Biosynthesis and Biodegradation of Cellulose

Biosynthesis and Biodegradation of Cellulose

edited by

Candace H. Haigler
Texas Tech University
Lubbock, Texas

Paul J. Weimer
E. I. du Pont de Nemours & Company, Incorporated
Wilmington, Delaware

Marcel Dekker, Inc.   New York • Basel • Hong Kong

Library of Congress Cataloging-in-Publication Data

Biosynthesis and biodegradation of cellulose/edited by Candace H.
Haigler, Paul J. Weimer.
    p.   cm.
  Includes bibliographical references.
  ISBN 0-8247-8387-5 (alk. paper)
  1. Cellulose--Biogradation. 2. Cellulose--Synthesis.
3. Microbial metabolism. I. Haigler, Candace H.
II. Weimer, Paul J.
  [DNLM: 1. Cellulose. 2. Biodegradation. 3. Cellulose-
-biosynthesis. 4. Microbiology. QU 83 B6156]
  QR160.B57   1991
  574.19'2482--dc20
  DNLM/DLC
  for Library of Congress   90-14143
                                             CIP

This book is printed on acid-free paper.

MARCEL DEKKER, INC.
270 Madison Avenue, New York, New York 10016

Current printing (last digit):
10 9 8 7 6 5 4 3 2 1

PRINTED IN THE UNITED STATES OF AMERICA

Preface

This book assembles into a single balanced volume the vast and widely dispersed literature on the biosynthesis and biodegradation of
cellulose, an abundant biopolymer of great natural and industrial
importance. Included are comprehensive summaries of the biosynthesis of cellulose in both in vitro and in vivo systems, along with
reviews of the phenomenon of cellulose degradation as practiced by
a variety of microbial groups and their component enzymes. Unlike
other books on the biology of cellulose, this volume aims to provide
an equal treatment of biosynthesis and biodegradation, which, unfortunately, are usually regarded as separate phenomena with little
in common. Thus, a major aim of the book is to broaden the horizons and knowledge of cellulose researchers who have until now
lacked a systematic and balanced source of information on aspects of
cellulose biology outside of their immediate research area. With respect to biosynthesis, all aspects of the field —including cell biology,
biophysics, biochemistry, and molecular biology —are covered to the
proportional level of knowledge in each area. Within the biodegradation domain, a broad canvas of microbial degradation has been included, instead of the narrow focus on fungal cellulolytic enzymes
that is normally encountered in technology-oriented books on biomass conversion or methods-oriented books on cellulose biochemistry.
Because its material is presented in the form of synthetic reviews

and commentaries on the most current research approaches, the book should be appreciated by both established and beginning researchers in the field.

Candace H. Haigler
Paul J. Weimer

Contents

v

Contributors

Edward A. Bayer   Weizmann Institute of Science, Rehovot, Israel

Astrid Beermann   Institut für Biochemie und Biotechnologie, Technische Universität Braunschweig, Braunschweig, Federal Republic of Germany

Pier Luigi Beltrame   Università di Milano, Milan, Italy

Moshe Benziman   Institute of Life Sciences, Hebrew University of Jerusalem, Jerusalem, Israel

R. Malcolm Brown, Jr.   University of Texas, Austin, Texas

Paolo Carniti   Università di Milano, Milan, Italy

Sean Carrington   University of the West Indies, Bridgetown, Barbados

C. Colson   Catholic University of Louvain, Louvain-la-Neuve, Belgium

Burk A. Dehority   Ohio Agricultural Research and Development Center, Ohio State University, Wooster, Ohio

D. P. Delmer   Hebrew University of Jerusalem, Jerusalem, Israel

Anne Mie C. Emons   Wageningen Agricultural University,
Wageningen, The Netherlands

D. E. Eveleigh   Rutgers University, New Brunswick, New Jersey

Bonaventura Focher   Stazione Sperimentale per la Cellulosa, Carta
e Fibre Tessili Vegetali ed Artificiali, Milan, Italy

David J. Frost   Rutgers University, New Brunswick, New Jersey

Candace H. Haigler   Texas Tech University, Lubbock, Texas

A. W. Khan   National Research Council of Canada, Ottawa, Ontario,
Canada

Shigenori Kuga   University of Tokyo, Tokyo, Japan

Raphael Lamed   George S. Wise Faculty of Life Sciences, Tel Aviv
University, Ramat Aviv, Israel

A. Lejeune   Catholic University of Louvain, Louvain-la-Neuve,
Belgium

Gordon Maclachlan   McGill University, Montreal, Quebec, Canada

Annamaria Marzetti   Stazione Sperimentale per la Cellulosa, Carta
e Fibre Tessili Vegetali ed Artificiali, Milan, Italy

Raphael Mayer   Institute of Life Sciences, Hebrew University of
Jerusalem, Jerusalem, Israel

D. H. Northcote   University of Cambridge, Cambridge, England

G. B. Patel   National Research Council of Canada, Ottawa, Ontario,
Canada

H. Quader   Universität Heidelberg, Heidelberg, Federal Republic
of Germany

Peter Rapp*   Institut für Biochemie und Biotechnologie, Technische
Universität Braunschweig, Braunschweig, Federal Republic of
Germany

*Current affiliation: Gesellschaft für Biotechnologische Forschung,
Braunschweig, Federal Republic of Germany.

S. M. Read   University of Melbourne, Parkville, Victoria, Australia

Danièle Reis   Laboratoire des Biomembranes et Surfaces Cellulaires Végétales, Ecole Normale Supérieure, Paris, France

Paul A. Richmond   University of the Pacific, Stockton, California

Peter Ross   Institute of Life Sciences, Hebrew University of Jerusalem, Jerusalem, Israel

Karl-Ludwig Schimz   Institut für Biotechnologie, Jülich, Federal Republic of Germany

Robert W. Seagull   Southern Regional Research Center, USDA/ARS, New Orleans, Louisiana

Margaret E. Sloan   Rutgers University, New Brunswick, New Jersey

Fred J. Stutzenberger   Clemson University, Clemson, South Carolina

Katherine Troyer   U.S. National Museum of Natural History, Smithsonian Institution, Washington, D.C.

Svein Valla   University of Trondheim, Trondheim, Norway

Brigitte Vian   Laboratoire des Biomembranes et Surfaces Cellulaires Végétales, Ecole Normale Supérieure, Paris, France

Bruce P. Wasserman   Rutgers University, New Brunswick, New Jersey

Paul J. Weimer*   E. I. du Pont de Nemours & Company, Incorporated, Wilmington, Delaware

Thomas M. Wood   Rowett Research Institute, Aberdeen, Scotland

---

Current affiliation: US Dairy Forage Research Center, USDA/ARS, Madison, Wisconsin.

Biosynthesis and Biodegradation of Cellulose

I

Biosynthesis

CANDACE H. HAIGLER

INTRODUCTION

This introduction provides an overview of Part I of the volume,
which deals with the biosynthesis of cellulose. Specific chapters are
mentioned where appropriate in a brief overview of the field. A
similar introduction can be found after Chap. 12 for Part II of the
volume on the biodegradation of cellulosic materials.

Because of the abundance of cellulose in the natural world
(Chap. 1), understanding the mechanism of its biosynthesis has
been a goal for several decades. Recently, interest in this process
has been stimulated by increasing evidence that plant cell walls have
critical roles in plant development and stress responses, the realization that microbial cellulose is a source of cellulose with special properties for commercial use, and the possibility of improving renewable
resources such as cotton and wood through genetic engineering. It
is clear to all who work in this field that there is a large deficiency
of basic knowledge without which practical improvements are impossible. Of equal importance to applied aims, understanding cellulose
synthesis will help illuminate how cells coordinate complex processes.
Even now we know or can predict that cellulose synthesis involves
differential gene expression, the endoplasmic reticulum, the Golgi
apparatus, the cytoskeleton, vesicle transport and fusion with the
plasma membrane, endocytosis of excess plasma membrane, specialized
fluidity domains in the plasma membrane, geometrical aggregation of
membrane proteins, localized ion concentrations, and a precisely

1

regulated biochemical pathway. After synthesis, the cellulose is integrated into walls by a biophysical process of which we have only rudimentary understanding (Chap. 2).

To the reader uninitiated in this field of research, the implication that there is so much left to know about the polymerization of glucose into a homopolymer probably seems strange. The deficiency can in part be explained by the complexity of the process. Not only is a molecule being made, but the molecule is being assembled under cellular control into a structural fibril at the cell surface (Chaps. 6 and 8). A major contributor to the deficiency, however, was the longstanding inability to synthesize cellulose in vitro from cell-free extracts of any organism (Chap. 9). Because of the intractable nature of the problem, many researchers turned to more productive areas of research. The few cellulose biochemists who , persevered have at last seen some rewards of their efforts as will be discussed below.

Meanwhile, beginning in 1976 microscopists discovered intriguing particle arrays in freeze-fracture replicas of plasma membranes of nearly all cellulose-synthesizing organisms prepared without the use of artifact-inducing cryoprotectants (Chaps. 3 and 4). Although conclusive assignment of a biochemical function to these particles is a major goal, most everyone now agrees that they are important participants in microfibril formation. Investigation of the relationship between polymerization and crystallization in *Acetobacter xylinum* (Chap. 5) established the theoretical basis for predicting that the different geometrical arrangements of the membrane particles determine the size and crystallinity of microfibrils. Similarly, microscopists observed a coincidence between the alignment of microtubules and cellulose microfibrils in certain cell types (Chap. 7). The extent of this coincidence in different cell types is still being examined, and a mechanism by which cytoplasmic microtubules might affect alignment of extracellular microfibrils is still being sought. The possible participation of other cytoskeletal components in cellulose deposition is largely unexamined. The development of cytochemical techniques based on enzymes (Chap. 2) or antibodies specific for one type of polysaccharide has begun to provide new insights into biophysical and cell biological aspects of wall assembly.

In 1981, the first major breakthrough in the area of biochemistry was realized through the ability to synthesize cellulose in vitro from membrane preparations and solubilized extracts of *A. xylinum* (Chap. 11). The key to this breakthrough was in maintaining a guanyl cyclase that converts GTP to an activator of the enzyme in the in vitro assay system. Unfortunately, application of the same strategies to higher plants has been unsuccessful, causing in vitro synthesis of cellulose from higher plant enzymes to remain as one of the great challenges in the field (Chaps. 8 and 9). The success with *A. xylinum* and very recent success with in vitro synthesis in *Dictyostelium discoideum* have renewed optimism that an elegant solution can be found to the problem in higher plants. Continuing efforts to find out more about the competing enzyme in higher plants, the callose synthase (Chap. 10), will certainly speed the resolution of the problem. The success with in vitro synthesis in *A. xylinum* has allowed research in cellulose biosynthesis to move into the molecular realm (Chap. 12). Polypeptides that participate in the process have now been tentatively identified in *A. xylinum* and higher plants. Some of the genes in the biosynthetic pathway in *A. xylinum* have been cloned.

This is an opportune time for other scientists to join in the research on cellulose biosynthesis. Part I of this book is designed to review the current status of the field as we move into the 1990s. In Part I, Chaps. 1 and 2 cover the occurrence and natural role of cellulose, Chaps. 3–7 deal with cell biological and biophysical factors in the cellulose synthesis, Chaps. 8–11 cover the biochemistry of cellulose synthesis, and Chap. 12 covers the newly emerging investigation of the molecular biology of cellulose synthesis. Some very recent developments briefly mentioned in this introduction are not covered in detail in the book chapters. This is a sign that a field that once moved very slowly is beginning to move more rapidly. We can expect that immunological and molecular probes will be developed to answer many longstanding questions about cellulose biochemistry and molecular biology. For example, antibodies to polypeptides involved in the process can be used in freeze-fracture labeling to test the identity of the organized membrane particles and as probes to investigate the molecular regulation of cellulose biosynthesis. This work is in its infancy in the public sector, but we can expect rapid progress in the future. In the excitement over biochemistry and molecular biology, we should make sure that the cell biological aspects of this process, which promise to reveal much about coordination of cell functions, are not ignored. As also indicated in the introduction to Part II on the biodegradation of cellulosic materials, research on cellulose biosynthesis will proceed most rapidly if a multidisciplinary approach is taken.

1

Occurrence and Functions of Native Cellulose

PAUL A. RICHMOND   *University of the Pacific, Stockton, California*

I. INTRODUCTION

Cellulose is synthesized by all higher plants as well as by a wide variety of other organisms. The amount of synthesis is enormous, making cellulose the most abundant biopolymer on earth (1). Most native cellulose is found in cell walls in the form of structural microfibrils. Although considerable effort has gone into investigating the various aspects of cellulose biosynthesis, and not without many exciting successes, much is yet to be understood about cellulose synthesis, microfibril structure, and microfibril deposition within cell walls.

Cellulose is a linear homopolymer composed of β-1,4-linked D-glucopyranosyl units (2). These residues form long chains with variable degrees of polymerization (DP; number of monomeric residues per polymer molecule). Although reliable measurements are difficult to obtain (3−5), DP values in higher plants appear to generally range between 7000 to 14,000 or more for secondary walls (3,6), whereas they are as low as 500 for primary walls (3,7). These extended glucan chains strongly associate by hydrogen bonding and van der Waals forces (8), laterally aggregating into crystalline cellulose microfibrils (4,9). Degrees of polymerization have not been correlated with native microfibril lengths. Length determination is difficult since microfibril ends are rarely seen in electron

micrographs of cell walls (7). It is possible that chain termination
occurs unevenly within each microfibril (7), resulting in imperfec-
tions that could account for the imperfect crystal structure of cellulose (10–13).

The cross-sectional substructure of cellulose microfibrils has not
been clarified for all organisms. One classical model proposed that
microfibrils are composed of thinner, elementary fibrils (9,14). The
crystalline portions of elementary fibrils were considered to be separated by paracrystalline regions (14); thus, a solid, crystalline
core would exist only for microfibrils that were equivalent to a
single elementary fibril. According to species and stage of development, microfibrils do vary in width (4,7,10), but sizes of the crystallite core vary accordingly (7,10). Current evidence, particularly
lattice imaging of algal microfibrils, supports the existence of a
single, central crystallite of variable width as the basis of the cross-
sectional organization of individual cellulose microfibrils (4,7,15;
also see Chap. 6).

Another longstanding controversy that relates to the supramolecular structure of cellulose microfibrils concerns the polarity of
the individual glucan chains within microfibrils. Native cellulose
was first proposed to contain parallel chains (16), an arrangement
in which the reducing ends of adjacent glucan chains are located at
the same end of a microfibril. Although diffraction data at the time
were insufficient to allow discrimination between parallel and antiparallel structures, subsequent theoretical considerations led most
workers during the 1930s to 1970s to favor an antiparallel model
(17). Refinements in computer analysis of X-ray crystallographic
data, however, have provided new evidence that lends support to
the original concept of a parallel arrangement of adjacent chains
(18–20). Using a completely different approach, Hieta and coworkers (21) corroborated this assessment. They showed by a silver-
labeling technique that the reducing groups of the glucan chains
are preferentially located only at one end of *Valonia* microfibrils
(see Chap. 6). Since chain polarity is likely determined by the
mechanism of biosynthesis, its importance to an understanding of
cellulose biosynthesis is apparent (7,12,22). On the other hand,
chain polarity would not appear to be critical for an appreciation of
the role of cellulose in cell wall structure.

Cellulose can exist in at least four different crystalline forms
(celluloses I–IV) as determined by X-ray crystallography (11,23).
The predominant native cellulose, designated cellulose I, is contrasted with its most common product of recrystallization, cellulose
II. Cellulose II is also formed by the process of mercerization,
which involves swelling (but not dissolving) cellulose in alkali, then

washing in water. Although rare, there is evidence that cellulose II is produced in nature (see Secs. II.B and II.D). Two other allomorphs of cellulose, III and IV, are reversibly produced by chemical treatments of either cellulose I or II (19,20). Cellulose IV has also been demonstrated by electron diffraction to occur in primary walls of higher plants (24,25). It is thought to exist as small microfibrils containing parallel chains with good longitudinal but poor transverse order.

The transition from cellulose I to II is irreversible. As a result it has been proposed that cellulose I is metastable, whereas cellulose II is the thermodynamically more stable form (4,19). Alternatively, solid state ^{13}C-NMR studies (based on the cross-polymerization/magic angle spinning technique) have led to the proposal that two different chain conformations, both stable, account for the differences between the two allomorphs (26). In addition, it has been argued that cellulose I occurs in nature due to the coupling of the processes of polymerization and crystallization during microfibril biogenesis so that adjacent chains crystallize with parallel orientation (7). This argument relies on cellulose II having an antiparallel chain orientation as based on computer-assisted analysis of X-ray data (19,20). With increasingly sophisticated computer programs, however, more recent analysis of X-ray data suggests that parallel and antiparallel models of cellulose II are equally probable (27,28). If cellulose I and II are of opposite chain polarity, then an apparent dilemma exists about how such a transition in polarity is possible, given that microfibrils only swell during mercerization. One model proposes that chains from adjacent microfibrils, which may lie in opposite orientation within the wall, interdigitate upon swelling to account for the change in polarity within the crystallite (20,22). On the other hand, if cellulose I and II both have parallel chains, the mechanism of mercerization is easier to explain, but another reason for the predominant synthesis of cellulose I in nature must be sought.

II. OCCURRENCE

A. Detection

Cellulose is found widely in the plant kingdom and in limited groups of fungi, bacteria, invertebrates, and protists. Evidence for the presence of native cellulose can be obtained by a number of physical and chemical methods. The most reliable of these methods is X-ray diffraction (4,9) due to the unique pattern produced by cellulose I. Typical results are seen in the X-ray diagrams from oriented ramie

(*Boehmeria nivea*) fibers and from fibers drawn from walls of the
green alga *Valonia* (4,19). The concentric rings of powder diagrams,
which derive from the interaction of X rays with randomly oriented
crystallites (4), are also sufficient to conclusively identify cellulose
(29,30). Similarly, the distinctive infrared spectrum of native cellulose (23) has been used for identification (31–33). Electron diffraction is another effective crystallographic technique. It is especially useful when only minute quantities of sample are available and
when it is advantageous to directly correlate the appearance and
crystallinity of the sample (34,35).

    Preliminary indications of cellulose can be obtained by chemical
and histochemical tests such as the presence of anthrone-positive
material after treatment with acetic-nitric reagent (36) and staining
with chlorozinc iodine (4,9,30,37) or Calcofluor White (38). Degradation by purified cellulase with a yield of only glucose and cellobiose
further indicates cellulose (39). Recently, highly purified cellulases
conjugated to colloidal gold (40–42) were used as probes to identify
cellulose by electron microscopy (43). In a related technique, antibodies to be used for immunolabeling were prepared against cellulose
derivatives. Antibodies that bind either cellobiose or three- to
eight-unit β-1,4-linked D-glucans, unfortunately, do not recognize
crystalline cellulose in higher plant walls (44). The above chemical
methods should be confirmed by crystallographic analysis or by
methylation analysis with either gas chromatography/mass spectrom-
etry (45,46) or proton magnetic resonance spectroscopy (47) to
establish the presence of β-1,4-linked D-glycopyranosyl residues.

B. Higher Plants and Algae

The cell walls of all higher plants (bryophytes and vascular plants)
contain cellulose in the form of crystalline microfibrils (9). The
percentage and crystalline form of cellulose within a wall varies according to cell type and developmental stage. For example, cellulose generally constitutes about 20–40% of wall dry weight in expanding, primary walls (4,48,49), whereas it commonly increases to
40–60% in secondary walls (4,50). This value approaches 100%,
however, for the secondary walls of cotton seed hairs (2,51). As
the cellulose content increases in secondary walls, so does the degree of microfibril ordering (4,7,9). Secondary wall microfibrils
have higher cellulose I crystallinity and may be thicker than primary wall microfibrils that exhibit cellulose IV crystallinity (9,24,
25,51). Higher plant secondary walls are utilized as the sole source
of industrial cellulose. Finished products are primarily formed or
derived from wood pulp and cotton (2).

    In contrast to higher plants, cellulose is distributed widely,
but nonuniformly, among the algae. Within a single division of

algae, wall composition may be highly variable. In the green algae (Chlorophyta) either cellulose, xylan, or mannan may serve as the structural wall polysaccharide (4,50,52,53). Microfibrillar composition may even vary among different cells within an individual alga. For example, the cyst walls of *Acetabularia* contain cellulose while the remaining walls of the plant utilize mannan (54). In addition to the Chlorophyta, cellulose is found in many or most of the algae in the Charophyta (stoneworts), Phaeophyta (brown algae), Chrysophyta (chryosophytes), Pyrrhophyta (dinoflagellates), and Rhodophyta (red algae) (4,50,55). However, detailed research on algal cellulose has focused primarily on the Chlorophyta.

Certain green algae produce exceptionally well-organized cellulose microfibrils. These microfibrils tend to be broad (about 25 nm wide), relatively straight, and highly crystalline (4,15; see Chaps. 3 and 6). Representative genera include members of both the Siphonocladales (e.g., *Valonia*, *Siphonocladus*, *Apjohnia*, and *Dictyosphaeria*) and the Cladophorales (e.g., *Cladophora*, *Chaetomorpha*, and *Rhizoclonium*). Cellulose accounts for 30–70% of wall dry weight in these algae (4), the highest amount being found in *Valonia* (56). *Oocystis* (57), a member of the Chlorococcales, and *Glaucocystis* (58), of uncertain affinity (55), are unicellular algae with large, orthogonally arranged microfibrils in their walls. The secondary walls of the desmids *Micrasterias* (59,60) and *Closterium* (61), as well as other Zygnematalean algae (59,62–64), contain well-ordered bands of broad microfibrils, although primary wall microfibrils are considerably thinner and less organized (59–61).

The remaining green algae with cellulosic walls produce microfibrils that lack the high degree of order of the previous types. The often studied charophyte *Nitella* (4,65) has typical 10-nm cellulose microfibrils (66,67) and is among this group. Another green alga, *Derbesia*, is unusual not only in its life history (55) but also in the composition of its wall. The skeletal polysaccharide of the sporophyte stage is mannan, whereas the wall of the gametophyte (*Halicystis* phase) contains cellulose in addition to other structural polysaccharides such as 1,3-linked xylan (4,53). Exceptionally, the cellulose appears to be cellulose II (68–70). A biosynthetic origin of cellulose II raises the question of mechanism. Any proposal must consider what is known about cellulose I biosynthesis as well as allomorph stabilities and chain polarities (19). If cellulose II has an antiparallel chain orientation, then its synthesis must be accomplished by a mechanism that is topologically different from the one that produces parallel chains in cellulose I. On the other hand, a parallel structure for cellulose II could conform to the synthase configuration for cellulose I. Given a parallel chain orientation for cellulose II, polymerization and crystallization might be uncoupled during its biosynthesis, whereas they are coupled for cellulose I (7; see Chap. 5).

Cellulose is apparently present, although often in small amounts, in all of the brown algae and most of the red algae. The walls of the brown algae *Laminaria*, *Fucus*, and *Pelvetia* contain about 20, 13, and 1% cellulose, respectively (71,72). Their microfibrils are about 20 nm wide (56). Cellulose content in the red algae ranges from 7% in *Rhodymenia* to 24% in *Ptilota* (71). Microfibrils in the red algae are about the same width as those in the brown algae (73,74), but are of noticeably lower crystallinity (4). Unlike other Rhodophyta, members of the subclass Bangiophycidae were originally thought not to contain cellulose (4,71). This small subclass and the much larger subclass Florideophycidae compose the sole red algal class Rhodophyceae in a number of taxonomic classifications (55). The walls of the macrothalli of *Bangia* and *Porphyra*, two bangiophycidean genera, contain xylan microfibrils and a nonfibrillar mannan cuticle (4,69,75), but no cellulose (59). The cell walls of their microthallic conchocelis phases, however, have now been shown to contain cellulose (76–78). This finding lends support to systematic schemes that unite the Rhodophyta into a single class and eliminate subclass designations (78). Cellulose distribution has also been of value in other taxonomic assignments within the algae (74).

Cellulose is the common structural polysaccharide in the golden algae (Chrysophyceae) and yellow–green algae (Xanthophyceae) within the division Chrysophyta. The number of species with cellulosic walls is limited, however. Many species either lack walls altogether or possess largely inorganic walls. Chrysophytes with cellulosic walls include *Tribonema* (79,80) and *Vaucheria* (79,81), both filamentous members of the Xanthophyceae. Cellulose microfibrils occur in the lorica of the chrysophycean flagellate *Dinobryon* (82), although chitin is found in the related species *Poteriochromonas* (83). *Pleurochrysis*, a chrysophyte in the class Prymnesiophyceae, is exceptional in producing Golgi-derived cellulosic scales that aggregate to form an external cell covering (31,84). In contrast, all other organisms are thought to synthesize cellulose at the plasma membrane (7,12,22,85; see Chap. 9). Putative synthase complexes (rosettes) (86; see Chap. 4) were detected in Golgi membranes of algae and higher plants (60,87), but these complexes are in the process of export and are likely inactive.

Lastly among the algae, certain dinoflagellates appear to synthesize cellulose. Specifically, the armored dinoflagellates are thought to utilize cellulose in their unique polygonal plates (88). The plates (theca) are enclosed within flattened vesicles that lie between two cellular membranes, each encompassing the cell (88,89). Consequently, the structural components of the wall are unusual in being positioned intracellularly. It has generally been accepted that thecal plates are composed of cellulose (55,88,90). Clear data, however, are lacking. Much of the evidence is based on histochemical treatments (88). Physical, chemical, and enzymatic tests on *Peridinium westii* (91) and *Cachonina niei* (92) yielded equivocal results, and typical microfibrils were not detected in thecal plates (93). On the other hand, microfibrils that were shown to be cellulosic by X-ray diffraction are present in the cyst wall of *Pyrocystis* (94), a nonmotile member of the Pyrrophyta. As seen in thin section, the wall is primarily constructed of multiple layers of orthogonally arranged microfibrils, which is reminiscent of the organization observed in the green alga *Oocystis* (57). Ribbonlike, 12-nm-wide microfibrils are observed by freeze-fracturing, and negatively stained preparations reveal these microfibrils to be composed of 3-nm fibrils.

C. Fungi and Protists

There is a general correspondence between the major taxonomic groupings of the fungi and wall polymer composition (95,96). Chitin and glucans are the chief wall constituents. Despite a degree of polymer heterogeneity among the fungal classes, wall ultrastructure is quite uniform (97). Fibrils of chitin, but occasionally cellulose, usually associate with a β-1,3/β-1,6-linked D-glucan to form an inner wall layer (98). An outer, usually alkali soluble layer completes the wall. Chitin and cellulose coexist in the Hyphochytridiomycetes (96,98), a small class of aquatic fungi (99,100), and in a few genera of the Ascomycetes (101). Convincing evidence for the occurrence of both chitin and cellulose in the same fungal wall was first given for the Hyphochytridiomycete *Rhizidiomyces* (30,102). Cellulose was demonstrated by X-ray diffraction (30) and chemical methods (102). Subsequently, the same techniques provided evidence for the presence of cellulose and chitin in 31 of 47 species of *Ceratocystis* and in four species of *Europhium*, related members of the Ascomycetes (103). Cellulose was also detected in *Ceratocystis* by infrared spectroscopy (32) and was localized ultrastructurally by labeling with an exoglucanase-gold complex (42).

In contrast to the simultaneous occurrence of chitin and cellulose in some fungal walls, chitin is completely replaced by cellulose in the Oomycetes (96,98). Cellulose is not the major component, however, accounting for about 15% of the wall dry weight (104). The principal wall polysaccharide is a branched β-1,3/β-1,6-linked D-glucan of low solubility (105,106). Cellulose I is revealed by X-ray diffraction only after chemical and enzymatic treatments to remove the majority of the branched glucan (81,106,107). The resulting X-ray diagram indicates a low order of crystallinity (107). Due to the difficulty of separating the branched glucan from cellulose, it was suggested that the two may be covalently linked (98, 105,106). A similar association was proposed for chitin and

β-1,3/β-1,6-linked D-glucan (98). A partial list of Oomycete genera with identifiable cellulose includes *Phytophthora* (105,108), *Pythium* (106,109), *Achlya* (81), *Saprolegnia* (81), *Dictyuchus* (81), *Atkinsiella* (107), and *Apodachlya* (104,110). Among these examples, the water mold *Saprolegnia* has been effectively utilized for the study of cellulose biosynthesis. Membrane preparations, with plasma membrane-localized glucan synthases, promote in vitro synthesis of β-1,4-linked D-glucan from micromolar concentrations of uridine-5'-diphosphate (UDP)-glucose (111).

In the protists cellulose appears to be confined to ameboid organisms, namely certain single-celled, nonfruiting amebae and some slime molds (Mycetozoa). Cellulose was first discovered in the stalk, or sorophore, of the cellular slime mold *Dictyostelium discoideum*. Identification was made by X-ray diffraction, chemical tests, and ultrastructural analysis (112–114). Subsequently, cellulose was shown to be the major component of the surface sheath that surrounds *Dictyostelium* aggregates and migrating slugs (39). Confusion initially existed about the crystalline form of *Dictyostelium* cellulose, but recent reexamination by electron diffraction shows that it is cellulose I (115). Other cellular slime molds with cellulosic coverings include *Acytostelium leptosomum* (116) and *Polysphondylium pallidum* (117). In addition, cellulose was detected by electron diffraction in the stalk of *Protostelium irregularis*, a protostelid slime mold (34). Cyst walls of the soil ameba *Acanthamoeba* contain about one-third cellulose (33,118). Cellulose is not present, however, in cultures of amebae prior to encystment (118). DP values of the glucan chains in the cyst wall are comparable to that of higher plant primary walls, about 2000–6000 (119). Based on chemical and cytochemical methods, cyst walls of the ameboflagellates *Schizopyrenus russelli* (120) and *Naegleria gruberi* (121) also contain cellulose.

D. Bacteria and Blue–Green Algae

The synthesis of cellulose in bacteria is restricted, with one exception, to certain gram-negative forms (122). Only *Acetobacter*, the vinegar bacteria, produces large amounts of cellulose. Most bacterial work has focused on *A. xylinum*, an organism that has served as a model system in studies of cellulose biosynthesis (7,12,43,123; See Chap. 5). X-ray diffraction of polymer from *A. xylinum* yields reflections corresponding to those for cellulose I (124–126). *Acetobacter* cellulose is unusual, however, in not being synthesized as a wall component. Instead it is secreted into the medium from pores along the length of the cell (127), forming a single cellulosic ribbon (126,128). The synthetic action of a culture of *A. xylinum* generates a thick, cellulosic pellicle that floats on the surface of the nutrient medium (124,128). The DP of the resulting cellulose varies according to the age of the culture, starting abound 2000 and increasing to nearly 4000 at steady-state growth (129). Surprisingly, some variants of *A. xylinum* appear to produce small amounts of cellulose II (70).

In addition to *Acetobacter*, native cellulose is produced by such gram-negative bacteria as *Pseudomonas*, *Rhizobium*, and *Agrobacterium* (122). During growth in liquid media these bacteria aggregate into flocs as the individual cells become enclosed in an irregular network of fibrils. X-ray diffraction analyses and succeptibility to cellulase indicate that the fibrils are formed of cellulose I. Cellulose constitutes up to 4% of floc dry weight. Based on chemical and spectroscopic criteria, cellulose is also synthesized by the gram-positive bacterium *Sarcina ventriculi* (130,131). Cells of this species aggregate into packets. The principal intercellular material of the packets is apparently cellulose, accounting for about 15% of the dry weight of the mass (131).

Cellulose also has been reported in the blue–green algae (Cyanophyta). Cellulase treatments, which were correlated with ultrastructural observations, together with Calcofluor White staining provided evidence that cellulose is a component of heterocyst envelopes of *Anabaena* (132). Microfibrils, presumably cellulosic, are present in the mucilaginous sheaths that surround *Nostoc* vegetative cells (133).

E. Animals

The production of cellulose in animals is limited chiefly to the ascidians, or tunicates, a common marine invertebrate found throughout the world. The adult body of these simple chordates (Urochordata) is invested with a special extracellular mantel, the tunic, which is composed of protein and polysaccharide (134). A cellulosic component of the tunic was first recognized in 1845 (135), following the discovery of cellulose in plants by less than a decade. The carbohydrate portion of the ascidian tunic consists of 50–100% cellulose depending on the consistency of the tunic (according to genus) (134). The similarities between plant and ascidian cellulose, called tunicin (135), are striking. X-ray powder diagrams and electron diffraction diagrams of tunicin are qualitatively the same as those for plant cellulose (29,136,137). In addition, mercerized tunicin produces cellulose II reflections identical to those of mercerized plant cellulose (29). DP values (about 4000) for tunicin (29,135) are similar to those of plant and bacterial celluloses. Examination of isolated tunicin by electron microscopy reveals microfibrils of somewhat variable size. Individual microfibrils from a number of species are around 12 nm wide (29,138,139), but other species appear to

have even thicker microfibrils (140,141). Ascidian cellulose is apparently synthesized by tunic epidermal cells (37,138,142), although
the evidence for this proposal is not conclusive (142).

The organization of cellulose within the ascidian tunic varies
with species. Some genera studied (e.g., *Boltenia, Molgula,* and
Perophora) possess tunics with a generally random, diffuse network
of microfibrils (37,143). On the other hand, cellulose organization
in at least one species, *Halocynthia papillosa,* resembles the helicoidal patterns found in numerous secondary plant cell walls (141;
see also Chap. 2). A third form is seen in *Pyura stolonifera* in
which freeze-fracturing reveals bundles of microfibrils that approach
400 nm in diameter (138). Such bundling is more typical of the aggregation of collagen fibrils that is often seen in vertebrate connective tissue (138,144).

Animal cellulose may not be confined to the tunicates. It has
been reported as a normal but minor fibrous component of connective tissue in mammals. Cellulose fibers have been detected in
bovine (145,146) and human (137,146,147) dermis as well as in human aorta (137). Additionally, increased fiber density has been
correlated with certain tissue pathologies (137,146,148). When
analyzed by X-ray (145–148) and electron diffraction (137), mammalian cellulose yields reflections that are essentially coincident with
those of other native cellulose. Biosynthesis of cellulose in mammals seems odd and consequently the source of this cellulose has
been questioned (149). It has been suggested that specimen contamination could have occurred during preparation, although care
was taken to avoid it (146,148). Mammalian cellulose could be of
fungal origin (149); however, mammalian cellulose (137,146), like
tunicin (137,150), is closely associated with protein. Cellulose-
protein complexes are not encountered in fungi that produce cellulose (98). Since mammalian cellulose is not needed as a reinforcing
fiber in connective tissue, its function is unclear. If mammalian
cellulose biosynthesis exists, it could be a vestige of chordate
evolution with no apparent function. Mammalian cellulose will remain enigmatic until it is reexamined by current methods.

III. FUNCTION

As seen in a wide variety of cell coverings, cellulose most commonly
functions as a reinforcing element. Cellulose microfibrils act to
strengthen the cell walls of higher plants, algae, and fungi, as well
as the tunics of ascidians and the stalks and cyst walls of some
protists. Fibrous reinforcing elements of various composition are
common to biological supportive structures. For example, chitin
(151) serves as a fibrillar reinforcer in supportive structures of

arthropods and many fungi, while collagen (152) is similarly utilized in vertebrate connective tissue.

Except for that of *Sarcina* (130,131), bacterial cellulose is not synthesized as a component of a cell covering and, accordingly, does not act as a structural reinforcer. In bacteria that form packets (130) or flocs (122), cellulose appears to function as a binding element which permits cell aggregation. On the other hand, the floating, cellulosic pellicle produced by *Acetobacter* apparently aids in maintaining the oxygen requirements of this nonmotile, strict aerobe (12).

Cellulosic walls are natural composite structures (see Chap. 2). They consist of microfibrils embedded in a relatively amorphous matrix (9,153) of stereoirregular polysaccharides, mainly hemicelluloses and pectins (5,7,85,154,155), and proteins (7,85,155,156). Due to their high modulus of elasticity (4,9), which is nearly one-half that of mild steel (157), cellulose microfibrils transform a gel-like matrix into a reinforced composite with great tensile strength (9). Although cellulose imparts tensile strength to the wall to resist turgor pressure (155,158), microfibrils bend easily, and wall mechanical strength is low under compression (9). High compressive strengths are only achieved in lignified walls of higher plants and ferns (9,155). Lignins are aromatic polymers that replace water in the matrix of these walls (159). Lignification greatly increases bonding within the wall (159), producing rigid, woody tissues able to withstand the compressive force of gravity.

The orientation of fibers within a composite material directly influences the material's mechanical properties (152,160–162). For example, a unidirectionally reinforced composite is much stronger parallel to the fibers than perpendicular to them where the matrix bears most of the load. Biased fiber reinforcement is a common feature of primary walls of plant cells that expand uniformly along their length (4,9,48,163,164). In growing cylindrical cells, preferential alignment of cellulose microfibrils in the circumferential (transverse) direction results in longitudinal expansion (163,164). An unbiased (random) arrangement of nascent microfibrils can be induced in the same cells by addition of microtubule-disrupting drugs (163,165–167; see Chap. 7). Under these conditions expansion shifts to the transverse in accord with the stresses on the wall (65,166,167). The effectivness of microfibrillar reinforcement depends on microfibril positioning within the thickness of the wall. As demonstrated in *Nitella* internodal cells and presumably applying to higher plant cells, only the inner portion (approximately one-fourth) of the expanding wall bears the stress (65,166,167). The role of cellulose in reinforcing this inner wall portion is directly demonstrated by blocking cellulose synthesis in elongating *Nitella* internodes (167). Matrix synthesis continues in these cells until

the inner one-fourth of the wall is cellulose-free. The cell then
expands until it bursts from the stress applied by turgor pressure
on the unreinforced inner wall. Thus, the angular distribution of
cellulose microfibrils in the inner portion of the wall in conjunction
with wall stress patterns determines the directionality of cell expansion and, consequently, cell shape (65,163,167,168). Further, the
summation of individual cellular elongation events results in the
initiation and extension of cylindrical plant organs (168,169). It is
therefore clearly evident that cellulose has a critical role in the control of plant development.

ACKNOWLEDGMENTS

I wish to express my appreciation to Dr. Candace H. Haigler and
Dr. Eric M. Roberts for their many valuable comments during the
preparation of this manuscript.

REFERENCES

1. J. R. Colvin, in *The Biochemistry of Plants*, Vol. 3 (J. Priess,
    ed.), Academic Press, New York, 1980, p. 543.
2. T. P. Nevell and S. H. Zeronian, in *Cellulose Chemistry
    and Its Applications* (T. P. Nevell and S. H. Zeronian,
    eds.), Ellis Horwood, Chichester, 1985, p. 15.
3. M. Marx-Figini, in *Cellulose and Other Natural Polymer Sys-
    tems* (R. M. Brown, Jr., ed.), Plenum Press, New York, 1982,
    p. 243.
4. R. D. Preston, *The Physical Biology of Plant Cell Walls*,
    Chapman and Hall, London, 1974.
5. P. M. Dey and K. Brinson, *Adv. Carboh. Chem. Biochem, 42*,
    265 (1984).
6. D. A. T. Goring and T. Timell, *Tappi, 45*, 454 (1962).
7. C. H. Haigler, in *Cellulose Chemistry and Its Applications*
    (T. P. Nevell and S. H. Zeronian, eds.), Ellis Horwood,
    Chichester, 1985, p. 30.
8. K. H. Gardner and J. Blackwell, *Biochem. Biophys. Acta, 343*,
    232 (1974).
9. A. Frey-Wyssling, *The Plant Cell Wall*, Gebrüder Borntraeger,
    Berlin, 1976.
10. H. Harada and T. Goto, in *Cellulose and Other Natural Polymer
    Systems* (R. M. Brown, Jr., ed.), Plenum Press, New York,
    1982, p. 383.

11. R. H. Marchessault and A. Sarko, *Adv. Carboh. Chem.*, *22*, 421 (1967).
12. D. P. Delmer, *Adv. Carboh. Chem. Biochem.*, *41*, 105 (1983).
13. F. Shafizadeh and G. D. McGinnis, *Adv. Carboh. Chem. Biochem.*, *26*, 297 (1971).
14. K. Mühlethaler, *Ann. Rev. Plant Physiol.*, *18*, 1 (1967).
15. J. Sugiyama, H. Harada, Y. Fujiyoshi, and N. Uyeda, *Planta*, *166*, 161 (1985).
16. K. H. Meyer and H. Mark, *Ber. Dtsch. Chem. Gesel.*, *61B*, 593 (1928).
17. K. H. Meyer and L. Misch, *Helv. Chim. Acta*, *20*, 232 (1937).
18. K. H. Gardner and J. Blackwell, *Biopolymers*, *13*, 1975 (1974).
19. J. Blackwell, in *Cellulose and Other Natural Polymer Systems* (R. M. Brown, Jr., ed.), Plenum Press, New York, 1982, p. 403.
20. A. Sarko, in *Cellulose: Structure, Modification and Hydrolysis* (R. A. Young and R. M. Powell, eds.), John Wiley and Sons, New York, 1986, p. 29.
21. K. Hieta, S. Kuga, and M. Usuda, *Biopolymers*, *23*, 1807 (1984).
22. D. P. Delmer, *Ann. Rev. Plant Physiol.*, *38*, 259 (1987).
23. R. H. Marchessault and P. R. Sundararajan, in *The Polysaccharides*, Vol. 2 (G. O. Aspinall, ed.), Academic Press, New York, 1983, p. 11.
24. H. Chanzy, K. Imada, and R. Vuong, *Protoplasma*, *94*, 299 (1978).
25. H. Chanzy, K. Imada, A. Mollard, R. Vuong, and F. Barnoud, *Protoplasma*, *100*, 303 (1979).
26. D. L. VanderHart and R. H. Atalla, in *The Structures of Cellulose: Characterization of the Solid States* (R. H. Atalla, ed.), American Chemical Society, Washington, D.C., 1987, p. 88.
27. A. D. French, in *Cellulose and Wood: Chemistry and Technology* (C. Schuerch, ed.), John Wiley and Sons, New York, 1989, p. 103.
28. A. Sakthivel, A. F. Turbak, and R. A. Young, in *Cellulose and Wood: Chemistry and Technology* (C. Schuerch, ed.), John Wiley and Sons, New York, 1989, p. 67.
29. B. G. Rånby, *Ark. Kemi*, *4*, 241 (1952).
30. M. S. Fuller and I. Barshad, *Am. J. Bot.*, *47*, 105 (1960).
31. D. K. Romanovicz, in *Cytomorphogenesis in Plants* (O. Kiermayer, ed.), Springer-Verlag, Wien, 1981, p. 27.
32. A. J. Michell and G. Scurfield, *Trans. Br. Mycol. Soc.*, *55*, 488 (1970).

33. G. Tomlinson and E. A. Jones, *Biochim. Biophys. Acta, 63*, 194 (1962).
34. R. L. Blanton and H. D. Chanzy, *J. Protozool., 32*, 740 (1985).
35. C. H. Haigler and H. Chanzy, *J. Ultrastruct. Mol. Struct. Res., 98*, 299 (1988).
36. D. M. Updegraff, *Anal. Biochem., 32*, 420 (1969).
37. J. D. Deck, E. D. Hay, and J.-P. Revel, *J. Morphol., 120*, 267 (1966).
38. P. J. Wood and R. G. Fulcher, *Cereal Chem., 55*, 952 (1978).
39. H. Freeze and W. F. Loomis, *J. Biol. Chem., 252*, 820 (1977).
40. H. Chanzy, B. Henrissat, and R. Vuong, *FEBS Lett., 172*, 193 (1984).
41. R. H. Berg, G. W. Erdos, M. Gritzali, and R. D. Brown, Jr., *J. Electron Microsc. Tech., 8*, 371 (1988).
42. N. Benhamou, H. Chamberland, G. B. Ouellette, and F. J. Pauze, *Can. J. Microbiol., 33*, 405 (1987).
43. F. C. Lin, R. M. Brown, Jr., J. B. Cooper, and D. P. Delmer, *Science, 230*, 822 (1985).
44. D. H. Northcote, R. Davey, and J. Lay, *Planta, 178*, 353 (1989).
45. B. Lindberg and J. Lönngren, *Meth. Enzymol., 50*, 3 (1978).
46. M. McNeil, A. G. Darvill, P. Åman, L.-E. Franzén, and P. Albersheim, *Meth. Enzymol., 83*, 3 (1982).
47. J. N. C. Whyte and J. R. Englar, *Can. J. Chem., 49*, 1302 (1971).
48. P. A. Roelofsen, *Adv. Bot. Res., 2*, 69 (1965).
49. K. W. Talmadge, K. Keegstra, W. D. Bauer, and P. Albersheim, *Plant Physiol., 51*, 158 (1973).
50. H. J. Rogers and H. R. Perkins, *Cell Walls and Membranes*, E. and F. N. Spon, London, 1968.
51. U. Ryser, *Eur. J. Cell Biol., 39*, 236 (1985).
52. E. Percival and R. H. McDowell, *Encycl. Plant Physiol. New Ser., 13B*, 277 (1981).
53. R. D. Preston, *Ann. Rev. Plant Physiol., 30*, 55 (1979).
54. W. Herth, A. Kuppel, and W. W. Franke, *J. Ultrastruct. Res., 50*, 289 (1975).
55. H. C. Bold and M. J. Wynne, *Introduction to the Algae*, 2nd ed., Prentice-Hall, Englewood Cliffs, 1985.
56. J. Cronshaw and R. D. Preston, *Proc. Roy. Soc. Lond. B, 146*, 37 (1958).
57. H. Sachs, I. Grimm, and D. G. Robinson, *Cytobiologie, 14*, 49 (1976).
58. D. G. Robinson and R. D. Preston, *J. Exp. Bot., 22*, 635 (1971).

59. M. Mix, *Arch. Mikrobiol.*, *55*, 116 (1966).
60. T. H. Giddings, D. L. Brower, and L. A. Staehelin, *J. Cell Biol.*, *84*, 327 (1980).
61. T. Hogetsu and H. Shibaoka, *Planta*, *140*, 7 (1978).
62. F. Buer, *Flora*, *154*, 349 (1964).
63. C. J. Dawes, *J. Phycol.*, *1*, 121 (1965).
64. M. Mix, *Ber. Dtsch. Bot. Ges.*, *80*, 715 (1968).
65. L. Taiz, J.-P. Métraux, and P. A. Richmond, in *Cytomorphogenesis in Plants* (O. Kiermayer, ed.), Springer-Verlag, Wien, 1981, p. 231.
66. M. C. Probine and R. D. Preston, *J. Exp. Bot.*, *13*, 111 (1962).
67. P. B. Green, *J. Biophys. Biochem. Cytol.*, *4*, 505 (1958).
68. W. A. Sisson, *Science*, *87*, 350 (1938).
69. E. Frei and R. D. Preston, *Nature*, *192*, 939 (1961).
70. E. M. Roberts, I. M. Saxena, and R. M. Brown, Jr., in *Cellulose and Wood: Chemistry and Technology* (C. Schuerch, ed.), John Wiley and Sons, New York, 1989, p. 689.
71. J. Cronshaw, A. Myers, and R. D. Preston, *Biochim. Biophys. Acta*, *27*, 89 (1958).
72. W. A. P. Black, *J. Mar. Biol. Assoc. UK*, *29*, 379 (1950).
73. A. Myers and R. D. Preston, *Proc. Roy. Soc. Lond. B*, *150*, 456 (1959).
74. B. C. Parker, *Ann. N.Y. Acad. Sci.*, *175*, 417 (1970).
75. E. Frei and R. D. Preston, *Proc. Roy. Soc. Lond. B*, *160*, 314 (1964).
76. M. R. Gretz, J. M. Aronson, and M. R. Sommerfield, *Science*, *207*, 779 (1980).
77. L. S. Mukai, J. S. Craigie, and R. G. Brown, *J. Phycol.*, *17*, 192 (1981).
78. M. R. Gretz, J. M. Aronson, and M. R. Sommerfield, *Photochemistry*, *23*, 2513 (1984).
79. R. Frey, *Ber. Schweiz. Bot. Ges.*, *60*, 199 (1950).
80. M. Cleare and E. Percival, *Br. Phycol. J.*, *7*, 185 (1972).
81. B. C. Parker, R. D. Preston, and G. E. Fogg, *Proc. Roy. Soc. Lond. B*, *158*, 435 (1963).
82. W. Herth, *Protoplasma*, *100*, 345 (1979).
83. W. Herth, A. Kuppel, and E. Schnepf, *J. Cell Biol.*, *73*, 311 (1977).
84. D. K. Romanovicz and R. M. Brown, Jr., *J. Appl. Polym. Sci.*, *Appl. Polym. Symp.*, *28*, 587 (1976).
85. D. P. Delmer and B. A. Stone, in *The Biochemistry of Plants*, Vol. 14 (J. Priess, ed.), Academic Press, San Diego, 1988, p. 373.
86. R. M. Brown, Jr., *J. Cell Sci.* (Suppl.), *2*, 13 (1985).

87. C. H. Haigler and R. M. Brown, Jr., *Protoplasma*, *134*, 111 (1986).
88. L. C. Morrill and A. R. Loeblich III, *Int. Rev. Cytol.*, *82*, 151 (1983).
89. R. Wetherbee, *J. Ultrastruct. Res.*, *50*, 58 (1975).
90. J. D. Dodge, *Bot. Rev.*, *37*, 481 (1971).
91. Z. Nevo and N. Sharon, *Biochim. Biophys. Acta*, *173*, 161 (1969).
92. A. R. Loeblich III, The Physiology, Morphology and Cell Wall of the Marine Dinoflagellate, *Cachonina niei*, Ph.D. dissertation, University of California, San Diego, 1971.
93. J. D. Dodge, *J. Mar. Biol. Assoc. UK*, *45*, 607 (1965).
94. E. Swift and C. C. Remsen, *J. Phycol.*, *6*, 79 (1970).
95. J. M. Aronson, in *The Fungi*, Vol. 1 (G. C. Ainsworth and A. S. Sussman, eds.), Academic Press, London, 1965, p. 49.
96. S. Bartnicki-Garcia, *Ann. Rev. Microbiol.*, *22*, 87 (1968).
97. D. Hunsley and J. H. Burnett, *J. Gen. Microbiol.*, *62*, 203 (1970).
98. J. G. H. Wessels and J. H. Sietsma, *Encycl. Plant Physiol. New Ser.*, *13B*, 352 (1981).
99. H. C. Bold, C. J. Alexopoulos, and T. Delevoryas, *Morphology of Plants and Fungi*, 5th ed., Harper and Row, New York, 1987.
100. F. K. Sparrow, in *The Fungi*, Vol. 4B (G. C. Ainsworth, F. K. Sparrow, and A. S. Sussman, eds.), Academic Press, New York, 1973, p. 71.
101. E. Barreto-Bergter and P. A. Gorin, *Adv. Carboh. Chem. Biochem.*, *41*, 47 (1983).
102. M. S. Fuller, *Am. J. Bot.*, *47*, 838 (1960).
103. T. R. Jewell, *Mycologia*, *66*, 139 (1974).
104. J. H. Sietsma, D. E. Eveleigh, and R. H. Haskins, *Biochim. Biophys. Acta*, *184*, 306 (1969).
105. L. P. T. M. Zevenhuizen and S. Bartnicki-Garcia, *Biochemistry*, *8*, 1496 (1969).
106. J. H. Sietsma, J. J. Child, L. R. Nesbitt, and R. H. Haskins, *J. Gen. Microbiol.*, *86*, 29 (1975).
107. J. M. Aronson and M. S. Fuller, *Arch. Mikrobiol.*, *68*, 295 (1969).
108. S. Bartnicki-Garcia, *J. Gen. Microbiol.*, *42*, 57 (1966).
109. B. A. Cooper and J. M. Aronson, *Mycologia*, *59*, 658 (1967).
110. C. C. Lin and J. M. Aronson, *Arch. Mikrobiol.*, *72*, 111 (1970).
111. V. Girard and M. Fèvre, *Planta*, *160*, 400 (1984).
112. K. B. Raper and D. I. Fennell, *Bull. Torrey Bot. Club*, *79*, 25 (1952).

113. K. Mühlethaler, *Am. J. Bot.*, *43*, 673 (1956).
114. K. Gezelius and B. G. Rånby, *Exp. Cell Res.*, *12*, 265 (1957).
115. E. M. Roberts, I. M. Saxena, and R. M. Brown, Jr., Does cellulose II occur in nature?, in *Proc. Electron Society of America* (47th Annual Meeting, 1989), Publisher: San Francisco Press, San Francisco; City where held: San Antonio, p. 780.
116. K. Gezelius, *Exp. Cell Res.*, *18*, 425 (1959).
117. M. A. Toama and K. B. Raper, *J. Bacteriol.*, *94*, 1150 (1967).
118. R. J. Neff and R. H. Neff, *Symp. Soc. Exp. Biol.*, *23*, 51 (1969).
119. W. E. Blanton and C. L. Villemez, *J. Protozool.*, *25*, 264 (1978).
120. A. K. Rastogi, A. C. Shipstone, and S. C. Agarwala, *J. Protozool.*, *18*, 176 (1971).
121. J. M. Werth and A. J. Kahn, *J. Bacteriol.*, *94*, 1272 (1967).
122. M. H. Deinema and L. P. T. M. Zevenhuizen, *Arch. Mikrobiol.*, *78*, 42 (1971).
123. C. H. Haigler and M. Benziman, in *Cellulose and Other Natural Polymer Systems* (R. M. Brown, Jr., ed.), Plenum Press, New York, 1982, p. 243.
124. J. Barsha and H. Hibbert, *Can. J. Res.*, *10*, 170 (1934).
125. B. G. Rånby, *Ark. Kemi*, *4*, 249 (1952).
126. K. Zaar, *Cytobiologie*, *16*, 1 (1977).
127. K. Zaar, *J. Cell Biol.*, *80*, 773 (1979).
128. R. M. Brown, Jr., J. H. M. Willison, and C. L. Richardson, *Proc. Natl. Acad. Sci. USA*, *73*, 4565 (1976).
129. M. Marx-Figini and B. G. Pion, *Biochim. Biophys. Acta*, *338*, 382 (1974).
130. E. Canale-Parola, R. Borasky, and R. S. Wolfe, *J. Bacteriol.*, *81*, 311 (1961).
131. E. Canale-Parola and R. S. Wolfe, *Biochim. Biophys. Acta*, *82*, 403 (1964).
132. U. Granhall, *Physiol. Plant.*, *38*, 208 (1976).
133. A. Frey-Wyssling and H. Stecher, *Z. Zellforsch. Mikroskop. Anat.*, *39*, 515 (1954).
134. M. J. Smith and P. A. Dehnel, *Comp. Biochem. Physiol.*, *40B*, 615 (1971).
135. S. Hunt, *Polysaccharide-Protein Complexes in Invertebrates*, Academic Press, London, 1970.
136. R. O. Herzog and H. W. Gonell, *Z. Physiol. Chem.*, *141*, 63 (1924).
137. D. A. Hall and H. Saxl, *Proc. Roy. Soc. Lond. B*, *155*, 202 (1961).

138. A. B. Wardrop, *Protoplasma, 70*, 73 (1970).
139. A. Frey-Wyssling and R. Frey, *Protoplasma, 39*, 657 (1950).
140. K. H. Meyer, L. Huber, and E. Kellenberger, *Experientia, 7*, 216 (1951).
141. D. Gubb, *Tissue and Cell, 7*, 19 (1975).
142. W. E. Robinson, K. Kustin, and G. C. McLeod, *J. Exp. Zool., 225*, 187 (1983).
143. A. K. Mishra and J. R. Colvin, *Can. J. Zool., 47*, 659 (1969).
144. D. W. Fawcett, *Bloom and Fawcett: A Textbook of Histology*, W. B. Saunders, Philadelphia, 1986.
145. D. A. Hall, P. F. Lloyd, H. Saxl, and F. Happey, *Nature, 181*, 470 (1958).
146. D. A. Hall, F. Happey, P. F. Lloyd, and H. Saxl, *Proc. Roy. Soc. Lond. B, 151*, 497 (1960).
147. A. J. Cruise and J. W. Jeffrey, *Nature, 183*, 677 (1959).
148. J. Toriumi, H. Shirasawa, K. Sano, and K. Komatsu, *Acta Pathol. Jap., 22*, 591 (1972).
149. L. Jurasek, J. R. Colvin, and D. R. Whitaker, *Adv. Appl. Microbiol., 9*, 131 (1967).
150. M. J. Smith and P. A. Dehnel, *Comp. Biochem. Physiol., 35*, 17 (1970).
151. J. F. V. Vincent, in *The Mechanical Properties of Biological Materials* (J. F. V. Vincent and J. D. Currey, eds.), Cambridge University Press, Cambridge, 1980, p. 183.
152. D. W. L. Hukins and R. M. Aspden, *Trends Biochem. Sci., 10*, 260 (1985).
153. R. D. Preston, in *Cellulose: Structure, Modification and Hydrolysis* (R. A. Young and R. M. Powell, eds.), John Wiley and Sons, New York, 1986, p. 3.
154. K. Kato, *Encycl. Plant Physiol.* (New Ser.), *13B*, 29 (1981).
155. A. Bacic, P. J. Harris, and B. A. Stone, in *The Biochemistry of Plants*, Vol. 14 (J. Priess, ed.), Academic Press, San Diego, 1988, p. 297.
156. D. T. A. Lamport, *Encyl. Plant Physiol.* (New Ser.), *13B*, 133 (1981).
157. M. C. Probine and N. F. Barber, *Aust. J. Biol. Sci., 19*, 439 (1966).
158. P. M. Ray, P. B. Green, and R. Cleland, *Nature, 239*, 163 (1972).
159. T. Higuchi, *Encyl. Plant Physiol.* (New Ser.), *13B*, 194 (1981).
160. T. M. Cornsweet, *Science, 168*, 433 (1970).
161. B. Harris, in *The Mechanical Properties of Biological Materials* (J. F. V. Vincent and J. D. Currey, eds.), Cambridge University Press, Cambridge, 1980, p. 37.

162. D. R. P. Hettiaratchi and J. R. O'Callaghan, *J. Theor. Biol.*, *74*, 235 (1978).
163. P. B. Green, in *Cytodifferentiation and Macromolecular Synthesis* (M. Locke, ed.), Academic Press, New York, 1963, p. 203.
164. L. Taiz, *Ann. Rev. Plant Physiol.*, *35*, 585 (1984).
165. T. Hogetsu and H. Shibaoka, *Planta*, *140*, 15 (1978).
166. P. A. Richmond, J.-P. Métraux, and L. Taiz, *Plant Physiol.*, *65*, 211 (1980).
167. P. A. Richmond, *J. Appl. Polym. Sci.: Appl. Polym. Symp.*, *37*, 107 (1983).
168. P. B. Green, *Ann. Rev. Plant Physiol.*, *31*, 51 (1980).
169. P. B. Green and R. S. Poethig, in *Developmental Order: Its Origin and Regulation* (S. Subtelny and P. B. Green, eds.), Alan R. Liss, New York, 1982, p. 485.

2

Relationship of Cellulose and Other Cell Wall Components: Supramolecular Organization

BRIGITTE VIAN and DANIÈLE REIS   *Laboratoire des Biomembranes et Surfaces Cellulaires Végétales, Ecole Normale Supérieure, Paris, France*

I. INTRODUCTION

Cellulose microfibrils, like the steel rods in reinforced concrete, are the reinforcing components in the plant cell wall. The mechanical properties of cell walls have often been discussed in terms of cellulose characteristics. However, cellulose is generally associated with an important embedding matrix that modulates its properties similarly to the binder in concrete. The present chapter is centered on the supramolecular organization of cell walls. Special attention is given to helicoidal organization, which appears widespread from the literature of the last decade (see Refs. 1 and 2 for complete bibliography). In fact, helicoidal walls do not constitute a unique example since numerous extracellular matrices of both animal and plant cells reveal such organization (3,4). The strong analogy often emphasized between helicoidal constructions and cholesteric systems will be taken into account in the final discussion of the morphogenesis of cell walls.

II. NONCELLULOSIC POLYMERS: THEIR RELATION WITH CELLULOSE

In recent years the complexity of the molecular composition of the cell wall has been greatly emphasized (5−9). The molecular composition varies with plant taxa, tissue type, and state of cell differentiation, so that it is impossible to establish a composition type. The structure of cellulose need not be described in detail here since it is the topic of other chapters (see Chaps. 1 and 4). We have chosen to focus on other polymers that have a close relationship to cellulose, giving special attention to hemicelluloses and their possible role in the morphogenesis of the helicoidal wall. Though further information is still necessary on the actual conformation of the different polymers within the wall, we will attempt to relate their chemical composition and microscopic structure in order to predict the in muro construction as closely as possible. Various organic molecules that are sometimes secondarily added to plant cell walls (e.g., lignin, phenols, and suberin) will not be discussed here.

A. Hemicelluloses: A Close Coating for Cellulose

Hemicelluloses represent a heterogeneous group made of several molecular types (glucans, xylans, and mannans). These molecules share the property of being tightly but noncovalently bound to the microfibrils of cellulose. Hemicelluloses all have a long β-1,4-linked linear backbone with short side chains, so that the whole molecule has the appearance of a pipecleaner (10). The association between cellulose and hemicelluloses is close; the ribbonlike conformation of both these molecular types is similar, thus favoring their alignment and packing by means of hydrogen bonds and other noncovalent interactions (5,11).

Xyloglucans, which are "cellulose-like" because of their glucose backbone (12–18), and xylans, which have a xylose backbone (19–24), have been well studied in recent years. Their primary sequences and structures are well known, but their molecular shapes and their behavior both in vitro and in muro are still open questions (24,25). Figure 1 shows isolated and regenerated glucuronoxylans and fucogalactoxyloglucans seen in the electron microscope after negative staining. Though what is observed corresponds to a dissociated state because of the extracting procedure, the similarity in appearance of both types of polymers is striking. They appear as 30- to 80-nm-long rodlets, which have a tendency to associate along their length forming spindles.

The spontaneous association of hemicelluloses with cellulose microfibrils has actually been shown in several cases: (1) cellulose-mannan and cellulose-glucomannan associations have been observed as shish kebabs after recrystallization and characterized by X-ray diffraction (26,27); (2) cellulose-xylan associations have been

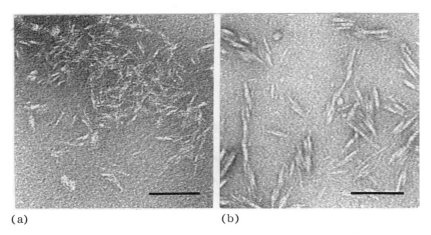

(a)   (b)

FIGURE 1   Negatively stained regenerated hemicellulose molecules.
(a) Fucogalactoxyloglucan from mung bean hypocotyl.   (b) Glucuro-
noxylan from linden tree wood.   Both types of subunits appear as
short and rigid rodlets which have a tendency to pack and join in
files.   Bar = 0.1 μm.

chemically analyzed, observed with electron microscope and shown to
exhibit a reversibility indicating weak bonds (28); and (3) cellulose-
xyloglucan association has been observed by biochemical analysis
(29).

B.   Pectins: A Plastifying Gel

Pectins are the most easily extractable polymers in the cell wall.
They are block copolymers of $1,4$-α-D-galacturonans that have
heavily branched zones alternating with unbranched ones (30).
When extracted by EDTA from the bulk of the wall, pectins constitute a loose network along which the abundance and regularity of
free acidic groups can be checked by using cytochemical markers
(Fig. 2).   The irregular molecular structure of pectins differs significantly from the periodic structure of cellulose and from the hemicelluloses.   Contrary to hemicelluloses, which align on cellulose because of their cellulose-like conformation, the buckled form of the
polygalacturonates confers the ability for strong cohesion with the
involvement of cations so that they form noncovalent gels (11).
Whether a percentage of the pectins are linked to other wall polymers,

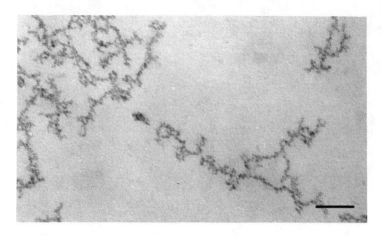

FIGURE 2   Regenerated molecules of pectin from mung bean hypocotyl
in which acidic groups are made visible by colloidal iron binding.
Nonlinear and flexible subunits form a loose network rich in car-
boxylic groups.   Bar = 0.1 μm.

like hemicelluloses, is still unknown (31).   It is clear that in the
wall heteromolecular assembly, the pectins constitute a network with
properties that can be modified by minute changes; they may be
esterified to varying degrees; they may have insertions of rhamnose
units leading to local kinks; and they may carry side chains of
varying length, complexity, and hydrophilic properties (11,30,32,33).

C.   Glycoproteins: An Interpenetrating Network

In 1960 Lamport and Northcote (34) first described extensin as a highly insoluble structural glycoprotein of the cell wall of dicotyledons
that is built from many hydroxyproline residues.   Since that time
many types of hydroxyproline-rich glycoproteins (HRPGs) have been
described in cell walls (e.g., arabinogalactan proteins, lectins, and
sexual agglutinins).   One feature that distinguishes extensin from
other HRGPs is that it is crosslinked in the wall, forming a rigid matrix that is theoretically inextensible, difficult to extract, and resistant to chemical or enzymatic attack.   In dicotyledons, extensin
accounts for 2–10% of primary cell walls (35).
   Numerous biochemical reviews have concerned the wall glycoproteins (35–38).   Extensin precursors have been purified from
cell walls and prepared for electron microscopic observations (39,

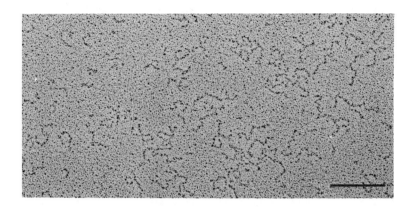

FIGURE 3   Rotary shadowing of precursors of extensin after spreading, showing elongated and flexuous subunits.   Bar = 0.1 μm. (Courtesy of D. T. A. Lamport.)

40).   They appear as elongated, flexuous, rodlike or wormlike molecules of 80 nm in average length (Fig. 3).   They become crosslinked in muro by peroxidase-catalyzed isodityrosine residues and, according to Lamport's working hypothesis, they form a covalent network with porosity that may allow penetration by cellulose microfibrils.

D.   Possible Interconnections

Molecular models have been proposed for the walls of both elongating cells and differentiated cells (37,41,42).   The main points of divergence concern the types of crosslinks between polymers.   A review has focused on the diversity and the importance of the polymer – polymer interactions implied in the building of the primary wall (31). Questions about the degree of precision of the crosslinks, the occurrence of multiple bonds, or the importance of physical entanglements are far from being solved.   It is likely that future developments will be based on the combinations of in vivo and in vitro approaches (29).   In experiments on association – disassociation between compatible molecular types, it is now possible to specifically label and precisely identify the molecular partners, owing to the development of new cytochemical probes (21,24,28,43,44).   This provides a promising tool to settle the problem of molecular interactions in the coming years.

III. SUPRAMOLECULAR CONSTRUCTION: THE
      HELICOIDAL CELL WALL

In the last decade an extensive bibliography has revealed that numerous cell walls have an ordered architecture revealing a helicoidal
pattern. The basic characteristics of such organization have been
described in detail in a review (2). Therefore we will briefly outline the helicoidal model, stressing its versatility and adaptability.

A. Basic Helicoidal Model

Typical helicoidal cell walls from both rapidly elongating and highly
differentiated cells have very similar construction (Fig. 4a,b).
They are multilayered and appear as a succession of nested arcs,
which are best defined when the section is oblique through the
specimen.
     The three-dimensional analysis of the arcs, through both elec-
tron microscopic observation with a goniometric stage and computer
modeling, has revealed their illusory character (3,45–49). The
arcs are actually made of short segments of microfibrils that lie
parallel in the plane of the laminae composing the helicoid. In
order to aid understanding the system, a model built from superimposed mobile disks is proposed (Fig. 5). On each disk the lines
correspond to the parallel microfibrils. The orientation of the microfibrils varies from one disk to the next. In many biological helicoids,
the layers are probably fictitious and the perceived orientations
likely represent only statistical orientations. In the present model,
the mutual angle between each successive sheet is uniform. The
whole gives the illusion of symmetrical arcs that are inverted on opposite sides. The model also shows the occurrence of characteristic
reversion points that are often encountered on sections.

FIGURE 4   Basic helicoidal texture of cell wall in (a) an elongating
epidermal cell of mung bean hypocotyl prepared by methylamine extraction and the PATAg test (periodic acid-thiocarbohydrazide-silver
proteinate), and (b) a stone cell of pear, a highly differentiated
cell, prepared by chlorite and subsequent methylamine extraction followed by the PATAg test. Both specimens are cut obliquely and
show the occurrence of a succession of nested arcs (underlined);
cy: cytoplasm. Bar = 0.5 μm.

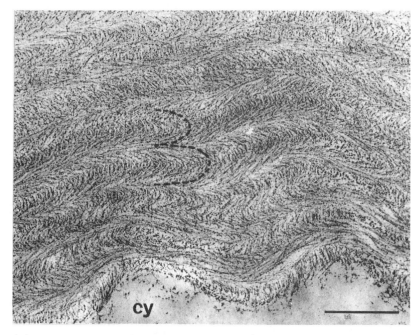

(a)

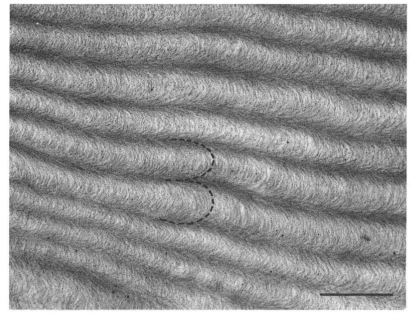

(b)

31

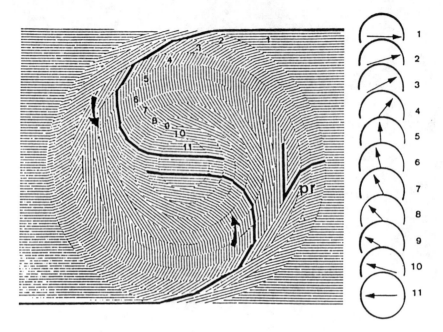

FIGURE 5   Schematic representation of the helicoidal pattern. Model-
ling made with 11 superimposed mobile disks rotating through a constant angle.   The whole gives illusion of arcs which are inverted on
opposite sides.   The hands of a clock mirror the changes in direction
of the microfibrils with time.

B.   Versatility of the Helicoidal
Model in Higher Plants

A systematic analysis shows that cell walls are not all constructed
with a regular arced pattern.   There are examples in which both a
regular helicoid and walls that at first sight do not present arced
textures coexist (Fig. 6).   A careful analysis indicates that both
appearances are probably relevant to the same helicoidal construction but that great textural variability exists depending on the differentiation state (2,44,50,51).

1.   Monotonous Helicoids

Monotonous helicoids occur when microfibrils pass periodically and
with great regularity through similar orientations (Fig. 4 and cell 1
of Fig. 6).   On a curve that expresses the direction of the microfibrils with time, they give rise to a regular sinusoid with a

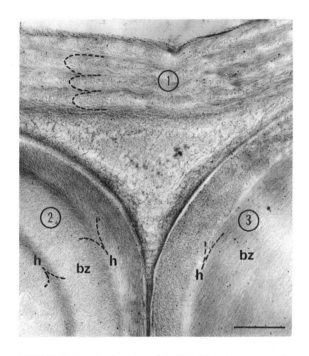

FIGURE 6   Variation of helicoidal texture with differentiation in papyrus stem prepared by methylamine extraction and the PATAg test.   The junction of three neighboring cells shows two types of textures in the walls: 1, parenchyma cell with regular arced pattern, 2 and 3, fibers with irregular pattern displaying transient helicoids (h) between blocked zones (bz).   Bar = 0.5 μm.

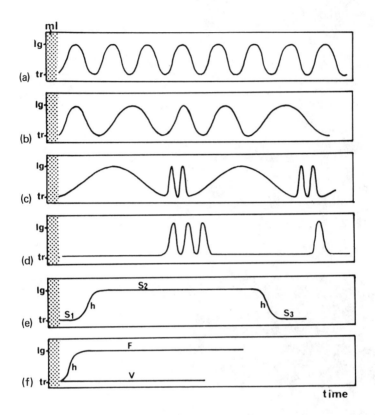

FIGURE 7 Diagram summarizing helicoidal wall expressions that vary according to cell differentiation. (a) Unvarying motion leading to a constant period and a monotonous helicoid as found in epidermis, parenchyma, and stone cells. (b) Varying motion leading to de-stabilized period as found in collenchyma and some sclerenchymas. (c) Bimodal pattern as found in certain sclereids. (d) Arrested motion and bursts of helicoids. (e) Arrested motion in three-layered walls. S1, S2, and S3 alternate with semihelicoids (h). (f) Arrested motion in which the rotation of the microfibrils is limited to a single semihelicoid (h) as found in some fibers, F, or is even lacking as in numerous vessels, V (ml, middle lamella; lg and tr, longitudinal and transverse orientation of the microfibrils, respectively.

constant period (Fig. 7a). Such monotonous helicoids, the pattern
of which can be maintained for a long time, are encountered in many
cells with thick and often lignified cell walls such as sclereids,
sclerocytes, and stone cells (Fig. 4b). They have been infrequently
described since they are difficult to handle for electron microscopy.
In particular, the exposure of the fibrillar framework of these compact walls often needs prolonged chemical or enzymic extractions,
which may have discouraged cytologists (see Refs. 2 and 52 for
discussion).

2. Destabilized Helicoids

Destabilized helicoids, in which no regular arced pattern is discernible, are encountered most frequently. In some cases the arcs
are irregular but still discernible. For example, in collenchyma
cells the period is variable, so that the width of the arcs changes
greatly throughout the thickness of the wall (Fig. 7b). In other
cases helicoids are only intermittent appearing as bursts between
zones where the rotation is either retarded (Fig. 7c) or arrested
(Fig. 7d). In Fig. 8 from a sclerocyte of birthwort stem, the
helicoids appear as bursts within a bulk of wall that apparently
shows no arced pattern. That the bursts do represent a helicoidal
construction can only be deduced when the typical inversion of arcs
is seen from goniometric analysis using a tilting stage. Conversely,
concerning the bulk of the wall where apparently no rotation occurs,
it is difficult to decide without a careful analysis whether it corresponds to a strong delay or an actual stop in the rotating motion.

Other systems exist where the rotation of the microfibrils may
be temporarily (Fig. 7e) or permanently (Fig. 7f) arrested. Fibers
of wood cells provide a well-studied example in which the arced
pattern is transient and limited to transitional semihelicoid zones between the S1, S2, and S3 layers (1,53). Due to the difficulty of
sharply visualizing such discrete helicoidal transitions, they have
not been widely observed before but are likely to be found in many
fibers.

3. Walls Without Apparent Helicoids

Though helicoids are likely widespread in various plant groups and
cell types and may be considered as a basic construction for cell
walls, many cases exist in which they are not actually seen. Different explanations for their absence can be proposed: (1) Typical
helicoidal walls have been built up by the cell but they have disappeared without any trace. For example, in highly expanding cells
it has been shown that the helicoidal texture constructed before the
onset of elongation is progressively and irreversibly lost during the
elongating process (54,55). (2) The rotating motion is arrested in

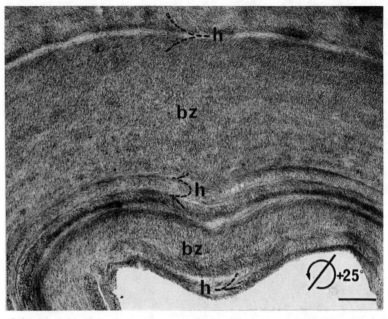

(a)

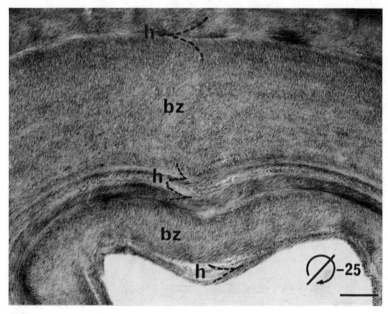

(b)

one predominant direction as soon as the wall is initiated. The helicoidal expression is then restricted to a unilayered wall (as is found in some vessels; see Ref. 56 and Fig. 7f). (3) The helicoid is not expressed. For example, in many organs helicoids are much more developed in the outer parts, which are constantly subjected to the highest strains, whereas they are not encountered in internal tissues. In the latter the walls are very thin and do not show any apparent order (1,57,58). (4) The helicoid does not exist, as is the case of suspension cultured cells, and probably in many apical cells (59).

C. Comparison with Cellulose-Rich Walls of Some Algae

Although the cell walls of algae have structural features in common with those of land plants, cellulose does not occupy a unique place as the skeletal substance in these diverse organisms (60,61). Xylans and mannans also often form crystalline microfibrils in their walls. When cellulose microfibrils are present, they may be found exclusively or in addition to microfibrils of other polysaccharides within the cell wall skeleton. Cellulose can constitute from 30 to 75% of the bulk of the wall, a proportion exceeding that found in most cellulosic secondary walls of higher plants. We will focus on the characteristics of cell walls of some of these cellulose-rich algae, which have received extensive attention in the literature. These algae have ordered walls either with criss-crossed textures which are likely not relevant to helicoidal organization, or with helicoids that are deposited during only part of the cell cycle.

The wall of *Oocystis* is the best known example of the typical criss-crossed pattern (62–64) (Fig. 9). The wall is composed of alternating layers in which large (6 nm average diameter), highly crystalline cellulose microfibrils lying in parallel run at a right angle to those of the adjacent layer. *Glaucocystis* provides another similar example with a crossed polylamellate wall (65). The case of *Micrasterias* is more complex because the secondary wall is formed from criss-crossed bands of many microfibrils associated in bundles

---

FIGURE 8 Irregular helicoidal texture in a sclerocyte of birthwort stem prepared by methylamine extraction and the PATAg test, and visualized with a tilting stage: (a) +25° tilt; (b) −25° tilt. Bursts of helicoids (h) alternate with blocked zones (bz) of varying size. Note the typical goniometric effect of arc inversion in the helicoid when the section is tilted. Bar = 0.5 μm.

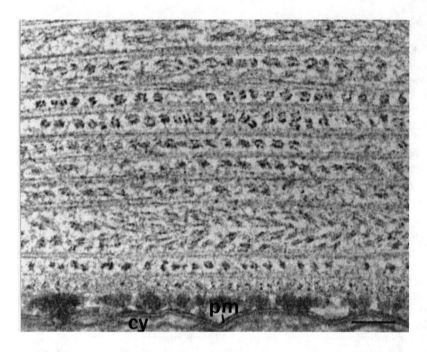

FIGURE 9   Criss-crossed texture of the mature cell wall of *Oocystis* visualized by positive staining.   Each layer of parallel microfibrils is orthogonally oriented to the succeeding one (pm, plasmalemma; cy, cytoplasm).   Bar = 0.1 μm.   (Courtesy of R. M. Brown.)

(66).   All these cases differ primarily in the degree of association of the microfibrils.

The actual texture of the wall of *Valonia* is less clear.   Analyzed in the pioneer studies on cell walls (67), it has been classically considered as having a "crossed" wall organization.   However, the existence of intermediate layers was recently proposed (68).   Conversely, the occurrence of typically arced patterns is unequivocal in many algal examples: embryos of *Pelvetia* (69), aplanospores of *Boergesenia* (70), zygotes of *Closterium* (71), filamentous cells and akinetes of *Pitophora* (72), and internode cells of *Nitella* (Fig. 10) (48).   The organization of the *Nitella* wall is highly similar to that of higher plants, showing a pattern of repetitive arcs (up to 60) corresponding to the helicoidal model described above.

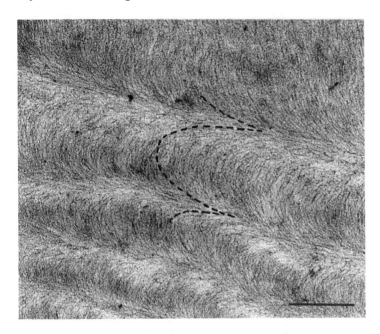

FIGURE 10   Helicoidal texture of the cell wall in a mature internode cell of *Nitella* visualized by positive staining.   Wall shows the typical succession of nested arcs.   Bar = 0.5 μm.   (Courtesy of S. Levy.)

IV. CELL MEDIATION OF ORDERED
WALL CONSTRUCTION

It has been well established biochemically that the secretion of the plant cell wall depends on coordinated and sequential synthesis involving the endomembrane system and the plasma membrane.   It is now important to determine what tools are at the cell's disposal to control the precisely ordered but adaptable pattern of helicoidal morphogenesis.   In the following discussion we hypothesize that the control of the wall supramolecular order is likely different for cells building simple criss-crossed walls with a high percentage of crystalline cellulose (like many algal cells) than for cells building helicoidal walls.   In the latter, cellulose has been shown from spectroscopic studies to be less crystalline (73).   It represents generally less than 50% of the bulk of the wall and is closely associated with a range of specific and complex matrix components.

A. Criss-Crossed Texture and Predominant Role
of the Transmembrane Machinery

Much has been written about the possible roles of the membrane-
synthesizing complexes (terminal complexes) and of the associated
cytoskeletal elements in determining microfibril orientation (see
Chap. 7). Hypothetical evolution schemes have been proposed that
try to integrate the changes in the putative cellulose synthase complexes and the corresponding organizations of cell walls (74–77).
Figure 11 illustrates examples chosen from the two main types of

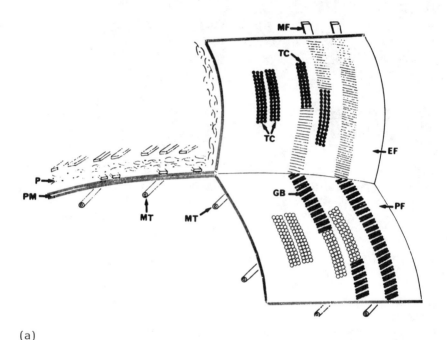

(a)

FIGURE 11   Models of cellulose microfibril deposition in algal systems.   (a) *Oocystis* model.   The synthesis and orientation of the
cellulose microfibrils (MF) appear as the result of the cooperation
between the terminal complexes (TC) and the granule bands (GB),
located in the exoplasmic (EF) and protoplasmic (PF) fracture faces
of the plasmalemma (PM), respectively.   Microtubules (MT) sub-
tending the plasmalemma may also orient the microfibrils.   (Reproduced from Ref. 63 by permission.)   (b) *Micrasterias* model.   The
oriented synthesis of the cellulose microfibrils (MF) appears as a
the result of the migration of arrays of rosettes (R) within the
plane of the plasmalemma (PM) (arrows).   (Reproduced from Ref.
66 by permission.)

described terminal complexes: the linear and the nonlinear rosette (the so-called rosettes) types.

What emerges from the literature is that spatially organized and consolidated complexes are encountered in algal cells that have ordered, but not helicoidal, walls. *Oocystis* provides the best example in that many observations and experimental treatments have supported the involvement of the linear terminal complexes, likely directed by microtubules, in the synthesis and the orientation of the crossed layers (62–64; see Chap. 3). Hexagonal arrays of rosettes (putative cellulose synthetases) are regularly found in algae building secondary walls made of regular bands of broad microfibrils (66; see Chap. 4). As emphasized by Herth (76), the variability in geometry of the terminal complexes is in good correlation with the variability in size, crystallinity, and the degree of association of the synthesized cellulose fibrils (see Chap. 4).

B. Helicoidal Walls and Importance of the Matrix

In helicoidal walls the problem then becomes to identify what regulates the gradual rotation of the cellulose microfibrils through different orientations. First we will consider that, although engaged in the synthesis of the cellulose, the plasmalemma-bound complexes acting in cooperation with the underlying cytoskeleton are an insufficient explanation for the complexity of the helicoidal ordering. Then, based on the analogy between helicoidal walls and cholesteric systems, we will hypothesize that the enrichment of the matrix in molecules capable of spontaneous association with cellulose may possibly help control such precise organization.

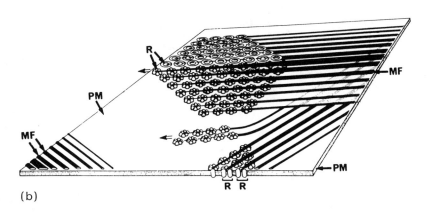

(b)

FIGURE 11   (Continued)

1. Involvement of Rosettes and Cytoskeleton

The cell wall of the alga *Closterium* provides a transitional example since it is built up from both criss-crossed and helicoidal layers. The freeze-fracture analysis of the plasma membrane during the cell cycle is significant: hexagonal arrays of rosettes are involved in the formation of the criss-crossed bundle layers, whereas a precise assembly of rosettes is never found during the formation of the helicoidal layers (71). In *Nitella*, often considered as a possible ancestor of vascular plants, the wall is unequivocally helicoidal. Recently, it was shown that its terminal complex morphology is indistinguishable from that observed in vascular plants (78). In the latter, recent progress in freeze-fracture studies on putative cellulose synthase complexes indicate the regular occurrence of rosettes (76,79,80). However, one never finds a pattern in the organization of the complexes that could correspond to the high degree of order encountered in helicoidal walls.

Whether the underlying microtubules and related compounds are directing the rosettes in orienting the microfibrils is still under debate (see Chap. 7). Recent developments of new techniques, particularly whole-cell indirect immunofluorescence, have provided new insights into the organization and behavior of cortical microtubules; they can wind around the cell describing helices of different pitches. This internal template is capable of changing its pitch in response to external signals. Such properties of microtubules have ruined the common assertion that cytoskeleton behavior is incompatible with helicoidal wall patterns (81–83). However, although a precise spatial organization of the cortical cytoskeleton occurs, it is probably not sufficient to explain the regular rotation of cellulose microfibrils observed in helicoidal walls. In such systems, the direct involvement of microtubules in wall ordering seems improbable (77, 84,85).

2. Helicoidal Walls and Cholesteric-Like Systems

For a more complete understanding of the present question, a new perspective is possible because of the striking similarity between helicoidal cell walls and cholesteric liquid crystals. The question is whether it is possible to transpose the mechanisms of assembly of molecules into cholesteric mesophases to the mechanisms of assembly of helicoidal walls.

In vitro and theoretical studies of macromolecular systems forming cholesteric mesophases indicate that, whatever the chemical nature of the molecules, their self-assembly is favored by some characteristics: elongated and stiff configuration; occurrence of short and flexible side chains; and possible helix conformation (86). Except for early studies on the spontaneous arrangement of cellulose

FIGURE 12   Cholesteric structure of the cellulose derivative.
Hydroxypropylcellulose visualized through crossed polarizers in the
light microscope. The sample is a fluid mesophase of 70% (w/v)
hydroxypropylcellulose in aqueous solution. Bar = 50 µm. (Courtesy
of R. Werbowyj.)

following partial hydrolysis (87), carbohydrates were only recently
analyzed in terms of self-assembly properties (88 – 92).   Cellulose
and hemicellulose are both chiral molecules theoretically capable of
constituting cholesteric mesophases.   When cellulose is too crystalline, the assembly becomes impossible (93).   Conversely, it has
been shown that certain cellulose derivatives, such as hydroxypropylcellulose, are capable of forming cholesteric and lyotropic
liquid crystals (Fig. 12; 93 – 97).   Such molecules possess side
chains added to the glucose backbone that prevent crystallization
and likely act as an internal plasticizer so that the molecules disen-
tangle and spontaneously orient themselves (93).

3. Helicoidal Assembly and Cellulose/
Hemicellulose Balance

Concerning cellulose in situ, it has been proven that the two steps
of polymerization and crystallization can be separated (74,98,99;
see Chap. 5).   The latter step occurs spontaneously and can be
interpreted as a form of homomolecular self-assembly.   In the helicoidal cell walls, a second degree of order exists which associates
the subunits in heteromolecular complexes.   That helicoidal walls
could self-assemble via a liquid crystal stage had already been
proposed from earlier studies (44,46,100 – 103).   Recent developments
strongly strengthen such a working hypothesis.   The change in

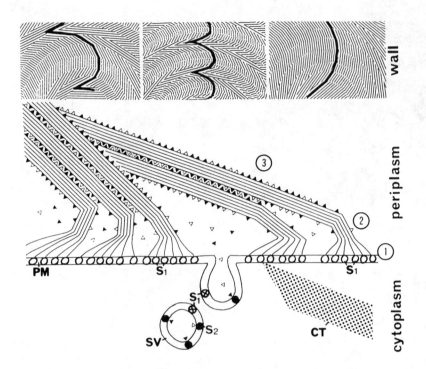

FIGURE 13 Hypothetical scheme of the successive steps in the construction of the helicoidal cell wall. 1, polymerization; 2, fibrillogenesis; 3, helicoidal assembly; △ ▼, pool of matrix subunits; ⊗, S1, inactivated cellulose synthetases located in the membrane of cytoplasmic secretory vesicles (SV); ○, S1, activated cellulose synthetases located in the plasmalemma (PM); ●, S2, matrix synthetases located in the membrane of secretory vesicles (SV). Matrix subunits are excreted into the periplasm and associate with the glucan chains in the course of crystallization, giving rise to a coating which allows the helicoidal construction. The different aspects of wall helicoids may depend on the cellulose-matrix balance (see text).

helicoidal order and even the lack of order obtained experimentally
under a change in hemicellulose composition or a deficit in cellulose
suggest that the helicoidal assembly is highly dependent on a precise balance, both qualitative and quantitative, between cellulose
and matrix components (104). Moreover, from the three-layered
cell walls of hardwoods, a correspondence has been shown between
an accumulation of glucuronoxylans, the major hemicellulose component of these walls, and the resuming of rotation of the cellulose
microfibrils (24).

Figure 13 is an attempt to schematically integrate the main
events. What emerges is that the association between cellulose and
compatible hemicelluloses likely occurs at the moment of fibrillogenesis. This close association could control the degree of crystallization of cellulose and provide the microfibrils a mobility that allows
a spontaneous orientation related to the preexisting wall layers.
The effect would be analogous to the formation of liquid crystals
when side chains are grafted on the glucan backbone in cellulose
derivatives (93). Indeed, a clear difference exists between liquid
crystals obtained from cellulose derivatives in which the molecules
are semirigid, and those obtained from microfibrillar suspensions in
which the polymeric whiskers are tremendously rigid (44,105). One
can hypothesize that in algal systems, which lack a close association
between cellulose and matrix, extensive homomolecular cellulose
crystallization occurs, whereas in higher plants the chiral hemicellulose molecules play the role of internal plasticizer or twisting
agents. To some extent the helicoidal texture seems dependent on
the balance between these rigid/flexible systems and a relation likely
exists between the type of matrix and the rate of rotation.

V. CONCLUDING REMARKS

We have focused on the characteristics of cellulose-rich cell walls,
emphasizing their three-dimensional organization. Many questions
are still unanswered concerning the morphogenesis of the ordered
construction. For example, what is the exact involvement of the
cytoskeletal elements? How is the processing of the matrix polymers controlled, particularly the grafting of the side chains? How
and when does the activation of both the synthesizing and the
breaking enzymes occur? This list is not complete and in conclud-
ing this chapter we would like to draw attention to the necessity
for a complete analysis of the sequence of events occurring within
the cell, so that some pieces still missing in the puzzle can be found.
We would like to note that this work was submitted to the publisher in
1987.

REFERENCES

1. A. C. Neville and S. Levy, in Biochemistry of Plant Cell Walls (C. T. Brett and J. R. Hillman, eds.), Cambridge University Press, New York, 1985, p. 99.
2. J. C. Roland, D. Reis, B. Vian, B. Satiat-Jeunemaitre, and M. Mosiniak, Protoplasma, 140, 75 (1987).
3. Y. Bouligand, Tissue and Cell, 18, 621 (1972).
4. A. C. Neville, Phys. Bull., 37, 74 (1986).
5. R. D. Preston, Ann. Rev. Plant Physiol., 30, 55 (1979).
6. R. R. Selvendran, B. J. H. Dube, and M. A. O'Neil, in Biochemistry of Plant Cell Walls (C. T. Brett, ed.), Cambridge University Press, New York, 1982, p. 39.
7. M. MacNeil, A. G. Darvill, S. C. Fry, and P. Albersheim, Ann. Rev. Bioch., 53, 625 (1984).
8. R. R. Selvendran, J. Cell Sci., 2, 51 (1986).
9. D. H. Northcote, J. Cell Sci., 4, 115 (1986).
10. A. C. Neville, J. Theor. Biol., 131, 243 (1988).
11. D. A. Rees, Polysaccharide Shapes, Chapman and Hall, London, 1977, p. 45.
12. G. Chambat, F. Barnoud, and J. P. Joseleau, Plant Physiol., 74, 687 (1984).
13. J. P. Joseleau and G. Chambat, Plant Physiol., 74, 694 (1984).
14. T. Hayashi and G. MacLachlan, Plant Physiol., 75, 596 (1984).
15. W. S. York, A. G. Darvill, and P. Albersheim, Plant Physiol., 75, 295 (1984).
16. S. C. Fry, Planta, 169, 443 (1986).
17. S. C. Fry, J. Exp. Bot., 40, 1 (1989).
18. T. Hayashi, Ann. Rev. Plant Physiol. Plant Mol. Biol., 40, 139 (1989).
19. D. Fengel and M. Przyklenk, Wood Sci. Technol., 10, 311 (1976).
20. N. C. Carpita, Plant Physiol., 72, 515 (1983).
21. K. Ruel, Etude en microscopie électronique des interrelations cellulose-hémicelluloses-lignine dans les parois végétales, Thesis, Grenoble, 1984.
22. Y. Kato and D. J. Nevins, Plant Physiol., 75, 759 (1984).
23. N. C. Carpita and D. Whittern, Carbohyd. Res., 146, 129 (1986).
24. B. Vian, D. Reis, M. Mosiniak, and J. C. Roland, Protoplasma, 131, 185 (1986).
25. M. C. Mc Cann, B. Wells, and K. Roberts, in Proc. 5th Cell Wall Meeting, Edinburgh, 1989, p. 79.
26. H. Chanzy, M. Dube, R. H. Marchessault, and J. F. Revol, Biopolymers, 18, 887 (1979).

27. H. Chanzy, A. Grosrenaud, J. P. Joseleau, M. Dube, and R. H. Marchessault, *Biopolymers*, *21*, 301 (1982).
28. F. Mora, K. Ruel, J. Comtat, and J. P. Joseleau, *Holzforschung*, *40*, 85 (1986).
29. T. Hayashi, M. P. F. Marsden, and D. P. Delmer, *Plant Physiol.*, *83*, 384 (1987).
30. M. C. Jarvis, *Plant Cell Env.*, *7*, 153 (1984).
31. S. C. Fry, *Ann. Rev. Plant Physiol.*, *37*, 165 (1986).
32. P. Albersheim, in *Plant Biochemistry* (J. Bonner and J. E. Varner, eds.), Academic Press, New York, 1976, p. 226.
33. R. Goldberg, *Physiol. Plant.*, *61*, 58 (1984).
34. D. T. A. Lamport and D. H. Northcote, *Nature*, *188*, 665 (1960).
35. L. G. Wilson and J. C. Fry, *Plant Cell Env.*, *9*, 239 (1986).
36. M. T. Esquerré-Tugaye and D. Mazau, *Physiol. Vég.*, *19*, 415 (1981).
37. D. T. A. Lamport, in *Cellulose: Structure, Modification and Hydrolysis* (R. A. Young and R. M. Rowell, eds.), John Wiley and Sons, New York, 1986, p. 77.
38. M. L. Tierney and J. E. Varner, *Plant Physiol.*, *84*, 1 (1987).
39. D. T. A. Lamport, E. P. Muldoon, M. Kieliszewski, J. J. Willard, B. Terhune, J. W. Heckman, and D. Everdeen, in *Cell Walls '86* (B. Vian, D. Reis, and R. Goldberg, eds.), Paris, 1986, p. 8.
40. K. M. M. Swords and L. A. Staehelin, in *Plant Fibers* (H. F. Linskens and J. F. Jackson, eds.), Springer-Verlag, New York, 1989, p. 79.
41. K. Keegstra, K. W. Talmadge, W. D. Bauer, and P. Albersheim, *Plant Physiol.*, *51*, 188 (1973).
42. D. Fengel and G. Wegener, *Wood*, Walter de Gruyter, Berlin, 1984, p. 227.
43. B. Vian and D. Reis, *Food Hydrocolloids*, *1*, 557 (1987).
44. J. C. Roland, D. Reis, B. Vian, and S. Roy, *Biol. Cell*, *67*, 209 (1989).
45. F. Livolant, *Biol. Cell.*, *31*, 159 (1978).
46. A. C. Neville, D. C. Gubb, and R. M. Crawford, *Protoplasma*, *90*, 307 (1976).
47. J. C. Roland and B. Vian, *Int. Rev. Cytol.*, *61*, 129 (1979).
48. A. C. Neville and S. Levy, *Planta*, *162*, 370 (1984).
49. D. Reis, B. Vian, B. Satiat-Jeunemaitre, and J. C. Roland, in *Cell Walls '86* (B. Vian, D. Reis, and R. Goldberg, eds.), Paris, 1986, p. 16.
50. J. C. Roland, in *Cell Walls '81* (D. G. Robinson and H. Quader, eds.), Wissenschaftliche Verlagsgesellschaft, Stuttgart, 1981, p. 162.

51. M. Mosiniak and J. C. Roland, *Ann. Sci. Natl. Bot.*, 7, 175 (1985).
52. M. Mosiniak-Bessoles, Contribution à l'étude des oscillations spontanées de l'assemblage cellulosique dans les parois des cellules végétales, Thesis, Paris, 1987.
53. J. C. Roland and M. Mosiniak, *IAWA Bull.*, 4, 15 (1983).
54. J. C. Roland, D. Reis, M. Mosiniak, and B. Vian, *J. Cell Sci.*, 56, 303 (1982).
55. B. Vian and J. C. Roland, *New Phytol.*, 105, 345 (1987).
56. H. Harada and W. A. Cote, in *Biosynthesis and Biodegradation of Wood Components* (T. Higuchi, ed.), Academic Press, New York, 1985, p. 1.
57. R. Prat and J. C. Roland, *Physiol. Vég.*, 18, 241 (1980).
58. B. Vian, in *Cellulose and Other Natural Polymer Systems* (R. M. Brown, ed.), Plenum Press, New York, 1982, p. 23.
59. M. M. A. Sassen, J. A. Traas, and A. M. C. Wolters-Arts, *Eur. J. Cell Biol.*, 37, 21 (1985).
60. E. Frei and R. D. Preston, *Nature*, 192, 939 (1961).
61. E. Perceval and R. H. McDowell, in *Plant Carbohydrate*, Vol. 2, *Extracellular Carbohydrates* (W. Tanner and F. A. Loewus, eds.), Springer-Verlag, New York, 1981, p. 277.
62. D. Montezinos and R. M. Brown, *J. Supramol. Struc.*, 5, 277 (1976).
63. D. Montezinos, in *Cellulose and Other Natural Polymer Systems* (R. M. Brown, ed.), Plenum Press, New York, 1982, p. 3.
64. H. Quader, *J. Cell Sci.*, 83, 223 (1986).
65. J. H. M. Willison and R. M. Brown, *J. Cell Biol.*, 77, 103 (1978).
66. T. H. Giddings, D. L. Brower, and A. Staehelin, *J. Cell Biol.*, 84, 327 (1980).
67. F. C. Steward and K. Mühlethaler, *Ann. Bot.*, 17, 301 (1953).
68. H. Harada and T. Goto, in *Cellulose and Other Natural Polymer Systems* (R. M. Brown, ed.), Plenum Press, New York, 1982, p. 383.
69. H. B. Peng and L. F. Jaffe, *Planta*, 133, 57 (1976).
70. S. Mizuta and S. Wada, *Bot. Mag. Tokyo*, 94, 343 (1981).
71. T. Noguchi and K. Ueda, *Protoplasma*, 128, 64 (1985).
72. N. L. Pearlmutter and C. A. Lembi, *J. Phycol.*, 16, 602 (1980).
73. R. H. Atalla and D. L. Vanderhart, *Science*, 223, 283 (1984).
74. R. M. Brown, C. H. Haigler, J. Suttie, A. R. White, E. Roberts, C. Smith, T. Itoh, and K. Cooper, *J. Appl. Polym. Sci.*, 37, 33 (1983).

75. R. M. Brown, in *The Cell Surface in Plant Growth and Development* (K. Roberts, A. W. B. Johnston, C. W. Lloyd, P. Shaw, and H. W. Woolhouse, eds.), Company of Biologists Limited, Cambridge, 1985, p. 13.
76. W. Herth, in *Botanical Microscopy 1985* (A. W. Robards, ed.), Oxford University Press, 1985, p. 285.
77. T. Itoh, in *Plant Cell Wall Polymers: Biogenesis and Biodegradation* (N. G. Lewis and M. G. Paice, eds.), American Chemical Society, Washington, D.C., 1989, p. 257.
78. T. Hotchkiss and R. M. Brown, *J. Phycol.*, *23*, 229 (1987).
79. W. Herth and I. Hauser, in *Structure, Function and Biosynthesis of Plant Cell Walls* (M. Dugger and S. Barnicki-Garcia, eds.), American Society of Plant Physiol., Riverside, 1984, p. 89.
80. R. L. Chapman and L. A. Staehelin, *J. Ultrastruc. Res.*, *93*, 87 (1985).
81. C. W. Lloyd, *Int. Rev. Cytol.*, *86*, 1 (1984).
82. I. N. Roberts, C. W. Lloyd, and K. Roberts, *Planta*, *164*, 439 (1985).
83. C. W. Lloyd, *Ann. Rev. Plant Physiol.*, *38*, 119 (1987).
84. P. A. Richmond, E. Byron, and M. Yousef, in *Proc. 5th Cell Wall Meeting*, Edinburgh, 1989, p. 19.
85. F. W. A. Wilms and J. Derksen, in *Proc. 5th Cell Wall Meeting*, Edinburgh, 1989, p. 48.
86. G. H. Brown and J. J. Wolken, *Liquid Crystals and Biological Structures*, Academic Press, New York, 1979.
87. R. H. Marchessault, F. F. Morehead, and N. M. Walter, *Nature*, *184*, 632 (1959).
88. G. Maret, M. Milas, and M. Rinaudo, *Polym. Bull.*, *4*, 291 (1981).
89. K. Van, T. Norisuye, and A. Teramoto, *Mol. Cryst. Liq. Cryst.*, *78*, 123 (1981).
90. H. Chanzy, B. Chumpitazi, and A. Peguy, *Carboh. Polym.*, *2*, 35 (1982).
91. F. Livolant, La structure cristalline liquide de l'ADN in vivo and in vitro, Thesis, Paris, 1984.
92. G. A. Geffrey, *Acc. Chem. Res.*, *19*, 168 (1986).
93. D. G. Gray, *J. Appl. Polym. Sci.*, *37*, 179 (1983).
94. R. Werbowyj and D. G. Gray, *Mol. Cryst. Liq. Cryst.*, *34*, 97 (1976).
95. R. Werbowyj and D. G. Gray, *Macromol.*, *13*, 69 (1980).
96. F. Fried, J. M. Gilli, and P. Sixou, in *Proc. 9th Cellulose Conf.*, Syracuse, 1982, p. 13.
97. G. Charlet and D. G. Gray, *Macromolecule*, *20*, 33 (1987).

98. C. H. Haigler, R. M. Brown, and M. Benziman, *Science, 210*, 903 (1980).
99. C. H. Haigler and M. Benziman, in *Cellulose and Other Natural Polymer Systems* (R. M. Brown, ed.), Plenum Press, New York, 1982, p. 273.
100. J. C. Roland, B. Vian, and D. Reis, *Protoplasma, 91*, 125 (1977).
101. D. Reis, *Ann. Sci. Nat., 19*, 163 (1978).
102. D. Reis, *Mol. Cryst. Liq. Cryst, 43* (1987).
103. J. H. M. Willison and R. M. Abeysekera, *J. Polym. Sci., 26*, 71 (1988).
104. B. Satiat-Jeunemaitre, *Biol. Cell, 59*, 89 (1987).
105. T. Folda, H. Hoffmann, H. Chanzy, and P. Smith, *Nature, 333*, 55 (1988).

3

Role of Linear Terminal Complexes in Cellulose Synthesis

H. QUADER   *Universität Heidelberg, Heidelberg, Federal Republic of Germany*

I. INTRODUCTION

Cellulose fibrils constitute the stiff backbone of the cell wall of higher plants and many algae, especially the Chlorophyceae. They are composed of β-1,4-linked glucan chains that are assembled into a crystalline entity, the microfibril. The number of polyglucan chains constituting a microfibril depends on the natural habitat and particular growth characteristics of the individual plants, resulting in extensive variation in microfibril width (1,2). The relatively thin 7- to 10-nm microfibrils observed in higher plant cells contrast with those found in some algae, which can be 12−30 nm wide. These thick microfibrils were the first to be individually visualized (see Figs. 1−3).

The notion that cellulose seems to be a chemical concept of unusual simplicity (3) certainly does not apply to its biogenesis. Besides the polymerization of β-1,4-linked glucan chains cellulose biosynthesis involves chain crystallization and, often, microfibril orientation (4). Ultrastructural studies have revealed within the plasma membrane (generally accepted as the site of microfibril assembly; see Chap. 9) particulate complexes that have been postulated to be involved in cellulose biogenesis or even to represent the cellulose synthase. This chapter concentrates on the particulate

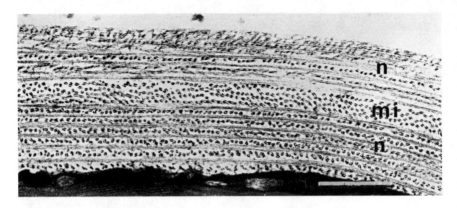

FIGURE 1   Cell wall segment of *Oocystis solitaria* showing alternating orientation of microfibril layers.   n, normal pattern of microfibril layers; mi, microfibril pattern caused by microtubule inhibitor
treatment (3 hr) in which all microfibrils apparently have similar
orientation.   Bar = 0.5 μm.

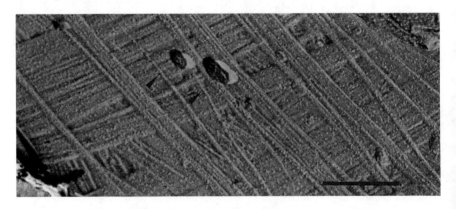

FIGURE 2   Alternating layers of cellulose fibrils of *Oocystis
solitaria* visualized by freeze-fracturing.   Note tapering at the ends
of two microfibrils (arrow heads)   Bar = 0.5 μm.

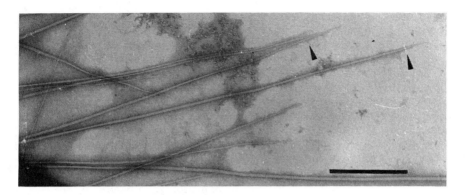

FIGURE 3   Isolated microfibrils of *Oocystis solitaria* visualized by freeze-fracturing.   Note tapering at the ends of two microfibrils (arrow heads).   Bar = 0.5 μm.   (Courtesy of Prof. E. Schnept.)

complexes, called "linear terminal complexes" (5), that have been observed in algae with thick solitary microfibrils.   These complexes are distinct from the rosette/globular type of terminal complex found in higher plants and some other algae (see Chap. 4).   The occurrence and possible function of linear terminal complexes in microfibril assembly will be discussed.

II. CELLULOSE BIOSYNTHESIS: SEARCH FOR THE ASSEMBLING UNIT

A. Historical Background

Isotope labeling and electron microscopy have significantly stimulated cellulose research and contributed to the understanding of cellulose biogenesis ever since the microfibrillar shape of cellulose was demonstrated by electron microscopy (see Ref. 1).   Thirty years ago, Roelofson (6) suggested a comprehensive idea regarding the biogenesis of cellulose fibrils.   The major aspects of his hypothesis were that the growth of cellulose fibrils is due to the action of an enzyme at the tip of the fibril; microfibril growth involves a nearly simultaneous process of polymerization and crystallization; and the synthesis of cellulose takes place on the exterior side of the plasma

membrane. A few years later this notion was supported by the
studies of Preston (7) and Moor and Mühlethaler (8). Preston
studied the formation of ordered microfibril lamellae in *Chaetamorpha*
and suggested that ordered granule complexes, which were fixed in
the membrane, participated in the formation of the microfibrils. At
the same time, Moor and Mühlethaler (8) improved the freeze-etch
technique by accomplishing quick freezing of the material. They
showed that closely packed particles occur in the plasma membrane
of yeast and that fibrils terminate in their vicinity. The model of
Preston (7) presumed that the cell surface was extensively occupied
by granules, which was supported by the observed particle aggregates on the yeast plasma membrane. Later, in two marine algae,
Cladophora and *Chaetamorpha* (9), and in the unicellular fresh
water alga *Oocystis* (10), particle aggregates were visualized on the
plasma membrane that were assumed to be involved in microfibril
production. Although these early observations were quite promising,
their significance was eventually questioned because of advancing
knowledge about molecular movement within the lipid bilayers of the
plasma membrane and because of the sometimes confusing terminology
used to identify the membrane faces in freeze-etch replicas at that
time.

B. The Terminal Complex Hypothesis

The model of fixed ordered granule complexes to explain the deposition of microfibril lamellae with different orientations had to be re-
evaluated after the fluid mosiac model for membranes was introduced
(11). The agreement regarding the nomenclature of the exposed
fracture faces of the cellular membranes by the freeze-etch technique (12) further clarified communication about membrane-associated
events. A major advance was made by Brown and Montezinos (5)
who investigated cell wall formation in the unicellular green alga
Oocystis apiculata. During microfibril deposition they observed long
particle aggregates in the outer leaflet of the plasma membrane bilayer (the EF face). They named these aggregates "terminal complexes" (TCs) because they are often found at the end of a microfibril imprint (see Fig. 4) and proposed that they were the cellulose-
synthesizing unit of *Oocystis*. The TCs are composed of three rows
of about 30–40 particles, each 7 nm in size, yielding a complex of
about 500 nm in length and 30–35 nm in breadth. A few years
previously, Robinson and Preston (10) had observed in the same
alga particle complexes called granule bands, which they erroneously allocated to the surface of the plasma membrane. Brown and
Montezinos (5) confirmed their occurrence, but according to the
nomenclature of Branton et al. (12) they are located on the PF face
of the plasma membrane.

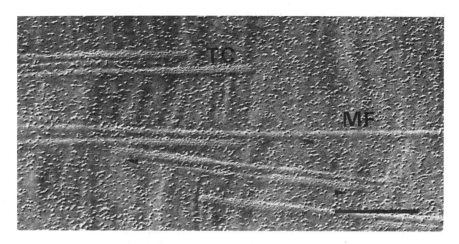

FIGURE 4   EF face of the plasma membrane of *Oocystis solitaria*. Linear terminal complexes (TC) in close association with microfibril imprints (MF). Arrow heads indicate direction of movement of the TCs.   Bar = 0.5 μm.

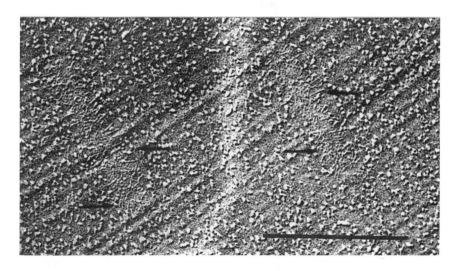

FIGURE 5   EF face of the plasma membrane of *Oocystis solitaria* treated with the calmodulin antagonist fluphenazine (10 μM).   Two doublets with altered morphology are indicated by arrows.   Appendices emanate from the subunits of the TC.   Compare the altered appearance of the TC with the control TC in Fig. 4.   Bar = 0.5 μm.

III. OCCURRENCE OF LINEAR TERMINAL
   COMPLEXES

All the organisms in which "linear terminal complexes" have been
demonstrated to date (Table 1) are characterized by cellulose fibrils
at least 12 nm wide and cell walls composed of alternating lamellae in
which the single microfibrils are aligned parallel to each other
(Fig. 1).

A. Complexes Composed of Single Rows
   of Particles

In *Pelvetia fastigiata* Peng and Jaffe (13) discovered single rows of
particles in the PF face of the plasma membrane. These particle
rows are often inferred to be cellulose fibril-synthesizing complexes
(14,15). The authors themselves did not suggest, however, that
they were involved in cellulose synthesis. They instead postulated
that these particles might orient the microfibrils because they are
found under and along microfibril imprints. Similar rows, but of only
only four to five particles, have been observed in cellulose stalk-
forming *Dictyostelium discoideum* (16), but no clear evidence could
be obtained regarding their function. At present, the evidence
that string-like TCs are involved in cellulose assembly is weak.

B. Complexes Composed of Two or More
   Rows of Particles

In *Pelvetia* the single strings of particles may occasionally be close
together, giving the impression of packed aggregates of two or even
more strings of particles (13). There is, however, no tight association between the particles of neighboring strings as is typical for
the linear TCs observed, for instance, in *Oocystis* (Fig. 4). The
involvement of these loosely aggregated strings in cellulose assembly
is as doubtful as in the case of single strings.

TCs similar to those demonstrated in *Oocystis apiculata* (5) or
Oocystis solitaria (17) have also been observed in several other
fresh water (Chlorococcales) and marine algae (Cladophorales and
Siphonocladales; Figs. 6 and 7). The linear TCs of the freshwater
and the marine algae are similar in their width of about 30 – 40 nm,
but differ in length, subunit size and alignment, and location in the
plasma membrane bilayer (Table 1). They are located in the outer
leaflet of the membranes of the freshwater algae *Oocystis*, *Glaucocystis*, and *Eremosphera*, but they traverse the plasma membrane in
marine algae (14,18,19,42) and go with either one of the membrane
leaflets during freeze-fracture (20). The granule bands observed in
the PF face of the plasma membrane of *Oocystis* have not been
detected in marine algae. The subunits of the TC of marine algae

TABLE 1   Organisms Shown to Have Linear Terminal Complexes in Their Plasma Membrane

Organism	Habitat	TC location	Length (nm)	Width (nm)	Particle size
Oocystis solitaria[a]	Fresh water	EF-face	350[f]	30−35[f]	7
Oocystis apiculata[b]	Fresh water	EF-face	500	30−35	7
Eremosphaera sp.	Fresh water	EF-face	?	?	?
Glaucocystis nostochinearum[c]	Fresh water	EF-face	350	35	7
Valonia macrophysa[d]	Sea water	EF-PF[d]	550	31	9
Valonia ventricosa	Sea water	EF-PF	600	30	8
Boergesenia forbesii	Sea water	EF-PF	665	33	9
Dictyosphaeria cavernosa	Sea water	EF-PF	400	33	8
Siphoncladus tropicus	Sea water	EF-PF	460	36	7
Struvea elegans	Sea water	EF-PF	520	36	7
Boodlea composita	Sea water	EF-PF	400	35	8
Microdyction	Sea water	EF-PF	?	?	?
Chaetomorpha aera	Sea water	EF-PF	250	32	8

[a]Ref. 23.
[b]Ref. 5.
[c]Ref. 41.
[d]Marine algae, Refs. 14, 18−20, 42.
[e]EF-PF indicates transmembrane location.
[f]Averaged values.

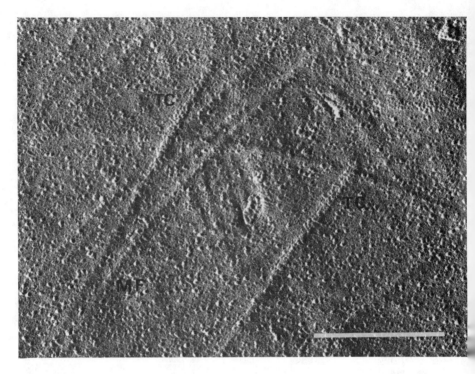

FIGURE 6   PF face of the plasma membrane of *Valonia macrophysa*.
Linear terminal complexes (TC), microfibril imprints (MF). Bar =
0.5 μm. (Courtesy of Dr. T. Itoh.)

are sometimes arranged in three rows as observed in *Oocystis*, but
they are also often randomly arranged within a TC and differ in
size between 9 and 12 nm (18−20,42).
     In several species of the Siphonocladales and in *Chaetomorpha*
the length of the TCs varies independently of the state of cell wall
formation. However, in some species the TCs visualized during
secondary wall formation are longer than those observed during primary wall deposition (19,20). Whereas the TCs of *Chaetamorpha*
consistently measure about 250 nm long, those of the Siphonocladales
range between 200 and 1000 nm long. With the onset of secondary
wall formation the TCs do not elongate further (19). Itoh (19)
suggests that a few subunits aggregate into small clusters that may
then act as nucleation centers for the TCs of the giant marine algae.
It is unknown where and how the TCs are assembled in *Oocystis*,
but the process may involve glycoproteins (21).

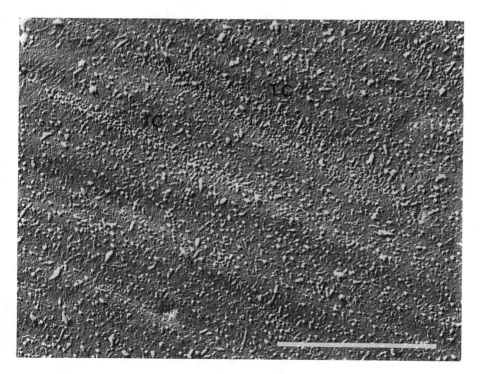

FIGURE 7   PF face of the plasma membrane of a protoplast of
Boergesenia forbesii (20 hr) regenerating a new cell wall. Linear
terminal complexes (TC) claimed to be transmembrane, but no microfibril imprints. Bar = 0.5 μm. (Courtesy of Dr. T. Itoh.)

IV.   EVIDENCE IN FAVOR OF A ROLE IN
CELLULOSE ASSEMBLY

A. Model Organisms

Cellulose assembly occurs during plant cell morphogenesis only during a defined phase of the cell cycle (22).   In order to investigate
cellulose biogenesis, the timing of the developmental phases must be
known in the organism of interest.   Otherwise the stage of cell wall
synthesis may be missed and/or the results obtained may be interpreted incorrectly.   Two good model organisms for investigating the
possible involvement of linear TCs in cellulose assembly are the unicellular green alga *Oocystis* (23) and the marine alga *Boergesenia* in
which protoplasts can be induced that regenerate a new wall (14).
Several lines of evidence from ultrastructural studies indirectly

support the hypothesis that the linear TCs observed in these algae
(Table 1) participate in cellulose biogenesis.

B. Terminal Complex-Microfibril Interaction

All the models put forward suggest an interaction between the intra-
membranous particle complexes and the microfibril during its synthesis. In *Oocystis* two types of TCs occur: TC doublets and single
TCs (5,23). The TCs first appear as doublets in the membrane,
then move apart as single TCs that are located at the end of membrane impressions resembling the contours of microfibrils. No gap
is discernible between the TC and the microfibril imprint, indicating
a close interaction between microfibril and TC (Fig. 4). This observation led Brown and Montezinos (5) to suggest that the TCs
were the sites of cellulose synthesis in *Oocystis*, although there was
(and still is) no supporting biochemical evidence. The close association between TCs and microfibrils was recently further demonstrated
in ultrathin sections of conventionally fixed and stained cells of
Boergesenia forbesii (24; Fig. 8).

Years before the TCs were discovered and related to cellulose
assembly, Schnepf and coworkers (25) observed that microfibrils
isolated from cell walls of *Oocystis* continuously taper down at one
end (Fig. 3). This is also detectable on shadowed wall preparations
that show the last deposited microfibril layer (Fig. 2). The length
along which the microfibril tapers approximately agrees with the
length of a TC, suggesting that microfibril assembly may occur
along the whole TC.

C. Changes in Linear Terminal Complexes
During Inhibited Microfibril Assembly

Several substances have been shown to disrupt cellulose biosynthesis (4,23,26). There are four major classes of these substances
that antagonize regular ionic functions, prevent protein synthesis
including the synthesis of glycoproteins, interfere with glucan chain
crystallization, and inhibit cellulose biosynthesis selectively.

1. Inhibition by Ionic Antagonists

The integrity and morphology of membranes and membrane potential
are very sensitive to ionic changes such as fluctuations in calcium,
magnesium, or potassium concentrations. Perturbation of calcium
and magnesium distribution by ionophores, cryptates, or the intracellular calcium chelator chlorotetracycline inhibits microfibril synthesis in *Oocystis* (23,27,28). The morphology of TC doublets
changes in the presence of these substances but recovers within a
short time if the agents are washed away. **Potassium-selective**

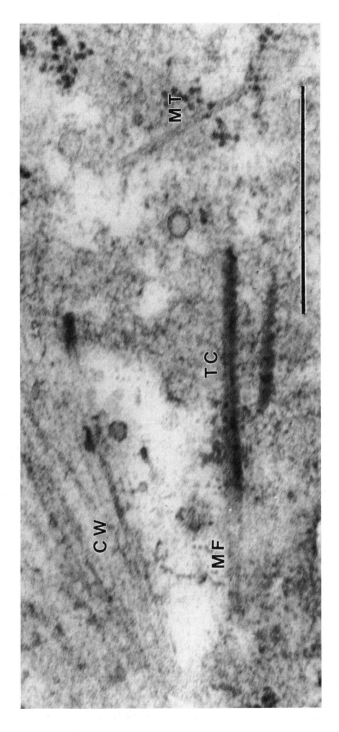

FIGURE 8   Oblique section of a *Boergesenia forbesii* protoplast regenerating its cell wall (CW).   A densely stained linear terminal complex (TC) is associated with a microfibril (MF).   Mt, microtubule. Bar = 0.5 µm.   (Courtesy of Prof. R. M. Brown, Jr.)

ionophores and cryptates do not affect microfibril assembly (29). A general effect on membranes of the calcium/magnesium-affecting agents rather than a specific effect on TCs cannot be ruled out. However, the coincidence between inhibited microfibril synthesis and changed TC morphology indirectly supports the notion that these complexes play a role in cellulose synthesis. Substances that antagonize the action of calmodulin such as fluphenazine or calmidazolium also inhibit microfibril assembly coordinately with alteration in morphology of TC doublets (29,30; Fig. 5) similar to that observed in the presence of the calcium ionophore A23187 (17) or EDTA, which extracellularly chelates magnesium (28). Again, it is not known whether these are specific effects on TCs or more generalized alterations.

2. Inhibition of Protein Synthesis and Glycosylation

The influence of cycloheximide, known to inhibit translation, on cellulose deposition was studied in *Oocystis* (23) and in *Boergesenia* protoplasts (31). In both organisms no microfibrils are synthesized in the presence of this drug. In *Oocystis*, TCs were not observed in the presence of cycloheximide, but they appeared within 1 hr after drug removal (17). In *Boergesenia* the TCs remained visible throughout the treatment. Robinson and Quader (17) suggested that in *Oocystis* each microfibril layer is assembled by a new set of enzymes, which is supported by the observation of two sets of TCs oriented perpendicularly to each other after certain inhibitor treatments [Fig. 9(d)]. TCs reversing their direction at the poles, as postulated by Montezinos (32), have never been observed in our laboratory. In the case of *Boergesenia* protoplasts, it has been suggested that an essential protein needs to be continually renewed for microfibril synthesis to continue (31). Tunicamycin, which blocks the formation of glycosylated lipid precursors involved in protein glycosylation, also inhibits microfibril assembly in *Oocystis* and no new TCs are formed in its presence (21).

3. Inhibition of Microfibril Crystallization

The subsequent crystallization of newly polymerized glucan chains into cellulose fibrils can be interferred with by dyes and fluorescent brighteners such as Congo red and Calcofluor white (33,34) that have a high affinity for 1,4-β-glucans. In *Oocystis* (17) these molecules cause a severe alteration of TC morphology and shape [see Fig. 9(a)(b)(c)(d)]. The pattern of microfibril imprints in the plasma membrane also deteriorates and only amorphous material (most likely noncrystalline 1,4-β-glucans; see Chap. 5) is deposited into the wall (35–37). TC morphology begins to change at the tip

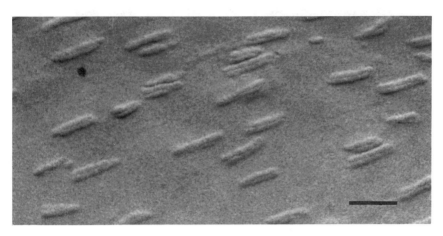

(a)

FIGURE 9   Linear terminal complexes (TCs) visualized at the EF face of the plasma membrane of *Oocystis solitaria* autospores treated with Congo red during cell wall production. Short treatment with Congo red causes a change in the appearance of TCs active in microfibril assembly (39; compare Fig. 4). (a) Treatment: 6 hr, TC doublets altered in shape and morphology. Bar = 0.5 μm. (b) Treatment: 4 hr, individual TCs of the doublet have begun to fragment. Bar = 0.5 μm. (c) Short treatment: 1 hr, heavily fragmented TC (fTC), microfibril imprints (MF) begin to fade. Bar = 0.5 μm. (d) Short treatment: 1 hr, fragmented single TCs (big arrow heads) perpendicularly oriented to TC doublets. The TC doublets have been very recently incorporated into the membrane because only the tips of the TCs have changed morphology. The change begins at opposite ends of each TC in the doublet (small arrow heads), corresponding to the normally opposite direction of movement of each individual TC. However, movement is inhibited in the presence of Congo red. Bar = 0.5 μm.

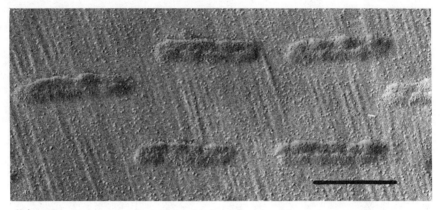

(b)

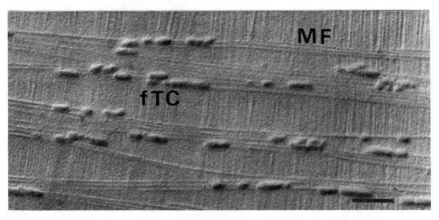

(c)

FIGURE 9   (Continued)

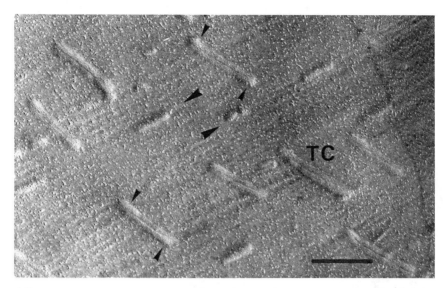

(d)

FIGURE 9   (Continued)

of the individual complexes [Fig. 9(d)] and fragmented TCs are
often observed after longer periods of treatment.   Early and advanced stages of TC fragmentation can be seen in Figs. 9(c) and
9(d), respectively.   It has been suggested that TCs move passively
through the plane of the membrane driven only by the force of the
elongating crystalline microfibril (34).   In the presence of these
dyes TC doublets do not separate by moving in opposite directions,
again supporting the relation between cellulose fibril assembly and
TCs (38,39).   *Boergesenia* TCs are also affected by dyes but, in
contrast to *Oocystis*, they disappear rather than adopting altered
morphology (40).   *Oocystis* cultures as well as *Boergesenia* protoplast cultures washed free of the administered dyes recover microfibril production and regain TCs with normal particle organization
after some time (36,40).   These observations are only understandable
if the TCs play an active role in glucan polymerization and microfibril crystallization.

4. Inhibition of Chain Polymerization

Coumarin and the herbicide 2,6-dichlorobenzonitrile prevent cellulose
synthesis by specific inhibition of the polymerization of 1,4-β-glucan

(4). In the presence of coumarin the TCs of *Oocystis* lose their
substructure (28), whereas in the presence of 2,6-dichlorobenzo-
nitrile the TCs neither change their morphology nor disappear as
revealed by freeze-fracturing (Quader, unpublished).

D. Alignment of Linear Terminal Complexes in the Presence of Microtubule Inhibitors

Microfibril imprints, single TCs, and TC doublets are aligned in
one parallel direction in *Oocystis* autospores with an undisturbed
array of microtubles (39). Substances that disassemble microtubules
in *Oocystis* such as colchicine or the herbicides amiprophosmethyl,
oryzalin, and trifluralin cause an aberrant micrifibril orientation.
In the presence of these drugs the normal crossed wall pattern due
to layers of microfibrils in alternating orientation changes to an uniform pattern characterized by microfibrils all oriented in parallel
(23; Fig. 1). The alternating orientation of microfibril layers corresponds well with the fact that successive sets of TCs are incorporated into the membrane in perpendicular orientation (17,39). In
the presence of anti-microtubule drugs, however, the TC doublets
are inserted in irregular orientation and microfibril imprints show a
strong tendency to bend rather than being straight as in control
cells (39). Therefore, the microtubules seem to have a dual function. They not only guide the aligned insertion of the TC doublets
(and thereby the change in orientation of microfibril deposition between successive lamellae), but they also regulate the parallel movement of the TCs during the formation of a microfibril layer (39).
The corresponding interruption of microfibril pattern and TC arrangement by microtubule-disassembling substances indicates a
close relation between microfibril assembly and TCs.

V. EVIDENCE AGAINST A ROLE IN CELLULOSE ASSEMBLY

It is now generally acknowledged that cellulose assembly is, with few
exceptions, a property of the plasma membrane. Biochemical procedures for membrane isolation and characterization through, for example, enzyme activity measurements are well established. However,
cellulose biosynthesis in vitro has not yet been achieved from isolated membranes of eukaryotic cells (see Chaps. 9 and 10). Not
even the proteinaceous nature of the TCs has been demonstrated
(26). The indirect evidence that linear TCs function in cellulose
fibril assembly is provocative, but electron micrographs can only

give static views. When cellulose synthesis is achieved in vitro in eukaryotes and the participating proteins are clearly identified, it should be possible to clearly associate these membrane structures with their cellular function.

VI. CONCLUDING REMARKS

The hypothesis by Brown and Montezinos (5) that the linear terminal complexes observed in algae with alternating layers of thick cellulose fibrils in their walls represent the cellulose-assembling machinery is supported by numerous ultrastructural observations. These observations focus mainly on the association between the terminal complexes and microfibrils and the observed conformity between induced alterations in microfibril deposition and changes in the presence or appearance of linear terminal complexes. Future researchers need to concentrate on finding conditions to demonstrate microfibril assembly, or at least glucan chain polymerization, by isolated terminal complexes.

REFERENCES

1. R. D. Preston, *The Physical Biology of Plant Cell Walls*, Chapman and Hall, London, 1974.
2. A. Frey-Wyssling, *The Plant Cell Wall*, Gebrüder Bornträger, Berlin, 1976.
3. J. R. Colvin, in *The Biochemistry of Plants, Vol. 3, Structure and Function* (J. Preiss, ed.), Academic Press, New York, p. 543 (1982).
4. D. P. Delmer, *Ann. Rev. Plant Phys.*, *38*, 259 (1987).
5. R. M. Brown and D. Montezinos, *Proc. Natl. Acad. Sci. USA*, *73*, 143 (1976).
6. P. A. Roelofson, *Acta Bot. Neerl.*, *7*, 77 (1958).
7. R. D. Preston, in *The Formation of Wood in Forest Trees* (M. H. Zimmermann, ed.), Academic Press, New York, p. 169 (1964).
8. H. Moor and K. Mühlethaler, *J. Cell Biol.*, *17*, 609 (1963).
9. D. G. Robinson, R. K. White, and R. D. Preston, *Planta*, *107*, 131 (1972).
10. D. G. Robinson and R. D. Preston, *Planta*, *104*, 234 (1972).
11. S. J. Singer and G. C. Nicholson, *Science*, *175*, 720 (1972).
12. D. Branton, S. Bullivant, N. B. Gilula, M. J. Karnovsky, H. Moor, K. Mühlethaler, D. H. Northcote, L. Packer, B. Satir, P. Satir, V. Speth, L. A. Staehelin, R. L. Steer, and R. S. Weinstein, *Science*, *190*, 54 (1975).

13. H. B. Peng and L. F. Jaffe, *Planta, 133,* 57 (1976).
14. R. M. Brown, *J. Cell Sci.,* Suppl. *2,* 13 (1985).
15. L. A. Staehelin and T. H. Giddings, in *Developmental Order: Its Origin and Regulation* (S. Subtelny, ed.), Alan R. Liss, New York, p. 133 (1982).
16. H.-J. Treede, Untersuchungen zur Differenzierung der Stielzelle bei *Dictyostelium discoideum,* Dipl. Thesis, Göttingen, 1984.
17. D. G. Robinson and H. Quader, *Eur. J. Cell Biol., 25,* 278 (1981).
18. T. Itoh and R. M. Brown, Jr., *Planta, 160,* 372 (1984).
19. T. Itoh, Microfibril assembly in giant marine algae, in *Proc. 1987 International Dissolving Pulps Conference,* TAPPI Press, Atlanta, 1987.
20. S. Mizuta and K. Okuda, *Botanica Marina, 20,* 205 (1987).
21. H. Quader, *Plant Physiol., 75,* 534 (1984).
22. D. G. Robinson and H. Quader, *J. Theor. Biol., 92,* 483 (1981).
23. H. Quader and D. G. Robinson, *Ber. Deutsch. Bot. Ges., 94,* 75 (1981).
24. K. Kudlicka, A. Wardrop, T. Itoh, and R. M. Brown, Jr., *Protoplasma, 136,* 96 (1987).
25. E. Schnepf, W. Koch, and G. Deichgräber, *Arch Mikrobiol., 55,* 149 (1966).
26. C. H. Haigler, in *Cellulose Chemistry and Its Applications* (T. P. Nevell and S. H. Zeronian, eds.), Ellis Horwood, Chichester, p. 30 (1985).
27. H. Quader and D. G. Robinson, *Eur. J. Cell Biol., 20,* 51 (1979).
28. D. Montezinos and R. M. Brown, Jr., *Cytobios, 23,* 119 (1979).
29. H. Quader, in *Molecular and Cellular Aspects of Calcium in Plant Development* (A. J. Trewavas, ed.), Plenum Press, London, p. 57 (1985).
30. H. Quader, in *Microtubules in Microorganisms* (P. Cappuccinelle and N. R. Morris, eds.), Marcel Dekker, New York, p. 313 (1982).
31. T. Itoh, R. L. Legge, and R. M. Brown, Jr., *J. Phycol., 22,* 224 (1986).
32. D. Montezinos, in *The Cytoskeleton in Plant Growth and Development* (C. W. Lloyd, ed.), Academic Press, London, p. 147 (1982).
33. C. H. Haigler, R. M. Brown, Jr., and M. Benziman, *Science, 210,* 903 (1980).
34. W. Herth, *J. Cell Biol., 87,* 442 (1980).

35. H. Quader, *Naturwissenschaften, 67,* 428 (1981).
36. H. Quader, D. G. Robinson, and R. van Kempen, *Planta, 157,* 317 (1983).
37. E. Roberts, R. W. Seagull, C. H. Haigler, and R. M. Brown, Jr., *Protoplasma, 113,* 1 (1982).
38. H. Quader, *Eur. J. Cell Biol., 32,* 174 (1983).
39. H. Quader, *J. Cell Sci., 83,* 223 (1986).
40. T. Itoh, R. M. O'Neil, and R. M. Brown, Jr., *Protoplasma, 123,* 174 (1984).
41. J. H. M. Willson and R. M. Brown, Jr., *J. Cell Biol., 77,* 462 (1978).
42. T. Itoh, *J. Cell Sci., 95,* 309 (1990).

4
Role of Particle Rosettes and Terminal Globules in Cellulose Synthesis

ANNE MIE C. EMONS   *Wageningen Agricultural University, Wageningen, The Netherlands*

The cell wall, an extracellular structure made by the cell, is biphasic; it consists of a skeleton of microfibrils held together, and apart, by a like matrix. In the higher plants the microfibrils are made of cellulosic 1,4-β-D-glucan chains. The composition and structure of the plant cell wall are discussed in Chap. 2 of this volume.

The cell wall components and/or their precursors are formed in the biosynthetic apparatus of the plant cell and have to pass the plasma membrane to become incorporated into the wall. The purpose of this chapter is to provide an overview of what is known about the role of the plasma membrane in cellulose microfibril formation. Other reviews of the topic include those by Mueller (1), Willison (2), Brown (3), and Herth (4).

I. THE PLASMA MEMBRANE

Like all biological membranes, the plant plasma membrane is composed of amphipathic lipids and proteins held together by noncovalent interactions. Amphipathic molecules in water assemble spontaneously, thus forming micelles or bimolecular layers in which the hydrophobic ends lie against each other and the hydrophilic ends orient toward the watery solvent. The lipid part of the biological membrane

self-assembles into a fluid bilayer structure in which the molecules
move rapidly, although they primarily stay within their own mono-
layer (5). The degree of mobility of the bilayer lipids depends on
the composition of their hydrophobic part and on temperature.

Membrane proteins are embedded in the membrane with their
hydrophobic domains in contact with the hydrophobic interior of the
lipid bilayer and their hydrophilic domains exposed to the watery
solvent. Many membrane proteins are transmembranous. Generally,
the lateral mobility of membrane proteins is less than the mobility of
the lipids. The lateral mobility of membrane proteins is inhibited if
proteins are aggregated or if they are attached to molecules outside
or inside the cell such as cell wall components or cytoskeletal elements, respectively. While the basic structure of the biological
membrane is determined by the lipid bilayer, its specific functions
are determined by the membrane proteins, which are receptors,
enzymes, or transporters.

A lipid bilayer is impermeable to large molecules and ions. Intact plasma membranes, however, are permeable to polar molecules
such as ions, sugars, amino acids, and nucleotides. Specific membrane proteins are responsible for part of the selective transport of
these substances through the plasma membrane. This transport may
be passive, the protein simply forming a channel in which the transported molecules have no direct contact with the hydrophobic part
of the lipids. The transport may also be active and require energy,
in which case the membrane transport protein has a specific binding
site for a specific chemical.

II. VISUALIZATION OF PLASMA
MEMBRANE PROTEINS

Freeze-fracture, which is described below, reveals the hydrophobic
inner layers of the plasma membrane. Without any prior treatment
with cryoprotectant or fixative, the material is placed between thin
gold or copper plates in a specimen holder and ultrarapidly frozen
in liquid ethane, liquid propane, liquid freon, or nitrogen slush.
The specimen holder is opened in the freeze-etch apparatus, causing
the biological material to fracture at the hydrophobic interface between the two phospholipid leaflets of membranes. Fracture and
platinum−carbon replication are performed at $\sim \pm 108°C$. Subsequently,
the replicas are floated on a cleaning solvent, washed in distilled
water, dried onto Formvar-coated grids, and examined in the transmission electron microscope. For details of freeze-fracture procedures, consult Robards and Sleytr (6).

By means of the freeze-fracture procedure many of the intramembrane proteins do not split but rather stay in one of the two lipid monolayers and remain visible as particles. In plasma membranes of plant cells, more particles fracture with the PF face (protoplasmic fracture face) of the membrane than with the EF face (exoplasmic fracture face) [terminology of Branton and coworkers (7)]. Particles may extend through the bilayer, although they partition with one of the membrane monolayers upon fracture. Since membrane transport particles are relatively large, it is probable that many of the visible intramembrane particles have a transport function.

III. DISCOVERY OF INTRAMEMBRANE PARTICLES RELATED TO MICROFIBRIL SYNTHESIS

Preston (8) postulated the ordered granule hypothesis for cellulose microfibril formation in 1974. In this model granules in the plasma membrane, lying in a specific pattern, constitute synthesizing enzymes. The ordered granule hypothesis was based on static membrane structure. In 1972 Singer and Nicholson (5) formulated the fluid mosaic model for the structure of cell membranes. According to this concept, the microfibril-synthesizing complexes would be allowed to move laterally in the plane of the membrane. In 1976 Brown and Montezinos (9) visualized complexes at microfibril termini in *Oocystis*. Chapter 3 of this volume deals with those complexes. In the same year Mueller and coworkers (10) visualized terminal complexes in plasma membranes of higher plant cells. More details of this complex were described by Mueller and Brown (11). The complex has two components: a PF face particle rosette that measures 25 ± 5 nm in diameter and consists of six particles with a diameter of 8 nm each (Figs. 1, 2a); and an EF face terminal globule that measures about 20 nm in diameter and in which a microfibril imprint terminates (Fig. 2b). Mueller and Brown (11) reported that rosettes are often observed at the ends of the visible imprints of wall microfibrils, but later reports show that many of them do not lie at a terminus of a microfibril imprint (12,13).

Particle rosettes have also been found in the PF face of the plasma membrane of the alga *Micrasterias* (14). In the plasma membrane adjacent to the secondary wall of *Micrasterias*, rosettes lie grouped in hexagonal patterns and imprints of bands of parallel microfibrils are attached to them. Figure 3a shows such an array of rosettes in the PF face of the plasma membrane of the alga *Spirogyra*; Fig. 3b shows an array of terminal globules in the EF

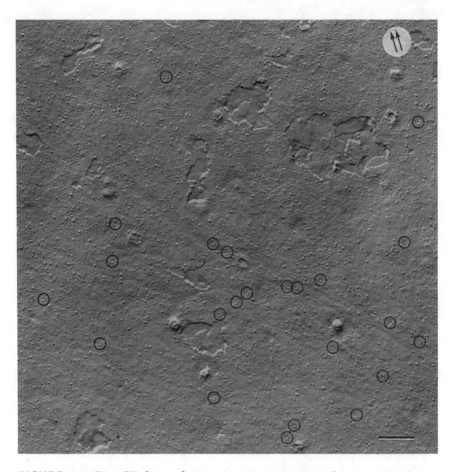

FIGURE 1   The PF face of the plasma membrane of a root cortical
cell of *Cucurbita pepo* showing microfibril imprints in the plasma
membrane and the distribution of particle rosettes (encircled).
Bar = 200 nm.   (Courtesy of A. W. Robards.)   *In this and in all
subsequent freeze-fracture micrographs, the direction of platinum
shadowing is indicated by the double arrow on white background.*

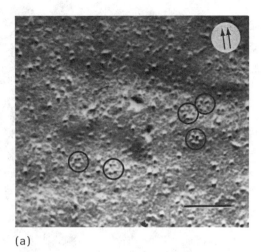

(a)

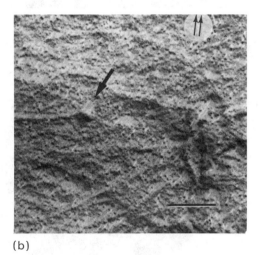

(b)

FIGURE 2 (a) The PF face of the plasma membrane of an *Equisetum hyemale* root hair showing several particle rosettes (encircled); a particle rosette consists of six particles arranged in a circle. Bar = 100 nm. (b) The EF face of the plasma membrane of an *E. hyemale* root hair showing a microfibril imprint terminating in a globule. The globule appears to be an imprint of material underlying the membrane and bears a normal EF face intramembrane particle. Bar = 100 nm.

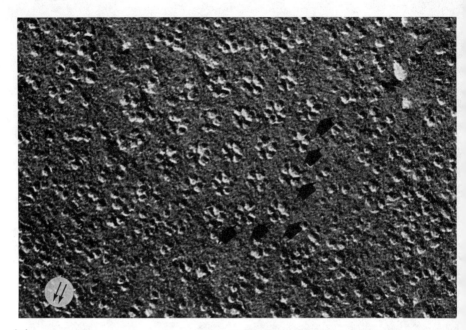

(a)

(b)

face of this alga. Imprints of microfibrils ending at globules can
be clearly observed.

Mueller and Brown (11) postulated that the terminal globule of
the EF face of the plasma membrane would fit between the constituting
particles of the PF face rosette and that this whole would be the
functional terminal complex of microfibril synthesis in higher plants.
However, the fact that the rosette of the PF face and the globule of
the EF face fit together forming one complex has not been proven by
double-replica technique (see also Sec. V). Plasma membrane
rosettes seem always to be observed if preparation conditions are
optimal, i.e., if uncryoprotected material and rapid-freezing methods
are used. Loss of cell turgor, indicated by lack of microfibril imprints, often also causes the particle rosettes to disappear (1,12).
Centrifuging soybean suspension-culture cells at low gravity forces is
enough to cause the disappearance of rosettes from the PF face of
the plasma membrane (15).

IV. OCCURRENCE OF PARTICLE ROSETTES
AND TERMINAL GLOBULES

Both rosettes and terminal globules have been found in the plasma
membranes of cellulose-synthesizing organisms from the algae
Zygnematales onward to the higher plants (Table 1). Rosettes always fracture with the PF face of the membrane; terminal globules
are found on the EF face of the plasma membrane.

V. EVIDENCE THAT ROSETTES AND GLOBULES
FUNCTION IN MICROFIBRIL SYNTHESIS

A. Particle Rosettes

A number of data clearly favor the hypothesis that rosettes have
some function in microfibril synthesis.

Rosettes now have been found in cellulose microfibril-depositing
cells of a great number of plant species (Table 1).

FIGURE 3   (a) The PF face of the plasma membrane of the alga
Spirogyra showing an hexagonal pattern of rosettes. Bar = 100 nm.
(Courtesy of W. Herth.)   (b) The EF face of the plasma membrane
of the alga *Spirogyra* showing imprints of microfibrils in the membrane terminating in a cluster of hexagonally arranged terminal
globules. Bar = 100 nm. (Courtesy of W. Herth.)

TABLE 1 Occurrence of Particle Rosettes and Terminal Globules in Plasma Membranes of Plant Cells

Plant species	EF structure	PF structure	Source	Microfibril width
Closterium sp.	—	Rosettes array	16	Thick
Closterium aterosum	—	Rosettes array	17	Thick
Micrasterias denticulata primary wall	—	Rosette	18	Thin
Micrasterias denticulata secondary wall	Globules array	Rosettes array	18	Thick
Micrasterias crux melitensis primary wall	—	Rosette	19	Thin
Micrasterias crux melitensis secondary wall	Globules array	Rosettes array	19	Thick
Mougotia	Globule	Rosette	10	Thin
Spirogyra sp. primary wall	—	Rosette	21	Thin
Spirogyra sp. secondary wall	Globules array	Rosettes array	21	Thick
Chara	Globule	Rosette	22	Thin
Nitella translucens	Globule	Rosette	20	Thin
Funaria hygrometrica	—	Rosette	23	Thin
Psilotum sp.	Globule	—	3	Thin
Equisetum hyemale	Globule	Rosette	12	Thin
Ophioglossum sp.	Globule	—	3	Thin
Adiantum capillus veneris	Globule	Rosette	24	Thin
Ginkgo biloba	Globule	—	3	Thin

Species				
Pinus taeda	Globule	Rosette	11	Thin
Lepidium sativum root hair	–	Rosette	Figure 9	Thin
Lepidium sativum xylem	–	Rosette	13	Thin
Raphanus sativus	Globule	–	25	Thin
Raphanus sativus	Globule	Rosette	2	Thin
Raphanus sativus	Globule	–	26	Thin
Cucurbita pepo root cortical cells	–	Rosette	Figure 1	Thin
Glycine max suspension cells	–	Rosette	15	Thin
Phaseolus aureus	Globule	Rosette	11	Thin
Phaseolus vulgaris	Globule	Rosette	2	Thin
Vigna radiata	–	Rosette	27	Thin
Acer pseudoplatanus suspension cells	–	Rosette	28	Thin
Daucus carota protoplasts	Globule	–	29	Thin
Daucus carota protoplasts	–	Rosette	28	Thin
Zinnia elegans suspension cells	–	Rosette	30	Thin
Hydrocharis morsus ranae	–	Rosette	Unpublished	Thin
Limnobium stoloniferum root hair	–	Rosette	Unpublished	Thin
Limnobium stoloniferum root cortex	–	Rosette	4	Thin
Allium sp.	–	Rosette	3	Thin
Lilium longiflorum pollen tube	–	Rosette	31	Thin

TABLE 1 (Continued)

Plant species	EF structure	PF structure	Source	Microfibril width
Avena sativa	Globule	Rosette	32	Thin
Hordeum vulgare	—	Rosette	2	Thin
Zea mays	Globule	Rosette	11	Thin
Zea mays xylem	—	Rosette	33	Thin

Note: Organisms with plasma membranes containing single rosettes and/or single globules possess thin, 5- to 10-nm-wide, microfibrils in shadowed preparations of the cell wall. Arrays of rosettes and hexagonal patterns of globules coexist with wider, 20- to 30-nm-diameter, microfibrils. Because actual microfibril width is not known, the designation "thin" is given for microfibrils that are 5–10 nm in diameter, "thick" for 20- to 30-nm-wide microfibrils both in shadowed preparations.

Because they disappear when cryoprotectants or fixatives
are applied (12,34), they were not found in earlier work on
plant cell membranes.

Rosettes are more abundant in those parts of the plasma membrane associated with extensive microfibril deposition (23).
The density of rosettes is high in systems producing a high
cellulose content, low in systems producing a low cellulose
content (4).

In *Funaria* protonema cells rosettes disappeared and microfibril
synthesis stopped after several hours of darkness; rosettes
reappeared quickly and cell growth and microfibril synthesis resumed when cells were reilluminated (35).

Higher plant cells, which have approximately 8-nm-wide cell wall
microfibrils in traditionally shadowed preparations, have
solitary rosettes.

In some algae, which have microfibrils in bundles that are thicker than in higher plants, it is clear that rows of rosettes
are terminally associated with microfibril imprints (16).

In algae such as *Micrasterias* that have rosettes grouped together
in hexagonal patterns within the plasma membrane, there is a
coincidence between the pattern of rosettes and the alignment of microfibrils. The center-to-center spacing of the
fibrils is equal to that of the rows of rosettes (16).

Microfibril diameter coincides with the number of rosettes in line
with the microfibril. In *Micrasterias*, the widest fibrils in
the center of a band of fibrils are formed by the longest
rows of rosettes in the center of the hexagonal array (16).
The primary wall of this alga contains thin microfibrils and
solitary rosettes, a correlation that is also observed in the
algae *Chara* (22) and *Nitella* (20) as well as in higher plants.

Rosettes are stage-specifically incorporated into the plasma membrane (14,18,30).

Considering the data on the occurrence, density, and arrangement of rosettes from Zygnematales to higher plants, there is considerable evidence for the putative function of rosettes in microfibril
synthesis. Organisms that have single rosettes possess cellulose
with a diameter of 5–10 nm in shadowed preparations (Table 1),
which is 3–4 nm in stained preparations (36). Organisms that have
hexagonal patterns of rosettes have wider microfibrils, the width depending on the number of rosettes in line with the microfibril. At
present there is much confusion on microfibril width. The traditional shadow-casting technique yields microfibril widths in higher plants
from 5 to 10 nm because shadow deposit varies greatly. The improved method of high-resolution shadowing (37), which has also
resolved the helix feature of DNA (38), shows higher plant microfibrils with a width of 3.68 nm. This figure is similar to that found

by staining with potassium permanganate of material from which wall matrix molecules had been dissolved (36).

B. Terminal Globules

Terminal globules on the EF face of the plasma membrane have also been suggested to function in microfibril synthesis. In a number of plant cells the globules clearly lie at microfibril termini (Figs. 2b and 3b). The terminal globule seems very labile and its structure is far less well defined than the structure of the particle rosette. Not all cells that have rosettes in the PF face of the plasma membrane have been shown to contain terminal globules in the EF face. Often the frequency of particle rosettes and terminal globules in one single cell is not the same (2). In *Equisetum hyemale* root haris, more rosettes than globules are always observed.

The terminal globule might represent the still uncrystallized product. Rather than being a particle in the EF face of the membrane, the globule would then be an imprint of material lying outside the membrane. In favor of this idea is the often observed phenomenon that one or more normal EF face particles lies on top of a terminal globule (Fig. 2b).

Though function cannot be assigned from static structures observed in the electron microscope, the evidence that rosettes and globules play a role in microfibril synthesis is compelling. Their relationship needs clarification. Proof of function in cellulose biosynthesis for rosettes and globules awaits their isolation in active form, their purification, and the raising of antibodies that specifically bind to them.

VI. INSERTION OF PARTICLE ROSETTES INTO THE PLASMA MEMBRANE

For the alga *Micrasterias* evidence has been shown for the insertion of rosettes into the plasma membrane by Golgi vesicles (18). This has been confirmed for the alga *Closterium* (39). Haigler and Brown (30) visualized rosettes in the plasma membrane, in Golgi vesicles, and in Golgi cisternae of differentiating xylem elements of *Zinnia elegans* mesophyll cells in suspension culture. Each Golgi vesicle contained only one rosette (Fig. 4). The rosettes present in Golgi cisternae and in Golgi vesicles have the same morphology, size, and symmetry as the well-formed rosettes of the plasma membrane. In the PF face of the plasma membrane rosettes are sometimes observed lying in a depression in the membrane (Fig. 5).

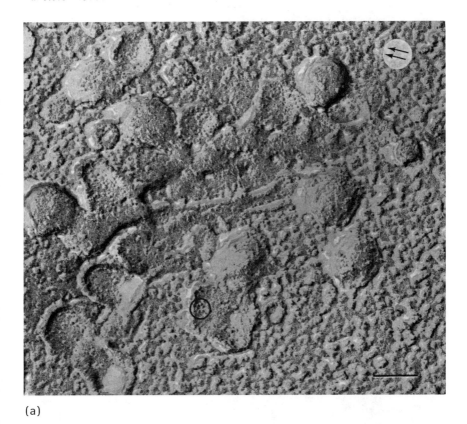

(a)

FIGURE 4   A freeze-fractured dictyosome of a *Zinnia elegans* suspension culture mesophyll cell with a particle rosette (encircled).
Bar = 100 nm. (Courtesy of C. H. Haigler.) (b) A Golgi vesicle of
a *Z. elegans* suspension culture mesophyll cell bearing a clear particle rosette (encircled). Bar = 100 nm. (Courtesy of C. H. Haigler.)

The rosettes that lie in plasma membrane impressions are always
well formed; they might represent the newly inserted rosettes.
Apart from well-formed rosettes the plasma membrane also possesses
malformed rosettes (23) and rosettes with larger diameters (12).
These observations were interpreted as indications of the degradation of rosettes after their functioning in microfibril synthesis (12).
Haigler and Brown (30) remark that a recycling mechanism from the

(b)

FIGURE 4   (Continued)

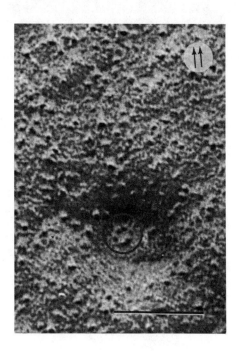

FIGURE 5   The PF face of
a root hair of *E. hyemale*
with a particle rosette lying
in an impression of the
plasma membrane. Bar =
100 nm.

plasma membrane cannot be completely excluded in which vesicles containing rosettes are transported inward and incorporated into Golgi cisternae.

VII. TURNOVER OF PARTICLE ROSETTES

Assuming that particle rosettes function in microfibril deposition, it is possible to calculate the rate of microfibril production per rosette. In root hairs of *Equisetum hyemale*, five to seven microfibrils intersect any micrometer line drawn across a completed microfibril lamella perpendicularly to the orientation of its microfibrils (Fig. 6). A 1-μm^2 area of plasma membrane of *E. hyemale* root hair contains 5–15 rosettes (12). Therefore, approximately 600 nm of microfibril is deposited per rosette, if all rosettes are active in forming the microfibrils of that lamella.

Reiss and coworkers (23) calculated the lifetime of rosettes to be in the range of about 10–20 min in *Funaria* caulonema tip cells, For xylem thickening Schneider and Herth (33) calculated a similar lifetime for rosettes. However, these calculations are both based on the unproven assumption that one rosette produces one 10-μm-long elementary microfibril. The length of microfibrils is not known; in the many cell wall preparations studied, microfibril ends have rarely been found. The degree of polymerization of cellulose varies from 500 to 14,000 glucose residues per molecule in protoplasts and in secondary walls, respectively (40). Microfibrils, however, may be much longer than the 7 μm predicted by a degree of polymerization of 14,000, because glucan chains may end randomly within microfibrils (40). If the assumptions are made for *E. hyemale* root hairs that one rosette produces a microfibril with a length of 10 μm and that only one lamella is under deposition at a time, only one out of 16 rosettes can be active, which is only 6%.

If microfibril content is 40% in *Lepidium* xylem (33) and one microfibril actually is 4 nm wide (33), then a 1-μm^2 area of wall lamella contains 100 μm of microfibril length. If one rosette makes a 10-μm microfibril, as Schneider and Herth (33) assume, this area will need 10 rosettes to produce the required amount of microfibril. However, in *Lepidium* xylem they found 135 rosettes per μm^2. Only 10 out of 135 rosettes, therefore, are active, which is 7%. Thus, also in *Lepidium* xylem, either not all rosettes are active or one rosette makes not more than 742 nm length of microfibril, approximately the same amount as found for *E. hyemale* root hairs assuming all rosettes are active. Since microfibrils are much longer than 500–700 nm it is more likely that many rosettes are inactive. This may explain why many rosettes are not associated with microfibril ends. From their study on the transport of rosettes from the

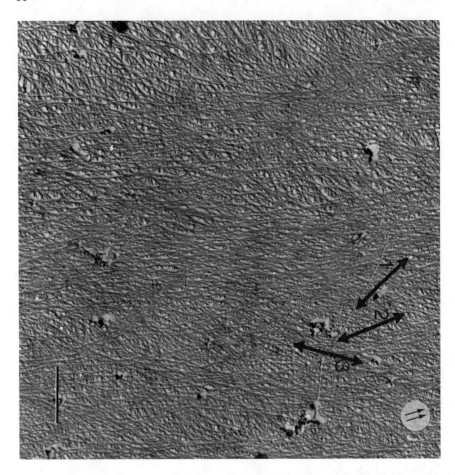

FIGURE 6   The platinum/carbon shadowed inner surface of the cell
wall of an *E. hyemale* root hair showing the last deposited microfibrils.   The microfibril lamella with microfibrils in orientation 1 is
under deposition.   The lamella with microfibrils in orientation 2 has
been completed.   The lamella with microfibrils in orientation 3 was
laid down before lamella 2.   A mean of six microfibrils intersecting
1 μm perpendicularly was counted from 60 wall preparations of completed lamellae.   The cell wall texture of the *E. hyemale* root hair is
of the helicoidal type; microfibrils lie parallel to each other in
lamellae, with the axis of parallel orientation changing from lamella
to lamella (60 – 62; see Chap. 2).   Bar = 500 nm.

Golgi apparatus to the plasma membrane, Haigler and Brown (30) concluded that to achieve cell wall thickening new rosettes must be continually inserted into the membrane. It may be deduced that intracellular transport and insertion of rosettes in the plasma membrane at specific sites requires a highly coordinated process. The most likely candidates to perform such a transport function are cytoskeletal components (Chap. 7).

VIII. THE HONEYCOMB PATTERN OF PARTICLES

Delmer (41) has argued that ordered arrays of plasma membrane particles have been seen in organisms that do not produce cellulose, such as in the alga *Chlamydobotrys* (42). The particle pattern found in *Chlamydobotrys* is similar to that found in yeast plasma membrane and has not been found in plasma membranes of higher plants.

The particle pattern of yeast plasma membrane (Fig. 7a) has been called a hexagonal array (43). In this view "hexagonal" is the term used in crystallography. Mathematically, the pattern is trigonal (Fig. 7b). A different pattern of particles that has been found in higher plant cells has also been called a hexagonal particle pattern. In this pattern particles pack less closely than in the hexagonal crystal lattice seen in yeast membranes. Mathematically, however, this pattern may be called hexagonal. The term hexagonal pattern should be reserved for the particle arrangement found in yeast plasma membrane because of its previous wide use in animal cell literature. For describing the pattern found in higher plant cells, the term honeycomb pattern has been introduced (44).

Particle rosettes (higher plants), hexagonal arrays of particle rosettes (some algae), linear arrays of particles (other Algae; see Chap. 3), and strings of particles (Bacteria; see Chap. 5) found in cellulose-synthesizing organisms have all been hypothesized to function in microfibril synthesis (3). Two functions have been attributed to the honeycomb pattern of particles: microfibril synthesis (45,46) and solute transport (44,47,48). However, it has also been hypothesized that particles arranged in honeycombs are an effect of plasmolysis (12,49). Volkmann (46) reported the presence of hexagonally ordered particles, clearly honeycombs, in *Lepidium* root hairs and the absence of particle rosettes in this plant material, but used glycerinated or sucrose-incubated roots. In *E. hyemale* root hairs, honeycomb patterns of particles were only seen during retraction of

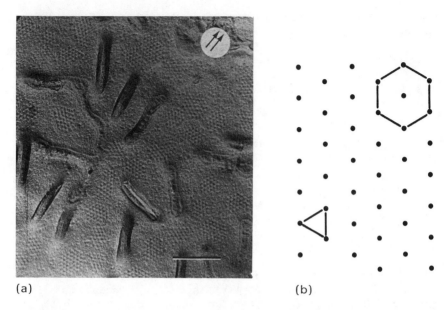

(a)   (b)

FIGURE 7   (a) The PF face of the plasma membrane of baker's
yeast showing a hexagonal particle pattern.   Bar = 200 nm.
(b) Schematic representation of the hexagonal particle configuration
found in the plasma membrane of yeast cells; the basic unit of this
pattern is a triangle.

the plasma membrane from the cell wall (Fig. 8).   Moreover, uncryoprotected *Lepidium sativum* root hairs grown in distilled water contain particle rosettes on the PF face of the plasma membrane (Fig. 9).
The honeycomb pattern of particles has also been found in protoplasts (49) and pollen tubes of tobacco (48) and lily (44).   These
pollen tubes, like the protoplasts, were cultured in a sucrose-con-
taining medium.   Reiss and coworkers (31) found single-particle
rosettes in lily pollen tubes, but conditions used for germination
were not reported.   Micrographs of PF faces from wheat leaf bases
dehydrated over 8 hr to a water content of 11% of that of the fully
turgid tissue published by Pearce (50) show particle patterns that
are clearly honeycombs.   In the same membrane particle-free patches
occurred which were hypothesized to be induced by cell dehydration (50).   In summary, the honeycomb pattern of particles has been
found in plant plasma membranes only if the medium contains sucrose
or glycerol or if the cells are stressed by plasmolysis or dehydration.

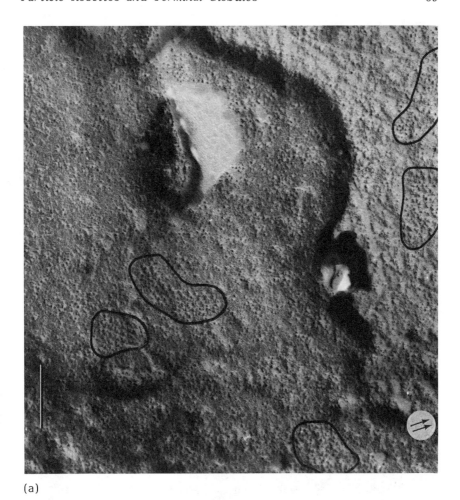

(a)

FIGURE 8   (a) The PF face of the plasma membrane of an
E. hyemale root hair showing a partly (lower right) retracted membrane.   The part that is not retracted from the cell wall (upper left)
shows microfibril imprints.   The honeycomb particle pattern is
visible in both parts (outlined).   Bar = 200 nm.   (b) A honeycomb
particle pattern visible in the retracting plasma membrane of the
cell in Fig. 8a.   Bar = 100 nm.

(b)

FIGURE 8   (Continued)

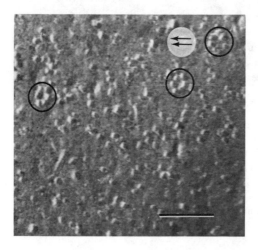

FIGURE 9   The PF face of the plasma membrane of an uncryoprotected root hair of *Lepidium sativum* grown in water showing particle
rosettes (encircled).   Bar = 100 nm.

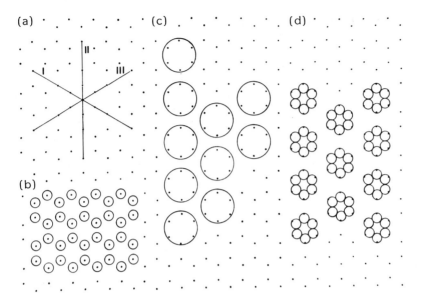

FIGURE 10   Schematic representation of the honeycomb pattern found in the plasma membranes of plant cells. Compared to the hexagonal pattern (Fig. 7b), particles are less closely packed because the basic unit is a hexagon, not a triangle. (a) shows the symmetry axes; (b) shows the honeycomb pattern of particles; (c) shows that the honeycomb pattern may be constructed from rosettes; (d) shows a hexagonal pattern of rosettes constructed from a honeycomb by converging the constituting particles of each hexamer.

Figure 10 shows the relation between the single rosette, the honeycomb pattern of particles, and the hexagonal pattern of rosettes. It is clear that the honeycomb pattern (Fig. 10b) can be constructed from particle rosettes without dropping out or adding particles (Fig. 10c). However, because adjacent hexamers in the honeycomb pattern share particles, 36 particles define 11 hexamers of the honeycomb (Fig. 10b), whereas 66 particles would be required to define 11 independent hexamers as in Fig. 10c. The hexagonal pattern of particle rosettes, which can be obtained by converging the particles of the hexamers (Fig. 10d), is similar to the hexagonal pattern of rosettes in the Zygnematales (cf. Fig. 3a).

IX. PUTATIVE FUNCTIONAL MODEL FOR
ROSETTES AND GLOBULES

Rosettes seem to function in microfibril synthesis, but there is a
lack of biochemical evidence that they are the microfibril synthesizing enzymes, as has been postulated (4,11).

During cell wall synthesis molecules must traverse the plasma
membrane. Cell wall matrix is thought to be brought through the
membrane by exocytosis of the Golgi-derived smooth vesicles in
which it is carried (51). The formation of cellulose microfibrils,
however, is different since it is a plasma membrane-associated
process (41; see Chap. 9). Mueller and Maclachlan (52) showed in
Pisum sativum that [^{3}H]glucose is polymerized into wall material at
the plasma membrane.

There seems to be sufficient evidence that UDP-glucose is the
major precursor to cellulose (41). Since UDP-glucose cannot be
supplied from the outside of the cell, the binding site for the substrate must be on the cytoplasmic side of the plasma membrane.
UDP-glucose cannot diffuse through the lipid bilayer of the plasma
membrane.

Membranes can regulate the passage of ions and/or small molecules by creating protein channels across the membrane. An example of such a protein channel is the one that forms the gap junction
in animal cells. Gap junctions permit diffusion from cell to cell of
low molecular weight, water-soluble components. The structural unit
is the connexon, a 6- to 7-nm-wide particle that is a hollow cylinder
of protein spanning the plasma membrane and joining end to end
with another connexon in the plasma membrane of an adjacent cell
(53). The connexon is made of six similar subunits arranged around
a 2-nm-wide water-filled channel. These connexons are clustered
together in well-defined areas. Regulation of passage of molecules
through the membrane is achieved by conformational changes of the
connexon proteins in response to chemical and electrical stimuli (54).

It could be that particle rosettes are such transport particles
representing sites in the plasma membrane where UDP-glucose
passes the membrane. 1,4-β-D-Glucan chains (cellulose) might be
polymerized during passage, the particles of the rosettes acting as
transporters and synthases. The assembly of particles in a rosette,
then, allows individual glucan chains to cocrystallize into a cellulose
microfibril and acts as a template determining width, crystallite size,
and substructure of the microfibril.

Figure 11 is a schematic representation of this new idea about
rosettes. Embedded and mobile in the fluid plane of the membrane
lie clusters of six particles (rosettes). Every particle is built up of
a number of protein molecules such that there is a channel in each
particle. In this hypothetical model the channel would be lined by

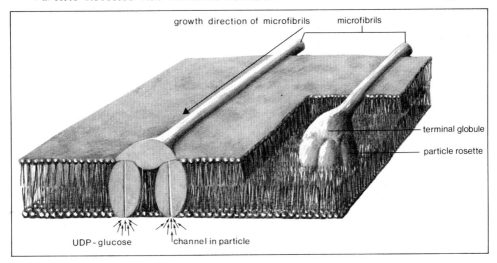

growth direction of microfibrils   microfibrils

terminal globule

particle rosette

UDP-glucose   channel in particle

FIGURE 11   Schematic representation of the microfibril terminal complex embedded in the lipid plane of the membrane. Rosettes are built of six particles; the proteins constituting one particle form a channel through which UDP-glucose passes the membrane while being polymerized into 1,4-β-glucan chains. The terminal globule on top of the rosette is pictured as the still uncrystallized product, which makes an imprint in the plasma membrane as the microfibril itself does.

amphipathic molecules with their hydrophilic faces bordering the channel. The other transmembrane domains interacting with the lipid components of the membrane could be largely hydrophobic. Through this channel cellulose precursors (UDP-glucose) pass. The passage is an active, energy-requiring process whereby UDP-glucose is specifically bound to the membrane transport proteins. This specific binding enables the formation of β-1,4 linkages. The rosette configuration is needed for microfibril synthesis because it brings together the required number of glucan chains to form a crystalline microfibril. In this scheme the glucan chains are not extruded through the central core, which might seem another possibility. From the analysis of many rosettes, the central core appears to be continuous with the rest of the lipid bilayer (12), which makes extrusion of glucan chains through it less plausible. Also Willison (2) concluded that rosettes do not have a hole in the middle for the extrusion of cellulose microfibrils.

The complex moves in the plasma membrane while generating cellulose. The cellulose microfibril grows by the continuous polymerization of supplied precursors. The movement of the complex in the membrane might be driven by the force generated during crystallization of the polymerized product into rigid microfibrils (55). Then, rosette movement does depend on the quality of fluidity of the membrane but does not need a driving force generated by the membrane or the underlying cytoskeleton.

In this concept the terminal globules might represent enzymes needed for crystallization, other proteins that hold the six particles of the rosette together in the required configuration, or the un-crystallized product of the synthases. If the terminal globule is an imprint of material lying outside the membrane, the number of terminal globules will coincide with the number of microfibrils under deposition. Few details on the number of distinct terminal globules are available. In *E. hyemale* root hairs the number of PF rosettes greatly exceeds the number of EF globules, and, as calculated above, there are many more rosettes than microfibrils under deposition. Based on counts of rosettes and terminal globules in relation to the number of microfibrils, Willison (2) concluded that terminal globules are better candidates for microfibril synthesizing enzymes than particle rosettes. Arguments in favor of the idea that terminal globules are imprints of material outside the membrane are the observations that normal EF particles may lie on top of the globule (Fig. 2b) and that terminal globules were not found in Golgi cisternae and in Golgi vesicles (Haigler, pers. commun.).

X. IMPLICATIONS OF THE MODEL FOR IN VITRO MICROFIBRIL FORMATION

Cellulose is only synthesized efficiently under natural conditions; β-1,3-linked glucose is synthesized predominantly from damaged cells or membrane preparations (51,56,57; Chap. 9 of this volume).

The model on the function of rosettes formulated here might hint at the reason for this phenomenon. Rosettes fall apart if cells are damaged (Fig. 12). The constituting particles of the rosettes prob-ably still produce glucan chains, but because the orientation of the constituting proteins of the particles is disturbed C-3 instead of C-4 becomes available for glycosylation, and β-1,3 instead of β-1,4 links are produced. This hypothesis is consistent with the previous sug-gestion that the same enzyme may catalyze both β-1,3 and β-1,4 linkages (58,59). Furthermore, only by grouping particles in a rosette configuration does crystallization of glucan chains into a microfibril become possible. It is also possible that cell disruption leads to dissociation of an important regulatory protein from the rosette complex, possibly causing the formation of β-1,3 rather than

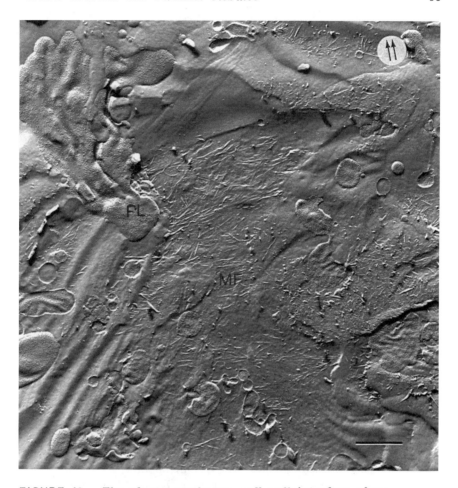

FIGURE 12   The plasma membrane –cell wall interface of an
E. hyemale root hair that was mechanically damaged just prior to
fast-freezing shows patches of PF face plasma membrane that are
always devoid of rosettes; PL: PF face of plasma membrane, MF:
microfibrils.   Bar = 200 nm.

β-1,4 linkages. A potential 18-kD regulatory protein has been
identified and proposed to correspond to the rosette particles,
which would be in association with the glucosyltransferase (41).
The 18-kD polypeptide might alternatively represent the factor that
holds the particles of the rosette together.

The biochemical nature of the particle rosettes involved in cellulose formation remains to be studied. Their isolation in the intact
functional state is difficult because it seems to depend on an as yet
unknown cytoplasmic factor, which might be the factor that holds
the particles of a rosette together in a special configuration.

ACKNOWLEDGMENTS

I am grateful to Dr. A. A. M. Van Lammeren for critical reading of
the manuscript and to Mr. A. B. Haasdijk for the artwork.

REFERENCES

1. S. C. Mueller, in *Cellulose and Other Natural Polymer Systems*
   (R. M. Brown, Jr., ed.), Plenum Press, New York, 1982,
   pp. 87–103.
2. J. H. M. Willison, *J. Appl. Polym. Sci.*, *37*, 91 (1983).
3. R. M. Brown, Jr., *J. Cell Sci.*, Suppl. 2, 13 (1985).
4. W. Herth, in *Botanical Microscopy 1985* (A. W. Robards, ed.),
   Oxford University Press, Oxford, 1985, p. 285.
5. S. J. Singer and G. C. Nicholson, *Science*, *175*, 720 (1972).
6. A. W. Robards and U. B. Sleytr, *Low Temperature Methods in
   Biological Electron Microscopy*, Elsevier, Amsterdam, 1985.
7. D. Branton, S. Bullivant, N. B. Gilula, M. J. Karnoosky,
   H. Moor, K. Muhlethaler, D. H. Northcote, L. Parker,
   B. Satir, P. Satir, L. A. Staehelin, R. L. Steere, and
   R. Weinstein, *Science*, *190*, 54 (1975).
8. R. D. Preston, *The Physical Biology of Plant Cell Walls*,
   Chapman and Hall, London, 1974.
9. R. M. Brown, Jr. and D. Montezinos, *Proc. Natl. Acad.
   Sci. USA*, *73*, 143 (1976).
10. S. C. Mueller, R. M. Brown, Jr., and T. K. Scott, *Science*,
    194, 949 (1976).
11. S. C. Mueller and R. M. Brown, Jr., *J. Cell Biol.*, *84*, 315
    (1980).
12. A. M. C. Emons, *Planta*, *163*, 350 (1985).
13. W. Herth, *Planta*, *164*, 12 (1985).

14. T. H. Giddings, Jr., D. L. Brower, and L. A. Staehelin, *J. Cell Biol.*, *84*, 327 (1980).

15. W. Herth and G. Weber, *Naturwissenschaften*, *71*, 153 (1984).

16. L. A. Staehelin and T. H. Giddings, *Developmental Order: Its Origin and Regulation*, Alan R. Liss, New York, 1982, p. 113.

17. T. Hogetsu, *Plant Cell Physiol.*, *24*, 777 (1983).

18. O. Kiermayer and U. B. Sleytr, *Protoplasma*, *101*, 133 (1979).

19. T. Noguchi, K. Tanaka, and K. Ueda, *Cell Struct. Funct.*, *6*, 217 (1981).

20. A. T. Hotchkiss and R. M. Brown, Jr., *J. Phycol.*, *23*, 229 (1987).

21. W. Herth, *Planta*, *159*, 347 (1983).

22. B. McLean and B. E. Juniper, *Planta*, *169*, 153 (1986).

23. H. D. Reiss, E. Schnepf, and W. Herth, *Planta*, *160*, 428 (1984).

24. M. Wada and L. A. Staehelin, *Planta*, *151*, 462 (1981).

25. J. H. M. Willison and B. W. W. Grout, *Planta*, *140*, 53 (1978).

26. J. H. M. Willison and R. M. Brown, *Protoplasma*, *92*, 21 (1977).

27. W. Herth, *Naturwissenschaften*, *71*, 216 (1984).

28. R. L. Chapman and L. A. Staehelin, *J. Ultrastruct. Res.*, *93*, 87 (1985).

29. C. Smith and R. M. Brown, Jr., *J. Appl. Polym. Sci.*, *37*, 33 (1983).

30. C. H. Haigler and R. M. Brown, Jr., *Planta*, *134*, 111 (1986).

31. H. D. Reiss, W. Herth, and E. Schnepf, *Naturwissenschaften*, *72*, 276 (1985).

32. D. S. Brown and R. M. Brown, Jr., *J. Cell Biol.*, *97*, 415a (1983).

33. B. Schneider and W. Herth, *Protoplasma*, *131*, 142 (1986).

34. S. C. Mueller and R. M. Brown, Jr., *Planta*, *154*, 489 (1982).

35. E. Schnepf, O. Witte, U. Rudolph, G. Deichgraeber, and H. D. Reiss, *Protoplasma*, *127*, 222 (1985).

36. A. M. C. Emons, *Acta Botanica Neerlandica*, *37*, 31−38 (1988).

37. G. C. Ruben and G. H. Bokelman, *Carboh. Res.*, *160*, 434 (1987).

38. G. C. Ruben, K. A. Marx, and T. C. Reynolds, *Ann. Proc. Elect. Microsc. Soc. Am.*, *39*, 440 (1981).

39. T. Noguchi and K. Ueda, *Protoplasma*, *128*, 64 (1985).

40. C. H. Haigler, in *Cellulose Chemistry and Its Applications* (R. P. Nevell and S. H. Zeronian, eds), Horwood, Chicester, UK, 1985, p. 30.

41. D. P. Delmer, *Ann. Rev. Plant Physiol.*, *38*, 259 (1987).

42. Y. Henry, M. Pouphile, T. Gulik-Krywicki, W. Wiessner, and M. Lefort-Tran, *Protoplasma*, *126*, 100 (1985).

43. H. Moor and K. Mühlethaler, *J. Cell Biol.*, *17*, 609 (1963).
44. A. M. C. Emons, M. Kroh, B. Knuiman, and T. Platel, in *Plant Sperm Cells as Emerging Tools for Crop Biotechnology* (H. J. Wilms and C. J. Keijzer, eds.), Pudoc, Wageningen, 1988, p. 41.
45. H. Robenek and E. Peveling, *Planta*, *136*, 135 (1977).
46. D. Volkmann, *Planta*, *162*, 392 (1984).
47. H. Schnabl, J. Vienken, and U. Zimmermann, *Planta*, *148*, 231 (1980).
48. M. Kroh and B. Knuiman, *Planta*, *166*, 287 (1985).
49. M. J. Wilkinson and D. H. Northcote, *J. Cell Sci.*, *42*, 401 (1980).
50. R. S. Pearce, *J. Exp. Bot.*, *36*, 1209 (1985).
51. D. H. Northcote, in *Soc. Exp. Biol. Sem. Series*, Vol. 28 (C. Brett and J. R. Hillman, eds.), Cambridge University Press, Cambridge, 1985, p. 177.
52. S. C. Mueller and G. A. Maclachlan, *Can. J. Bot.*, *61*, 1266 (1983).
53. J. D. Pitts and M. E. Finbow, *J. Cell Sci.*, Suppl. 4, 239 (1986).
54. P. N. T. Unwin and P. D. Ennis, *Nature*, *307*, 609 (1984).
55. W. Herth, *J. Cell Biol.*, *87*, 442 (1980).
56. D. H. Northcote, *J. Cell Sci.*, Suppl. 4, 115 (1986).
57. D. P. Delmer, *Adv. Carboh. Chem. Biochem.*, *41*, 105 (1983).
58. S. R. Jacob and D. H. Northcote, *J. Cell Sci.*, Suppl. 2, 1 (1985).
59. D. P. Delmer, *Rec. Adv. Phytochem.*, *11*, 45 (1977).
60. A. M. C. Emons and A. M. C. Wolters-Arts, *Protoplasma*, *117*, 68 (1983).
61. A. M. C. Emons and N. Van Maaren, *Planta*, *170*, 145 (1987).
62. A. M. C. Emons, *Can. J. Bot.*, *67*, 2401 (1989).

5

Relationship Between Polymerization and Crystallization in Microfibril Biogenesis

CANDACE H. HAIGLER   *Texas Tech University, Lubbock, Texas*

I. INTRODUCTION

The control of cellulose I microfibril formation by living organisms has been of intense interest ever since such microfibrils were visualized in shadowed specimens in the electron microscope (1,2). Subsequent microscopic and crystallographic examination of cellulose microfibrils from a variety of organisms demonstrated that the width and crystallite size of microfibrils varied greatly between organisms and even within organisms at different stages of development (3). Furthermore, the crystalline polymorph formed predominantly in nature, cellulose I, is different from cellulose II, which is formed when cellulose I is swollen in alkali or precipitated from solution (4). As long as four decades ago, Frey-Wyssling et al. proposed that the uniformity of microfibril size in plant and bacterial cellulose "speaks for a control of cellulose crystallization by the living cytoplasm" (2) and Rånby argued, based on the preferential formation of cellulose II under acellular conditions, that cellulose I formed not from glucan chains crystallizing acellularly but rather "from chains under the influence of a specific enzyme system or by direct synthesis in the cell" (5).

This idea of cellular control was not initially extended to bacterial cellulose microfibrils, which were widely believed to crystallize acellularly (6–9). Two predominant hypotheses were developed to explain acellular microfibril formation in *Acetobacter xylinum*:

(1) extracellular tip growth in which a glucose carrier was proposed to diffuse away from the cell and add a residue onto one
chain at the end of a growing microfibril so that polymerization and
crystallization occurred simultaneously (9); and (2) the intermediate
high polymer hypothesis in which preformed cellulose chains dif-
fused away from the cell and crystallized spontaneously into microfibrils so that polymerization and crystallization were widely separated in space and time (6). Neither of these hypotheses can explain how uniformly sized microfibrils are nucleated acellularly and
there is no evidence that the intermediate compounds required for
either mechanism exist in *A. xylinum* cultures (10). The work of
Brown and coworkers (11) and Zaar (12,13) provided compelling
microscopic evidence that microfibril formation was not acellular in
A. xylinum but rather occurred in close association with the cell.

Arguments for and against various relationships between polymerization and crystallization in microfibril formation in plants and
bacteria were based largely on theoretical considerations or indirect
evidence until 1979 when it was discovered that chemicals could be
used to alter cellulose microfibril assembly in vivo (14). These
altering chemicals (fluorescent brightening agents, direct dyes, and
cellulose derivatives) allowed direct testing of the temporal relationship between polymerization and crystallization in cellulose biogenesis
and provided insight into how organisms regulate the size of microfibrils. This chapter will summarize how these chemicals have been
used to investigate the control of cellulose biogenesis in *A. xylinum*
and other organisms. As will be elaborated below, major conclusions
derived from initial studies of *A. xylinum* have been borne out by
further studies of higher plants and other organisms that synthesize
both cellulose and chitin. These altering agents have therefore allowed elucidation of basic principles that govern the biogenesis of at
least two major structural polysaccharides.

II. *ACETOBACTER XYLINUM* AS A MODEL ORGANISM

The utility of *A. xylinum* as a model organism was recognized in the
last century as evidenced by A. J. Brown in 1886: "The production
of cellulose by a simple cell plant (*A. xylinum*), and its use as a
cell connecting medium, seems of great interest in view of the important part which cellulose plays in a similar manner in the more
highly organized forms of the vegetable kingdom; and it appeared
that any information that could be gained . . . might perhaps assist
in understanding the complex reactions which go on in higher plants.

To this end my first experiments were made. . . ." (15). Subsequently it has been argued that *A. xylinum* is an unsuitable model for biogenesis of higher plant cellulose and that reference to it in developing unifying hypotheses has led to unnecessary controversy (16). This opinion is largely contradicted by the clearly demonstrated utility of *A. xylinum* for elucidating the control of cell-associated microfibril biogenesis as discussed below. Even though the same regulatory mechanism for glucan chain polymerization recently discovered for *A. xylinum* does not seem directly applicable to higher plants (see Chaps. 9 and 11), the ability to achieve efficient in vitro synthesis with solubilized cellulose synthetase from any organism has renewed optimism that similar success can be achieved with higher plants. Although thoughtful consideration and experimental evidence are necessary to determine how extensively conclusions about prokaryotic cellulose synthesis can be generalized to eukaryotes, it is clear that *A. xylinum* remains a fundamentally important model organism in cellulose research.

Acetobacter xylinum is a gram-negative bacterium that synthesizes an extracellular fibril of cellulose (often called a ribbon; Fig. 1) in association with a row of pores and particles that can be

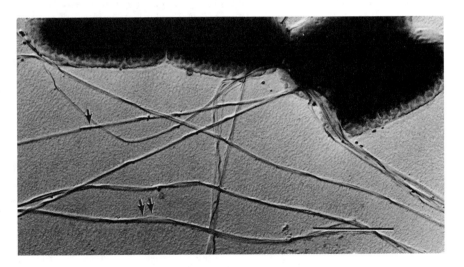

FIGURE 1   Tungsten/tantalum shadowing of several normal ribbons synthesized by *A. xylinum*. Points where a twist occurs and where smaller fibrillar subunits are visible within a ribbon are marked by single and double arrows, respectively. Bar = 1 μm.

observed after freeze-fracture in the outer membrane of the cell
(11,13). Recent cell fractionation studies demonstrated that the
cellulose synthetase activity is associated with the cytoplasmic membrane (17); the outer membrane pores and particles probably function in extrusion of the polymer chains. [The particles could be
artifacts of freeze-fracturing caused by deformation of transmembrane pore proteins (18).] The ribbon, which is a composite of
hundreds of glucan chains extruded along the row of pores, is
formed parallel to the longitudinal cell axis at a rate of about 2 μm/
min (11). Ribbon synthesis causes the cell to move forward and
rotate about its axis. Normal ribbons are 40–60 nm wide, they
twist with a periodicity of 0.6–0.9 μm, and they contain smaller
fibrillar subunits (3). Diffraction analysis demonstrates that the
ribbons contain cellulose I with a crystallite size of about 7.0 nm,
which is greater than in higher plant cellulose but lower than in
many algae (19). Recent analysis of celluloses from different
sources with ^{13}C NMR suggests that A. xylinum cellulose has
greater crystallographic similarity to the cellulose in algae than in
higher plants (20), although the significance of these differences is
not yet understood. Investigations with altering agents (see below)
suggest that the assembly of the typical A. xylinum ribbon is highly
dependent on the intact cell, which is confirmed by the observation
that in vitro preparations synthesize small disorganized fibrils (21)
with cellulose II crystallinity (17).

   Acetobacter xylinum uses the glucose in a complex growth medium (22) to synthesize cellulose. The cellulose microfibrils accumulate into a pellicle that floats on top of the culture medium under
laboratory conditions and entraps the cells (Fig. 2). Acetobacter
xylinum isolated from nature on rotting fruit does not usually show
extensive accumulation of cellulose until it is cultured in the laboratory (R. M. Brown, Jr., pers. comm.), although A. xylinum
growing on vats of vinegar or wine will form a floating pellicle
(15,23). Variants that differ in rate and extent of cellulose accumulation, which are characterized by different colony types on agar,
can be isolated after serial transfer or shaking of one strain from
the American Type Culture Collection (Fig. 3a–c). Therefore,
caution is appropriate in generalizing results obtained with one isolate of A. xylinum because it is subject to great variation in culture (24). All experiments described herein were performed on
isolates from ATCC 23769.

   Although some strains of A. xylinum may synthesize soluble
polysaccharides under some conditions (25), only cellulose accumulates in the pellicle. This constitutes one of the main advantages of
A. xylinum as a model organism; cellulose assembly can be studied
without uncontrollable interference of other polysaccharides that
accumulate in the plant cell wall. The cells and the fibril assembly

FIGURE 2   Pellicles of cellulose floating on top of the culture medium of *A. xylinum*. The thin pellicle on the left, which did not
thicken further, was synthesized by a thin-pellicle variant. The
thick pellicle on the right, which filled the flask after the photo-
graph was taken, was synthesized by a thick-pellicle variant.

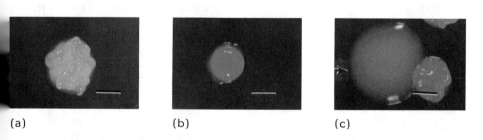

(a)   (b)   (c)

FIGURE 3   *A. xylinum* variants with different colony types, which
correlate with differences in cellulose synthesis, were isolated from
ATCC 23769.   Colonies characteristic of the thick-pellicle variant (a),
the thin-pellicle variant (b), and a variant that does not synthesize
a pellicle (c, arrow) are shown.   The colonies of the cellulose-producing variants are cohesive and will come off the agar intact when
touched with an inoculating needle.   Non-pellicle-producing variants
have frequently been called "celluloseless," but some of these synthesize small amounts of cellulose II (100).   Bar = 1 mm.

process are, however, readily accessible to the experimental addition of drugs or other altering chemicals without diffusion limitations that can hinder investigations on cellulose biogenesis in complex tissues.

III. AGENTS THAT ALTER CELLULOSE MICROFIBRIL ASSEMBLY IN VIVO

A. Direct Dyes and Fluorescent Brightening Agents

Direct dyes and fluorescent brightening agents (FBAs) share the structural characteristic of being flat molecules containing conjugated aromatic rings, which gives them a high propensity to bind to flat, ribbonlike β-1,4-linked glucan chains via dipolar and dispersion interactions (26). Both types of molecules bind with their long axes parallel to the cellulose chains (27,28) due to the juxtaposition of the pyranose rings of cellulose and the aromatic rings and heterocycles of the planar dyes. Hydrogen bonds between the dyes and cellulose may stabilize the interaction, but they are not essential (26). Direct dyes, which are dyes of the azo type, are used industrially to color cellulosic textiles and FBAs, which are based on 4,4'-diaminostilbene, are used to whiten them (they absorb ultraviolet light and emit visible light). Both types of molecules have been used extensively in biological and medical histology (29,30), although staining does not distinguish between various glucose-containing polymers with β-1,3 or β-1,4 linkages (31–33). The staining properties of FBAs continue to be exploited both qualitatively and quantitatively in (1) studies of wall regeneration in protoplasts (34–36); (2) analysis of secretory systems in plants (37,38); (3) preliminary characterization of cell wall composition and growth patterns (39,40); (4) cytopathologic detection of fungi in tissues (41,42); (5) analysis of cell viability (43); and (6) assays for polysaccharide hydrolases (44).

The first report that a direct dye could alter cell growth (probably via alteration of cell wall assembly) was published in 1896. Georg Klebs, Professor in Basel, reported that Congo red (CI 22120) caused cell swelling, inhibition of extension, and abnormally directed cross-walls when included in the growth medium of a filamentous alga (45). Alterations in cell shape were also observed when a red alga (46) and root tips (31) were grown in dyes, but no explanations were offered when these results were published. It is now known, beginning with the investigations on *A. xylinum* (14,47–49), that these types of molecules can bind to polysaccharide chains immediately upon their extrusion at the cell surface of prokaryotes and eukaryotes, thereby preventing normal assembly of microfibrils and cell walls. Many, but not all, direct dyes and FBAs prevent *A. xylinum* microfibril crystallization in vivo (50). Variance in effectiveness of the different dyes is likely due to their different affinities for cellulose under the experimental conditions, which would be affected by the kind, number, and position of substituents on the planar molecular backbone (51).

Users of these compounds should also be aware that the real concentration of dye may differ between commercial preparations. For example, most work done with *A. xylinum* was performed with FBA 28 {4,4'-bis[4-anilino-6-bis(2-hydroxyethyl)amino-s-triazin-2-ylamino]-2,2'-stilbenedisulfonic acid; CI 40622} either in the form of Calcofluor White ST or Tinopal LPW. Calcofluor White ST was a gift many years ago from American Cyanamid (Bound Brook, NJ), but they no longer manufacture this compound. Tinopal LPW was a gift from Ciba Geigy (Greensboro, NC) in 1980; it is also now obsolete. FBA 28 is now available from Polysciences under the name Cellofluor and from Calbiochem under the name Bioglo. Comparison of the absorbance at 350 nm of both forms showed that Tinopal LPW contains five to six times more active FBA 28 per unit weight than Calcofluor White ST (52). Tinopal LPW is at least 95% active FBA 28 (Robert Rummage, Ciba Geigy, pers. commun.) but may contain up to 5% salts and inert ingredients. FBAs must also be protected from light if the effective concentrations are maintained because absorption of light at 350 nm causes isomerization of the trans isomer (A_{max} = 350 nm) to the cis isomer (A_{max} = 270 nm), which is nonplanar and unable to bind or alter cellulose. When FBA 28 is completely isomerized to equilibrium, only 6–7% effective trans isomer remains (Doug Parkes, Ciba Geigy, pers. commun.). Solutions of FBAs may be made and used under red light to avoid reduction of effective concentration by isomerization. The problem of isomerization is not encountered with direct dyes.

Normal ribbons were synthesized by *A. xylinum* in the presence of all basic, acid, ingrain, polymeric, and reactive dyes that we tested and in urea, dimethylsulfoxide, and glycerol. The core molecule of FBAs, diaminostilbene, and nonplanar cis isomer of FBA 28 were also ineffective (52). These controls indicate the importance of the highly planar molecular structure of direct dyes and FBAs in inducing altered cellulose assembly. Simple competition for hydrogen bonding sites as expected from urea, dimethylsulfoxide, and glycerol is insufficient.

B. Cellulose Derivatives

High molecular weight cellulose molecules with substitutions on some or all of the hydroxyl groups are used in industry as soluble thickeners, protective colloids, binders, stabilizers, and suspending

agents. Carboxymethylcellulose (CMC) was previously added by
Colvin and Beer (53) to cultures of *A. xylinum* at 5% (w/v) to test
the intermediate high-polymer hypothesis, based on the expectation
that CMC would prevent diffusion of polymers and fibril formation if
this mechanism were operative. Fibrils were observed, which was
erroneously interpreted as evidence that microfibrils must instead
form by extracellular tip growth, but no comment was made on
whether or not the fibrils were altered. Ben-Hayim and Ohad (54)
observed a 30% increase in the rate of polymerization of cellulose
and increased orientation of the pellicle when *A. xylinum* was grown
in 0.02% CMC. They interpreted their results as evidence that
acidic polysaccharides could function to orient microfibrils in plant
cell walls.

Of the cellulose derivatives we tested, carboxymethylcelluloses
(cellulose gums 4M6F, 7LF, 7MF, 7HF, and 12MFP; gifts from
Hercules, Inc., Wilmington, DE) were the most effective in altering
cellulose microfibril assembly. All cellulose derivatives alter cellulose assembly at a higher level of organization than the smaller direct dyes and FBAs (55). Observation of the extent of alteration
in the electron microscope suggested that a CMC with degree of substitution (DS) = 0.7 was more effective than those with DS = 0.4 or
1.2, which is consistent with kinetic data of Ben-Hayim and Ohad
(54). Comparing three CMCs with DS = 0.7 but different molecular
weights indicated that lower molecular weight molecules (MW = 90,000)
were more effective than those with higher molecular weights (MW =
250,000 – 750,000). Some examples of the neutral cellulose derivatives,
hydroxypropyl-, hydroxymethyl-, hydroxypropylmethyl-, and
methylcellulose (gifts from Hercules, Inc., Wilmington, DE or Dow
Chemical, Midland, MI) induced slightly altered ribbons. No alteration was observed when cells were incubated in xanthan gum (a
bacterial polysaccharide in which a β-1,4-linked backbone is shielded
by numerous substituents composed of three sugar molecules; a gift
from Kelco Co., Clark, NJ), starch, laminarin, or agar. This illustrates the importance of a partially unshielded cellulose backbone to
allow hydrogen bonding of the derivatives with the surfaces of the
fibrillar subunits of the normal ribbon, thereby preventing their
normal association. CMC may alter most effectively because its negative charge causes the coated microfibrillar subunits to repel each
other more extensively than neutral cellulose derivatives (54). In
addition, the neutral derivatives tested had higher degrees of substitution and larger substituents than the CMCs (except for methyl
cellulose), which could have hindered their association with native
cellulose. Methyl cellulose was the most effective of the neutral
cellulose derivatives in inducing alteration.

IV. ALTERATION OF CELLULOSE ASSEMBLY IN *ACETOBACTER XYLINUM*

A. Alteration by Fluorescent Brightening Agents and Direct Dyes

Because this work has been reported in several original papers (47–50,55,56) and two reviews (3,57), only the most important points will be summarized below. When low concentrations (1–4 μM; A_{350} = 0.18–0.30) of FBA 28 or low to high concentrations (1–250 μM) of various direct dyes are added to the culture medium of *A. xylinum*, bands of fine fibrils are extruded perpendicularly to the longitudinal cell axis (Fig. 4) (47,50). The appearance of the fibrils differs somewhat between different altering agents, probably reflecting different affinities of the dyes for cellulose (50). The smallest fibrils measure 1.5 nm wide on high-resolution grids internally calibrated with ferritin; this has been proposed to be the basic unit of synthesis that is extruded from one pore at the cell surface (57). The instantaneous reversibility between altered and normal assembly (or vice versa) when the dye is washed away from (or added to) cells attached to grids suggests that the alteration is

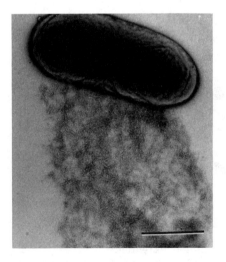

FIGURE 4 Negative staining of the highly fibrillar altered cellulose synthesized at low concentrations of FBA 28. Bar = 1 μm.

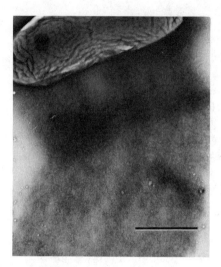

FIGURE 5   Bands of altered cellulose that lack fibrillar substructure
are synthesized at higher concentrations of FBA 28.   Higher magnification micrographs show that these bands are composed of many
randomly oriented small sheets (56).   Bar = 1 µm.

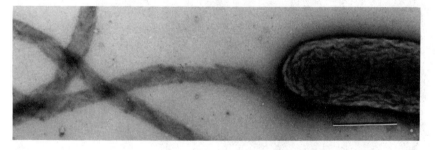

FIGURE 6   Some cells synthesize a single extended sheet (or tube;
see Fig. 7) at high concentrations of FBA 28.   All of the material
shown in this micrograph was synthesized by the included cell; the
single extended sheet passes in and out of the frame of the figure.
Bar = 1 µm.

induced extracellularly as the dye binds to the surface of each
small fibril and prevents their normal association (47).

When the concentration of FBA 28 is raised to between 8 and
250 μM (A_{350} > 0.40), the fibrillar substructure in the altered
cellulose as visualized by negative staining or tungsten/tantalum
shadowing is completely lost (49,56). The altered cellulose is synthesized as a band of many small sheets (Fig. 5) or as one extended sheet (Fig. 6). This sheetlike cellulose is never induced by
direct dyes under our experimental conditions, possibly reflecting a
particularly high affinity of FBA 28 for cellulose (50,56). It has
been shown that FBA 28 is more specific for β-1,4-linked glucan
than the direct dye Congo red (58). Different isolates of *A. xylinum*
differ in the precise concentrations where transitions between
fibrillar altered cellulose, bands of sheets, and single extended
sheets occur. The trends are generally the same, however, except
with some thin pellicle isolates that synthesize fibrillar bands over
the whole concentration range of FBA 28 (52). Why these differences between isolates occur is not understood.

Uptake of radioactive glucose, measurement of the degree of
polymerization, and crystallographic analysis demonstrate that the
altered material is high molecular weight cellulose that is noncrystalline in the wet state (57). One researcher contended that a
strong crystalline reflection in this type of altered cellulose is due
to the 020 plane of the cellulose lattice (59). Evidence obtained by
selected area electron diffraction indicates that this reflection at
$d = 0.399$ nm is due to crystalline stacking between the dye molecules when they are precipitated onto cellulose in high concentration, not to any crystallinity of the cellulose itself (56). Drying or
washing, however, causes the altered material to crystallize into
cellulose I with reduced crystallite size compared to that of controls (47,56).

In the case of the former extended sheets, the microfibrils generated by washing form a helix (Fig. 7), indicating that the
extended sheet was really a tube that probably formed by rolling of
a flat sheet of glucan chains stabilized by precipitated dye molecules
(56,60). The pitch of the helix is about 1 μm, which is similar to
the 0.6- to 0.9-μm twist in normal ribbons. This similarity and the
uniform handedness of the helices suggests that an inherent feature
of the cellulose molecule may cause this helical conformation (60).
Supporting this possibility, it has recently been demonstrated that
cellulose and/or cellulose derivatives can adopt helical conformations
under some conditions (61,62). It was previously proposed that
relief of thermodynamic strain during crystallization might account
for the twist of normal ribbons (57), but these data show that
crystallization is not required to generate a helical, or twisting,

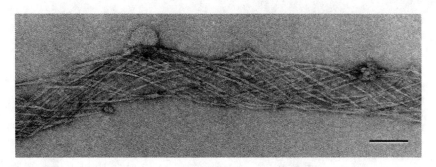

FIGURE 7   A helix of microfibrils formed by gentle washing of an
extended sheet with water at pH 3.0, which caused the dye to dissociate from the cellulose. The microfibrils contain cellulose I crystallites with reduced size compared to controls. All of these microfibrils would have been part of one normal ribbon. Bar = 100 nm.

arrangement of glucan chains. It should be emphasized that the
helix of washing-induced microfibrils has much greater pitch than
3.3- to 3.6-nm twist claimed to exist in *A. xylinum* and plant microfibrils on the basis of electron microscopic observations of shadowed
specimens (63). These observations are likely artifacts of the
shadowing method (64); they are not supported by diffraction data
and are inconsistent with the straight glucan chains observed in
lattice images (see Chap. 6).
      The rate of incorporation of radioactive glucose in the presence
and absence of FBA 28 demonstrates that the rate of polymerization
increases up to fourfold when the constraint of crystallization is released by the presence of the dye (48). The rate of glucose oxidation
to CO_2 increases only 10 – 15%; this slight increase is probably because
the cells do not aggregate while synthesizing noncrystalline cellulose,
so that more oxygen is available. When the free dye is bound to
newly synthesized cellulose so that its concentration falls below the
threshold level required to induce altered cellulose assembly, the
rate of polymerization returns immediately to that of the control.
The rate of polymerization increases with increasing concentration of
FBA 28 in the low range and then plateaus, corresponding to the
progressive disruption of fibril assembly as concentration increases
in the low range followed by the formation of sheets over a wide
range of higher concentrations (56).
      FBA 28 had no effect on in vitro synthesis of glucans in par-
ticulate membrane preparations of *A. xylinum* (M. Benziman, pers.
commun.), but this experiment was done before the optimized,
solubilized in vitro system was developed. Since the product of the

optimized in vitro system is cellulose II (17), which probably results from uncoupled polymerization and crystallization (see Chap. 1),
it is unlikely that inclusion of FBA 28 would affect in vitro cellulose
synthesis in a physiologically relevant way.

B. Alteration by Cellulose Derivatives

When 1–40 μM CMC is added to the culture medium of *A. xylinum*,
the cellulose that would normally comprise one ribbon is splayed
into separate fibrillar subunits averaging around 10 nm wide (55;
Figs. 8 and 9). These subunits leave the cell surface at average
intervals of 150 nm along the row of extrusion sites. The crystallinity of CMC-altered cellulose is not changed compared to controls,
implying that the normal crystallite is contained within these separate fibrillar subunits. The average 10 nm width (6–12 nm range)
of the subunits separated by CMC is consistent with the average
7.0-nm crystallite size in *A. xylinum* cellulose (19). Reversals in
direction of cellulose assembly that occur in the presence of CMC

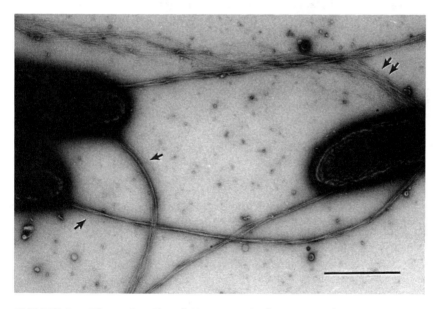

FIGURE 8   The subunits of the normal ribbon remain separated
when synthesized in the presence of CMC. They may lie closer
together, thereby resembling normal ribbons, but close inspection
shows separated (single arrows) or sometimes highly splayed
(double arrows) subunits. Bar = 1 μm.

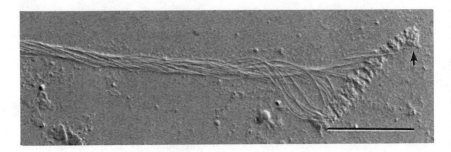

FIGURE 9   Platinum/carbon shadowed subunits of one ribbon that
have been separated by CMC. Note that the subunits originate from
separated groups of disorganized material (arrow). This form of
disorganized material (called native band) is sometimes synthesized
without addition of altering agents just before normal fibril assembly
begins. Bar = 1 μm.

reveal how these separate fibrillar subunits are formed (Fig. 10).
Extrusion sites seem to be grouped along the cell surface; the
spatial separation between the groups leads to crystallization of
fibrillar subunits that then associate only along their surfaces to
form the composite ribbon. The average number of grouped extrusion sites probably accounts for the average crystallite size of
A. xylinum cellulose. CMC is too large to prevent association of
the small 1.5-nm bundles of glucan chains originating at each pore,
but it can coat the surfaces of the larger fibrillar subunits thereby
preventing their normal association. We were able to confirm the
result of Ben-Hayim and Ohad that addition of CMC increases the
rate of cellulose polymerization about 30% (compared to at least 400%
in the presence of FBA 28). Therefore, the final assembly of the
ribbon after crystallization has occurred imposes part of the constraint on the rate of polymerization, but not nearly so great a constraint as the formation of the crystallites themselves. The evidence that fibrillar subunits containing the crystallites are part of
the normal ribbon is supported by the observation that the ribbon
first splays into 6- to 12-nm fibrillar subunits during degradation
with cellulases (65).

C. Implications of Altered Cellulose Assembly

Several conclusions can be drawn from alteration of cellulose assembly by direct dyes and FBAs (3,57): (1) there must be at least a
slight temporal gap between polymerization and crystallization in

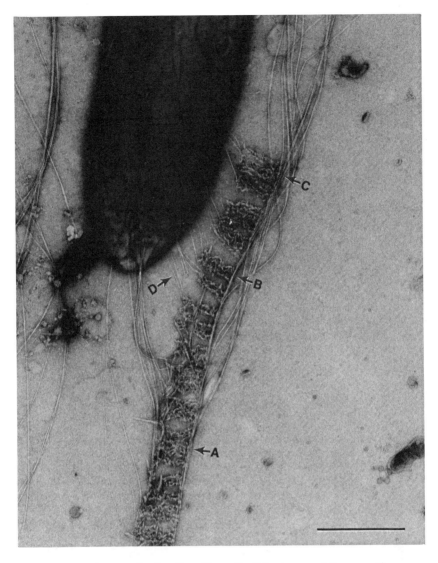

FIGURE 10   Reversals in direction of ribbon assembly sometimes
occur in the presence of CMC. This cell was synthesizing about 10
separate fibrillar subunits, progressing toward the bottom of the
figure.   It then stopped, synthesized a small amount of native band
(see Fig. 9), and resumed synthesis of fibrillar cellulose progressing
toward the top of the page.   Clusters or groups of disorganized
cellulose 60 – 200 nm wide (e.g., arrows A – C) reflect grouped extrusion sites that account for synthesis of separate fibrillar subunits
within normal ribbons.   Fine fibrils originating from one or a few
extrusion sites within each group are visible near the cell surface
(e.g., arrow D).   This fortuitous micrograph provides almost a dynamic view of ribbon assembly.   Bar = 500 nm.   (Reproduced by
permission of John Wiley and Sons, Inc., from Ref. 50.)

A. *xylinum* since FBAs and direct dyes do not disrupt a preformed
crystalline lattice (staining occurs by binding to surfaces of microfibrils). (2) Polymerization and crystallization must be coupled
processes since disrupting crystallization affects the rate of polymerization. (3) Crystallization must be the rate-limiting step in the
biogenesis of cellulose I. (4) Also, even though crystallization is
prevented in the presence of altering agents, the glucan chains
must be maintained in an ordered state so that cellulose I forms when
the dye is displaced by washing or drying. The ordered condition
of the wet, noncrystalline material is confirmed by its birefringence
(57). Whatever feature of the biogenetic process determines the
preferential formation of cellulose I in nature is preserved in the
noncrystalline material. If cellulose I is parallel and cellulose II is
antiparallel (4), preservation of parallel chain order (as dyes bind
with and stabilize chains extruded in parallel) would be sufficient to
explain the generation of cellulose I from this noncrystalline material.
If cellulose II also has parallel chain orientation with cellulose I and
cellulose II being differentiated by slight differences in conformation
of anhydroglucose units (66), the conformation characteristic of
native cellulose I must be preserved in the altered cellulose despite
the absence of crystallization. It seems possible that the biosynthetic mechanism could determine a particular molecular conformation
that cannot be recovered once disrupted by swelling or solvation.

   Alteration of fibril assembly at a higher level of organization by
CMC suggests that the organization of the synthesizing machinery in
A. *xylinum* accounts for the formation of the unique ribbon (3,57).
Because of the gap that occurs between polymerization and crystallization, fibrillar subunits about 1.5 nm wide from closely spaced
extrusion sites cocrystallize to form a larger fibrillar subunit containing one crystallite. Fibrillar subunits from adjacent groups of extrusion sites then associate by hydrogen bonding along their surfaces to form the composite ribbon. This mechanism of fibril formation has been called cell-directed self-assembly because the size and
subunit composition of the fibril, which self-assembles extracellularly,
is determined by the arrangement of synthesis/extrusion sites at the
cell surface.

   The results with CMC also show that other polysaccharides can
interact directly with cellulose to control microfibril size. This is
analogous to the proposed interaction of hemicelluloses with cellulose
microfibrils in the plant cell wall (see Chap. 2). It has been shown
that a purified hemicellulose, xyloglucan, can alter cellulose microfibril assembly in A. *xylinum* similarly to CMC when the pH of the
medium is less than 6 (67). Therefore, it is quite plausible that
microfibril size in higher plants can be partly controlled by coincident extrusion of hemicellulosic polysaccharides and control of ionic
conditions that affect binding of hemicelluloses to cellulose.

V. ALTERATION OF CELLULOSE ASSEMBLY IN OTHER ORGANISMS

Direct dyes, especially Congo red, and FBA 28 (or its equivalent, Calcofluor White ST) have been used to alter cellulose microfibril assembly and/or cell wall deposition in many eukaryotes including (1) root hairs of radish, corn, *Trianea bogotensis*, and *Ceratopteris thalictroides* (50,68,69); (2) protoplasts of tobacco, the moss *Physcomitrella*, and the alga *Boergesenia forbesii* (70−72); (3) autospores of *Oocystis apiculata* (73,74); (4) semicells of the desmid *Micrasterias denticulata* (68); (5) pollen tubes of *Lillium longiflorum* (68); (6) hyphae of the fungus *Saprolegnia monoica* (75); and (7) microcysts of the cellular slime mold *Polysphondylium pallidum* (76). Altered walls appear disorganized without distinct microfibrils and sometimes are deposited with abnormally thickened bulges. The tip-growing pollen tubes, root hairs, and fungal hyphae exhibit bulbous expansion at the tip and slowed elongation in the presence of altering agents. Both dyes have also been shown to prevent the formation of the extracellular cellulosic spikes of the chlorococcalean alga *Acanthosphaera zachariasi* (77). In *A. zachariasi* as in *A. xylinum*, FBA 28 has been demonstrated to be more effective in altering microfibril assembly than Congo red, possibly because of its higher specificity for β-1,4-linked glucans (58).

It has been suggested that the altered material in eukaryotes is inappropriately called cellulose because of its noncrystalline nature (78). This crystallinity-based definition of cellulose is more restrictive than the biochemical definition that is used in most standard references (79,80). The biochemical definition of cellulose as a high molecular weight homopolymer of β-1,4-linked glucan is to be preferred with the crystalline state clarified when necessary.

In all eukaryotes studied so far, it has been possible to find a concentration of dye or brightener that causes alteration in fibril assembly but does not cause cell death. Plant protoplasts, which are notoriously delicate, can be regenerated into plants when cultured in the presence of FBAs (70). Effective but nonlethal concentrations have generally ranged between 0.001 and 1 mg/ml (about 1−1000 μM), with the lower concentrations effective for protoplasts where no preexisting wall can hinder diffusion or bind added dye. It is noteworthy that this lower limit is the same as the threshold concentration required to alter *A. xylinum* cellulose assembly where the microfibrils are readily accessible to the dye as they are in protoplasts. In most eukaryotes the alteration is reversible, just as in *A. xylinum*, at least for incubation times of a few hours (72). In *O. apiculata*, complete disruption of wall assembly by 800 μM Calcofluor White ST (about 130 μM active FBA 28) causes new autospores to die (74). The cell death may be due to formation of an

abnormal wall, but whether a lower effective concentration of dye
might have been nonlethal and whether the lethal effect was due to
impurities in the dye were not tested.

In each of the organisms mentioned above (except possibly
Polysphondylium pallidum; see below), cellulose microfibril assembly
occurs at the surface of the plasma membrane. Dyes can penetrate
an existing cell wall to reach the exterior of the plasma membrane,
but they do not penetrate inside living cells (31,43). Therefore,
the cellulosic scales of the alga *Pleurochrysis scherffelii*, which are
formed intracellularly within the cisternae of the Golgi apparatus
and then exocytosed, are not altered by the inclusion of dyes in the
culture medium (68). It has been claimed, based on Calcofluor
staining, that *P. pallidum* amoebae also synthesize cellulose in intracellular vesicles and transport it to the cell wall during microcyst
formation (76). Despite this putative intracellular cellulose synthesis and in contrast to *P. scherffelii*, microcyst wall assembly is still
altered by the presence of the dye. Given the general impermeability
of living cells to Calcofluor (31) and its ability to stain the cytoplasm and nuclei of many types of dead cells (43) as well as other
polysaccharides (31–33), it is clear that the authors have over-
interpreted intracellular Calcofluor staining as definitively indicative
of cellulose. (They try to support this claim by use of commercial
cellulases, which are notorious for contamination with other degradative enzymes.) Phagocytic amoebae might take up Calcofluor
whereas most cells do not, but the intracellular presence of cellulose
must be established by other methods to establish *P. pallidum* as
the second exception to the rule that cellulose synthesis is a cell
surface phenomenon.

In two organisms where the assembly of cellulose microfibrils is
elaborate, the ability to alter assembly depends on the stage at
which the altering agent is added: (1) the spikes in *A. zachariasi*
are only effectively altered if the dye is present when they are
initiated (77); and (2) the primary wall of the desmid *M. denticulata*,
which contains small single microfibrils, is always alterable whereas
the secondary wall, which contains aggregates of microfibrils, is
only altered if dye is added before onset of its synthesis (68).
The significance of this atypical absence of disruption cannot presently be understood since dyes do prevent formation of large microfibrils in other algae such as *O. apiculata*.

There is little published information about the effect of FBAs or
dyes on the rate or cellulose synthesis in vivo in eukaryotes. The
synthesis of very thick altered walls in many eukaryotes suggests
that the rate of polymerization may increase. A preliminary report
(text only with no graphic data) indicates that the rate of cellulose
synthesis does increase at least 40% when hyphae of the fungus
Saprolegnia monoica are incubated in low concentrations of Congo

red (25 µg/ml), whereas a slight inhibition is observed at high concentrations (400 µg/ml) (75). In contrast, it has been reported that up to 100 µM Calcofluor White ST causes no increase in rate of cellulose synthesis in suspension cultures of tobacco (81). However, 100 µM of this commercial formulation contains only about 18 µM active dye (see Sec. III.A), which is only 10 times higher than the threshold concentration required to induce altered cellulose assembly in *A. xylinum* cells with no preexisting cellulose. It is possible that the walls of the suspension cells bound the added dye so that insufficient free dye remained to prevent crystallization of newly synthesized cellulose. More extensive investigation of the rate of cellulose synthesis while crystallization is disrupted should be performed with eukaryotes.

Similarly, no reports of the effect of cellulose derivatives on microfibril assembly in eukaryotic organisms have been published. Problems with accessibility of these high molecular weight molecules to the active site of cellulose synthesis in organisms with existing cell walls would be expected. It should be possible, however, to study the effect of cellulose derivatives on initial wall regeneration in protoplasts.

The ability to alter cellulose microfibril assembly in a variety of eukaryotes demonstrates that polymerization and crystallization are also consecutive processes in these higher organisms and that final microfibril assembly occurs outside the plasma membrane. It should be emphasized, however, that the glucan chains are still attached to their polymerizing enzymes when crystallization occurs; the significance of this distinction will be discussed in Section VII. It has also been possible to investigate other aspects of cell wall assembly in eukaryotes by use of direct dyes and FBAs to alter cellulose assembly and/or to mark terminal complexes present when the dye is added. For example, it has been possible to show that (1) terminal complexes are subject to turnover and are moved by the force of microfibril crystallization in the membrane of the alga *O. apiculata* (82,83); (2) microtubules probably control terminal complex alignment in *O. apiculata* (84); (3) assembly of crystalline cellulose microfibrils and/or hydrogen bonding between microfibrils in the innermost wall layer is required for gravitropic bending in plants (85); and (4) a *rigid* cell wall is not a prerequisite for cell division in plant protoplasts although synthesis of wall material is required (70).

VI. ALTERATION OF CHITIN ASSEMBLY

After the ability to alter cellulose assembly was discovered, many researchers were able to demonstrate that the same types of molecules

also alter chitin assembly and/or growth habit in (1) the loricae of
the chrysoflagellate alga *Poterioochromonas stipitata* and the
folliculinid ciliate *Eufolliculina uhligi* (86,87); (2) walls and bud
scars of the yeasts *Saccharomyces cerevisiae* and *Rhodotorula
glutinis* (88–92); (3) protoplasts, yeast cells, and mycelia of the
yeast *Candida albicans* (88,93); (4) hyphae of the fungi *Aspergillus
niger* and *Geotrichum lactis* (91,94); (5) extracellular spikes of the
centric diatom *Thalassiosira fluviatilis* (68); and (6) the periotrophic
membranes of insects (95; there is no direct evidence that chitin
assembly has been altered in this case, but the permeability of the
membrane is increased). It has been demonstrated that altered material produced in vitro has no crystallinity when wet (normal
in vitro chitin is crystalline) but that it dries into typically crystalline, antiparallel chain, α-chitin (95).

In many cases, the growth habit of the organism is changed by
the addition of the altering agents with the specific changes being
determined by the location of chitin in the normal wall. For example, Congo red halts cell proliferation in *R. glutinis* (90), which
contains large amounts of chitin in its entire wall, whereas it halts
cytokinesis only in *S. cerevisiae*, which contains chitin predominantly in the bud scars (89). Congo red has no effect on the
growth of *Schizosaccharomyces pombe*, which has only 1,3-β-glucan
in its wall (89). In filamentous fungi, altering agents cause abnormally thickened septa and often cause lysis at the hyphal tips,
presumably due to weakening of the newly synthesized cell wall (91).
These results suggest that chitin polymerization and crystallization
are also consecutive processes in vivo, which has been confirmed
with an in vitro system from *Schizophyllum commune* that produces
crystalline chitin (96). In the in vitro system extensive crystallization begins within 2.5 min after polymerization of chitin. Use of
altering agents has provided direct evidence that the wall at the
tip of a fungal hypha is viscoelastic due to the synthesis of noncrystalline chitin, which becomes more crystalline further from the
hyphal apex (96).

Chitin polymerization and crystallization have also been demonstrated to be coupled processes because the rate of polymerization
increases when crystallization is disrupted. This conclusion has
frequently been suggested by synthesis of abnormally thickened
walls in the presence of altering agents and directly confirmed by
an increase in incorporation of [^{14}C]glucose into chitin in mycelia,
protoplasts, and permeabilized cells of yeasts (91,92). Incorporation enhancement ranged from a minimum of 1.5-fold to a maximum
of 32-fold in different organisms compared to the controls. From
the study on permeabilized cells (92), it has been suggested that
Calcofluor induces activation of chitin synthase zymogens by an
unknown mechanism that depends on protein synthesis. Altering

agents have been shown to inhibit chitin polymerization in vitro, but these results are highly variable and probably not physiologically relevant (91,92,96 – 98).

VII. SUMMARY

The proof that polymerization and crystallization are consecutive but coupled processes in cellulose I biogenesis in *A. xylinum* provides the theoretical basis for the mechanism of regulation of microfibril size that is suggested by freeze-fracture observations of terminal complexes with different sizes and shapes (see Chaps. 3 and 4). The evidence is compelling that the linear arrangement and grouping of cellulose synthesis/extrusion sites in *A. xylinum* regulate the fibrillar substructure and crystallite size of the cellulose synthesized by this prokaryote. The ability of FBAs and dyes to interfere with fibril assembly in eukaryotes demonstrates that polymerization and crystallization are also consecutive processes in higher organisms. A preliminary report suggests that the two processes are coupled in cellulose-synthesizing eukaryotes. This possibility is supported by the correlation between different types of eukaryotic terminal complexes and different sizes and shapes of microfibrils (see Chap. 6), but definitive kinetic data are needed to prove that coupling also occurs in eukaryotes. Chitin polymerization and crystallization have been directly proven to be consecutive, coupled processes in eukaryotes, but chitin microfibril terminal complexes have not been observed. Until there is more evidence regarding the organization of chitin synthases in membranes, we cannot know how far the similarity with cellulose-synthesizing organisms extends.

The process of cell-directed self-assembly is likely to be a universal, or at least a very common, mechanism by which organisms regulate the size and crystallinity of structural microfibrils composed of polysaccharides. The potential size and shape of a cellulose microfibril can be regulated by the size and geometry of terminal complexes in the plasma membrane. In organisms such as *M. denticulata*, the ordered packing of many terminal complexes further regulates the arrangement of microfibrils in the wall. It is the temporal gap between polymerization and crystallization that allows glucan chains extruded by different enzymes within one terminal complex to coalesce before crystallization to form the typical microfibril occurs. Microfibril size can be further regulated by the coincident extrusion of hemicellulosic polysaccharides such as xyloglucan that can bind to cellulose.

It is inappropriate to say that inhibition of microfibril crystallization by FBAs and dyes proves the intermediate high-polymer hypothesis (86,89,98,99); this hypothesis implies that crystallization occurs acellularly from chains that have been released from their polymerizing enzymes. In contrast, coupled polymerization and crystallization implies that the chains remain attached to their polymerizing enzymes until crystallization occurs. The importance of this coupling to the concept of cell-directed self-assembly is emphasized by the observation that the normal crystallite size or fibril structure of *A. xylinum* cellulose cannot be recovered when the normal coupling in association with the cell is disrupted even if removal of excess dye allows formation of cellulose I.

The wide distribution of cellulose in nature (see Chap. 1) is correlated with significant diversity in microfibril size and crystallinity (see Chap. 6). The large microfibrils in the walls of many algae, which are woven together into very strong cell walls, are probably an adaptation to the physical stresses of existing as unicells or filaments in aquatic environments. The unique ribbon of *A. xylinum* seemingly allows a floating pellicle to be formed; other cellulose-synthesizing bacteria such as *Rhizobium* and *Alcaligenes* lack the linear array of extrusion sites, synthesize less organized fibrils, and only flocculate in their medium (R. M. Brown, Jr., pers. commun.). The rosette terminal complex in some algae and all higher plants may allow production of a more extensible wall as smaller, dispersed microfibrils interact with hemicelluloses. The mechanism of coupled polymerization and crystallization, which provides the basis for microfibril formation by cell-directed self-assembly, has allowed different organisms to synthesize microfibrils with different sizes and crystallinities and, thereby, to be better adapted to different environments.

REFERENCES

1. R. D. Preston, E. Nicolai, R. Reed, and A. Millard, *Nature*, *162*, 665 (1948).
2. A. Frey-Wyssling, K. Mühlethaler, and R. W. G. Wyckoff, *Experientia*, *4*, 475 (1948).
3. C. H. Haigler, in *Cellulose Chemistry and Its Applications* (T. P. Nevell and S. H. Zeronian, eds.), Ellis Horwood Ltd., Chichester, 1985, p. 30.
4. J. Blackwell, in *Cellulose and Other Natural Polymer Systems* (R. M. Brown, Jr., ed.), Plenum Press, New York, 1982, p. 403.
5. B. G. Rånby, *Acta Chem. Scand.*, *6*, 101 (1952).
6. A. Frey-Wyssling and K. Mühlethaler, *J. Polym. Sci.*, *1*, 172 (1946).

7. M. Aschner and S. Hestrin, *Nature*, *157*, 659 (1946).
8. M. Schramm and S. Hestrin, *J. Gen. Microbiol.*, *11*, 123 (1954).
9. J. R. Colvin, S. T. Bayley, and M. Beer, *Biochem. Biophys. Acta*, *23*, 652 (1957).
10. D. P. Delmer, *Adv. Carboh. Chem. Biochem.*, *41*, 105 (1983).
11. R. M. Brown, Jr., J. H. M. Willison, and C. L. Richardson, *Proc. Natl. Acad. Sci. USA*, *73*, 4565 (1976).
12. K. Zaar, *Cytobiologie*, *16*, 1 (1977).
13. K. Zaar, *J. Cell Biol.*, *80*, 773 (1979).
14. C. H. Haigler and R. M. Brown, Jr., *J. Cell Biol.*, *83*, 70a (1979).
15. A. J. Brown, *J. Chem. Soc. (Lond.)*, *49*, 432 (1886).
16. J. H. M. Willison, in *Cellulose and Other Natural Polymer Systems* (R. M. Brown, Jr., ed.), Plenum Press, New York, 1982, p. 105.
17. T. E. Bureau and R. M. Brown, Jr., *Proc. Natl. Acad. Sci. USA*, *84*, 6985 (1987).
18. U. B. Sleytr and A. W. Robards, *J. Microsc.*, *126*, 101 (1982).
19. I. Nieduszynski and R. D. Preston, *Nature*, *225*, 274 (1970).
20. R. H. Atalla and D. L. Vanderhart, *Science*, *223*, 283 (1984).
21. F. C. Lin, R. M. Brown, Jr., J. B. Cooper, and D. P. Delmer, *Science*, *230*, 822 (1985).
22. S. Hestrin and M. Schramm, *Biochem. J.*, *58*, 345 (1954).
23. J. Barsha and H. Hibbert, *Can. J. Res.*, *10*, 170 (1934).
24. J. L. Shimwell and J. G. Carr, *Nature*, *201*, 1051 (1964).
25. R. O. Couso, L. Ielpi, and M. A. Dankert, *J. Gen. Microbiol.*, *133*, 2123 (1987).
26. I. D. Rattee and M. M. Breuer, *The Physical Chemistry of Dye Adsorption*, Academic Press, New York, 1974.
27. T. Vickerstaff, *The Physical Chemistry of Dyeing*, Interscience, New York, 1954.
28. W. Herth and E. Schnepf, *Protoplasma*, *105*, 129 (1980).
29. R. D. Lillie, *H. J. Conn's Biological Stains*, 9th ed., Williams and Wilkins, Baltimore, 1977.
30. M. A. Darken, *Science*, *133*, 1704 (1961).
31. J. Hughes and M. E. McCully, *Stain Tech.*, *50*, 319 (1975).
32. P. J. Wood and R. G. Fulcher, *Cereal Chem.*, *55*, 952 (1978).
33. P. J. Wood, R. G. Fulcher, and B. A. Stone, *J. Cereal Sci.*, *1*, 95 (1983).
34. T. Itoh, R. L. Legge, and R. M. Brown, Jr., *J. Phycol.*, *22*, 224 (1986).
35. M. G. Meadows, *Anal. Biochem.*, *141*, 38 (1984).

36. D. W. Galbraith, *Physiol. Plant.*, *53*, 11 (1981).
37. J. Heslop-Harrison and Y. Heslop-Harrison, *Is. J. Bot.*, *34*, 187 (1985).
38. J. D. Walton and E. D. Earle, *Planta*, *165*, 407 (1985).
39. P. V. Sengbusch, M. Mix, I. Wachholz, and E. Manshard, *Protoplasma*, *111*, 38 (1982).
40. G. Kritzman, I. Chet, Y. Henis, and A. Hütlermann, *Is. J. Bot.*, *27*, 138 (1978).
41. J. G. Monheit, G. Brown, M. M. Kott, W. A. Schmidt, and D. G. Moore, *Am. J. Clin. Pathol.*, *85*, 222 (1986).
42. R. L. Sautter and H. G. Kwee, *Am. J. Clin. Pathol.*, *87*, 295 (1987).
43. J. M. C. Fischer, C. A. Peterson, and N. C. Bols, *Stain Tech.*, *60*, 69 (1985).
44. P. J. Wood, J. D. Erfle, and R. M. Teather, *Meth. Enzymol.*, *160*, 59 (1988).
45. G. Klebs, *Die Bedingungen der Fortpflanzung bei einigen Algen und Pilzen*, Gustav Fischer, Jena, 1896, p. 338.
46. S. D. Waaland and J. R. Waaland, *Planta*, *126*, 127 (1975).
47. C. H. Haigler, R. M. Brown, Jr., and M. Benziman, *Science*, *210*, 903 (1980).
48. M. Benziman, C. H. Haigler, R. M. Brown, Jr., A. R. White, and K. M. Cooper, *Proc. Natl. Acad. Sci. USA*, *77*, 6678 (1980).
49. R. M. Brown, Jr., C. H. Haigler, and K. Cooper, *Science*, *218*, 1141 (1982).
50. R. M. Brown, Jr., C. H. Haigler, J. Suttie, A. R. White, E. Roberts, C. Smith, T. Itoh, and K. Cooper, *J. Appl. Polym. Sci.: Appl. Polym. Symp.*, *37*, 33 (1983).
51. D. G. Duff and C. H. Giles, in *Water: A Comprehensive Treatise, Vol. 4, Aqueous Solutions of Amphiphiles and Macromolecules* (F. Franks, ed.), Plenum Press, New York, 1975, p. 169.
52. C. H. Haigler, Alteration of cellulose assembly in *Acetobacter xylinum* by fluorescent brightening agents, direct dyes, and cellulose derivatives, Ph.D. dissertation, University of North Carolina, Chapel Hill, 1982.
53. J. R. Colvin and M. Beer, *Can. J. Microbiol.*, *6*, 631 (1960).
54. G. Ben-Hayim and I. Ohad, *J. Cell Biol.*, *25*, 191 (1965).
55. C. H. Haigler, A. R. White, R. M. Brown, Jr., and K. M. Cooper, *J. Cell Biol.*, *94*, 64 (1982).
56. C. H. Haigler and H. Chanzy, *J. Ultrastruct. Mol. Struct. Res.*, *98*, 299 (1988).
57. C. H. Haigler and M. Benziman, in *Cellulose and Other Natural Polymer Systems* (R. M. Brown, Jr., ed.), Plenum Press, New York, 1982, p. 273.

58. P. J. Wood, *Carbohy. Res.*, *95*, 271 (1980).
59. A. Kai, *Makromol. Chem. Rapid Commun.*, *5*, 307 (1984).
60. C. H. Haigler and H. Chanzy, in *Cellulose and Wood: Chemistry and Technology* (C. Schuerch, ed.), John Wiley and Sons, New York, 1989, p. 493.
61. P. Zugenmaier, *J. Appl. Polym. Sci.: Appl. Polym. Symp.*, *37*, 223 (1983).
62. A. M. Ritcey, J. Giasson, J. F. Revol, and D. G. Gray, in *Cellulose and Wood: Chemistry and Technology* (C. Schuerch, ed.), John Wiley and Sons, New York, 1989, p. 189.
63. G. C. Ruben and G. H. Bokelman, *Carboh. Res.*, *160*, 434 (1987).
64. J. H. M. Willison and A. J. Rowe, *Replica, Shadowing, and Freeze-Etching Techniques*, North-Holland, New York, 1980.
65. A. R. White and R. M. Brown, Jr., *Proc. Natl. Acad. Sci. USA*, *78*, 1047 (1981).
66. R. H. Atalla, in *The Structure of Cellulose: Characterization of the Solid State* (R. H. Atalla, ed.), American Chemical Society, Washington, D.C., 1987.
67. T. Hayashi, M. P. F. Marsden, and D. P. Delmer, *Plant Phys.*, *83*, 384 (1987).
68. W. Herth and I. Hausser, in *Structure, Function, and Biosynthesis of Plant Cell Walls* (W. M. Dugger and S. Bartnicki-Garcia, eds.), American Society of Plant Physiology, Rockville, MD, 1984, p. 89.
69. H. T. H. M. Meekes, *J. Exp. Bot.*, *37*, 1201 (1986).
70. G. Hahne, W. Herth, and F. Hoffman, *Protoplasma*, *115*, 217 (1983).
71. J. Burgess and P. J. Linstead, *Micron*, *13*, 185 (1982).
72. T. Itoh, R. M. O'Neill, and R. M. Brown, Jr., *Protoplasma*, *123*, 174 (1984).
73. H. Quader, *Naturwissenschaften*, *67*, 428 (1981).
74. E. Roberts, R. W. Seagull, C. H. Haigler, and R. M. Brown, Jr., *Protoplasma*, *113*, 1 (1982).
75. P. Nodet, A. Capellano, and M. Fevre, Effects of Congo red on polysaccharide synthesis and glucan synthase activities of the fungus *Saprolegnia*, in *Cell Walls '86, Conference Proceedings*, Paris, 1986.
76. A. H. C. Choi and D. H. O'Day, *J. Bacteriol.*, *157*, 291 (1984).
77. E. Schnepf, G. Deichgräber, and W. Herth, *Protoplasma*, *110*, 203 (1982).
78. H. Quader, D. G. Robinson, and R. Van Kempen, *Planta*, *157*, 317 (1983).
79. A. L. Lehninger, *Biochemistry*, 2nd ed., Worth, New York, 1975.

80. R. D. Preston, *The Physical Biology of Plant Cell Walls*, Chapman and Hall, London, 1974.

81. W. Blaschek, U. Semler, and G. Franz, *J. Plant Physiol.*, *120*, 457 (1985).

82. D. G. Robinson and H. Quader, *Eur. J. Cell Biol.*, *25*, 278 (1981).

83. H. Quader, *Eur. J. Cell Biol.*, *32*, 174 (1983).

84. H. Quader, *J. Cell Sci.*, *83*, 223 (1986).

85. D. B. Folsom and R. M. Brown, Jr., in *Physiology of Cell Expansion During Plant Growth* (D. J. Cosgrove and D. P. Knievel, eds.), American Society of Plant Physiologists, Rockville, MD, 1987, p. 58.

86. W. Herth, *J. Cell Biol.*, *87*, 442 (1980).

87. M. Mulisch, W. Herth, P. Zugenmaier, and K. Hausmann, *Biol. Cell.*, *49*, 169 (1983).

88. V. Elorza, H. Rico, and R. Sentandreu, *J. Gen. Microbiol.*, *129*, 1577 (1983).

89. G. L. Vannini, F. Poli, A. Donini, and S. Pancaldi, *Plant Sci. Lett.*, *31*, 9 (1983).

90. S. Pancaldi, F. Poli, G. Dallolio, and G. L. Vannini, *Caryologia*, *38*, 247 (1985).

91. C. Roncero and A. Duran, *J. Bacteriol.*, *163*, 1180 (1985).

92. C. Roncero, M. H. Valdivieso, J. C. Ribas, and A. Duran, *J. Bacteriol.*, *170*, 1945 (1988).

93. H. Rico, F. Miragall, and R. Sentandreu, *Exp. Mycol.*, *9*, 241 (1985).

94. S. Pancaldi, F. Poli, G. Dallolio, and G. L. Vannini, *Arch. Microbiol.*, *137*, 185 (1984).

95. D. Zimmerman and W. Peters, *Comp. Biochem. Physiol.*, *86B*, 353 (1987).

96. C. A. Vermeulen and J. G. H. Wessels, *Eur. J. Biochem.*, *158*, 411 (1986).

97. C. P. Selitrennikoff, *Exp. Mycol.*, *8*, 269 (1984).

98. C. P. Selitrennikoff, *Exp. Mycol.*, *9*, 179 (1985).

99. J. R. Colvin and D. E. Witter, *Protoplasma*, *116*, 34 (1983).

100. E. Roberts, I. M. Saxena, and R. M. Brown, Jr., in *Cellulose and Wood: Chemistry and Technology* (C. Schuerch, ed.), John Wiley and Sons, New York, 1989, p. 689.

6

Physical Structure of Cellulose Microfibrils: Implications for Biogenesis

SHIGENORI KUGA   *University of Tokyo, Tokyo, Japan*

R. MALCOLM BROWN, JR.   *University of Texas, Austin, Texas*

I. INTRODUCTION

At present, the science of cellulose seems to be entering a new and dynamic stage of activity as a result of recent progress in structural and biological studies. While no cellulose scientist has doubted the fundamental role of the biogenic mechanism in determining the structure of native celluloses, our knowledge in both of these fields has been frustratingly limited. A series of findings in the last few years, however, provided clues toward understanding the basic features of the molecular construction of cellulose microfibrils in diverse organisms. In this chapter we will attempt to describe and organize these new findings in such a way as to give an overview of this rapidly evolving branch of cellulose research.

II. CRYSTAL STRUCTURE AND BIOGENESIS

A remarkable fact about cellulose is that D-glucose, which is a typical hydrophilic molecule, polymerizes into a highly insoluble macromolecule. It seems quite likely that the physicochemical stability of cellulose has its origin in the high symmetry of β-1,4-linked glucan structure, giving rise to its tendency to crystallize. As a result, glucan chain molecules aggregate to form the microfibril almost as soon as they are polymerized. Therefore, the crystal structure is regulated by both the mechanism of polymerization and the structure of the enzyme complex. Microfibril biogenesis could be defined in terms relating to "biocrystallization."

A. Parallel or Antiparallel

One of the central questions about cellulose structure is the mode of the polymer chain arrangement in the crystal lattice, i.e., whether the glucan chains are oriented parallel or antiparallel to each other. The antiparallel structure in native cellulose, necessitating either a chain-folding process after polymerization or the involvement of two types of cellulose-polymerizing enzymes, has caused much confusion and controversy, particularly among biochemists. For example, a biochemist has written, "For those who study biosynthesis, it is therefore a relief to learn that the weight of evidence favors the parallel model" (1).

The "weight of evidence" at that time had been obtained by X-ray diffraction studies combined with molecular modeling analysis, which were carried out in the 1970s (2,3). Recently, evidence for the extended chain, parallel packing structure has emerged from electron microscopic studies of *Valonia* cellulose. Hieta et al. (4) showed that disrupted *Valonia* microfibrils were stained asymmetrically by selective staining of reducing ends of cellulose (cf. Fig. 1). Also, Chanzy and Henrissat (5) showed that cellulose molecules in the microfibrils of *Valonia* are eroded unidirectionally by cellobiohydrolase, which removes cellobiose from the nonreducing end of cellulose (Figs. 2 and 3).

Recently, Sugiyama et al. (6–8) achieved lattice imaging of cellulose, which was a significant technical achievement since cellulose is a highly electron beam-sensitive material. By this technique, they could show that *Valonia* microfibrils are essentially a single crystal with a cross-sectional size of about 20 × 20 nm (Figs. 4 and 5). Though this observation cannot tell anything about the chain polarity, it directly showed the lattice fringes corresponding to single molecular sheets of cellulose, and excluded the possibility of regular disruption of crystalline order, such as chain folding, along the fiber axis.

Through these electron microscopic studies, the extended parallel chain structure was firmly established for *Valonia* cellulose. Judging from the universal occurrence of cellulose I in nature, it is quite likely that the basic parallel chain structure is shared by cellulose-producing organisms other than *Valonia*. Therefore, the recent results of a refined X-ray study with ramie cellulose (9) is

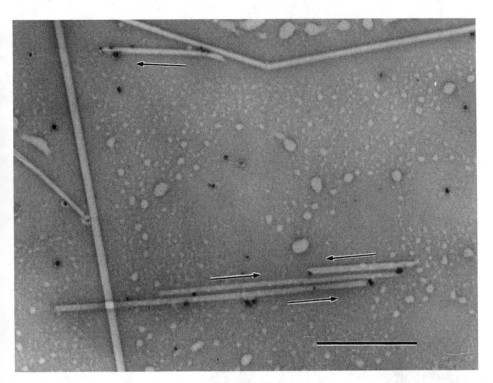

FIGURE 1   Fragments of cellulose microfibrils of *Valonia macrophysa* labeled with silver at the reducing ends. Silver is deposited on silver proteinate particles, which are attached to the reducing groups of cellulose. The asymmetrical staining demonstrates that the cellulose molecules are arranged with the same sense in the microfibrils. Arrows indicate the direction of the reducing end in each microfibril. Bar = 500 nm.

encouraging in that it reversed the previous conclusion that ramie cellulose is antiparallel (10). Also, the reducing end staining technique was recently applied to bacterial cellulose, showing the parallel structure again (11) (Fig. 6). Thus, accumulating evidence points to the universality of parallel chain structure in native celluloses, providing more relief for biochemists. In turn, it is likely that the particular structure of cellulose I is the unique result of biosynthesis, because another polymorph, cellulose II, is irreversibly

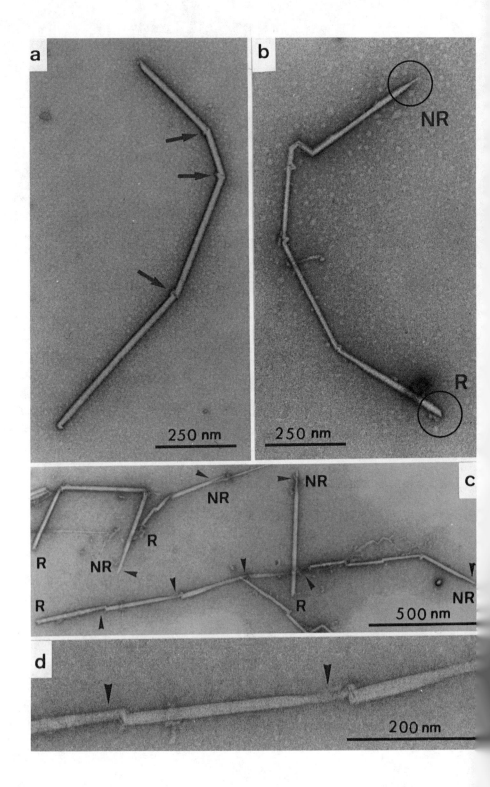

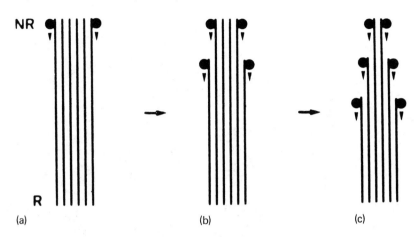

FIGURE 3   Schematic representation of enzymatic degradation of *Valonia* cellulose. In (a) the exo enzyme (●) attacks accessible non-reducing ends (NR) at the crystal surface. In (b) degradation of the outer layer exposes a new surface of nonreducing ends to the enzymes. As degradation proceeds, a sharpened tip morphology (c) appears at the nonreducing end of the microcrystal. (Reproduced by permission from Ref. 5.)

formed when native cellulose I is swollen in alkali or regenerated from solution (see Chap. 1).

B. "Up or Down" and "Head or Tail"

Given the above basic structure, two questions arise in the context of the correlation of the structure with biosynthesis:

(1)   There are two possible structures in the parallel chain model for cellulose I, denoted as "parallel up" and "parallel down" by Gardner and Blackwell (2). These structures refer to the way in which the monoclinic unit cell is skewed from the corresponding (imaginary) orthorhombic unit cell (with rectangular cross-section),

FIGURE 2   Mode of enzymatic degradation of microfibrils of *Valonia ventricosa*. (a) Mechanically destrupted microfibrils. Arrows indicate kinked defects. Bar = 250 nm. (b) After digestion with a cellobiohydrolase (CBH). Circled areas: R, reducing ends; NR, nonreducing ends. Arrowheads denote the areas of CBH attack. Bar = 500 nm. (d) Enlargement of an area of (c). Bar = 200 nm. (Reproduced by permission from Ref. 5.)

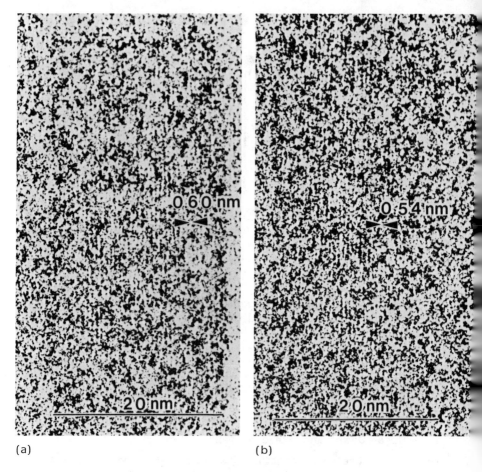

(a)   (b)

FIGURE 4   Lateral lattice images of *Valonia* cellulose obtained with
the disintegrated microfibril.   Lattice fringes of (a) (110) plane and
(b) (110) plane.   Arrowheads denote the single-lattice spacings.
The indexing of crystallographic axes is based on Ref. 3.   (Reproduced by permission from Ref. 7.)

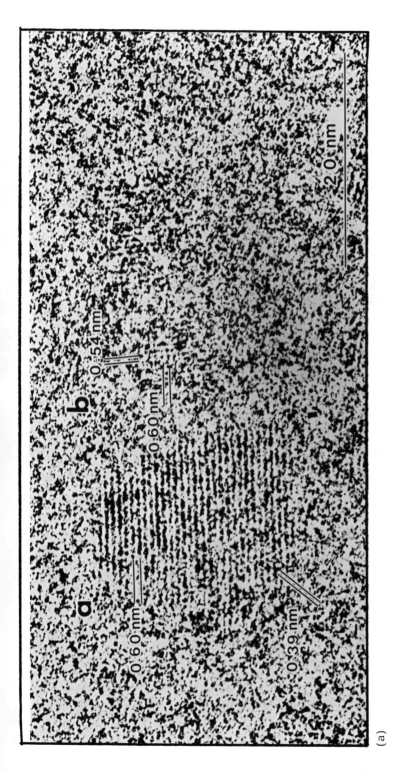

(a)

FIGURE 5 (a) Cross-sectional lattice images obtained with the ultrathin section of the cell wall of *Valonia macrophysa*. Two neighboring microfibrils are visualized. (b) Schematic drawing of the cross-section of the microfibril. The direction parallel to the cell wall surface is horizontal in both (a) and (b). (Reproduced by permission from Ref. 8.)

38-42 cellulose chains

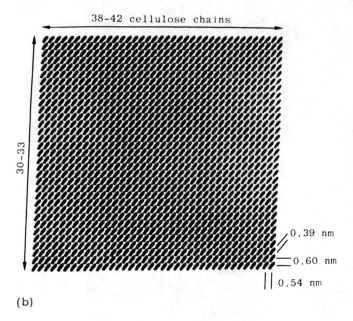

30-33

0.39 nm

0.60 nm

0.54 nm

(b)

FIGURE 5   (Continued)

while keeping the molecular polarity (see Fig. 7). The difference
between up and down structures is a subtle one compared with that
between parallel and antiparallel packing, and is more difficult to
distinguish experimentally. The situation is somewhat analogous to
the case of chirality in natural organic molecules, but up and down
structures are not physically equivalent as in the case of D and L
structures, however subtle the difference may be. The latest X-ray
analysis with ramie (9) favored up structure, agreeing with Gardner
and Blackwell's model for *Valonia*. Since we have no a priori reason
to exclude the possibility of the down structure, both confirming the
structure for variety of native celluloses and discovering the reason
of the choice of up (or down) structure by nature are the next

---

FIGURE 6   (a and b) Fragment of bacterial cellulose ribbon treated
in the same way as for Fig. 1. Silver is deposited on silver proteinate particles, which are attached to the reducing groups of cellulose.
The asymmetrical staining demonstrates the parallel arrangement.
(Reproduced by permission from Ref. 11.)

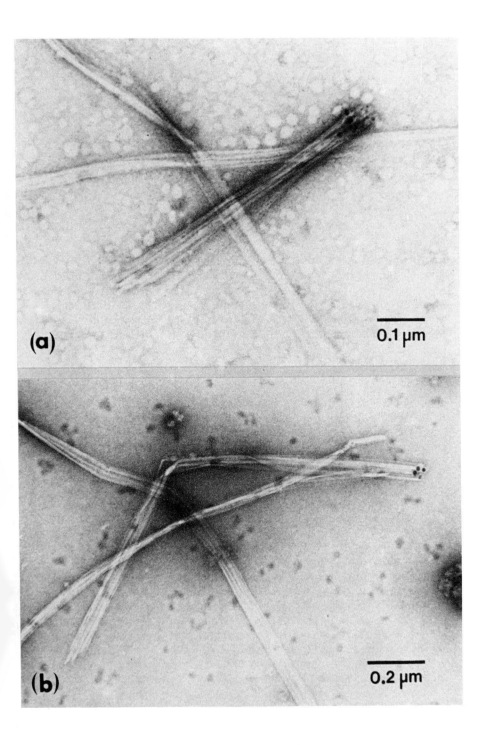

(a)

0.1 μm

(b)

0.2 μm

133

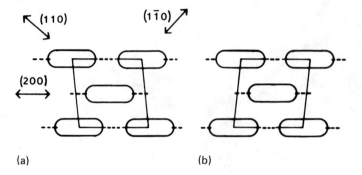

(a)   (b)

FIGURE 7   Unit cells for the (a) "parallel-up" and (b) "parallel-down" models for cellulose I (2).   In both (a) and (b), all molecules have their reducing ends directed toward the viewer.   Broken lines show the direction of hydrogen bondings, which lie in the (200) planes.   The senses of "skew" in (a) and (b) are opposite to each other.   The indexing of crystallographic axes is based on Ref. 3.

challenges.   Also, there is a logical possibility that the both structures actually occur in nature, though it seems rather unlikely.

(2)   To date, we have no knowledge about the direction of polymerization of cellulose.   Obviously, there are two possibilities in the mode of polymerization of polysaccharides; the reducing end ahead (pointing away from the synthesizing enzyme) or behind.   Though both are known to occur in nature, the latter mechanism seems to be characteristic of complex polysaccharides such as O antigen (12).   A typical glucan, glycogen, is known to be synthesized with the non-reducing end ahead (13; also see 14).   Thus, we have good reasons to expect that cellulose is synthesized with the reducing end ahead; yet the experimental evidence is still lacking.

When we find answers to these questions, we will be able to tell the direction in which a cellulose microfibril was synthesized by knowing the sense of "skew" of the unit cell seen in the cross-section such as the one in Fig. 5b.   If the parallel-up structure and reducing-end-ahead synthesis are correct, the particular microfibril shown in Fig. 5b would be synthesized in such a way that the cellulose-synthesizing enzyme complex would project toward the viewer (see also Sec. IV.B).

C.   Native Cellulose II or IV

So far in this chapter, we have used the words "native cellulose" and "cellulose I" as synonyms.   However, there have been several reports claiming crystalline forms of native cellulose other than cellulose I. The claim of native cellulose II in the cell wall of a green alga *Halicystis* (currently *Derbesia tenuissima*) (15,16) understandably has been controversial. This material was subsequently reinvestigated several times (17,18) and it became well established that the cell wall contained cellulose not crystallized in the form of cellulose I. A recent reinvestigation clearly showed that cellulose II does form part of the wall of this alga, but the mechanism of biosynthesis of this atypical native allomorph has not yet been determined (19). For prokaryotic native cellulose II, consult (20).

Several studies claimed the existence of cellulose IV_I in nature (20–23). The difference between the diffraction patterns of cellulose I and IV is rather small, especially when the crystallinity is not very high, which is the case for the primary wall cellulose examined in these studies. It has been argued that the native cellulose IV is a laterally disorganized cellulose I, which results from remarkably small microfibril widths (22). The existence of native cellulose IV_I also raises intriguing question about its biogenic mechanism which, as for native cellulose II, can only be completely answered once the structures of these allomorphs have been clearly established.

D. Cellulose Iα and Iβ, Unit Cell Size

Another recent finding is a variation in cellulose I structure as revealed by solid state ^{13}C-NMR, denoted as Iα and Iβ; the former is predominant in algal and bacterial celluloses whereas the latter is predominant in higher plant cellulose (24). So far ^{13}C-NMR is the only method capable of detecting this difference. This finding seems important because it is the first unequivocal crystallographic difference between the two classes of native cellulose.

On the other hand, it has long been known that *Valonia* cellulose requires an eight-chain unit cell for the full assignment of electron diffraction reflections (25). While no reliable diffraction data have been available for other materials to determine the true unit cell size, it is possible that the difference between cellulose Iα and Iβ is related to a difference in the unit cell size. Also, this difference might result from the difference in the morphology of cellulose-synthesizing complexes in the two groups of organisms (see below).

III. VARIETY OF MICROFIBRILLAR MORPHOLOGY AND ITS ORIGIN

A. "Elementary Fibril" Concept

Since Frey-Wyssling proposed the "elementary fibril" concept (26), the idea has been accepted as a major working hypothesis about the fundamental fibrillar structure of native celluloses and even of regenerated celluloses by some researchers. However, in light of the above discussion about the primary importance of the nature of the organism in determining the structure of cellulose, there seems to be no plausible reason to expect a universal fibrillar unit for cellulose. In fact, lattice images of Valonia cellulose (Figs. 4 and 5) directly showed the single-crystalline nature of the microfibrils and thus disproved the existence of a smaller structural unit in the microfibril.

Additional support has been obtained in a lattice-imaging study of the microfibrils of Boergesenia forbesii and bacterial cellulose (27). The former is a green alga closely related to Valonia and was found to have highly crystalline microfibrils very similar to those of Valonia, but with a flattened cross-section in the direction of cell surface.

The bacterial cellulose of Acetobacter xylinum is known to have a thin and flat ribbon morphology, a single ribbon being produced from a single cell. Lattice imaging of this material showed the presence of one undisrupted crystal lattice as wide as 25 nm, revealing that large crystallites like those from Valonia or Boergesenia are sometimes present in bacterial cellulose. (The observation of a lattice image 25 nm wide does not contradict the average crystallite size of 7 nm as estimated by X-ray diffraction of bacterial cellulose because the lattice image can be selectively obtained from a single large crystallite whereas X-ray diffraction provides size estimates based on averaging all crystallites in the sample.)

In view of these observations, we should not perpetuate the use of the term "elementary fibril" in its original sense. It seems more reasonable to denote the smallest natural fibrillar units just as "microfibrils," which will have varying size and shape inherent to the different organisms producing them.

B. Terminal Complex

There has also been remarkable progress in the study of ultrastructure of the cell surface where cellulose microfibrils are produced. Though its true role in the biogenesis of cellulose is still unknown, a membrane-associated complex known as the "terminal complex" (TC) is now believed to represent the cellulose-synthesizing enzyme complex. Terminal complexes have been seen in the freeze-fractured plasma membranes of various cellulose-synthesizing organisms including many algae, a bacterium, and higher plants (28).

Notably, terminal complexes have different morphologies in different classes of organisms; this variation corresponds to different types of cellulose microfibrils found in nature. Green algae belonging to Siphonocladales and other algae such as Oocystis, Eremosphera, and Glaucocystis have "linear" TCs with long rectangular shapes (see Chap. 3), producing highly crystalline microfibrils with square or rectangular cross-sections. In contrast, the smaller microfibrils of certain advanced algae (28), ferns, and higher plants are associated with a small hexagonal structure called a "rosette" (see Chap. 4). Some green algae belonging to Zygnematales are known to have a characteristic array of rosettes and produce microfibrils with an intermediate size between those of the Siphonocladales and higher plants (29). In contrast, one member of the Zygnematales, *Mougeotia*, has only unorganized rosettes typical of higher plants (30).

Acetobacter *xylinum* has a linear type of TC, but its structure is distinctly different from the algal linear TCs (see Chap. 5). The ribbonlike bacterial microfibril is correspondingly different from the rectangular algal microfibrils. The bacterial TC is also immobilized on the cell surface and secretes cellulose into the medium, whereas the algal TC travels over the cell surface depositing a microfibril on the inner surface of the cell wall.

Based on these observations, it seems highly likely that the size and shape of the microfibril is determined by the structure of the TC, which is a characteristic of the organism. There is an interesting correlation between the shapes of linear TCs and the cross-sectional shape of the microfibril. While microfibrils of *Valonia* have a nearly square cross-section of 20 × 20 nm and linear TCs of about 350 × 25 nm (31), the microfibrils of *Boergesenia* have an average cross-section of 25 × 12 nm (27) and a TC size of about 510 nm × 25 nm (32). If the subunits in these TCs have similar properties, the TC length may be related to the width of the microfibril (Fig. 8).

C. Crystallization Mechanism

Probably the most challenging question about these algal celluloses is how the glucan chains become aligned in such a precise way as to assemble a particular size and shape of microfibril. It is therefore important to note the basic feature of the crystal structure of cellulose I.

In all models for cellulose I derived from X-ray diffraction and packing analyses (2,3,10), the glucose residue is arranged with its ring plane nearly parallel to the bc [or (200)] plane (see Fig. 5. The indexing of the crystallographic planes is based on Ref. 3). This molecular geometry is characterized by a remarkable anisotropy in the mode of attraction between neighboring molecules. Molecules within the (200) plane are strongly hydrogen-bonded with each other, whereas there is no (2), or at most weak (3), hydrogen bonding across the (200) plane.

This arrangement does not necessarily mean either that the cohesive force within the (200) planes is substantially stronger than those across it or that the (200) plane is actually formed in the

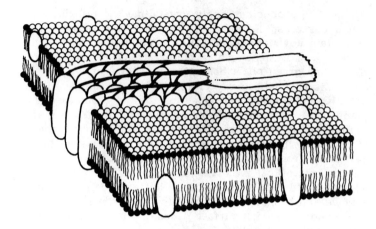

FIGURE 8   Model for the microfibril formation by the terminal
complex (TC) of *Valonia macrophysa*. Each subunit of the TC contributes to the polymerization of one or more glucan chains. The
forward movement of the TC in the fluid membrane may be the result of coupled polymerization and crystallization processes. (Reproduced by permission from Ref. 31.)

early stage of the microfibril formation. Rather it means that the
(imaginary) (200) molecular sheets in the crystal are bound together mainly by the van der Waals force. This imples that the
surface of the (200) plane should be essentially hydrophobic, and
it would be highly unfavorable to expose this plane in aqueous environments. (The hydrophobic nature of the glucose residue in
the direction perpendicular to the ring plane is demonstrated by
the formation of inclusion compounds between amylose and iodine,
or cyclodextrins and nonpolar guest molecules. See, e.g., Ref.
33.)

    It is possible that this hydrophobicity of the (200) plane is re-
sponsible for the formation of the rectangular microfibrils of *Valonia*
or *Boergesenia*, which have their (110) and (1$\bar{1}$0) planes exposed.
Interestingly, the cellulose ribbon of *A. xylinum* has been found to
have a thin sheet structure with the sheet plane parallel to the
(1$\bar{1}$0) plane (27). Thus, cellulose molecules might have a tendency
to form a sheet structure in the direction parallel to the (1$\bar{1}$0)
plane, and these are subsequently stacked with each other to form
rectangular microfibrils in algal celluloses.

    Algal celluloses have another important feature in that their
microfibrils are disposed in the cell wall with the (1$\bar{1}$0) lattice plane

lying parallel to the cell surface. This feature, called the "preferred orientation" or, more specifically for this case, "uniplanar orientation" (34), is another puzzle to be solved. It may simply be a result of the rectangular shape of the microfibril, which would tend to lie flat on the surface of the plasma membrane. The idea seems to apply to the microfibril of *Boergesenia* with a wider ($1\bar{1}0$) plane but cannot explain the same feature with that of *Valonia* with a nearly square cross-section. Therefore, we have to seek a mechanism to cause some affinity between the ($1\bar{1}0$) plane and the surface of the plasma membrane.

D. Rosette Systems

Microfibrils of higher plant cellulose are fundamentally different from those of many algae and bacteria in the following ways:

1. The lateral dimension of the microfibril is much smaller, typically being 3–4 nm in negatively stained preparations.
2. The crystallites are much smaller and seem to be separated from each other by amorphous or paracrystalline regions.
3. There are no preferred orientations with (110) and ($1\bar{1}0$) planes.
4. The microfibrils have a characteristic pattern of ^{13}C-NMR spectrum (discussed above).
5. They are much more readily affected by alkali or acid, resulting in mercerization or hydrolytic disintegration, respectively.

As a whole, higher plant cellulose seems to be less strictly controlled by the cellular mechanisms in the process of microfibril formation, or biocrystallization. It seems likely that these differences reflect the difference in the functions of linear TCs and rosettes.

There seems to be an important difference in the modes of constructions of linear TCs and rosettes. Freeze-fracture studies suggested that linear TCs are assembled on the plasma membrane from their subunits, which are transported there via Golgi vesicles as isolated particles (35). In contrast, rosettes, which consist of six subunits similar to those of linear TCs, seem to be preassembled in the Golgi apparatus and transferred to the plasma membrane (36).

Thus, the morphology and assembly of terminal complexes seems to reflect two distinct groups in cellulose-synthesizing organisms. Probably these groups are widely separated from each other in phylogeny. In this context, the highly ordered array of rosettes found in the green alga *Micrasterias* (29) deserves special attention. Unfortunately, nearly nothing is known about the property of cellulose microfibrils produced by this organism. Studying this material may give some clue to understanding the molecular process of biocrystallization in both rosette and linear TC systems.

IV. OTHER PROBLEMS

A. Degree of Polymerization

Though many intensive studies have been done about degree of
polymerization (DP) and its distribution of native celluloses (see,
e.g., Ref. 37), the biological mechanism regulating DP is totally unknown. It may be the case that cellulose synthesis continues in-
definitely, and only accidental interruptions in polymerization result
in a statistical distribution of DP, with a finite average value. Still
there is a significant variation in the average DP of celluloses of
various origins, demanding an interpretation.

A recent study on in vitro cellulose synthesized by isolated
cytoplasmic membrane of *A. xylinum* was found to have a very sharp
DP distribution ($\overline{DP}_w/\overline{DP}_n$ = 1.1) (38), as contrasted with in vivo
product ($\overline{DP}_w/\overline{DP}_n$ = 2 − 4) (39). Further studies of in vitro synthesis may give a clue to understanding the origin of DP control.

B. Morphogenesis of the Cell Wall

Biosynthesis of cellulose, microfibril formation, and deposition of
microfibrils into the cell wall are essential parts of the morphogenesis
of the plant cell wall. In particular, much attention has been paid
to the mechanisms controlling the TC movement and therefore determining the longitudinal orientation of microfibrils (see Chap. 7).

Revol (40) found that a single lamella of the *Valonia* cell wall,
composed of parallel microfibrils, contained two types of microfibrils
with regard to the directions of the c axis of the cellulose crystal.
This feature was also directly visualized through lattice imaging (8).
With the confirmed parallel chain structure of *Valonia* cellulose and
the assumption of the occurrence of only one of the "up" and "down"
structures as discussed above, the direction of the c axis should
indicate the way in which the TC traveled leaving the microfibril.
Therefore, the double directionality of the c axis seems to correspond to the observation of TCs traveling in opposite directions (31).
Thus more detailed information about the distribution of the direction of the c axis will be useful in elucidating the mode of TC
movement.

V. CONCLUSION

Because of the unique physical and chemical properties of cellulose,
research on its structure and biogenic mechanism provides an excellent opportunity for interactions between physical chemistry,

biochemistry, and cell biology. Recent progress in each of these areas is bringing about a more unified picture of the biogenesis and resulting structure of cellulose.

REFERENCES

1. D. Delmer, *Adv. Carboh. Chem. Biochem.*, *41*, 105 (1984).
2. K. H. Gardner and J. Blackwell, *Biopolymers*, *13*, 1975 (1974).
3. A. Sarko and R. Muggli, *Macromolecules*, *7*, 486 (1974).
4. K. Hieta, S. Kuga, and M. Usuda, *Biopolymers*, *23*, 1807 (1984).
5. H. Chanzy and B. Henrissat, *FEBS Lett.*, *184*, 285 (1985).
6. J. Sugiyama, H. Harada, Y. Fujiyoshi, and N. Uyeda, *Mokuzai Gakkaishi*, *30*, 98 (1984).
7. J. Sugiyama, H. Harada, Y. Fujiyoshi, and N. Uyeda, *Mokuzai Gakkaishi*, *31*, 61 (1985).
8. J. Sugiyama, H. Harada, Y. Fujiyoshi, and N. Uyeda, *Planta*, *166*, 161 (1985).
9. D. P. Miller and W. A. Roughead. Detailed conformational refinements on cellulose I (ramie) and cellulose II (Fortisan), in *Proc. Int. Dissolving Pulps Conference* (Tappi, Geneva, 1987), p. 177.
10. A. D. French, *Carboh. Res.*, *61*, 67 (1978).
11. S. Kuga and R. M. Brown, Jr., *Carboh. Res.*, *180*, 345–350 (1988).
12. P. W. Robbins, D. Bray, M. Dankert, and A. Wright, *Science*, *155*, 1536 (1967).
13. L. F. Leloir, J. M. Olavarria, S. H. Goldemberg, and H. Carminatti, *Arch. Biochem. Biophys.*, *81*, 508 (1959).
14. J. F. Robyt, *Trends Biochem. Sci.*, *4*, 47 (1979).
15. W. A. Sisson, *Science*, *87*, 350 (1938).
16. W. A. Sisson, *Contrib. Boyce Thompson Inst.*, *12*, 171 (1941).
17. P. A. Roelofsen, V. Ch. Dalitz, and C. F. Wijnman, *Biochem. Biophys. Acta*, *11*, 344 (1953).
18. E. Frei and R. D. Preston, *Nature*, *192*, 939 (1961).
19. E. Roberts Ph.D., Dissertation, Univ. of Texas at Austin (1990).
20. E. Roberts, F. C. Lin, M. R. Gretz, S. Kuga, and R. M. Brown, Jr., *J. Bacteriol*, in Press (1990).
21. R. D. Preston, *Polymer*, *3*, 511 (1962).
22. H. Chanzy, K. Imada, A. Mollard, R. Vuong, and F. Barnoud, *Protoplasma*, *100*, 303 (1979).
23. H. Nishimura, K. Okano, and I. Asano, *Mokazai Gakkainshi*, *28*, 484 (1982).

24. D. L. VanderHart and R. H. Atalla, *Macromolecules*, *17*, 1465 (1984).
25. G. Honjo and M. Watanabe, *Nature*, *181*, 326 (1958).
26. A. Frey-Wyssling, *Science*, *119*, 80 (1954).
27. S. Kuga and R. M. Brown, Jr., *J. Electron Microsc. Tech.*, *6*, 349 (1987).
28. R. M. Brown, Jr., *J. Cell Sci. (Suppl.)*, *2*, 13 (1985).
29. T. H. Giddings, Jr., D. L. Brower, and L. A. Staehelin, *J. Cell Biol.*, *84*, 327 (1980).
30. A. T. Hotchkiss and R. M. Brown, Jr., *J. Phycol.*, *23*, 229 (1987).
31. T. Itoh and R. M. Brown, Jr., *Planta*, *160*, 372 (1984).
32. T. Itoh, R. M. O'Neil, and R. M. Brown, Jr., *Protoplasma*, *123*, 174 (1984).
33. M. L. Bender and M. Komiyama, *Cyclodextrin Chemistry*, Springer-Verlag, Berlin, 1978, Chap. 3.
34. J. A. Howsmon and W. A. Sisson, in *Cellulose and Cellulose Derivatives* (E. Ott, H. M. Spurlin, and M. W. Grafflin, eds.), Part I.B., John Wiley and Sons, New York, 1954, pp. 291–294.
35. T. Itoh, Microfibril assembly in giant marine algae, in *Proc. Int. Dissolving Pulps Conference*, Tappi, Geneva, 1987, p. 117.
36. C. H. Haigler and R. M. Brown, Jr., *Protoplasma*, *134*, 111 (1986).
37. M. Marx-Figini, in *Cellulose and Other Natural Polymer Systems* (R. M. Brown, Jr., ed.), Plenum Press, New York, 1982, p. 243.
38. T. E. Bureau and R. M. Brown, Jr., *Proc. Natl. Acad. Sci. USA*, *84*, 6985 (1987).
39. G. J. F. Ring, in *Cellulose and Other Natural Polymer Systems* (R. M. Brown, Jr., ed.), Plenum Press, New York, 1982, p. 299.
40. J.-F. Revol and D. A. I. Goring, *Polymer*, *24*, 1547 (1983).

7
Role of the Cytoskeletal Elements in Organized Wall Microfibril Deposition

ROBERT W. SEAGULL   *Southern Regional Research Center,*
USDA/ARS, New Orleans, Louisiana

I. INTRODUCTION

Specific morphogenesis (changes in cell shape) requires precise co-ordination between environmental signals, cytoplasmic response, and cell wall development. Each cell can react to its environment by modifying expansion characteristics in response to specific stimuli. The organization of reinforcing cellulosic microfibrils in the wall determines the polarity characteristics of turgor-pressure-driven cell expansion and thus cell morphogenesis (1,2).

The cell controls the organization of the wall which surrounds it. Since their discovery in plant cells in 1963, cortical cytoplasmic microtubules (MTs) have been considered the most likely candidates to influence the organization of microfibrils in the cell wall. While the vast majority of research on the subject of microfibril orientation in plant cells seems to be consistent with the involvement of MTs in wall organization, there is a growing list of examples where MTs are reported not to be involved. In addition, other cytoskeletal elements have also been invoked as having a possible role in this process (3–5).

Many reviews have been written on this subject (6–11). To date most evidence is consistent with the hypothesis that MTs are somehow involved in the organized deposition of wall microfibrils;

144   *Seagull*

however, difficulty arises in proving that the cytoskeleton is essential for the organized deposition of wall fibrils. The purpose of this
chapter is to outline the evidence for and against a role for the cytoskeleton in regulating wall microfibril orientation in plants.

II. CELL WALL SYNTHESIS

For the purposes of this chapter I will be discussing cell wall synthesis only with respect to laying down arrays of organized wall
microfibrils. This type of deposition can occur during either primary or secondary wall synthesis. I will not discuss the specific
chemical composition of various walls (see Chap. 2) or the biochemical synthesis of cellulose microfibrils (see Chaps. 8–11).
    The components of the wall-synthesizing machinery that are im-
portant for this chapter are the cellulose synthetases and the fluid
plasma membrane. I will assume the validity of the fluid mosaic
model for membrane structure (12) and that cellulose microfibrils
are synthesized at or on the plasma membrane by membrane-bound
synthetic complexes (13; see Chaps. 3, 4, and 9).

III. COMPONENTS OF THE CYTOSKELETON

The following discussion will be limited to the interphase state of
the cell cycle, when organized microfibrils are being deposited. With
the introduction of glutaraldehyde fixation, it became immediately apparent that MTs are a major element of the cortical cytoplasm of most
plant cells. More recently, it became clear that actin-like microfilaments also form a major element of the plant cytoskeleton.
    Microtubules were first described as hoops around the cell cir-
cumference (14). These structures were later described as over-
lapping short segments around the cell (15) or as multistart helical
arrays around the cell (16). The precise organization of the MT
arrays is far from being clear. Some authors report the MTs to be
long (17–19), while others report them to be predominantly short
(3,15). These observations were made using different cell systems
and techniques. While it is often assumed so, there is no direct
evidence to suggest that MT arrays from various cell types necessarily need be composed of similarly sized MTs. Also, all the techniques used to observe MTs carry the risk of technique-induced
artifacts (3).
    Certain characteristics of MTs seem essential if their possible
role in wall deposition is to be established. First, MTs must be

close to the plasma membrane. The plasma membrane is thought to be the site of cellulose microfibril assembly (13), and if MTs are to influence their deposition, then it would seem reasonable for them to be located near the site of synthesis. Second, there must be some type of interaction of the MTs with the plasma membrane. Connections between MTs and the plasma membrane have been observed by many authors (3,15,20−27). The precise meaning of this interaction and the mechanism by which it is achieved remains to be determined.

The presence of actin-like microfilaments in plant cells has long been established. In 1974, algal cells were shown to contain actin-like filaments (28). The evidence for the existence and possible role(s) of microfilaments in plant cell function was reviewed several times (7,29,30). During interphase, a population of microfilaments other than those involved in streaming was proposed (3,31,32). Subsequent observations substantiated the idea that there may be a cortical network of actin associated with the MTs and plasma membrane (33−41). The possible role of actin-like microfilaments in cell wall microfibril orientation will be discussed later.

IV. LOCALIZED WALL DEPOSITION

Several model systems have been very useful in the examination of the interaction between the cytoskeleton and wall microfibril deposition. Systems that undergo localized wall deposition have been especially useful. Both tracheary elements and stomatal guard cells deposit localized arrays of organized microfibrils. The deposition of secondary cell wall in tracheary elements is characterized by the laying down of organized arrays of cellulose microfibrils in discrete bands. As a result, tracheary element morphogenesis has been the most intensively studied cell system with respect to determining the biological mechanism controlling the specific patterns of wall deposition. This subject was reviewed previously (7,42). In this chapter I will only emphasize the crucial observations that are used to synthesize current concepts of cytoskeletal involvement in wall organization.

A. Tracheary Elements

The investigation of tracheary element development resulted in many observations that form the foundation of our current beliefs on how MTs may be involved in wall deposition. In the early investigations of MT arrays, it was observed that clusters of MTs appeared adjacent to the localized deposits of wall material in developing tracheary elements (43−46). MTs were only located in regions where secondary wall bands were being deposited (43,45,47). Clustering of MTs seemed to initiate localized wall deposition (48). The orientation of wall microfibrils around the cell circumference was precisely predicted by the orientation of MTs (43,45,49).

Work with disrupting agents further substantiated a role for MTs in microfibril (MF) orientation. It has been shown that colchicine, while having dramatic effects on cell wall organization, did not affect either cell wall synthesis or protein secretion (50,51). Colchicine treatment resulted in a dramatic reduction in the numbers of MTs, which in turn resulted in a loss of organized wall patterns (52). Colchicine had little effect when applied late in tracheary element development, with the removal of MTs having no detectable effect on wall formation (52). If MTs were disrupted before tracheary element differentiation (20,53), then wound-induced xylogenesis was inhibited. However, if differentiation had started before MT disruption, then tracheary elements developed but wall thickenings were smeared (less discrete) and less organized.

The distribution of MTs seemed to change during tracheary element development. Early in differentiation the MTs were evenly distributed along the cell length; however, before wall thickenings developed, MTs clustered into discrete bands (48). These bands were maintained until later in tracheary element development when MTs were observed between the wall bands (54).

In colchicine-treated cells, wall microfibrils were deposited in swirled patterns (53) indicative of groups of microfibrils interacting together or being synthesized as a unit. Later observations showed that the walls of tracheary elements that developed in the absence of MTs had a patchy birefringence pattern (55), indicating localized arrays of parallel microfibrils but absence of organization over the whole cell. Thus, inherent in the microfibrils themselves may be a mechanism for maintaining order among closely associated fibrils. MTs may only be needed to maintain order between groups of microfibrils.

The location of MTs altered the properties of the plasmalemma. When developing tracheary elements were plasmolyzed, the membrane associated with the bands of MTs remained more firmly attached to the cell wall (49). When more severely plasmolyzed, the shrunken protoplast maintained undulations that coincided with the location of MT groups (54). These observations stressed the interaction between MTs, the plasma membrane, and the location of wall deposition.

More recently, tracheary element development and the role of MTs was studied using single-cell suspension cultures of *Zinnia* leaf mesophyll and immunocytochemical staining of MTs (43,44,56−61). These studies showed that MTs form bands before any wall material can be detected (56,57). The first deposited wall material appeared more firmly attached to the plasma membrane than to the cell wall. Most of the developmental changes that occurred in the MT arrays were completed before any signs of wall differentiation were detected. Changing the cellular patterns of MT arrays had dramatic effects on wall patterns. The development of specific patterns of MTs may involve actin-mediated reorientation (44,61) and MT bundling (57). While drug treatment resulted in MT patterns normally not found in control cells, in all cases the wall patterns that developed were precisely predicted by the MT arrays (44,57,58,61). The pattern of wall deposition in developing tracheary elements is thus a direct consequence of the array of MTs present before wall differentiation.

B. Stomatal Guard Cells

Stomatal guard cells, although less well studied, also provide an excellent model system for the study of localized wall deposition. Stomatal guard cells undergo localized wall thickening, which appears to involve MTs. Before aperture formation, MTs became concentrated in the central region of the cell (62–65). Microtubules were linked to the plasma membrane (66) and were found up to four layers deep in the cortical cytoplasm (64). Cell wall thickening occurred after MT relocation, with wall microfibrils laid down in arrays that were parallel to the MTs. As the wall thickened the number of MTs increased (63). In mature guard cells the MTs were no longer clustered, but were more evenly dispersed in the cytoplasm (62).

In the grasses, the guard cells changed shape during their development to take on an elongate or "bone-shaped" conformation (26). This change in shape was the result of the deposition of MFs in an axial orientation. The cortical MTs shifted their orientation in advance of the shift in MF orientation (26,67).

As with tracheary elements, the application of MT-disrupting agents to developing guard cells had a dramatic affect on their morphology (69). The application of colchicine early in guard cell development resulted in the elimination of MTs and deposition of smeared cell walls (63). Cells that were further along in their development were less affected by colchicine. Colchicine did not inhibit localized cell wall deposition in *Zea mays*, but rather caused thickenings to be smeared and not as compact (68). It was argued (68) that these observations support the idea that other factors, in addition to MTs, must be acting in the regulation of where wall thickenings will occur.

Further support for MT involvement in localized microfibril deposition is seen in seed hairs of *Cobaea scandens* which produce

localized cell wall deposits composed of organized arrays of cellulose
microfibrils (70). Prior to the development of a cellulosic thread
around the seed hairs, the cortical MTs were evenly distributed
along the plasmalemma. Deposition of the thread was signaled by the
the clustering of MTs into bands containing 10-18 MTs. Individual
MTs were 20-30 μm long. The location of MT clusters precisely
corresponded to the location of the wall band of cellulose.

V. POLYLAMELLATE WALLS

If the cytoskeleton is involved in directing the orientation of parallel
arrays of microfibrils, then one must explain the production of
polylamellate cell walls. These walls have multiple layers of parallel
microfibrils, with each layer having a different orientation. Polylamellate walls are common in many cell types (71; see Chap. 2).

A. Walls with Two Microfibril Orientations

The green algae Oocystis has proven to be an extremely useful
model system for the analysis of polylamellate cell walls (for review,
see Ref. 8). Early observations clearly showed that Oocystis produces a cell wall composed of 20-30 layers of microfibrils (each layer
is one microfibril thick) that are oriented at right angles to each
other (72,73). The cortical cytoplasm contained MTs that were associated with the plasma membrane and were oriented parallel to one
of the microfibril orientations. In 90% of the cases the MTs were
parallel to the fibril layer which was in the process of being synthesized. Due to the normal shifts in fibril orientation, MTs were
sometimes found (10% of the time) oriented perpendicular to the
innermost layer of microfibrils. This was interpreted as reflecting
instances in which the MTs had shifted in anticipation of the synthesis of a new microfibril later with a different orientation (72).
The time required for MTs to shift orientation was found to be comparable to the times required for MTs to depolymerize and re-
polymerize, thus indicating that the mechanism regulating the shifts
in MT pattern may be polymerization/depolymerization (72,74).

Pharmacological experiments done on Oocystis indicated that MTs
may function to establish orientation of the microfibril array and
then are no longer needed until a new orientation of fibrils is required (75-77). Disruption of MTs with cholchicine had no effect
on the orientation of wall microfibrils but rather resulted in the
loss of fibril switching. Recovery from colchicine exposure resulted
in a resumption of microfibril layering within the wall. Colchicine
treatment did not decrease wall synthesis (74), nor did it have an
effect on the membrane-associated terminal complexes (78).

Numerous other MT disrupting agents (such as griseofulvin, oryzalin, colcemid, isopropyl-N-phenylcarbamate, trifluralin, podophyllotoxin, oncodazol, and amiprophos-methyl) were also applied to *Oocystis* (76). These drugs had a variety of effects on MTs and microfibril orientation, but common to all was the observation that when MTs were affected so was wall microfibril deposition. The authors claimed that because of the diversity of chemical agents used, the variety of modes of action, and the consistent relationship between MTs and microfibrils, these studies proved that MTs were the causal agent responsible for controlling microfibril orientation in *Oocystis* (77). Subsequent analyses of terminal complexes (cellulose synthetases) after exposure to various MT-disrupting agents indicated that MTs may function in establishing the alignment of complexes in the plasma membrane (79).

B. Walls with Many Microfibril Orientations

Polylamellate walls of higher plant tissue often consist of many lamellae, each containing parallel microfibrils deposited at a different angle relative to the cell's long axis (see Chap. 2). In lettuce hypocotyl and other tissues, MTs often were seen parallel to the innermost layer of wall microfibrils (80,81). In other cells that deposit polylamellate walls the cortical MTs were seen with various orientations relative to the cell long axis (80–83). These variations were believed to be responsible for the deposition of the various orientations of the wall fibrils. When MTs were observed not to be parallel to the innermost wall microfibrils, they were generally displaced toward the predicted new microfibril orientation (81).

The ability of MTs to establish a new orientation during wall deposition has been convincingly shown. The precise dynamics of the reorganization and the mechanism that controls it remain unclear. The most likely models to accomplish this reorganization are (1) disassembly/reassembly and (2) shifting of intact MT arrays. While it is clear that MTs do undergo assembly/disassembly during the cell cycle, there is little or no evidence that large-scale MT turnover occurs during interphase reorganization (82). Data have been produced that are consistent with disassembly/reassembly (discussed above), but to date no studies have proven it.

In a series of articles by Lloyd and coworkers (reviewed in Refs. 16 and 84), a dynamic helical model for MT reorganization was proposed to explain how orientation can be changed over the whole cell. MTs in an orientation that is transverse to the long axis of the cell are thought to be arranged into a shallow pitched helix. To accomplish more axial MT orientations, the helical array is stretched or extended by some type of dynamic MT–MT interaction (sliding). All the orientations of MTs (transverse, oblique, and

axial) that are found in cells that deposit polylamellate walls can be
explained via this model. There are researchers who believe, however, that other factors besides microtubules are also involved in
determining microfibril orientation in polylamellate walls (see Chap.
2).

C. Effect of Hormones on Microtubule/ Microfibril Orientation

Plant hormones may play a role in controlling and changing MT
arrays in the cell. The effect of plant hormones on MTs and cell
wall organization is well documented (80,81,85−89). Exposure to
GA_3 (gibberellin) increased cell elongation by altering MT orientation
to predominantly transverse to the cell long axis (89−91). This
resulted in large quantities of MFs being deposited with a transverse
orientation to the long axis, thus promoting cell elongation (90).

  Hormone-induced cell elongation could be reduced or eliminated
by treatment with colchicine (80,85,87). This inhibition was not
due to a drug-induced prevention of hormone penetration into the
cell (85) or to an inhibition of cellulose synthesis (51). Although
colchicine-treated cell lacked MTs, they maintained the deposition of
parallel arrays of microfibrils (87). Like *Oocystis*, these cells
lacked the ability to change the orientation of microfibril deposition
when MTs were absent. Alternatively, colchicine-induced cell swelling
was reduced by subsequent treatment with GA_3 (89). The observed
antagonism between GA_3 and colchicine with respect to cell elongation indicated that the two agents may have targets in common.

  Other plant hormones have also been shown to affect MT organization and thus microfibril deposition. Kinetin and benzimidazole
have been shown to promote cell swelling in *Azukia angularis* by inducing MTs to orient parallel to the long axis of the cell (86).
Colchicine prevented these hormone-induced changes. Exposure to
ethylene also induced axially oriented MTs in pea (81) and mung
bean (82). The rapidity of MT reorientation (within 15 min of exposure to ethylene) indicated that intact MTs were most likely shifting orientation rather than depolymerizing and repolymerizing in a
new orientation (82). In all cases the interaction between MTs and
microfibrils was maintained since along with a new orientation for
MTs there was a shift in orientation of newly deposited microfibrils.

VI. HAIR AND FIBER CELLS

In this section I wish to discuss systems which develop highly
elongated cells. Cell systems such as root hairs, cotton fibers, and
pollen tubes elongate partially or exclusively via tip synthesis and

have cell walls that are composed of parallel arrays of microfibrils (92). Many of these systems also deposit polylamellate walls. This cell type contains examples that strongly support the consensus that the cytoskeleton does play a crucial role in cell wall organization (discussed in this section) as well as some of the most convincing examples that MTs are not involved in organized cell wall deposition (discussed in the next section).

In radish root hairs, MTs were found in an axial orientation in the cell. When parallel arrays of microfibrils were deposited, they were oriented parallel to the MTs (93). A good correlation exists between MT and microfibril orientation in these cells, with variability in MT orientation being reflected by the wall fibrils (3). Later observations on radish, using indirect immunofluorescence microscopy and dry cleaving (94,95), confirmed that axial orientation of the MTs and the coalignment with microfibrils.

Not all root hairs deposit axially oriented wall fibrils. Root hairs of *Allium*, *Equisetum*, *Lepidium*, and others have polylamellate or helicoidal walls (94-96). In *Allium*, MT arrays form helical patterns and thus could function in the deposition of helical arrays of microfibrils (11,95). The other root hair systems seem to contain only axial MT arrays (94), indicating that there was no correlation between MT and microfibril alignment (for further discussion, see subsequent section).

Cotton fibers develop a polylamellate wall that is composed of various arrays of parallel microfibrils (97). Unique to cotton fibers is the development of microfibril reversals (regions in the wall where spiraling wall fibrils exhibit abrupt changes in gyre). The cortical cytoplasm of developing cotton fibers was shown to contain arrays of MTs that were associated with the plasma membrane (98). The initiation of secondary wall development was signaled by a large increase in cell wall synthesis and increased numbers of MTs (98,99). Coincident orientation between MTs and microfibrils has been observed in cotton fibers (83,99-101). Disruption of MT arrays with colchicine, as in other systems, resulted in abnormal wall development (101,102). During wall development, the shifts in MT orientation coincided with the shifts in microfibril orientation (83,103). The location of wall reversals was evidenced by abrupt changes in the orientation of cytoplasmic MTs (83,101). The coincident spiraling and reversing of gyre exhibited by the cytoplasmic MTs and the cellulosic microfibrils outside the cell would seem highly improbable if the two were not mutually connected in some way (101).

The developing lorica of the algae *Poteriochromonas* is another excellent elongating system to study the possible involvement of cytoskeletal elements in the deposition of parallel arrays of wall microfibrils. The number of "primary" microfibrils corresponded to the number of MTs, MTs were bridged to the plasma membrane (the

site of fibril synthesis), the microfibrils precisely paralleled the
MTs, the width of the wall fibrils was comparable to that of the
MTs, and the disruption of MTs with colchicine resulted in the abnormal formation of the lorica (104). As microfibril synthesis was
not restricted to the region of MTs, it was concluded that MTs were
needed to initiate fibril orientation and spacial organization, but
once deposition was started MTs were no longer needed (104). Although the microfibrils that were deposited are composed of chitin
and not cellulose (105), it is believed that the regulatory mechanisms involved were at least comparable (106).

VII. EVIDENCE AGAINST CYTOSKELETAL INVOLVEMENT IN WALL ORGANIZATION

Thus far I have discussed evidence that is consistent with a role
for cytoskeletal elements in wall organization. However, as alluded
to above, there are numerous citations in the literature that can be
used to support the contention that cytoskeletal elements are not involved in regulating microfibril organization in the wall.

In discussing the mechanism for the deposition of parallel arrays
of microfibrils, it is essential that we concentrate only on the examples in which wall fibrils are deposited in such a manner. Thus,
examples of randomly organized microfibril arrays being deposited
over parallel arrays of MTs should not be considered as evidence
against the involvement of MTs in organized wall deposition. In recent reviews of this information (8,107), numerous examples of this
type were used to dismiss the involvement of MTs. One of the
earliest examples of this type was from radish root hairs (93), in
which it was reported that in the tip region of the hair (3 – 25 μm
from the tip) randomly organized microfibrils were deposited despite
the presence of parallel arrays of axially oriented MTs. Subsequent
work showed that in the tip region the MT populations were incomplete and thus perhaps unable to function properly in microfibril
orientation (3). In older regions of the hairs the MT –MF parallelism
was maintained. Similar types of observations and arguments can be
made for pollen tubes. Although observations of MT arrays in pollen
tubes have yet to be done to the same level of detail as for radish
root hairs, it was clear the MTs were axially oriented in pollen tubes
(108), yet in the tip region the microfibrils had a random organization (109).

A variety of root hair systems have been shown to develop
helicoidal (or polylamellate) cell wall textures (94,110 – 113). Analyses
of MT organization in these systems illustrated that MTs were arranged with a predominantly axial orientation. The authors believed

that they had examined enough cells to be confident that MTs were
not parallel to microfibrils in these systems and thus were not involved in determining wall organization.

The data published by these authors cannot be simply dismissed
as artifact or misinterpretation. The polylamellate nature of the
wall allows one to suggest that when MTs are not seen to parallel
the innermost wall microfibrils it may be due to the MTs having
been reoriented in anticipation for the deposition of the next layer
of fibrils. The authors argue that given the numbers of cell examined they should have found examples of MTs paralleling microfibrils. If MTs were involved in the deposition of these polylamellate walls, then one should see MTs with orientations other than
axial. In many of the root hair systems examined the MT orientation was reported as only axial (94). This group of workers provided the most consistent data to support the idea that, at least in
some root hairs, MTs are not involved in the deposition of organized
arrays of wall microfibrils.

There were many early observations of MT arrays in elongating
cells that deposited polylamellate walls where the MTs were not
parallel to the innermost fibrillar later in all cases (80,114–119).
These have been cited as evidence against the involvement of MTs
in wall organization (7,8). It has now become clear that in this
type of cell system, rather than being static, MT arrays are very
dynamic, capable of changing orientation and organization in response to biological stimuli (see previous discussion of polylamellate
wall deposition). When MTs are not parallel to the innermost layer
of microfibrils, they are usually displaced to the predicted next
orientation of fibril deposition (81).

The use of MT-disrupting agents has increased tremendously
our understanding of the role of MTs in wall organization. Concur-
rent loss of MTs and disruption of normal wall organization remains
one of the cornerstones for the support of MT involvement. Loss
of MTs, however, does not result in the deposition of randomly organized wall microfibrils. In the presence of colchicine, wall thickenings (in either developing tracheary elements or guard cells) become smeared but the microfibrils within the band remain parallel,
often taking on a swirled appearance (53,63). Similar observations
have been made in the alga *Oocystis* where, in the absence of MTs,
highly organized arrays of microfibrils are deposited (see previous
discussion). Superficially it could be argued that MTs do not play
a role in regulating fibril orientation since parallel arrays of microfibrils can be deposited in the absence of MTs. Parallel arrays of
microfibrils are deposited in the absence of MTs only if the parallelism was initiated before MTs were depolymerized. We know that
microfibrils maintain a certain degree of rigidity due to their

crystalline nature. Interaction between adjacent fibrils during or
just after synthesis could maintain the relative orientation of microfibrils over localized areas. If order in the fibrillar array was
established before MTs were disrupted, then order could be maintained to a certain extent by the physical characteristics of the
microfibrils themselves. In the case of *Oocystis*, the order can be
maintained in the absence of MTs for a long time. The need for the
establishment of microfibril order before it can be maintained in the
absence of MTs was shown using wound-induced xylogenesis (21,53).
If MTs were removed (using colchicine) before differentiations
started, then no wall thickenings developed. However, if differentiation had started before MTs were removed, then the wall thickenings were only modified.

Several marine algae have been studied with respect to the involvement of MTs in ordered wall deposition. In *Valonia* and
Boergesenia, cells in the early stages of wall regeneration (aplanospores) contain randomly organized MTs yet deposit arrays of parallel
microfibrils (120,121). Developing rhyzoids contain arrays of MTs
that parallel the long axis of the cell, yet wall microfibrils exhibit
three different orientations (transverse, oblique, and longitudinal).
These findings are consistent with the hypothesis that microfibril
orientation is not controlled by MTs (121). Similar evidence was
reported in thallus cells of *Chaetomorpha* (122) and *Boodlea* (123).
Thus some algal systems appear not to use MTs for the deposition
of organized arrays of wall microfibrils.

The development of peristome in the moss *Rhacopilum tomentosum*
has been used as evidence against the direct involvement of MTs in
microfibril orientation (104). While MTs were concentrated in the
areas of wall thickening during the early stages of development, in
older cells (which are still depositing organized wall fibrils) MTs
were fewer in number and less organized (104). These observations
were used to argue against a direct involvement of MTs in the movement of cellulose synthetases, as proposed by Heath (124). The
authors did not exclude the possibility that MTs may have been involved in initiating microfibril orientation. Thus, this report does
not totally discount any role for MTs in wall organization, but
rather discounts only a very specific characteristic of the possible
interaction.

VIII. MODELS FOR THE INVOLVEMENT OF
       CYTOSKELETAL ELEMENTS IN
       ORGANIZED CELL WALL DEPOSITION

There have been numerous models proposed to explain how the cytoskeleton can influence the organization of cellulose microfibrils

outside the plasma membrane. This subject has been reviewed in detail (4) and unfortunately not many data have been published since to help clarify the precise mechanism of MT involvement. To date, sufficient data have not been collected to prove or disprove any of these hypotheses. Details of current evidence and future data needed to discriminate between the hypotheses have been reviewed (4). The models can be broadly broken down into direct and indirect categories, depending on whether or not the cytoskeleton is proposed to have a direct connection with the cellulose-synthesizing machinery.

A. Direct Hypotheses

All direct hypotheses postulate some type of direct linkage between the wall fibril synthetic machinery (usually cellulose) and the cytoskeleton. These hypotheses differ in the proposed nature of this linkage (force-generating or static) and the mechanism for generating the movement of the synthetic complexes.

Heath proposed the first of the direct hypotheses (124). In brief, it states that the cellulose synthetic complexes are moved along the stationary, tracklike MTs via force-generating cross-bridges between the two. The precise path or orientation of the microfibrils is thus directly predicted by the positioning of the MTs.

The other direct model was proposed in response to observing microfilaments associated with cortical MTs of varying lengths (3). This hypothesis states that the synthetic complexes are bridged directly to short MTs via some type of static connection and that movement is generated between adjacent MTs via actin-like microfilaments. The direction of movement and thus microfibril orientation is governed by positioning of the long MTs in the array.

Predictions can be made based on these models; however, the data to support them do not exist. Actual MT length profiles and the identity of the MT bridge are still unknown. The association between MTs and microfilaments has been observed in a number of systems (3,4,18,27,37–41,94,125). However, the widespread occurrence of actin-like filaments associated with MTs has yet to be fully documented. The evidence from microfilament disruption experiments using cytochalasin B is contradictory. A number of reports show that treatment with cytochalasin does not affect ordered microfibril deposition (104,126,127). More recently, however, it was shown that cytochalasin treatment dramatically affects microfibril deposition in developing cotton fibers (Seagull, in preparation). This most recent observation may support a role for microfilaments in wall deposition but does not specifically support their functioning as proposed by Seagull and Heath (3). The fact that cytochalasin treatment also disrupted MT organization may indicate that the affect on microfibrils may be indirect (via MTs).

B. Indirect Hypotheses

The indirect hypotheses differ from the direct ones in one fundamental aspect. Indirect hypotheses propose that there is not a
direct connection between cytoskeletal elements and the microfibril
synthetic machinery. These models can be further subdivided into
active and passive, depending on whether or not some type of
force-generating mechanism is proposed to cause a shear force in
the plasma membrane.

An indirect, active hypothesis was first proposed by Hepler and
coworkers (7,53). Briefly, it states that MTs interact with each
other, with the plasma membrane or with both the plasma membrane
and each other via force-generating cross-bridges to create a flow
or shear force in the membrane. This flow could be transmitted to
the microfibril synthetic machinery causing it to move in the direction of the flow (i.e., parallel to the MTs). No direct interaction
between the cellulose synthetic machinery and the cytoskeleton is
needed. Once again, all that can be said is that the literature is
consistent with, but does not specifically support, this hypothesis.

A passive indirect model was recently proposed by Herth (106).
Elegant in its simplicity, this hypothesis uses the force of microfibril formation to move the synthetic complexes. The interaction of
MTs with the plasma membrane changes membrane fluidity, thereby
restricting the movement of synthetic complexes (and thus microfibril orientation) to parallel to the MTs. This somewhat less precise control of fibril orientation would explain the variability in orientation that is often observed in cell walls (for discussion, see
Ref. 4). A very attractive feature of this model is that it uses only
one force-generating system, that of microfibril formation. It has
been shown that microfibril formation generates enough force to
move an entire bacterium (128); thus it seems reasonable that a
membrane-bound enzyme complex could be moved by the same mechanism. The simple fact that this mechanism has been shown to
exist for microfibril synthesis, whereas other force-generating
mechanisms (active MT cross-bridges or actin interactions) have not,
makes this model very appealing. To conclusively prove this model
as accurate will still require intensive investigation.

The last of the indirect passive models differs substantially from
all the others in that it has application only to the localized wall
deposition that occurs in tracheary elements and guard cells. Proposed by Schnepf (129), the hypothesis states that MTs, via their
association with the plasma membrane, create an extracytoplasmic
space between the wall and the membrane. The mechanism for this
resides in the ability of MTs to interact with each other to form a
type of contractile ring to pull the membrane away from the wall.
The major flaw with this model is its limited application. It seems

unlikely that cells would have one mechanism for depositing localized fibrillar arrays and another for depositing arrays over the entire cell surface. This model also requires the same as yet unknown force-generating mechanism as the other models (except the Herth model).

The exercise of model building has extreme value for biological systems in that it allows one to focus on the important questions left to be answered. All of the above models make testable predictions concerning the structural and chemical characteristics of the cytoskeleton. For example, all models predict some type of force-generating mechanism being involved in microfibril orientation. Chemical identification of the protein components (perhaps through the production of specific monoclonal antibodies) may provide data on the nature of any force-generating mechanism.

This author tends to believe that the Herth model may most closely resemble the in vivo mechanism by which plant cells control MF orientation. Elegant in its simplicity and requiring only components that we know to exist (i.e., MTs bridged to the plasma membrane and movement via a mechanism known to be associated with fibril formation), this model incorporates all the characteristics essential for proper functioning in microfibril orientation. Precision of fibril orientation (or the degree of parallelism) may be regulated in either of two ways. By increasing or decreasing the MT numbers in the system the cell can modify the degree of parallelism among MTs. The more parallel the MTs, the more parallel the microfibrils. Alternatively, by changing the fibrils themselves, perhaps by changes in the degree of polymerization of the cellulose, microfibrils could be made more rigid. Once an orientation is established in the rigid fibril, it could be maintained without being "guided" by the MTs.

IX. CONCLUDING REMARKS

The precise involvement of cytoskeletal elements in the regulation of oriented microfibril deposition still remains unclear. The cytoskeletal elements are involved in most if not all systems is firmly established. Many of the examples of apparent lack of coalignment can be explained in light of the more recent observations on MT dynamics and reorientation. Root hair systems seem to be a true discrepancy in this generalization. These cells grow via mechanisms different from those of most plant cells (tip vs. intussusceptive growth) and may in fact possess other mechanisms for regulating microfibril organization. I do feel, however, that it is unwarranted to dismiss all the supportive data for MT involvement simply because a specific cell type does not appear to behave consistently.

158   Seagull

The cell has the ability to change shape by changing the pattern
of cell wall organization. Changes in cell shape that are induced by
hormonal stimulation occur via changes in MT organization. Clearly
the cytoskeleton (MTs, microfilaments, and other as yet unidentified
regulatory proteins) is capable of detecting and responding to these
stimuli. How this response, which is essential to proper morphogenesis, occurs remains one of the central unresolved questions in
plant cell development. Recently, several MT-binding proteins were
isolated from carrot tissue (130). These proteins were shown to induce MT bundling, a characteristic which may be important for the
regulation of MT organization and disposition in the cell.

We are continually changing the level at which we are asking
the question. Plant morphology is determined by the shape of the
individual cells. Cell shape is controlled via microfibril orientation.
The orientation of microfibrils is controlled by the positioning of
the MT arrays. The question now may then be stated as: What
controls the positioning of the MT arrays? While this question goes
well beyond the scope of this chapter, it is clear that it must be
answered before we can truly understand how morphogenesis is
controlled.

REFERENCES

1. P. B. Green, Science, 138, 1404 (1962).
2. P. B. Green, Ann. Rev. Plant Physiol., 31, 51 (1982).
3. R. W. Seagull and I. B. Heath, Protoplasma, 103, 205 (1980a).
4. I. B. Heath and R. W. Seagull, in The Cytoskeleton in Plant
    Growth and Development (C. W. Lloyd, ed.), Academic Press,
    New York, 1982.
5. S. C. Mueller and R. M. Brown, Planta, 154, 501 (1982).
6. E. H. Newcomb, Ann. Rev. Plant Physiol., 20, 253 (1969).
7. P. K. Hepler and B. A. Palevitz, Ann. Rev. Plant Physiol.,
    25, 309 (1974).
8. D. G. Robinson and H. Quader, in The Cytoskeleton in Plant
    Growth and Development (C. W. Lloyd, ed.), Academic Press,
    New York, 1982, p. 109.
9. B. E. S. Gunning and A. R. Hardham, Ann. Rev. Plant
    Physiol., 33, 651 (1982).
10. C. W. Lloyd, in Developmental Biology (L. W. Brosder, ed.),
    Plenum Press, New York, 1986, p. 31.
11. C. W. Lloyd, Ann. Rev. Plant Physiol., 38, 119 (1987).
12. S. J. Singer and G. L. Nicolson, Science, 175, 720
    (1972).

13. D. Montezinos, in *The Cytoskeleton in Plant Growth and Development* (W. Lloyd, ed.), Academic Press, New York, 1982, p. 147.
14. M. C. Ledbetter and K. R. Porter, *J. Cell Biol.*, *19*, 250 (1963).
15. A. R. Hardham and B. E. S. Gunning, *J. Cell Biol.*, *77*, 14 (1978).
16. C. W. Lloyd, *Int. Rev. Cytol.*, *86*, 1 (1984).
17. C. W. Lloyd, A. R. Slabis, A. J. Powell, and G. W. Peace, *Cell. Biol. Int. Rep.*, *6*, 171 (1982).
18. J. A. Traas, *Protoplasma*, *119*, 212 (1984).
19. C. R. Hawes, *Eur. J. Cell Biol.*, *38*, 201 (1985).
20. J. Cronshaw, *Can. J. Bot.*, *43*, 1401 (1965).
21. L. W. Roberts and S. Baba, *Plant and Cell Physiol.*, *9*, 315 (1968).
22. H. J. Marchant, *Protoplasma*, *98*, 1 (1979).
23. C. W. Lloyd, S. B. Lowe, and G. W. Peace, *J. Cell Sci.*, *45*, 257 (1980).
24. B. Galatis and K. Mitrakos, *Am. J. Bot.*, *67*, 1243 (1980).
25. P. Van der Valk, P. J. Rennie, J. A. Connolly, and L. C. Fowke, *Protoplasma*, *105*, 27 (1980).
26. B. A. Palevitz, in *The Cytoskeleton in Plant Growth and Development* (C. W. Lloyd, ed.), Academic Press, New York, 1982, p. 345.
27. R. W. Seagull, *J. Ultrastruct. Res.*, *83*, 168 (1983).
28. B. A. Palevitz, J. F. Ash, and P. K. Hepler, *Proc. Natl. Acad. Sci. USA*, *71*, 363 (1974).
29. R. E. Williamson, *Can. J. Bot.*, *58*, 766 (1980).
30. W. T. Jackson, in *The Cytoskeleton in Plant Growth and Development* (C. W. Lloyd, ed.), Academic Press, New York, 1982, p. 3.
31. R. W. Seagull and I. B. Heath, *Eur. J. Cell Biol.*, *20*, 184 (1979).
32. R. W. Seagull and I. B. Heath, *Protoplasma*, *103*, 231 (1980b).
33. M. Parthasarathy, *Eur. J. Cell Biol.*, *39*, 1 (1985).
34. M. Parthasarathy, T. D. Perdue, A. Witztum, and J. Alvernaz, *Am. J. Bot.*, *72*, 1318 (1985).
35. J. Derksen, J. A. Trass, and T. Oostendorp, *Plant Science*, *43*, 77 (1986).
36. R. W. Seagull, M. M. Falconer, and C. A. Weerdendurg, *J. Cell Biol.*, *104*, 995 (1987).
37. J. A. Traas, J. H. Doonan, D. J. Rawlins, P. J. Shaw, and C. W. Lloyd, *J. Cell Biol.*, *105*, 387 (1987).
38. H. Kobayashi, H. Fukuda, and H. Shibaoka, *Protplasma*, *143*, 29 (1988).

39. H. Kobayashi, H. Fukuda, and H. Shibaoka, *Protoplasma, 143,* 29 (1988).
40. A.-C. Schmit and A.-M. Lambert, *J. Cell Biol., 105,* 2157 (1987).
41. T. Noguchi and K. Ueda, *Protoplasma, 143,* 188 (1988).
42. P. K. Hepler, in *Cytomorphogenesis in Plants* (O. Kiermayer, ed.), Springer-Verlag, New York, 1982, p. 327.
43. P. K. Hepler and E. H. Newcomb, *J. Cell Biol., 20,* 529 (1964).
44. F. B. P. Wooding and D. H. Northcote, *J. Cell Biol., 23,* 327 (1964).
45. J. Cronshaw and G. B. Rouch, *J. Cell Biol., 24,* 415 (1965).
46. K. Esau, V. I. Cheadle, and R. H. Gill, *Am. J. Bot., 53,* 765 (1966).
47. S. C. Maitra and D. N. De, *J. Ultrastruct. Res., 34,* 15 (1971).
48. J. D. Pickett-Heaps and D. H. Northcote, *J. Cell Sci., 1,* 109 (1966).
49. A. W. Robards and P. Kidwai, *Cytobiologie, 6,* 1 (1972).
50. M. J. Chrispeels, *Planta, 108,* 283 (1972).
51. T. Hogetsu, H. Shibaoka, and M. Shimokoriyama, *Plant Cell Physiol., 15,* 389 (1974).
52. J. D. Pickett-Heaps, *Dev. Biol., 15,* 206 (1967).
53. P. K. Hepler and D. E. Fosket, *Protoplasma, 72,* 213 (1971).
54. L. Goosen-de Roo, *Acta Bot. Neerl., 22,* 279 (1973).
55. D. L. Brower and P. K. Hepler, *Protoplasma, 87,* 91 (1976).
56. M. M. Falconer and R. W. Seagull, *Protoplasma, 125,* 190 (1985).
57. M. M. Falconer and R. W. Seagull, *Protoplasma, 128,* 157 (1985).
58. M. M. Falconer and R. W. Seagull, *Protoplasma, 133,* 140 (1986).
59. M. M. Falconer and R. W. Seagull, *Protoplasma, 143,* 10 (1987).
60. H. Fukuda, *Plant Cell Physiol., 28,* 517 (1987).
61. H. Fukuda and H. Kobayashi, *Development, Growth and Differentiation, 31,* 9 (1989).
62. P. B. Kaufman, L. B. Petering, C. S. Yocum, and D. Baic, *Am. J. Bot., 57,* 33 (1970).
63. B. A. Palevitz and P. K. Hepler, *Planta, 132,* 71 (1976).
64. B. Galatis and K. Mitrakos, *Am. J. Bot., 67,* 1243 (1980).
65. B. A. Palevitz and J. B. Millinax, *Cell Motil. Cytoskel., 13,* 170 (1989).
66. M. E. Doohan and B. A. Palevitz, *Planta, 149,* 389 (1980).

67. A. L. Cleary and A. R. Hardham, *Protoplasma, 149*, 67 (1989).
68. B. Galatis, *Can. J. Bot., 60*, 1148 (1982).
69. B. Galatis, *Planta, 136*, 103 (1977).
70. H. Quadar, G. Deichgraber, and E. Schnepf, *Planta, 168*, 1 (1986).
71. J.-C. Roland and B. Vian, *Int. Rev. Cytol., 61*, 129 (1979).
72. H. Sachs, I. Grimm, and D. G. Robinson, *Cytobiologie, 14*, 49 (1976).
73. D. G. Robinson, I. Grimm, and H. Sachs, *Protoplasma, 89*, 375 (1976).
74. D. G. Robinson and H. Quadar, *Eur. J. Cell Biol., 21*, 229 (1980).
75. I. Grimm, H. Sachs, and D. G. Robinson, *Cytobiologie, 14*, 61 (1976).
76. D. G. Robinson and W. Herzog, *Cytobiologie, 15*, 463 (1977).
77. H. Quadar, I. Wagenbreth, and D. G. Robinson, *Cytobiologie, 18*, 39 (1978).
78. D. G. Robinson and H. Quadar, *Eur. J. Cell Biol., 25*, 278 (1981).
79. H. Quadar, *J. Cell Sci., 83*, 223 (1986).
80. K. V. Sawhney and L. M. Srivastava, *Can. J. Bot., 53*, 824 (1975).
81. J. M. Lang, W. R. Eisinger, and P. B. Green, *Protoplasma, 110*, 5 (1982).
82. I. N. Roberts, C. W. Lloyd, and K. Roberts, *Planta, 164*, 439 (1985).
83. R. W. Seagull, *Can. J. Bot., 64*, 1373 (1986).
84. C. W. Lloyd and R. W. Seagull, *Trends Biochem. Sci., 10*, 467 (1986).
85. H. Shibaoka, *Plant Cell Physiol., 13*, 461 (1972).
86. H. Shibaoka, *Plant Cell Physiol., 15*, 225 (1974).
87. K. Takeda and H. Shibaoka, *Planta, 151*, 385 (1981).
88. T. Mita and H. Shibaoka, *Plant Cell Physiol., 24*, 109 (1983).
89. T. Mita and H. Shibaoka, *Protoplasma, 119*, 100 (1984).
90. T. Mita and M. Katsumi, *Plant Cell Physiol., 27*, 651 (1986).
91. T. Skashi and H. Shibaoka, *Plant Cell Physiol., 28*, 339 (1987).
92. J. C. O'Kelley and P. H. Carr, *Am. J. Bot., 14*, 261 (1954).
93. E. H. Newcomb and H. T. Bonnett, *J. Cell Biol., 27*, 575 (1965).
94. J. A. Traas, P. Braat, A. M. C. Emons, H. Meekes, and J. Derksen, *J. Cell Sci., 76*, 303 (1985).
95. C. W. Lloyd and B. Wells, *J. Cell Sci., 75*, 225 (1985).

96. M. M. Sassen, A. Pluymaekers, H. J. Meekes, and A. M. C. Emons, in *Cell Walls '81. Proc. 2nd Cell Wall Meeting* (D. Robinson and H. Quadar, eds.), Wissenschaftliche Verlagsgesellschaft, 1981, p. 189.

97. L. Waterkeyn, *Technical Monograph from the Belgian Cotton Research Council*, Int. Inst. Cotton, Manchester, 1985, p. 17.

98. J. M. Westafer and R. M. Brown, Jr., *Cytobios*, *15*, 111 (1976).

99. R. W. Seagull, in *Proc. 47th Annual EMSA Meetings* (G. W. Bailey, ed.), p. 760.

100. J. H. M. Willison and R. M. Brown, Jr., *Protoplasma*, *92*, 21 (1977).

101. L. L. Yatsu and T. J. Jacks, *Am. J. Bot.*, *68*, 771 (1981).

102. R. W. Seagull, in *Cellulose and Wood-Chemistry and Technology. Proc. 10th Cellulose Conf.*, Syracuse, N.Y., 1988. (C. Schuerch, ed.), Wiley Interscience, 1989, p. 811.

103. H. Quadar, W. Herth, U. Ryser, and E. Schnepf, *Protoplasma*, *137*, 56 (1987).

104. E. Schnepf, G. Roderer, and W. Herth, *Planta*, *125*, 45 (1975).

105. W. Herth, A. Kuppel, and E. Schnepf, *J. Cell Biol.*, *73*, 311 (1977).

106. W. Herth, *J. Cell Biol.*, *87*, 442 (1980).

107. A. M. C. Emons and A. M. C. Wolters-Arts, *Protoplasma*, *117*, 68 (1983).

108. W. W. Franke, W. Herth, W. J. Van Der Woude, and D. J. Moore, *Planta*, *105*, 137 (1972).

109. M. M. A. Sassen, *Acta Bot. Neerl.*, *13*, 175 (1964).

110. A. M. C. Emons, *Protoplasma*, *113*, 85 (1982).

111. H. J. Pluymaekers, *Protoplasma*, *112*, 109 (1982).

112. A. M. C. Emons and J. Derksen, *Acta Bot. Neerl.*, *35*, 311 (1986).

113. A. M. C. Emons and N. van Maaren, *Planta*, *170*, 145 (1987).

114. L. M. Srivastava and T. P. O'Brien, *Protoplasma*, *61*, 257 (1966a).

115. L. M. Srivastava and T. P. O'Brien, *Protoplasma*, *61*, 277 (1966b).

116. A. W. Robards, *Protoplasma*, *65*, 449 (1968).

117. J. D. Pickett-Heaps, *Aust. J. Bot.*, *21*, 255 (1968).

118. S. C. Chafe and A. B. Wardrop, *Planta*, *92*, 13 (1970).

119. A. N. T. Heyn, *J. Ultrastruct. Res.*, *40*, 433 (1972).

120. T. Itoh and R. M. Brown, Jr., *Planta*, *160*, 372 (1984).

121. S. Hayano, T. Itoh, and R. M. Brown, Jr., *Plant Cell Physiol.*, *29*, 785 (1988).

122. K. Okuda and S. Mizuta, *Plant Cell Physiol.*, *29*, 785 (1988).

123. S. Mizuta and K. Okuda, *Bot. Gaz.*, *28*, 461 (1987).
124. I. B. Heath, *J. Theor. Biol.*, *48*, 445 (1974).
125. R. M. Brown, Jr., *J. Cell Sci.*, *2*(Suppl), 13 (1985).
126. E. Schnepf and G. Diechgraber, *Z. Pflanzenphysiol.*, *94*, 283 (1980).
127. B. Palevitz, *Can. J. Bot.*, *58*, 773 (1980).
128. R. M. Brown, Jr. and T. J. Colpitts, *J. Cell Biol.*, *79*, 157a (1978).
129. E. Schnepf, *Portugaliae Acta Biologia, Series A*, *14*, 451 (1974).
130. R. J. Cyr and B. A. Palevitz, *Planta*, *117*, 245 (1989).

8

Site of Cellulose Synthesis

D. H. NORTHCOTE   *University of Cambridge, Cambridge, England*

I. INTRODUCTION

The sites of polysaccharide synthesis are at membranes and for cell wall polysaccharides this is usually the membranes of the Golgi apparatus. Membrane constituents formed or placed at the Golgi membranes may also appear at the plasma membrane since there is a shuttle of vesicles to and from the organelle and the cell surface that involves a fusion of the vesicles at the cell membrane (1).

Pectin and hemicellulose molecules when compared to cellulose have have relatively short linear chains, which are branched, and together with water they form the matrix of the growing wall composite. This matrix material is packed into the wall from vesicles by fusion with the plama membrane; nevertheless, orientation of these polysaccharide molecules does occur within the wall (2; see Chap. 2).

The fibrillar material of the wall composite is made up of microfibrils, which are in turn constructed from linear molecules of β-1,4-linked glucan chains held in an organized manner in the microfibrils by intra- and intermolecular hydrogen bonds. These cellulose microfibrils give the composite wall a high tensile strength.

During formation of the wall the matrix is first deposited and into this the microfibrils are woven together with a continued deposition of matrix materials (3). The microfibrils are spun out and take on a definite orientation within the wall. Thus although it is possible to envisage the presynthesis and packaging of hemicellulose and pectin away from the site of the wall before its deposition, it is difficult to consider the synthesis of cellulose other than at the cell surface so that it can be spun directly into the wall and quickly organized. Direct evidence for the site of synthesis of cellulose microfibrils is difficult to obtain, but there is much indirect evidence that supports the idea of a mobile enzyme system located within or on the surface of the plasma membrane that synthesizes microfibrils of cellulose (2,4,5).

For the purpose of this chapter it will be assumed that the system for cellulose synthesis is at the plasma membrane. The site of synthesis will be discussed from two points of view: what is known about the site and what is required at a site for cellulose synthesis. A model system will be constructed by making analogies with the mechanisms now known for the glycosylation reactions that occur in the Golgi apparatus during the formation and transport of glyco-proteins (6,7).

Proteins present at the membranes of the cell can be seen as particles by freeze-etch and freeze-fracture studies (8–11) and in some cases organized groups of particles in the form of linear arrays or rosettes and globules can be seen in association with the micro-fibrils (12–14; see Chaps. 3 and 4). These could represent the complex of proteins necessary for cellulose synthesis, i.e., the formation of β-1,4-glucan chains and their assembly into microfibrils.

It is possible to consider the minimum number of activities nec-essary for this synthesis. These are probably carried out by pro-teins although more than one activity may be brought about by a single protein at various sites on the molecule. All the proteins making up the complex must be arranged so that the polysaccharide chains are synthesized at a definite place with a definite orientation with respect to one another. This necessitates that the constituent parts of the system, including the growing polysaccharide chains, be held in a definite relationship one to another. This would be achieved by holding the protein components in definite arrays, which are probably represented by those seen in the freeze-etch studies.

II. THE ENZYME COMPLEX

During cellulose synthesis and microfibril formation, the plasma mem-brane is obviously an important part of the synthetic system since any disruption of the membrane causes an inhibition of the synthe-sis of β-1,4-glucan chains and an increase in the synthesis of β-1,3-glucans (15; see Chap. 10). It will be argued below that this is caused in part by the consequent loss of organization in the donor and recipient components of the transglycosyl activities that make up the synthetic complex for cellulose.

A. Components of the Complex

The activities necessary for cellulose synthesis can be summarized as follows and they will be for the most part similar to those for the synthesis of any cell wall polysaccharide (16): (1) Glucosyltransferases. (2) Transporters. These are necessary for the transport of uridine diphosphate glucose, the donor molecule, across the membrane from the site of its synthesis in the cytoplasm to the transglucosylase at the outer surface of the membrane. (3) Separate structural proteins, which may be part of the glucosyltransferase activity and which serve to bind the donor and acceptor compounds and allow a correct orientation of these molecules so that the transglucosylase acts in a specific way. (4) Transmembrane proteins. These would associate the complex with the internal cytoskeletal elements in order to bring about directed movement and the consequent organization of the microfibrils in the wall as they are spun out from the synthetic complex. (5) Acceptor molecules which are necessary for the transglucosyl reactions. These will finally be the growing polysaccharide chain. (6) Subsidiary proteins may also be present which could act in the assembly of the complex and its location on the membrane. In addition to these more obvious components which are necessary for the production of the growing cellulose chains, other proteins acting as modulators may be needed either alone or in association with smaller molecules such as cations present at both sides of the membrane (17). All these functions assembled within and on the membrane emphasize the role of the membrane for the complete synthesis.

B. Arrangement of the Components

The way in which the various parts of the system are distributed about the membrane can best be discussed by making analogies with the glycosylation systems for glycoproteins found in the Golgi apparatus (7). A model system for transfucosylation from GDP-fucose has been proposed. The synthesis of GDP-fucose occurs in the cytoplasm and its entry into the lumen of the Golgi apparatus is by means of a specific transporter that spans the membrane and recognizes the nucleotide and the sugar portions of the donor. Once inside, the fucose is transferred by a fucosyltransferase to acceptors bound to the membrane. The transferase is also bound to the membrane and is in close proximity to the acceptor molecules. For

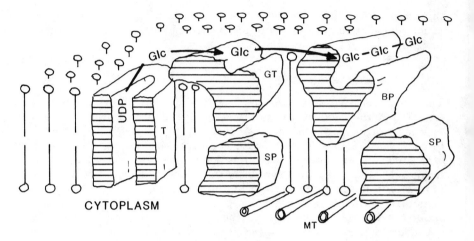

FIGURE 1   Diagramatic representation of the proposed complex of proteins that brings about the synthesis of β-1,4-glucan chains at the plasma membrane. T, transporter for UDP-Glucose; GT, glucosyltransferase; BP, binding protein for holding the acceptor molecule (this is the growing oligosaccharide chain); SP, subsidiary proteins that may act as assembly units, as modulators in conjunction with cations, and/or as links to cytoskeletal elements in the cytoplasm; MT, microtubules.

cellulose synthesis when UDP-glucose is the donor molecule for β-1,4-glucan synthesis, transport from the cytoplasm occurs probably by a specific transporter to the corresponding position to that of the lumen of the Golgi apparatus, i.e., to the outer surface of the plasma membrane. It is likely that the UDP-glucose never becomes free at this stage, but is held by the transporter so that the glucose can be transferred by an adjacent glucosyltransferase to an acceptor also located in close proximity to the donor and glucosyltransferase (Fig. 1). The UDP formed by the transglucosylation will then be liberated back to the cytoplasm either as UDP or UMP if a phosphatase is also involved.

III.   FUNCTIONS OF THE COMPLEX

A.   Specificity of Donor and Acceptor Molecules

The idea that has grown up from a study of the synthesis of glycoproteins in which there is a specific sequence of sugars is that there is a definite specificity of the transglycosylase not only for the donor molecule but for the acceptor molecule of the transferred sugar as well (18). These definite sequences would then be built up because the addition of any one new sugar depends on the structure of the acceptor, which changes in a stepwise manner incorporating individual sugars combined by varying linkages. This idea is not so appropriate for polysaccharide formation when a major proportion of the polymer is made up by a single monosaccharide and branching by insertion of different sugars can occur randomly and sometimes infrequently (2). During the synthesis of this type of polymer the donor compounds are probably specific for the appropriate transglycosylase, but the acceptor molecule is not so specific since the branching by single glycosyl radicals can be added at random along the main chain (19). Hydrolases can also act under appropriate conditions as transglycosylases; thus in this type of reaction the glycosyl group is transferred either to the hydroxyl group of water or to that of a sugar depending on the relative concentrations (20). This reflects the availability rather than the specificity of the particular hydroxyl groups onto which the transfer occurs.

Some polysaccharides carry mixed linkages or sugars in their main chains; laminarin has linkages other than β-1,3 scattered in the chain (21) and glucomannans have varying lengths of glucose and mannose residues distributed in the main chain all joined by β-1,4 linkages. It is tempting to assume that in the synthesis of these polysaccharides a single transglycosyl complex can form the mixed polymer. In the case of glucomannan synthesis it has been shown that the presence of GDP-mannose alters the product formed by a membrane preparation of pine cells when GDP-glucose is supplied (22). In the absence of GDP-mannose, a mixed β-1,3- and β-1,4-glucan is formed, but when GDP-mannose is supplied in addition to GDP-glucose the glucan production is inhibited and a glucomannan is formed. The two activities for the transfer of glucose and mannose must be closely associated so that the functioning of one can influence the other, or more probably the same transglycosylation function can handle both GDP-glucose and GDP-mannose and give different products in the presence and absence of GDP-mannose (22). If the transglycosylase gives a specific linkage onto the acceptor molecule, then presumably this is lined up and oriented in a definite manner either by the transglycosylase molecule itself or by a protein in close proximity to it. If the same transglycosylase can give two or more types of linkages onto the same sugar-acceptor residue, then the orientation of the acceptor molecule can be altered independently of the transglycosylase and thus there are probably at least two proteins involved: one for transfer and holding the donor and another for holding the acceptor (Fig. 1).

B. Protein Binding of Olgiosaccharides

Proteins are known to be able to bind sugar and oligosaccharide
molecules in a definite groove within the conformation of the protein
(23). It is possible therefore that part of the complex for cellulose
synthesis ensures that the acceptor molecule is held in such a manner that the transglycosylase takes place onto an available and specific hydroxyl group of the receiving sugar. The other nonbonding
hydroxyl groups could be held by hydrogen bonds within the groove
of the holding protein by particular amino acids. An analogy can be
made with the hexokinase reaction, for which the conformation and
molecular mechanism have been described in detail. It has been
shown that the glucose fits into a cleft of the enzyme, which closes
as the glucose is bound, and is held in such a way that its
6-hydroxyl is presented to the γ-phosphate of the bound ATP (24).
     That such specific associations can occur between proteins and
oligosaccharides and sugars is well known. In addition to hexokinase, proteins such as lysozyme and taka-amylase are known to
hold oligosaccharides in a ribbon-like configuration by hydrogen bonding and van der Waals forces within a binding site groove (25).
Precise and stereospecific interactions are formed and maintained by
the orientation of hydrogen-bonding residues, which are in turn
fixed by complex hydrogen bond networks with other residues within
the binding sites. Extensive networks of hydrogen bonds are thereby formed which extend outward from the carbohydrate molecule to
at least three shells of amino acid residues in the protein with the
hydroxyl groups of the sugar serving as hydrogen bond donors and
acceptors (23). During binding, conformational changes can occur
that allow the carbohydrate to be oriented for the binding to progress
and result in specific interaction between the protein and the carbohydrate, Hexa- and heptasaccharides are known to be held in this
way.

C. The Acceptor Molecule

During the transglucosylase reaction, the glucosyl radical could first
be transferred to water within an associated binding protein and the
resultant glucose could then be extended by subsequent transfers
to the nonreducing end of the growing chain (22) (Fig. 1). It can
be seen that it is possible to have the acceptor sugar precisely
oriented to receive the new glycosyl residues produced by the
transglucosylase at the particular hydroxyl presented to the donor
molecule. Once the chain of sugars extends beyond the channel of
the protein that is holding it in a particular direction, then hydrogen
bonds between adjacent carbohydrate chains can be formed. These
intermolecular hydrogen bonds would stabilize the conformation of the
glucan chains and help to hold them in the crystalline structure of

the microfibrils. For the microfibril to be assembled, some 30 glucan chains must be held in close proximity by intermolecular H bonds and van der Waals forces. These chains, each one of which must be the product of a complete transglucosylase system, must therefore be synthesized close to one another. This requires an assemblage of sets of the protein complexes that synthesize the individual chains and imposes a further organizational complexity on the system.

D. The Multienzyme System for Callose and Cellulose Synthesis

Multienzyme systems in various states of organization occur for many metabolic processes, e.g., biological oxidation in mitochondria, glycogen synthesis and degradation on membranes, fatty acid synthesis in the cytoplasm, and protein synthesis on ribosomes. Some are organized on membranes and some are free protein complexes and aggregates. The coordinated synthesis of single cellulose chains certainly needs the cooperation of a set of proteins and these must be organized close to one another in correct orientation for the sequence of the synthesis to occur in an economical and rapid manner. It follows that the complex has to be held in this coordinated way and, therefore, the membrane on and in which the proteins are held becomes an important part of the synthetic process. Once the membrane is disrupted and the organization of the proteins is lost, the precise transglucosylation reactions could be partially or completely disrupted, although the transport of the donor UDP-glucose would be no longer a problem since it would then leak through the damaged membrane. The glucose might then be transferred by the transglucosylase to other available hydroxyl groups of the acceptor. In this way β-1,3 and even some β-1,2 links could be established (26). This mechanism has an advantage for the plant since on wounding the lesion can be plugged quickly by the rapid formation of callose, β-1,3-glucan, which can be equally quickly removed when the damaged wall has been repaired (27).

It has been shown that with the same membrane preparation using UDP-glucose as a donor the proportion of the synthesis of β-1,3 to β-1,4 links in the glucans formed by the preparation can be increased by freezing and thawing the membranes, i.e., damage and consequent disorientation of the protein complexes brings about changed synthetic capabilities of the membrane (26). It is therefore possible that there is a transglucosylase on the membrane capable of transferring glucose to the secondary hydroxyl groups of another glucose molecule and that the particular linkage that is formed depends on the orientation of the preexisting polysaccharide, oligosaccharide, or monosaccharide held on the membrane by the protein system described previously. When the membrane is damaged the orientation is disturbed and more β-1,3 linkages are formed. It is of interest to note that with the cellulose-synthesizing system of *Acetobacter xylinum* some β-1,2 linkages can be formed when an in vitro membrane preparation is used. If in plant tissues it is further postulated that the breakdown of callose is continually taking place and that this degradation is inhibited by tissue damage, then it is possible to reconcile radioactive cold chase experiments that indicate that callose is a possible precursor of cellulose (28,29); any glucose incorporated into callose could be recycled and appear in cellulose (26). The theory also accounts for the small amounts of callose that are sometimes found at the inner surface of the cell wall immediately next to the plasma membrane (30). The rate of cellulose and callose synthesis at any particular time will therefore not only be controlled by the amount of glucosyltransferase at the plasma membrane but by the state of the membrane and the organization of the rest of the complex.

The increase in callose synthesis at the sieve plate pore is correlated with the close proximity of the endoplasmic reticulum to the plasma membrane at the sites of callose deposition (31). This endoplasmic reticulum that is closely opposed to the plasma membrane could bring about the necessary changes in orientation and binding to proteins by the various parts of the acceptor and donor system even by the regulation of the concentrations of divalent cations such as Ca^{2+}, which could act as bridges between various parts of the system or act as modulators of the various activities (17,32).

IV. ASSEMBLY OF THE COMPLEX

Given that such a complex of proteins is necessary for cellulose synthesis, then the site and method of its assembly and the way in which the complex is targeted to specific sites on the plasma membrane can be considered. In some haptophycean algae, cellulose can be synthesized as part of structured scales which serve as an exoskeletal covering of the cells. These scales are synthesized by enzymes contained within the Golgi cisternae and vesicles; therefore, in these unusual organisms the cellulose and microfibril synthetic system is present and functioning at the Golgi apparatus (33). The rosettes that may represent the complexes for cellulose synthesis in higher plants (see Chap. 4) have also been found both at the plasma membrane and in the membranes of the Golgi apparatus (34) making it possible that the enzyme complex is constructed in the cytoplasmic endomembrane system and transported on vesicle membranes to the plasma membrane. Proteins are formed and inserted into the membrane of the endoplasmic reticulum by specific signal-and-stop sequences. They can be transferred to the Golgi apparatus where they are modified and sorted so that they can be segregated within the compartments of the Golgi cisternae for transport to particular cell sites. In certain cases the sorting mechanism is known to be achieved by the formation on the protein of recognition groups such as mannose-6-phosphate (6,35,36). In this way they are directed to specific tubules of the Golgi apparatus probably in the mid- or trans region of the organelle where they are incorporated into vesicles destined for the various regions of the cell.

In the case of cell wall formation, both the polysaccharides of the matrix (hemicellulose and pectin), which are probably free in the lumen of the vesicles, and the membrane-bound enzymes necessary for the synthesis of cellulose are transported from the Golgi apparatus by means of vesicles. It is not known whether particular polysaccharides such as the xylans of the hemicellulose and the arabinogalactans of the pectins are transported in separate vesicles or together in one vesicle, nor is it known if the rosettes are transported by vesicles that carry polysaccharides. Concentrations of rosettes can be seen at the sites of secondary thickening of developing xylem cells where both matrix material and cellulose are actively being laid down (37). Both the deposition of the polysaccharides and the concentration of the rosettes can be explained in part by the directed movement of the Golgi vesicles to these areas. Microtubles that underlie areas of cell wall thickening in developing xylem cells may help determine preferred areas of vesicle fusion. However it is possible, especially if the vesicles carrying the polysaccharides and membrane-bound enzymes are different (38), that other signals and receptors at the membrane surfaces may be involved for the recognition of the sites for incorporation.

A. Movement and Turnover of the Complex

It is known that proteins can move laterally within the lipid bilayer of the plasma membrane; therefore the complete enzymatic system could move in this way. The enzymic system that gives rise to the microfibrils is thought to be mobile, thereby accounting for the changes in organization of the microfibrils that are seen in the wall at the various stages of development. Since the microfibrils are laid down in definite directions, the movement of the synthesizing complex must be directed and there must be a motile force. The complex is possibly guided by some direct association with the microtubules that underlie the plasma membrane and may be moved via a similar association or by the force of crystallization itself (39; see Chap. 7).

The arrays of rosettes are difficult to detect on the plasma membrane, and great care has to be taken with regard to the stage of growth of the cell and the preservation of the material in order to visualize them. Kinetic studies based on the rate of growth of cells, length and degree of polymerization of glucan chains, number of rosettes per unit area of the surface, and the assumption that one rosette produces one microfibril indicate that the rosettes have a turnover time of 10−20 min (37; see Chap. 4). In addition, incomplete rosettes are frequently seen. These observations lead to the conclusion that the rosettes are unstable. If they are continually broken down and removed from the surface it would seem, from economical considerations for protein conservation by the cell, that the components would be recycled back to the surface by a mechanism rather similar to that for the protein receptors at the surface of coated pits and vesicles during endocytosis (40,41). These ideas are only of relevance for the site of cellulose synthesis if it is assumed that the rosettes represent the enzyme complexes necessary for the synthesis.

V. CONCLUSIONS

The synthesis of cellulose is brought about by a complex of proteins with different enzymic and other functional activities. A minimum number of three activities is necessary. These are (1) a transporter for UDP-glucose, (2) a glucosyltransferase, and (3) a binder protein to hold the growing β-1,4-linked glucan chain, which is the acceptor molecule. In addition to these three proteins, there are probably subsidiary proteins present that are necessary to link the system to cytoskeletal elements and to act as modulators of the other activities in response to the presence of cations such as Ca^{2+} (Fig. 1). The complex is assembled in the Golgi apparatus and transferred to the plasma membrane where it is mobile within the surface of the bilayer. It is probably recycled back to the cytoplasm after its synthetic function is completed. Disturbance of the membrane and consequent disorganization of the complex may result in changes in orientation of the acceptor molecule, thereby allowing the same transglucosylase activity to form β-1,3 in addition to β-1,4 links.

REFERENCES

1. D. H. Northcote, in *Biosynthesis and Biodegradation of Wood Components* (T. Higuchi, ed.), Academic Press, London, 1985, p. 87.
2. D. H. Northcote, *Ann. Rev. Plant Physiol.*, *23*, 113 (1972).
3. D. H. Northcote, in *Polysaccharides in Food* (J. M. V. Blanshard, and J. R. Mitchell, eds.), Butterworths, London, 1979, p. 3.

4. M. J. Chrispeels, *Ann. Rev. Plant Physiol.*, *27*, 19 (1976).
5. C. H. Haigler, in *Cellulose Chemistry and Its Applications* (T. P. Nevell and S. H. Zeronian, eds.), Ellis Horwood, Chichester, 1985, p. 30.
6. R. Kornfeld and S. Kornfeld, *Ann. Rev. Biochem.*, *54*, 631 (1985).
7. C. B. Hirschberg and M. D. Snider, *Ann. Rev. Biochem.*, *56*, 63 (1987).
8. J. P. Sequest, T. Gulik-Krzywicki, and C. Sandet, *Proc. Natl. Acad. Sci. USA*, *71*, 3294 (1974).
9. W. J. Vail, D. Papahadjopoulos, and M. A. Moscarello, *Biochim. Biophys. Acta*, *345*, 463 (1974).
10. G. L. Nicolson, *Biochim. Biophys. Acta*, *457*, 57 (1976).
11. P. Pinto da Silva, B. Kachar, M. R. Torrisi, C. Brown, and C. Parkinson, *Science*, *213*, 230 (1981).
12. T. H. Giddings, D. L. Brower, and L. A. Staehelin, *J. Cell Biol.*, *84*, 327 (1980).
13. S. C. Mueller and R. M. Brown, *J. Cell Biol.*, *84*, 315 (1980).
14. H.-D. Reiss, E. Schnepf, and W. Herth, *Planta*, *160*, 428 (1984).
15. D. P. Delmer, *Ann. Rev. Plant Physiol.*, *38*, 259 (1987).
16. D. H. Northcote, *Soc. Exp. Biol. Sem. Series*, *28*, 177 (1985).
17. H. Kauss, *Ann. Rev. Plant Physiol.*, *38*, 47 (1987).
18. W. M. Watkins, *Carboh. Res.*, *149*, 1 (1986).
19. L. F. Leloir, *Biochem. J.*, *91*, 1 (1964).
20. R. Dedonder, in *PAABS Symposium 2* (R. Piras and H. G. Pontis, eds.), Academic Press, New York, 1972, pp. 21–78.
21. E. Percival, *Chemistry and Enzymology of Marine Algal Polysaccharides*, Academic Press, New York, 1967, pp. 53–72.
22. G. Dalessandro, G. Piro, and D. H. Northcote, *Planta*, *175*, 60 (1988).
23. F. A. Quiocho, *Ann. Rev. Biochem.*, *55*, 287 (1986).
24. T. A. Steitz, M. Shoham, and W. S. Bennett, *Phil. Trans. Roy. Soc. Lond.*, *B293*, 43 (1981).
25. Y. Matsuura, M. Kunsunoki, W. Harada, and M. Kakudo, *J. Biochem.*, *95*, 697 (1984).
26. S. R. Jacob and D. H. Northcote, *J. Cell Sci. (Suppl. 2)*, 1 (1985).
27. W. Eschrich, in *Encyclopeadia of Plant Physiology, New Series 1* (H. Zimmerman and J. A. Milburn, eds.), Springer-Verlag, Berlin, 1975, pp. 40–56.
28. Ch. Pillonel and R. Meier, *Planta*, *165*, 76 (1985).
29. H. Meier, L. Bucks, A. J. Buchala, and T. Homewood, *Nature*, *289*, 821 (1981).
30. L. Waterkeyn, *Protoplasma*, *106*, 49 (1981).

31. D. H. Northcote and F. B. P. Wooding, *Proc. Roy. Soc. Lond.*, *B163*, 524 (1966).
32. T. Hayashi, S. M. Read, J. Bussell, M. Thelen, F.-C. Lin, R. M. Brown, and D. P. Delmer, *Plant Physiol.*, *83*, 1, 54 (1987).
33. D. M. Allen and D. H. Northcote, *Protoplasma*, *83*, 389 (1975).
34. C. H. Haigler and R. M. Brown, *Protoplasma*, *134*, 111 (1986).
35. K. von Figura and A. Hasilik, *Ann. Rev. Biochem.*, *55*, 167 (1986).
36. S. R. Pfeffer and J. E. Rothman, *Ann. Rev. Biochem.*, *56*, 829 (1987).
37. B. Schneider and W. Herth, *Protoplasma*, *131*, 142 (1986).
38. L. R. Giffing, B. G. Mersey, and L. C. Fowke, *Planta*, *167*, 175 (1986).
39. I. B. Heath and R. W. Seagull, in *The Cytoskeleton in Plant Growth and Development* (C. W. Lloyd, ed.), Academic Press, London, 1982, pp. 163–182.
40. M. S. Brown, R. G. W. Anderson, and J. L. Goldstein, *Cell*, *32*, 633 (1983).
41. C. R. Hopkins, *Nature*, *304*, 684 (1983).

9
Biochemistry and Regulation of Cellulose Synthesis in Higher Plants

S. M. READ   *University of Melbourne, Parkville, Victoria,*
Australia

D. P. DELMER   *Hebrew University of Jerusalem, Jerusalem, Israel*

I. INTRODUCTION

Microfibrils of cellulose are a fundamental component of the cell walls of higher plants, but the process by which $1,4-\beta$-glucan chains are synthesized and then assembled into the cellulose microfibrils of the wall remains one of the major mysteries of plant biochemistry. It is still not possible to make preparations from higher plant cells that are able to synthesize, convincingly and reproducibly, either true microfibrillar cellulose or even appreciable quantities of any $1,4-\beta$-glucan shown to be related to cellulose. To a reader from outside this field it may therefore seem that nothing has altered in the last decade, for a review one of us (D. P. D.) wrote 12 years ago (1) contains these same comments. However, in spite of the continuing inability to measure any activity of higher plant cellulose synthase in vitro, there is now a considerably clearer understanding of the regulation of the synthesis of this polysaccharide in vivo and of the polymeric products that plant membranes do make when provided with the substrate uridine diphosphoglucose (UDP-glucose). In this chapter we present the current state of understanding of plant cellulose synthesis and, in addition, show that a knowledge of how plants change the nature of the polysaccharide synthesized at their cell surface is directly relevant to achieving the synthesis of cellulose in vitro.

Recent reviews on cellulose synthesis include those of Delmer
(2,3), Haigler (4), Franz and Heiniger (5), Buchala and Meier on
the cotton fiber system (6), and a general review on the biosynthesis
of plant cell walls by Delmer and Stone (7). Much of the present
enthusiasm for understanding the biochemical mechanism of cellulose
synthesis in higher plants has come from several recent advances in
work with bacteria, fungi, cellular slime molds, and algae, which
are outside the scope of this article but which will be briefly reviewed below. These and other topics relevant to cellulose synthesis, including the structure of the cellulose microfibril, the nature
of other polysaccharides present in plant cell walls and capable of
interacting with cellulose, the location of cellulose synthesis at the
plasma membrane, and the association of cytoskeletal elements with
regions of active cellulose deposition, are covered in detail elsewhere in this volume.

One of the most fruitful systems for studying cellulose synthesis has been the gram-negative bacterium *Acetobacter xylinum*,
which secretes a ribbon of cellulose into its growth medium (see
Chap. 5). The key to achieving high in vitro activities of synthesis of alkali-insoluble 1,4-β-glucan from UDP-glucose came through
the discovery of a protein factor that was needed for enzyme activity, but which was easily dissociated from the membrane-bound
synthase in the absence of the protective agent polyethylene glycol
(8,9; see Chap. 8). This protein factor is a guanyl cyclase enzyme
that converts two molecules of GTP to one of cyclic diguanylic acid,
the true activator of the synthase (10). Active cellulose synthase
was subsequently purified from trypsinized membranes of *A. xylinum*
by solubilization in digitonin followed by product entrapment, and
preliminary analyses of its polypeptide composition have been reported (11,12). Synthesis of "membrane-bound oligosaccharides"
(substituted and branched periplasmic 1,2-β-glucans) by *Escherichia
coli* membranes also requires a protein from the soluble fraction, and
this has been identified as the acyl carrier protein essential for fatty
acid synthesis (13); its molecular role in 1,2-β-glucan synthesis remains fascinatingly obscure (14). These results from *A. xylinum*
and *E. coli* are good examples of the very close link often found between studies on the mechanism of synthesis of a polymer and studies
on the regulation of its synthesis, and suggest immediately that a
complex set of effector requirements and regulatory proteins may
similarly be required in the higher plant system; a candidate for
such a regulatory polypeptide is in fact already identified (15; see
Sec. III).

Cellular slime molds, such as *Dictyostelium discoideum*, synthesize cellulose during fruiting body formation; the cellulose forms

part of the stalk and the sorocarp, as well as the spore walls. Initial work was able to demonstrate only synthesis of a glycogen-like 1,4-α-glucan polymer from UDP-glucose in vitro (16), but the synthesis of significant quantities of apparently authentic cellulose by *D. discoideum* prestalk membranes incubated with UDP-glucose, Mg^{2+}, and cellobiose was recently reported (17). Many independent physical and chemical techniques (see Sec. III) were used to confirm that a fibrillar, microcrystalline 1,4-β-glucan was being made. However, *Dictyostelium* is not a plant, and differs from higher plants in not containing any measurable 3-linked glucan, although the *D. discoideum* membranes did make small amounts of 1,3-β-glucan in vitro.

     Chitin, a 1,4-β-linked polymer of N-acetylglucosamine, is a structural component of the walls of many groups of fungi, being present at low levels in yeast and hyphal cell walls, and as a major component of the septa separating dividing yeast cells (18−20). It can be considered a substituted 1,4-β-glucan and, like cellulose, forms highly insoluble crystalline microfibrils due to extensive hydrogen bonding between adjacent chains. However, there are important differences between the two polymers, with the chains in α-chitin running antiparallel to each other (21), whereas in native cellulose they are probably parallel (22) with all the reducing ends of the glucan chains being at one end of the microfibril. In addition, the degree of polymerization of cellulose from higher plants is much higher, at 2000−14,000 residues per chain (2,23), than that of chitin chains extracted from the septa (dividing walls) of yeast cells (24). Chitin synthase enzymes have been purified from the basidiomycete *Coprinus cinereus* (25) and the yeast *Saccharomyces cerevisiae* (24), and the purified enzymes are large, multimeric complexes. There are at least two, apparently unrelated, chitin synthases in yeast (26,27), with differing responses to digitonin, divalent cations, and activation by proteolysis, but neither of these enzymes appears to have any requirement for an additional lipid intermediate or primer for in vitro synthesis of chitin from UDP-N-acetylglucosamine. Analysis of the mechanism of polymerization catalyzed by the fungal chitin synthases is relevant to an understanding of the synthesis of all polysaccharides by membrane-bound enzymes.

     While these bacterial and fungal systems have yielded progress at the molecular level, algae and higher plants have provided ultrastructural data on the process of deposition of cellulose microfibrils (3,4,28−31). Complexes of particles, in linear groups or hexameric rosettes, have been visualized in the plasma membranes of cellulose-synthesizing organisms, as described in detail in Chaps. 3 and 4.

Higher plants seem to have exclusively the hexameric rosette type
of structure. Although there is as yet no direct biochemical evidence linking these rosettes with the cellulose synthase enzyme,
their association with areas of the wall where the rate of cellulose
deposition is high is striking. In addition, the lability of the rosette structures in the plasma membrane of suspension-cultured cells
(32) parallels the ready shutdown of cellulose synthesis that occurs
in such systems (see Sec. III). It appears possible, therefore,
that one way in which the cell can control the rate and location of
cellulose synthesis is through the macromolecular organization of the
cellulose synthase enzyme complex and the subcellular sites of insertion of this complex into the plasma membrane.

   Many other aspects of the process of cellulose synthesis are also
precisely controlled in vivo. The degree of polymerization of cellulose is altered, for example, from 2000−6000 to 14,000 residues per
chain between the primary and secondary walls of cotton fibers (23),
and the rate of cellulose deposition increases up to 100-fold during
this period (1). There is also a range in observed cellulose fibril
diameters or crystallite sizes from 1.5-nm "subelementary fibrils" to
20-nm or larger fibrils (2−4), with this parameter also being fixed
for a particular cell type or stage of differentiation. The pattern
and direction in which cellulose fibrils are laid down in the cell wall
can determine cell shape and is also tightly controlled (4,33), possibly sometimes by cytoskeletal microtubules (7; see Chap. 7).
Cellulose deposition has even been shown in cotton fibers to vary on
a diurnal rhythm (34). Lastly, even in intact cells the synthesis of
cellulose is a labile process, ceasing rapidly under a great variety
of disrupted or perturbed conditions when the plasma membrane begins to synthesize high levels of the 1,3-β-glucan callose (3,7,34).

   It is known that synthesis of 1,4-β-glucan chains can be uncoupled from their subsequent crystallization into microfibrils (4; see
Chap. 5), and in this sense the term "cellulose synthesis" is am-
biguous. In intact cells, the bringing together of newly synthesized 1,4-β-glucan chains into oriented cellulose microfibrils is a
highly controlled step (3,4) and may depend on the degree to which
the nascent glucan chain is bound by the other wall polysaccharides,
such as xyloglucan (35) or glucuronoarabinoxylan (36). Haigler (4)
suggested "cellulose biogenesis" to describe the postsynthetic formation of structural microfibrils; equally, until the enzyme systems
have been properly characterized it may be better to talk about
"1,4-β-glucan synthase" rather than cellulose synthase.

   The concept that the activities of the various polysaccharide
synthases vary in a tightly controlled manner through cellular differentiation is attributable to Northcote (37). However, the mechanism of action of these enzymes is quite unknown, nor have any been
purified. It is therefore not known whether control of synthase

activity is exerted through changes in the total amount of enzymes, in their state of association in the membrane, in their interaction with regulatory proteins, or in altered amounts of primer. In these respects, then, cellulose synthase is not alone in its mystery. The enzyme that synthesizes 1,3-β-glucan, callose synthase, has been extensively studied because 1,3-β-glucan was found to be the product formed in assays for 1,4-β-glucan (cellulose) synthesis, and callose synthase is thus the best model we have for a higher plant cell wall polysaccharide synthase. This enzyme will be described in Sec. V, and it is possible that understanding how the cell regulates the ability of its plasma membrane to make these two different β-glucans will help us prevent or reverse the change to callose synthesis that occurs in vitro, and allow substantial cellulose synthesis to take place.

II. IDENTIFICATION OF GLUCAN PRODUCTS

When intact cells or tissues are fed with $^{14}CO_2$ or [^{14}C]sucrose, the radioactive cellulose product has to be separated from all the other labeled cell constituents and cell wall components. However, when plant extracts are supplied with UDP-[^{14}C]glucose, only three ethanol-insoluble radioactive polysaccharides are generally synthesized: 1,4-β-glucan, 1,3-β-glucan, and a mixed-link 1,3;1,4-β-glucan. The mixed-link glucan is only produced in grasses and related monocots (38). Other possible products include xyloglucan, glucomannan, and starch. Xyloglucan contains a 1,4-β-glucan backbone, but the activity of xyloglucan synthesis is very low in the absence of added UDP-xylose (7,35,39,40). Similarly, glucomannan, with a backbone containing 1,4-β-linked glucose and mannose residues, is not synthesized unless GDP-glucose (and GDP-mannose) are supplied as substrates (7,41). The amount of α-linked glucose residues produced from UDP-glucose is generally also very low if dark-grown tissues or cotton fibers, which contain no starch (42), are used; digestion by α-amylase has been used to confirm this or quantitate starch levels as necessary.

Much confusion has been caused by the lack of adequate product analysis in the early literature on cellulose synthesis (1–3). More recently, too, difficulties have been encountered in the characterization of products synthesized in vitro (43,44). It is now generally realized that no single technique is sufficient for identification and quantitation of the three types of β-glucan produced (3,4,41), and reaction products must be analyzed by some combination of solubility properties, degradation by acid hydrolysis, acetolysis, periodate treatment or enzymic digestion (with determination of the products in each case), and linkage analysis after methylation.

Crystalline cellulose microfibrils are insoluble in the Updegraff acetic-nitric reagent (45) but, while the residue after Updegraff treatment of whole cell walls contains only 1,4-β-linked glucose, 1,4-β-glucan synthesized in vitro is generally acetic-nitric-soluble and thus not crystalline (46). The ultimate proof of synthesis of crystalline cellulose polymers comes from physical techniques such as X-ray or electron diffraction, and this was recently successful in the *Acetobacter* and *Dictyostelium* systems (17,47,48). Cellulose is also insoluble in strong alkali (e.g., 4 M KOH), but alkali-insoluble 1,3-β-glucan fibrils have been found in pollen tube walls (49), and both 1,3-β-glucan and 1,4-β-glucan synthesized in vitro from UDP-glucose can have a range of solubility in alkali depending on the conditions in which they were synthesized (see, for example, Ref. 50).

Limited acid hydrolysis followed by chromatographic separation and identification of the oligosaccharides produced is a classical way to distinguish 1,4-β-glucan from 1,3-β-glucan, and acetolysis can be used similarly (51). Cleavage of polymers with purified specific glucan hydrolases can, however, provide detail about sugar linkages simply by measuring the amount of polymer degraded (as radioactivity or reducing sugar released). For example, the enzymes 1,3-β-glucan exohydrolase from *Euglena gracilis*, 1,3;1,4-β-glucan endohydrolase from *Bacillus subtilis*, and 1,4-β-glucan endohydrolase from *Streptomyces* sp. (which also hydrolyzes 1,3;1,4-β-glucan) enabled Henry and Stone to measure the levels of 1,4-β-glucan, 1,3;1,4-β-glucan, and 1,4-β-glucan produced by membranes from rye grass cells (52). These and similar enzymes are extracted from a range of sources, but commercial preparations generally require further purification and characterization before use. The accuracy of the technique is enhanced by analysis of the oligosaccharides released during enzymatic digestion. Treatment of polysaccharides with periodate specifically degrades 1,4-β-glucan residues but not 1,3-β-glucan residues (53). Susceptibility to periodate is one of the characteristics required of 1,4-β-glucan polymers; the erythritol produced can also be quantitated following acid hydrolysis.

The method of choice for linkage analysis of glucan polymers is, however, methylation (54). Carbohydrates are isolated and their hydroxyl groups methylated under strongly alkaline conditions; acid hydrolysis and reduction then produces new hydroxyl groups, which are acetylated, and the resulting partially methylated alditol acetates are separated and identified by GLC or HPLC. Each sugar linkage, such as 3-substituted glucose or 2,5-substituted arabinose, gives a unique derivative. The technique is quantitative and reliable and coupling the GLC to a mass spectrometer allows unambiguous identification of the products. However, the analysis of radioactive products requires that the profile of radioactive peaks eluted be accurately overlaid on the GLC elution profile.

Taken together, the three techniques of enzymic digestion, periodate treatment, and methylation analysis permit the accurate determination of the amounts of 1,3-β- and 1,4-β-linked glucose residues in a sample. Treatment with the 1,3;1,4-β-glucan endohydrolase from *B. subtilis*, with analysis of oligosaccharides produced under various chemical and enzymic hydrolysis conditions, is required to show whether the two different linkages are contained in two different polymers, or whether a mixed-link 1,3;1,4-β-glucan is also present.

III. CELLULOSE SYNTHESIS IN INTACT AND SEMI-INTACT SYSTEMS

Feeding intact plants with $^{14}CO_2$ or cut branches with [^{14}C]sugars provides the least perturbed systems for studying cellulose synthesis. However, these kinds of experiments can give only limited mechanistic information, which is why the normal next step for the biochemist is to attempt to demonstrate the reaction of interest occurring in a cellular extract. It is at this stage that research on cellulose synthesis is blocked, as tissue homogenates are found to have little ability to polymerize UDP-glucose (the probable substrate for cellulose synthesis in vivo; see below) into 1,4-β-glucan, instead making 1,3-β-glucan at high rates (see Secs. IV and V). In a search for a successful model, semi-intact systems such as protoplasts, tissue slices, or excised and cultured cotton ovules have been investigated, with the general finding that the loss of ability to synthesize 1,4-β-glucan varies directly with the degree of disruption inflicted on the tissue.

Cotton fibers have been used for much of the work on cellulose synthesis (1,6), especially since intact ovules with their attached fibers can be maintained and grown in culture throughout their developmental period (1,55). The secondary wall of the fiber is almost pure cellulose, and a very high rate of deposition of 1,4-β-glucan is attained, peaking at about 26−28 days postanthesis (1,56). The cotton fiber is, however, somewhat unusual in that a significant amount of 1,3-β-glucan, callose, is also present in developing fibers, unlike most other somatic cell types. The level of 1,3-β-linked glucose reaches a maximum of about 5% of the wall by weight at the beginning of secondary wall deposition, and then drops both in absolute terms and relative to the total wall weight as cellulose synthesis progresses (42,55−57; see also 58). When $^{14}CO_2$ or [^{14}C]glucose is fed to intact plants or whole fruit capsules, the rate of synthesis of 1,3-β-glucan observed is much higher than expected from the 1,3-β-glucan content of fibers, with 30−35% of the radioactivity in cell wall glucans being found initially in callose (59). In isolated, cultured ovules, even when handled very gently, half the radioactivity from [14C]glucose is in callose at early stages of labeling (42), and in isolated seed clusters fed [14C] sucrose or [14C] glucose the majority of the radioactivity is found initially in callose, the exact amount depending on the conditions used (6,60). Detached fibers, isolated in the absence of polyethylene glycol as a protective agent, synthesize essentially only 1,3-β-glucan from UDP-[14C]glucose (61), the same result as obtained with isolated membrane preparations (62−64).

It appears as though two factors are at work here. One is that cell damage or wounding is promoting the synthesis of 1,3-β-glucan at the expense of 1,4-β-glucan. The second is that the initial rate of synthesis of callose, even that observed when whole and presumably quite undamaged branches are fed $^{14}CO_2$, is higher than expected on the basis of the 1,3-β-glucan content of the cotton fibers. This suggests that constant callose turnover is occurring and led to the idea that callose might be an intermediate in cellulose synthesis (59). Some data in favor of turnover of 1,3-β-glucan were presented by Meier and coworkers (6,59), although performing pulse-chase experiments with whole plants or isolated fruit clusters is technically difficult, especially when the total amount of 1,3-β-glucan is decreasing with time as the fibers mature. Maltby and coworkers were unable to demonstrate callose turnover in cultured ovules (42). The situation is still unresolved, but it is clear that a demonstration of movement of radioactivity from callose to cellulose is no proof of any obligate intermediate status for callose; ordinary breakdown or turnover of 1,3-β-glucan would release glucose units that could then be incorporated into cellulose. Cotton fiber walls have high levels of exo-1,3-β-glucanase, as required for callose metabolism, as well as a transglucosylase activity (65). However, this transglucosylase appears to synthesize 1,6-β-glucosyl linkages in acceptor molecules and therefore to be more suited to introducing the occasional 1,6-β-linked branch found in cotton fiber callose (50) than to transferring glucose units from callose to a growing 1,4-β-linked cellulose chain.

The case for UDP-glucose being the cytoplasmic precursor for synthesis of cellulose in higher plants (2,5,7) is probably established as well as can be considering the current inability to perform the reaction in vitro. The two enzymes that synthesize UDP-glucose, UDP-glucose pyrophosphorylase and sucrose synthase, both have high activities in plant tissues, and sucrose synthase is a major sink enzyme for sucrose (7). Franz (66) and Carpita and Delmer (67) showed that there is a substantial pool of UDP-glucose in cotton fibers; this was labeled readily with exogenous [14C]glucose and turned over sufficiently rapidly to be a competent precursor for cellulose (67). Lastly, UDP-glucose is a very efficient substrate for in vitro synthesis of cellulose in *A. xylinum* (8,9) and of 1,3-β-glucan and 1,3;1,4-β-glucan in plant extracts. The levels of GDP-glucose, the only other nucleotide sugar precursor to have been suggested for cellulose (1,2), were very low in cotton (67), and although GDP-glucose pyrophosphorylase is present in plant tissue, its activity is also low, and sucrose synthase prefers uridine nucleotides to guanine nucleotides as substrate (68). These data suggest against a role for GDP-glucose in cellulose synthesis and, in fact, the GDP-glucose:1,4-β-glucan synthase assayed in plants has been suggested to synthesize glucomannan (2,7,41) or possibly a xyloglucan precursor. In cotton fibers this activity is detectable only during primary wall synthesis and declines to zero as secondary wall cellulose deposition is initiated (1).

It is now generally accepted that nucleotide sugars are not capable of crossing plasma membranes, and there is no consistent evidence for extracellular glycosyltransferases using nucleotide sugars in plants. Incorporation by cell or tissue preparations of sugar residues from added UDP-[14C]glucose can only occur if the cell membrane and permeability barrier has been damaged, or if this substrate is broken down to [14C]glucose which is then taken up into the cells. Thus, although detached cotton fibers use UDP-glucose to synthesize exclusively 1,3-β-glucan and no 1,4-β-glucan (62,63), feeding of UDP-glucose to intact fibers still attached to their ovules during the period of rapid secondary wall growth was much less efficient (60). Carpita and Delmer (61) showed that incubation in polyethylene glycol (PEG) preserved the ability of detached cotton fibers to make cellulose from [14C]glucose. PEG promoted resealing of the cut ends of the fibers, and is known to protect membranes from damage as well as maintaining the integrity of multisubunit complexes (see also Ref. 8). PEG also prevented the incorporation of UDP-[14C]glucose into noncellulosic glucans (mostly 1,3-β-glucan), consistent with incorporation from UDP-glucose occurring only across a disrupted membrane barrier.

Cotton fibers are of course not the sole system that has been used for studying cellulose synthesis. Raymond and coworkers (69) cut tissue slices from pea seedlings in an attempt to find a preparation that has retained the cellulose synthase activity lost by tissue homogenates, but concluded that radioactive polysaccharide was formed from UDP-[14C]glucose only at the surfaces of cut cells and not by intact cells. This was subsequently confirmed by autoradiography (70). Anderson and Ray (71), also using pea stem slices, found that intact cells were capable of incorporating [14C]-glucose into cellulose in a cyanide-sensitive manner, whereas incorporation from UDP-[14C]glucose occurred at the plasma membrane, was cyanide-insensitive, and yielded solely 1,3-β-glucan. Cooper showed that when radish roots were incubated in solutions of

UDP-[^{3}H]glucose, only visibly damaged root hairs synthesized radioactive polysaccharide, and this was concluded to be callose (72).
Brett found that the incorporation of UDP-[^{14}C]glucose into suspension-cultured soybean cells was very low unless the cells were
vigorously stirred, a treatment that stimulated 30-fold synthesis of
alkali-insoluble 1,3-β-glucan as a wound response (73). The UDP-
glucose was concluded not to penetrate the plasma membrane of intact cells. Even gentle centrifugation was found to cause loss of
plasma membrane rosettes (putative cellulose synthase complexes)
from suspension-cultured cells (32), and a range of such treatments,
including centrifugation, cold shock, or gentle permeabilization with
dimethylsulfoxide (DMSO), can cause the appearance of callose (as
indicated by fluorescence with the synthetic aniline blue fluoro-
chrome; D. P. Delmer, unpublished observations).

   Protoplasts made from plant cells are able to regenerate a cell
wall, and the deposition of new cellulose microfibrils on the membrane surface can be followed microscopically as well as biochemically (74). However, the new cell walls and the polysaccharide secreted into the culture medium by regenerating protoplasts often
contain a much larger fraction of noncellulosic glucose (mostly
1,3-β-glucan) than is seen in the parent tissue (74). When [^{14}C]-
glucose was fed to protoplasts it was incorporated efficiently into
cellulose as well as into 1,3-β-glucan, but the products of labeling
with UDP-[^{14}C]glucose were mostly 1,3-β-glucans (75,76), suggesting that the incorporation of exogenous UDP-glucose was occurring
mainly in damaged protoplasts. Preparation of protoplasts using
highly purified cell wall degrading enzymes is a more rapid and less
damaging procedure, and such protoplasts synthesize very little
callose (77).

   It therefore appears clear that one of the factors retained in
intact cellulose-synthesizing systems but lost when cells are damaged
is the permeability barrier of the plasma membrane. However, it is
not yet clear which, if any, of the several known effects of membrane disruption is the direct cause of the cessation of cellulose
synthesis. First, when suspension-cultured soybean cells are exposed to chitosan or other polycations, a leakage of electrolytes
from the cells occurs, and Ca^{2+} moves in from the external medium
(78, 79, and references therein). This elevated Ca^{2+} level activates the 1,3-β-glucan synthase directly. Second, a disrupted
membrane will become depolarized, and it has been suggested that a
membrane potential (positive inside the cell) is essential for cellulose
synthesis in A. xylinum cells (80). However, establishment of a
membrane potential in cotton fiber membrane vesicles did not alter
the balance of glucose linkages formed in favor of 1,4-β-glucan (46).
Last, it is possible that lipid breakdown products (81), fluidity
changes in the lipid bilayer, an altered conformation of multienzyme

complexes, activation of a protease (82,83), or even loss of regulatory proteins (3,15) may result in inhibition of 1,4-β-glucan synthesis and activation of 1,3-β-glucan synthesis. A better understanding of which of these changes is involved in the mechanism of loss of cellulose synthesis would undoubtedly be of assistance in attempts to resurrect 1,4-β-glucan synthase activity in vitro.

One other use of intact systems is to assay the effectiveness and specificity of inhibitors of cellulose synthesis (Ref. 7 reviews the small molecules that perturb cell wall polysaccharide synthesis in higher plants). The herbicide dichlorobenzonitrile (DCB) is an effective and specific inhibitor of cellulose synthesis in higher plants at 10 μM (15,84,85), but does not inhibit 1,4-β-glucan synthesis in vitro, which is consistent with the idea that the small amount of 1,4-β links formed in cell-free systems is not related to cellulose. Dichlorophenylazide, a photoreactive analog of DCB, is also an inhibitor of cellulose synthesis and binds specifically to an 18-kD polypeptide in cotton fiber extracts (15,64). The levels of this protein increase in parallel with the rate of secondary wall cellulose synthesis in the fibers. Since it is mainly found in the supernatant fraction of tissue homogenates, the DCB-binding protein has been suggested to play a regulatory rather than a catalytic role in cellulose synthesis (3,15).

IV. IN VITRO STUDIES ON CELLULOSE SYNTHESIS

Since the first report by Barber and coworkers in 1964 (86), there have been many attempts to synthesize in vitro a 1,4-β-glucan polymer that could be related to cellulose (early work is reviewed in Refs. 1, 2, and 5). Undoubtedly, some 1,4-β-glucan has been regularly made in a number of preparations from higher plants, but several considerations have combined to confuse the situation. First, 1,4-β-glucan synthase activity, with UDP-glucose or GDP-glucose as substrate, is generally associated with Golgi membranes (87,88) rather than the plasma membrane where cellulose is made. This activity is best seen using low concentrations of UDP-[^{14}C]glucose (0.5–10 μM) of high specific activity, and the amount of product then made is very small. Second, poor or insufficient linkage analysis has put some data in doubt, especially considering the high 1,3-β-glucan synthase activities generally found, and it has not always been possible to assay 1,4-β-glucan synthase activity reproducibly (43,44). Last, polymerization from GDP-glucose, the substrate first used by Barber and coworkers (86), is probably related to synthesis of polysaccharides other than cellulose (2,41).

The Golgi-localized 1,4-β-glucan synthase discovered in peas by Ray was named glucan synthase I (GSI) to distinguish it from the plasma membrane-bound glucan synthase II (GSII) that synthesizes 1,3-β-glucan (88). GSI is assayed with 0.5 μM UDP-glucose in the presence of 10 mM or higher concentrations of Mg^{2+}, and is noticeably labile to freezing and thawing or pelleting and resuspension of membranes. Since the enzyme is not assayed at saturating levels of UDP-glucose, and since Golgi membranes prepared by sucrose gradient centrifugation contain sufficient contaminating plasma membrane for 1,3-β-glucan synthase activity to become predominant as UDP-glucose concentrations are increased, it is difficult to calculate the total 1,4-β-glucan synthase activity. Micromolar levels of unlabeled UDP-glucose stimulate the incorporation of UDP-[^{14}C]xylose into polysaccharide, which suggests that GSI could provide the 1,4-β-glucan backbone required for synthesis of xyloglucan in the Golgi (35,40,89). By treatment of reaction products with highly purified 1,3-β-glucanase to remove the large amount of 1,3-β-glucan made, Brummell and MacLachlan were able to show definitively this relationship between synthesis of 1,4-β-glucan and synthesis of the backbone of xyloglucan (90). However, Hayashi and Matsuda found differences between synthesis of alkali-insoluble 1,4-β-glucan and incorporation of glucose from UDP-[^{14}C]glucose into alkali-soluble xyloglucan, and proposed that some of the measured GSI activity represented a "proenzyme" form of cellulose synthase present in Golgi bodies on its route to the plasma membrane (35,39,40). Although only Golgi cisternae, and not Golgi-derived secretory vesicles, have GSI activity (91), Haigler and Brown recently observed rosettes, representing putative cellulose synthase complexes, in both these membrane types in differentiating *Zinnia* mesophyll cells (92). In addition, there is precedent for an intracellular activity of cellulose synthase, as the alga *Pleurochrysis* is capable of assembling cellulosic wall components in its Golgi bodies (93). In sum, however, the low level of activity of GSI, the association of much of this activity with xyloglucan synthesis, and the fact that it is not localized at the plasma membrane means that there is probably little connection between GSI and cellulose synthesis.

It has been frequently observed, both with semi-intact systems (69) and with isolated membranes (52,62,88,94), that products formed from low (micromolar) concentrations of UDP-glucose contain a high proportion of 1,4-β-glucan linkages, whereas higher (millimolar) concentrations of UDP-glucose promote formation of 1,3-β-glucan linkages. The phenomenon of "substrate activation" observed for 1,3-β-glucan synthesis (where the activity is negligible at very low concentrations of UDP-glucose and rises steeply with increasing UDP-glucose levels; see, for example, Ref. 62) is now known to occur only if this enzyme is assayed under suboptimal conditions (see Sec. V). Fully activated 1,3-β-glucan synthase displays Michaelis–Menten kinetics (50). The proportion of 1,4-β links made is therefore an elastic number depending on the concentration of UDP-glucose, the degree of activation of the 1,3-β-glucan synthase, and the lability of the Golgi GSI in each preparation. Use of concentrations of UDP-glucose above 0.1 mM is generally to be recommended, as this more closely resembles the physiological condition, prevents interference from competing reactions (such as degradation or steroyl glucosylation), and allows rates of polysaccharide synthesis achieved (in nmol/min/mg protein) to be meaningfully compared with the in vivo rate and between different laboratories.

There are, however, few such studies on optimizing the absolute amount of 1,4-β-glucan made from higher plant systems. Unfortunately, the subcellular location of the 1,4-β-glucan synthase activity is also not always determined; the possibility of a plasma membrane-bound enzyme therefore remains in some cases. Jacob and Northcote (95) found that freezing and thawing membranes from celery petioles severely reduced the proportion of 1,4-β-glucan to 1,3-β-glucan that was synthesized from 1 mM UDP-glucose; this observation is reminiscent of the lability of Golgi GSI. The pH optimum for synthesis of 1,4-β-glucan from 1 mM UDP-glucose by rye grass membranes was 8.0, higher than the optimum pH (7.0) for 1,3-β-glucan synthase activity (52); these activities were distributed throughout the endomembrane system (96). Bacic and Delmer showed that establishment of a potential across cotton fiber vesicle membranes stimulated the synthesis of both 1,3-β-glucans and 1,4-β-glucans (46) but did not shift the balance of linkages synthesized in favor of 1,4-β-glucans. In conclusion, it seems that high rates of 1,4-β-glucan synthesis have yet to be reproducibly observed in vitro, and we are still a long way from showing the identity of this product with cellulose.

In vitro studies have also been used to investigate possible mechanisms of polymerization, even though particular intermediates in the reaction may not be found if these are the essential components for cellulose synthesis that are lost or inactivated on tissue homogenization. Lipid-linked intermediates have a well-established role in the assembly of oligosaccharide chains of plant and animal glycoproteins and bacterial cell walls, but there is very little evidence for their requirement in cellulose or callose synthesis (2,5,7, 97). In general, all systems with a requirement for a specific lipid intermediate need that lipid to be added back for the solubilized system to be active; the observation that the digitonin-solubilized *Acetobacter* cellulose synthase is fully active without added lipid (9) thus implies that glucose is transferred directly from UDP-glucose onto protein or the growing glucan chain. The requirement of fungal chitin synthase and higher plant 1,3-β-glucan synthase for

positively charged phospholipids to retain activity (24,98—100) probably reflects the need for a boundary layer of phospholipid around
these integral membrane proteins rather than a specific involvement
of lipid intermediates in polysaccharide biosynthesis.

The storage carbohydrates starch (101) and glycogen (102) are
known to have protein primers involved in the initiation of new
glucan chains. Mung bean membranes are capable of forming glycoproteins when incubated with UDP-[^{14}C]glucose (103) and labeling
of a 44-kD polypeptide has been demonstrated (104). However, the
role of these components in glucan synthesis remains to be determined. The other high molecular weight precursor that has been
suggested for cellulose is callose, as discussed in Sec. III. Isolated
membrane preparations always contain a certain amount of glucan
chains, which have been suggested to function as a primer for
glucan synthesis (105,106), but partially purified 1,3-β-glucan
synthase retains high activity without any added primer (99).

V. CALLOSE SYNTHASE AS A MODEL ENZYME

Callose is found at a variety of places and at various developmental
stages in whole plants (reviewed in 7,34,38), including cotton fibers
(6,42,55—58), cell plates, sieve plates, and plasmodesmata, pollen
mother cells (34), pollen tubes (107), and gravitropically responding
cells (108). In most of these situations, with the exception of pollen
tubes, 1,3-β-glucan is only a transient wall component. However,
the best known cause of deposition of callose, as "wound callose,"
is stress, whether physiological, mechanical, chemical, or pathogen-
induced (34,79,108). In several cases, the rapid induction of 1,3-β-
glucan synthesis has been related to changes in membrane permeability (see Sec. III) and elevated Ca^{2+} levels (78,79).

The enzyme that synthesizes 1,3-β-glucan is relatively easy to
work with because, unlike most cell wall polysaccharide synthases,
it exhibits high activity in tissue homogenates supplied with the
correct effectors, and the substrate UDP-[^{14}C]glucose is readily
available. The enzyme, called glucan synthase II (GSII) by Ray
(88), is localized at the plasma membrane (88,99,109) and synthesizes a linear 1,3-β-glucan (although callose extracted from plants
has detectable levels of 1,6-β-linked branches; see Ref. 50). Full
in vitro activity of 1,3-β-glucan synthase requires the presence of
both micromolar levels of Ca^{2+} and millimolar levels of cellobiose or
a similar molecule as activators (50,64,83,110—112). The activated
enzyme then has a K_m for UDP-glucose of 0.2—0.3 mM. Enzyme
preparations with lower affinity for this substrate have generally
been assayed under nonoptimal conditions (52,62,63,69,88,94,96).

A variety of compounds will replace cellobiose in activating 1,3-β-glucan synthase (50,111). n-Alkyl-β-D-glucosides, such as octylglucoside, are particularly potent, but even sucrose and glycerol are activators at high concentrations. These compounds do not appear to serve as primers for callose synthesis (50,105) and may instead be replacing endogenous, heat-stable, low molecular weight activators ("supernatant factors") that are lost from membranes on homogenization (111,112). The endogenous activator appears to contain an aglycone that is much smaller, with only five carbon atoms (113), than the dioleoyldiglyceride originally reported (112).

Mg^{2+} has a variety of effects on 1,3-β-glucan synthase, partially activating the enzyme in the absence of Ca^{2+}, and reducing the concentration of Ca^{2+} required for half-activation (50); polyamines stimulate in a similar fashion (109). Mg^{2+} also confers alkali insolubility on the 1,3-β-glucan made in its presence (50), prevents the enzyme from being extracted from membranes by the detergent CHAPS (99), and increases the sedimentation velocity of the enzyme through glycerol gradients (50); the molecular basis for these effects is unknown. The dependence on Ca^{2+} of callose synthase activity in vitro presumably reflects the role that elevated Ca^{2+} levels play in activating this enzyme in wounded cells (78,79). Lysophosphatidylcholine, unsaturated fatty acids, and other amphipathic compounds which stimulate 1,3-β-glucan synthase in vitro have also been suggested to be involved, as lipid breakdown products, in the regulation of the enzyme in vivo (81). Digitonin, which is generally included in 1,3-β-glucan synthase assays, also increases enzyme activity; this may be due in part to it acting as a detergent and allowing the substrate UDP-glucose to penetrate membrane vesicles that have formed with the enzyme active site facing inward (114). Alternatively, digitonin may have a direct effect on the fluidity of the lipid bilayer near the enzyme (81), and such a complex effect of digitonin is also suggested by its stimulatory effect on 1,3-β-glucan synthase solubilized in the detergent CHAPS (99; S. M. Read, unpublished observations).

Several laboratories have attempted to purify 1,3-β-glucan synthase, but progress has been hampered by instability of the activity and by the difficulty of eluting the enzyme from columns. It now appears that solubilization of activity in CHAPS rather than digitonin gives a preparation that is easier to handle, and 40-fold purification has been reported (99). Retention of activity requires that phospholipid be added, and it is known that disruption of boundary layer lipid inhibits the enzyme (81,98,100). The solubilized enzyme is very large: it barely migrates into polyacrylamide gels (110), sediments rapidly upon centrifugation on glycerol gradients (50), is

excluded from a Sepharose 4B gel filtration column (100), and is
visible in negatively stained preparations in the electron microscope
(50). The difficulty of purifying 1,3-β-glucan synthase has stimulated other approaches to analysis of its structure, including active
site-directed radiolabeling with the substrate analogs UDP-pyridoxal
(114) and 5-azido-UDP-glucose (115), and production of monoclonal
antibodies capable of binding the enzyme (D. P. Delmer, unpublished observations). The 1,3-β-glucan synthase enzyme is inhibited by reaction with the substrate-analog UDP-pyridoxal, and so
has been concluded to have an amino group at its active site (114).

What, then, do studies on callose synthase tell us about the
possible nature of cellulose synthase and the changes that occur in
the plant plasma membrane on conversion from cellulose synthesis to
callose synthesis? The cellulose synthase enzyme is probably large
and complex, with many constituent polypeptides, particularly if the
hexameric rosettes visualized in the electron microscope (29–31)
represent its active form. Sensitivity to the lipid environment in
the membrane could cause problems in attempts to assay 1,4-β-
glucan synthesis, either because the lipid composition changes after
tissue disruption or because the lipid conformation is altered in
detergent-permeabilized vesicles. It is also possible that enzyme
activators may dissociate from isolated membranes; these could include small molecules such as cyclic dinucleotides (10) or the possible "supernatant factor" represented by cellobiose (112), or else
regulatory proteins such as guanyl cyclase (8,9), acyl carrier protein (13,14), or the DCB-binding protein (15). The Ca^{2+} concentration may need to be buffered at some low level, and a membrane
potential could also be necessary. Kauss and coworkers found that
1,3-β-glucan synthase from soybean could be activated independently of Ca^{2+} by treatment with trypsin or an endogenous protease
(83), although this has not been reported for the enzyme from other
systems. It is also possible that a rapidly acting protease might be
rendering 1,4-β-glucan synthase inactive on homogenization (82),
with a similar process occurring on perturbation of intact cells and
requiring insertion of new rosettes (enzyme complexes) into the
plasma membrane before cellulose synthesis could be resumed. Lastly, synthesis of cell wall polysaccharides may depend on maintaining
the precise physical relationship of the various parts of the system,
including the different glycosyltransferases spanning the membrane
and the newly made portions of the cell wall just outside (see Chap.
9). Removal of the cell wall may cause significant changes in the
conformation of the membrane-bound enzymes, changes that will need
to be understood and reversed before cellulose synthesis can be assayed in vitro.

The possibility has been suggested that cellulose and callose
synthesis are catalyzed by the same enzyme, with the different

activities being regulated in opposing ways (1,3,37,116). This hypothesis is based mainly on the observation that the synthesis of
these two β-glucans is never known to occur simultaneously, and it
suggests that when the plasma membrane rosettes (presumed to represent cellulose synthase enzyme complexes) dissociate on cell disruption, the individual subunits make 1,3-β-glucan. It is consistent
with this hypothesis that both activities are found at the plasma
membrane, use UDP-glucose as substrate, and the maximal rate of
callose synthesis achieved in vitro is similar to the calculated rate of
cellulose deposition in vivo.

Although it is generally believed that synthesis of each different
linkage in glycoproteins or polysaccharides requires a separate glycosyltransferase, there is a precedent in the Lewis blood group
antigens for one enzyme being able to transfer a sugar residue, in
this case fucose from GDP-fucose, onto an acceptor in either 1,3-α
linkage of 1,4-α linkage (117). In the case of mammalian galactosyl-
transferase, binding of the regulatory protein α-lactalbumin alters
the nature of the galactose acceptor from N-acetylglucosamine residues on glycoproteins to free glucose, which results in the synthesis of the milk disaccharide lactose (118). A 1,4-β-galactosyl linkage is formed in each case, however, and so the situation is not directly analogous to the case of cellulose and callose, but it does
illustrate that binding of a regulatory protein can alter the specificity
of a glycosyltransferase. Stone has suggested that a single enzyme
may be responsible for synthesis of both the 1,3-β links and the
1,4-β links of the mixed-link 1,3;1,4-β-glucan of grasses (38). The
active site of this synthase would have to bind at least three adjacent
glucosyl residues in the growing polymer and incorporate the next
glucose unit in either 1,3-β linkage or 1,4-β linkage depending on
the linkage of the previous residues. There is, however, no direct
evidence as yet in favor of the identity of the 1,3-β-glucan synthase
and 1,4-β-glucan synthase enzymes of plant plasma membranes, and
this proposal remains just an interesting idea.

VI. CONCLUSION

The regulation of cellulose synthesis may be considered from two aspects. First, intact cells can control the quantity and the type
(microfibril size, direction of polymerization, and so on) of the cellulose deposited into their walls as well as restricting this deposition
to certain subcellular locations. We are still far from a molecular
understanding of these processes and can only speculate on the types
of control mechanisms that may occur. Second, cellulose synthesis
is rapidly shut down in a number of situations in which plasma
membrane perturbations and permeability changes occur. This is

the phenomenon responsible for the inability to assay cellulose synthesis in vitro with isolated membrane preparations. Several possibilities for the mechanism of this off switch for cellulose synthesis
have been considered in this chapter, and it could be related to
the mechanism of activation of callose synthesis that occurs in these
situations. We are also still far from understanding the developmental regulation of callose synthesis in intact cells, such as at the cell
plate, sieve plates, or plasmodesmata.

Work on cellulose synthesis in microbial systems, particularly
Acetobacter, has moved ahead of research on higher plants (see
Chap. 8), and data from simpler systems can often be used to understand more complex ones. Achievement of high in vitro activities
with *Acetobacter* cellulose synthase came with the discovery of a
soluble regulatory protein; similarly, a soluble protein is needed for
in vitro synthesis of 1,2-β-glucan by *E. coli*. It is therefore encouraging that a DCB-binding protein has been identified as a potential regulatory protein in higher plants (15). Recent results
with *Dictyostelium* (17) do suggest that true eukaryotic cellulose
synthesis is also possible in vitro, but it is likely that further
direct attempts to synthesize cellulose with higher plant membranes
will continue to fail until the precise nature of their lesion in in vitro
cellulose synthesis is known.

Steady progress is being made on purification of higher plant
callose synthase and identification of its constituent polypeptides
(99,114). However, it is probable that genetic and immunological
probes developed from *Acetobacter* cellulose synthase (9,11,12) will
greatly accelerate the molecular analysis of the cellulose and callose
synthase enzymes of higher plants by allowing the detection of
homologous genes and proteins. It is the possibility of developments in these areas that suggests that a review of the mechanism
of higher plant cellulose synthesis written in 10 years time may be
considerably different from this one.

ACKNOWLEDGMENTS

We thank Prof. A. E. Clarke, Plant Cell Biology Research Centre,
School of Botany, Melbourne University, Australia, in whose laboratory some of this work was written.

REFERENCES

1. D. P. Delmer, in *Recent Advances in Phytochemistry*, Vol. 11
    (F. A. Loewus and V. C. Runeckles, eds.), Plenum Press,
    New York, 1977), p. 45.

2. D. P. Delmer, *Adv. Carboh. Chem. Biochem.*, *41*, 105 (1983).
3. D. P. Delmer, *Annu. Rev. Plant Physiol.*, *38*, 259 (1987).
4. C. H. Haigler, in *Cellulose Chemistry and Its Applications* (R. P. Nevell and S. H. Zeronian, eds.), Ellis Horwood, Chichester, 1985, p. 30.
5. G. Franz and U. Heiniger, in *Encyclopedia of Plant Physiology, New Series, Vol. 13B, Plant Carbohydrates II. Extracellular Carbohydrates* (W. Tanner and F. A. Loewus, eds.), Springer-Verlag, New York, 1981, p. 47.
6. A. J. Buchala and H. Meier, in *Society for Experimental Biology Seminar Series, Vol. 28, Biochemistry of Plant Cell Walls* (C. T. Brett and J. R. Hillman, eds.), Cambridge University Press, Cambridge, UK, 1985, p. 221.
7. D. P. Delmer and B. A. Stone, in *The Biochemistry of Plants, Vol. 14, Carbohydrates* (J. Preiss, ed.), Academic Press, New York, 1988, p. 373.
8. Y. Aloni, D. P. Delmer, and M. Benziman, *Proc. Natl. Acad. Sci. USA*, *79*, 6448 (1982).
9. Y. Aloni, R. Cohen, M. Benziman, and D. Delmer, *J. Biol. Chem.*, *258*, 4419 (1983).
10. P. Ross, H. Weinhouse, Y. Aloni, D. Michaeli, P. Weinberger-Ohana, R. Mayer, S. Braun, E. de Vroom, G. A. van der Marel, J. H. van Boom, and M. Benziman, *Nature*, *325*, 279 (1987).
11. F. C. Lin and R. M. Brown, Jr., *J. Appl. Polym. Sci.; Appl. Polym. Symp.*, *43*, 38 (1989).
12. R. Mayer, P. Ross, H. Weinhouse, D. Amikam, G. Volman, P. Ohana, and M. Benziman, in *Proc. Fifth Cell Wall Meeting* (S. C. Fry, C. T. Brett, and J. S. G. Reid, eds.), Scottish Cell Wall Group, Edinburgh, 1989, Abstr. 38.
13. H. Therisod, A. C. Weissborn, and E. P. Kennedy, *Proc. Natl. Acad. Sci. USA*, *83*, 7236 (1986).
14. H. Therisod and E. P. Kennedy, *Proc. Natl. Acad. Sci. USA*, *84*, 8235 (1987).
15. D. P. Delmer, S. M. Read, and G. Cooper, *Plant Physiol.*, *84*, 415 (1987).
16. C. Ward and B. E. Wright, *Biochemistry*, *4*, 2021 (1965).
17. R. L. Blanton, D. H. Northcote, C. H. Haigler, and E. M. Roberts, in *Proc. Fifth Cell Wall Meeting* (S. C. Fry, C. T. Brett, and J. S. G. Reid, eds.), Scottish Cell Wall Group, Edinburgh, 1989, Abstr. 140.
18. E. Cabib, in *Encyclopedia of Plant Physiology, New Series, Vol. 13B, Plant Carbohydrates II. Extracellular Carbohydrates* (W. Tanner and F. A. Loewus, eds.), Springer-Verlag, New York, 1981, p. 395.
19. J. Ruiz-Herrera, in *Cellulose and Other Natural Polymer Systems* (R. M. Brown, Jr., ed.), Plenum Press, New York, 1982, p. 207.

20. J. G. H. Wessels, *Int. Rev. Cytol.*, *104*, 37 (1986).
21. D. Carlstrom, *J. Biophys. Biochem. Cytol.*, *3*, 669 (1957).
22. A. D. French, in *Cellulose Chemistry and Its Applications* (R. P. Nevell and S. H. Zeronian, eds.), Ellis Horwood, Chichester, 1985, p. 84.
23. M. Marx-Figini, *J. Polym. Sci.*, *Part C*, *28*, 57 (1969).
24. M. S. Kang, N. Elango, E. Mattia, J. Au-Young, P. W. Robbins, and E. Cabib, *J. Biol. Chem.*, *259*, 14966 (1984).
25. G. W. G. Montgomery, D. J. Adams, and G. W. Gooday, *J. Gen. Microbiol.*, *130*, 291 (1984).
26. S. J. Silverman, A. Sburlati, M. L. Slater, and E. Cabib, *Proc. Natl. Acad. Sci. USA*, *85*, 4735 (1988).
27. P. Orlean, *J. Biol. Chem.*, *262*, 5732 (1987).
28. L. A. Staehelin and T. H. Giddings, in *Developmental Order: Its Origin and Regulation* (S. Subtelny and P. B. Green, eds.), Alan R. Liss, New York, 1982, p. 133.
29. T. H. Giddings, D. L. Brower, and L. A. Staehelin, *J. Cell Biol.*, *84*, 327 (1980).
30. W. Herth, *Planta*, *164*, 12 (1985).
31. B. Schneider and W. Herth, *Protoplasma*, *131*, 142 (1986).
32. W. Herth and G. Weber, *Naturwissenschaften*, *71*, 153 (1984).
33. P. B. Green, *Ann. Rev. Plant Physiol.*, *31*, 51 (1980).
34. J. R. Gipson, in *Cotton Physiology*, Vol. 1, The Cotton Foundation Reference Book Series (J. R. Mauney and J. McD. Stewart, eds.), The Cotton Foundation Publisher: Memphis, TN, 1986, p. 47.
35. T. Hayashi, *Ann. Rev. Plant Physiol. Plant Mol. Biol.*, *40*, 139 (1989).
36. B. Vian, D. Reis, M. Mosiniak, and J. C. Roland, *Protoplasma*, *131*, 185 (1986).
37. D. H. Northcote, in *Society for Experimental Biology Seminar Series*, Vol. *28*, *Biochemistry of Plant Cell Walls* (C. T. Brett and J. R. Hillman, eds.), Cambridge University Press, Cambridge, UK, 1985, p. 177.
38. B. A. Stone, in *Structure, Function and Biosynthesis of Plant Cell Walls* (W. M. Dugger and S. Bartnicki-Garcia, eds.), American Society of Plant Physiologists, Rockville, MD, 1984, p. 52.
39. T. Hayashi and K. Matsuda, *Plant Cell Physiol.*, *22*, 1571 (1981).
40. T. Hayashi and K. Matsuda, *J. Biol. Chem.*, *256*, 11117 (1981).
41. C. T. Brett, *J. Exp. Bot.*, *32*, 1067 (1981).
42. D. Maltby, N. C. Carpita, D. Montezinos, C. Kulow, and D. P. Delmer, *Plant Physiol.*, *63*, 1158 (1979).

43. T. Callaghan and M. Benziman, *Nature*, *311*, 165 (1984).
44. T. Callaghan and M. Benziman, *Nature*, *314* 383 (1984).
45. D. M. Updegraff, *Analyt. Biochem.*, *32*, 420 (1969).
46. A. Bacic and D. P. Delmer, *Planta*, *152*, 346 (1981).
47. F.-C. Lin, R. M. Brown, Jr., J. B. Cooper, and D. P. Delmer, *Science*, *230*, 822 (1985).
48. T. E. Bureau and R. M. Brown, Jr., *Proc. Natl. Acad. Sci. USA*, *84*, 6985 (1987).
49. W. Herth, W. W. Franke, H. Bittiger, A. Kuppel, and G. Keilich, *Cytobiologie*, *9*, 344 (1974).
50. T. Hayashi, S. M. Read, J. Bussell, M. Thelen, F.-C. Lin, R. M. Brown, Jr., and D. P. Delmer, *Plant Physiol.*, *83*, 1054 (1987).
51. B. Lindberg, J. Lonngren, and S. Svensson, *Adv. Carboh. Chem. Biochem.*, *31*, 185 (1975).
52. R. J. Henry and B. A. Stone, *Plant Physiol.*, *69*, 632 (1982).
53. I. J. Goldstein, G. W. Hay, B. A. Lewis, and F. Smith, in *Methods in Carbohydrate Chemistry*, Vol. 5 (R. L. Whistler, ed.), Academic Press, New York, 1965, p. 361.
54. P. J. Harris, R. J. Henry, A. B. Blakeney, and B. A. Stone, *Carboh. Res.*, *127*, 59 (1984).
55. M. C. Meinert and D. P. Delmer, *Plant Physiol.*, *59*, 1088 (1977).
56. H. R. Huwyler, G. Franz, and H. Meier, *Planta*, *146*, 635 (1979).
57. H. R. Huwyler, G. Franz, and H. Meier, *Plant Sci. Lett.*, *12*, 55 (1978).
58. L. Waterkeyn, *Protoplasma*, *106*, 49 (1981).
59. H. Meier, L. Buchs, A. J. Buchala, and T. Homewood, *Nature*, *289*, 821 (1981).
60. C. Pillonel, A. J. Buchala, and H. Meier, *Planta*, *149*, 306 (1980).
61. N. C. Carpita and D. P. Delmer, *Plant Physiol.*, *66*, 911 (1980).
62. D. P. Delmer, U. Heiniger, and C. Kulow, *Plant Physiol.*, *59*, 713 (1977).
63. U. Heiniger and D. P. Delmer, *Plant Physiol.*, *59*, 719 (1977).
64. D. P. Delmer, G. Cooper, D. Alexander, J. Cooper, T. Hayashi, C. Nitsche, and M. Thelen, *J. Cell Sci. (Suppl. 2)*, 33 (1985).
65. P. Bucheli, M. Durr, A. J. Buchala, and H. Meier, *Planta*, *166*, 530 (1985).
66. G. Franz, *Phytochemistry*, *8*, 737 (1969).
67. N. C. Carpita and D. P. Delmer, *J. Biol. Chem.*, *256*, 308 (1981).

68. T. ap Rees, in *The Biochemistry of Plants*, *Vol. 14, Carbo-hydrates* (J. Preiss, ed.), Academic Press, New York, 1988, p. 1.
69. Y. Raymond, G. B. Fincher, and G. A. Maclachlan, *Plant Physiol.*, *61*, 938 (1978).
70. S. C. Mueller and G. A. Maclachlan, *Can. J. Bot.*, *61*, 1266 (1983).
71. R. L. Anderson and P. M. Ray, *Plant Physiol.*, *61*, 723 (1978).
72. K. M. Cooper, in *Cellulose and Other Natural Polymer Systems* (R. M. Brown, Jr., ed.), Plenum Press, New York, 1982, p. 167.
73. C. T. Brett, *Plant Physiol.*, *62*, 377 (1978).
74. J. H. M. Willison, in *Plant Protoplasts* (L. C. Fowke and F. Constabel, eds.), CRC Press, Boca Raton, 1985, p. 77.
75. A. S. Klein, D. Montezinos, and D. P. Delmer, *Planta*, *152*, 105 (1981).
76. D. Haass, W. Blaschek, H. Koehler, and G. Franz, in *Cell Walls '81, Proc. 2nd Cell Wall Meeting, Göttingen, April 1981* (D. G. Robinson and H. Quader, eds.), Wissenschaftliche Verlags GmbH, Stuttgart, 1981, p. 119.
77. E. M. Shea and N. C. Carpita, *Plant Physiol. Suppl.*, *86*, Abstr. 198 (1988).
78. H. Kohle, W. Jeblick, F. Poten, W. Blaschek, and H. Kauss, *Plant Physiol.*, *77*, 544 (1985).
79. H. Kauss, *Ann. Rev. Plant Physiol.*, *38*, 47 (1987).
80. D. P. Delmer, M. Benziman, and E. Padan, *Proc. Natl. Acad. Sci. USA*, *79*, 5282 (1982).
81. H. Kauss and W. Jeblick, *Plant Physiol.*, *85*, 131 (1987).
82. V. Girard and G. A. Maclachlan, *Plant Physiol.*, *85*, 131 (1987).
83. H. Kauss, K. Kohle, and W. Jeblick, *FEBS Lett.*, *158*, 84 (1983).
84. D. Montezinos and D. P. Delmer, *Planta*, *148*, 305 (1980).
85. W. Blaschek, U. Semler, and G. Franz, *J. Plant Physiol.*, *120*, 457 (1985).
86. G. A. Barber, A. D. Elbein, and W. Z. Hassid, *J. Biol. Chem.*, *239*, 4056 (1964).
87. P. M. Ray, T. L. Shininger, and M. M. Ray, *Proc. Natl. Acad. Sci. USA*, *64*, 605 (1969).
88. P. M. Ray, in *Plant Organelles* (E. Reid, ed.), Ellis Horwood, Chichester, 1979, p. 135.
89. P. M. Ray, *Biochim. Biophys. Acta*, *629*, 431 (1980).
90. D. A. Brummel and G. A. Maclachlan, in *Plant Cell Wall Polymers, Biogenesis and Biodegradation* (N. G. Lewis and

M. G. Paice, eds.), American Chemical Society, Washington, D.C., 1989, p. 18.

91. L. Taiz, M. Murry, and D. G. Robinson, *Planta*, *158*, 534 (1983).

92. C. H. Haigler and R. M. Brown, Jr., *Protoplasma*, *134*, 111 (1986).

93. D. K. Romanovicz, in *Cellulose and Other Natural Polymer Systems* (R. M. Brown, Jr., ed.), Plenum Press, New York, 1982, p. 127.

94. S. Amino, T. Yoshihisa, and A. Komamine, *Physiol. Plant*, *65*, 67 (1985).

95. S. R. Jacob and D. H. Northcote, *J. Cell Sci. Suppl.* *2*, 1 (1985).

96. R. J. Henry, A. Schibeci, and B. A. Stone, *Biochem. J.*, *209*, 627 (1983).

97. G. Maclachlan, in *Society for Experimental Biology Seminar Series*, Vol. *28*, *Biochemistry of Plant Cell Walls* (C. T. Brett and J. R. Hillman, eds.), Cambridge University Press, Cambridge, UK, 1985, p. 199.

98. B. P. Wasserman and K. J. McCarthy, *Plant Physiol.*, *82*, 396 (1986).

99. M. E. Sloan, P. Rodis, and B. P. Wasserman, *Plant Physiol.*, *85*, 516 (1987).

100. M. E. Sloan and B. P. Wasserman, *Plant Physiol.*, *89*, 1341 (1989).

101. S. Moreno, C. E. Cardini, and J. S. Tandecarz, *Eur. J. Biochem.*, *157*, 539 (1986).

102. C. Smythe, F. B. Caldwell, M. Ferguson, and P. Cohen, *EMBO J.*, *7*, 2681 (1988).

103. G. Franz, *J. Appl. Polym. Sci.; Appl. Polym. Symp.*, *28*, 611 (1976).

104. S. M. Read, M. Thelen, and D. P. Delmer, in *Cell Walls '86, Proc. 4th Cell Wall Meeting, Paris, September 1986* (B. Vian, D. Reis, and R. Goldberg, eds.), Groupe Parois, Paris, 1986, p. 308.

105. G. Maclachlan, *Plant Physiol.*, *84*, 327 (1987).

106. W. Blaschek, D. Haass, H. Koehler, U. Semler, and G. Franz, *Z. Pflanzenphysiol.*, *111*, 357 (1983).

107. P. J. Harris, M. A. Anderson, A. Bacic, and A. E. Clarke, in *Oxford Surveys of Plant Molecular and Cell Biology*, Vol. 1 (B. J. Miflin, ed.), Clarendon Press, Oxford, 1984, p. 161.

108. M. J. Jaffe and A. C. Leopold, *Planta*, *161*, 20 (1984).

109. J. Fink, W. Jeblick, W. Blaschek, and H. Kauss, *Planta*, *171*, 130 (1987).

110. M. P. Thelen and D. P. Delmer, *Plant Physiol.*, *81*, 913 (1986).

111. T. Callaghan, P. Ross, P. Weinberger-Ohana, and
      M. Benziman, *Plant Physiol.*, *86*, 1104 (1988).
112. T. Callaghan, P. Ross, P. Weinberger-Ohana, G. Garden,
      and M. Benziman, *Plant Physiol.*, *86*, 1099 (1988).
113. P. Ohana, G. Volman, R. Mayer, P. Ross, D. Delmer, and
      M. Benziman, in *Proc. Fifth Cell Wall Meeting* (S. C. Fry,
      C. T. Brett, and J. S. G. Reid, eds.), Scottish Cell Wall
      Group, Edinburgh, 1989, Abstr. 174.
114. S. M. Read and D. P. Delmer, *Plant Physiol.*, *85*, 1008
      (1987).
115. B. P. Wasserman, D. J. Frost, A. Wu, and S. M. Read, in
      Proc. Fifth Cell Wall Meeting (S. C. Fry, C. T. Brett, and
      J. S. G. Reid, eds.), Scottish Cell Wall Group, Edinburgh,
      1989, Abstr. 35.
116. D. H. Northcote, *Proc. Natl. Acad. Sci. USA*, *79*, 5282
      (1982).
117. J.-P. Prieels, D. Monnom, M. Dolmans, T. A. Beyer, and
      R. L. Hill, *J. Biol. Chem.*, *256*, 10456 (1981).
118. R. L. Hill and K. Brew, in *Advances in Enzymology and
      Related Areas of Molecular Biology*, Vol. 43 (A. Meister, ed.),
      John Wiley and Sons, New York, 1975, p. 411.

10

Molecular Approaches for Probing the Structure and Function of Callose and Cellulose Synthases

BRUCE P. WASSERMAN, DAVID J. FROST, and
MARGARET E. SLOAN   *Rutgers University, New
Brunswick, New Jersey*

I. INTRODUCTION

Elucidation of the biochemical mechanism of cell wall polysaccharide
biosynthesis has attracted considerable attention over the past 30
years (1–5). The earliest studies, focusing on the characterization
of polysaccharide biosynthesis in isolated membrane preparations,
raised an important and longstanding question. Namely, is callose
synthase activity (1,3-β-D-glucan synthase) measured in vitro catalyzed by the same enzyme complex responsible for cellulose biosynthesis (1,4-β-D-glucan) in vivo, i.e., is it a component of the
particle complexes demonstrated morphologically to be associated with
microfibril deposition? If so, what factors cause cellulose biosynthesis to switch over to callose biosynthesis upon the disruption of intact plant cells? Alternatively, if cellulose and callose synthesis are
mediated by separate enzymes, each specific for one linkage type,
why does cellulose biosynthesis cease to occur in vitro? The fact
that these questions have persisted for so many years represents a
serious gap in our understanding of plant growth and development.
   To answer these questions, a number of laboratories have been
focusing on the purification and molecular characterization of callose
and cellulose synthases. Purification and identification of the polypeptide subunits involved in the biosynthesis of (1,3)- and(1,4)-β-
linked D-glucans would enable generation of appropriate molecular

probes to study the genetic regulation of cell wall biopolymer formation. Two recent advances in this area are the identification of a
57-kD UDP-Glc-binding polypeptide of callose synthase by photoaffinity labeling (6) and the substantial purification of cellulose synthase from *Acetobacter xylinum* using product entrapment (7).

This chapter summarizes past and recent efforts to purify and
identify membrane-bound callose and cellulose synthases from both
higher plants and microorganisms, and to identify key protein subunits. Perhaps the most troublesome aspect of purifying these
enzymes has been to maintain enzyme activity through a series of
purification steps. Important factors that must be considered in
the design of any purification protocol for membrane-bound enzymes
include (1) choice of detergent for solubilization, (2) stability of
the solubilized enzyme, (3) phospholipid requirements of the solubilized enzyme, (4) finding effective separation methods, and (5)
unequivocal demonstration that the enriched polypeptide(s) represent actual components of the enzyme complex.

II. SOLUBILIZATION OF CALLOSE
    AND CELLULOSE SYNTHASES

A. Solubilization in Digitonin

1. Solubilization

Since callose and cellulose synthases are membrane-bound, the first
step in their purification must be solubilization. Many different
detergents have been effectively used to extract membrane enzymes,
but until recently, digitonin was the only detergent known to effectively release callose (8–14) or cellulose synthase (15) activity
without significant levels of inactivation occurring during the solubilization step. The structure of digitonin is shown in Fig. 1.
Digitonin is a nonionic steroidal glycoside with a critical micelle
concentration of approximately 0.02% and a micelle molecular weight
of 70,000 (16). To obtain reasonable levels of solubilization
digitonin levels of 1%, well above its critical micelle concentration,
have generally been used.

Enzyme stability in digitonin has been a major concern. Some
digitonin-solubilized callose synthases such as those from rye grass
(12), potato (13), and as well as cellulose synthase from *Acetobacter*
(15) were rapidly inactivated upon overnight storage at 4°C. Inactivation of the potato callose synthase could be partially delayed
by the addition of lipids to the incubation medium. Red beet callose
synthase was stable for at least a week at 4°C in digitonin (14).

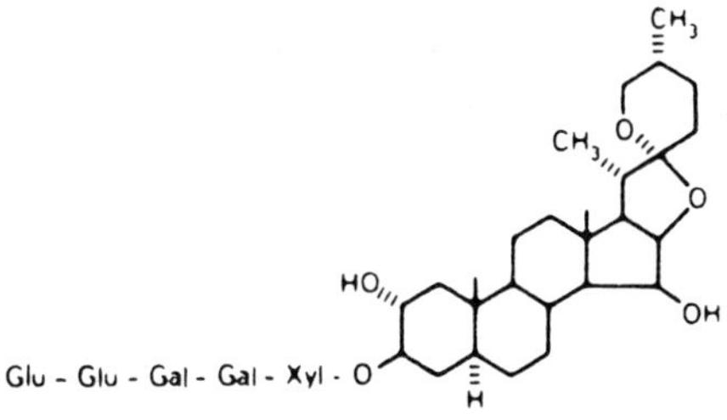

DIGITONIN

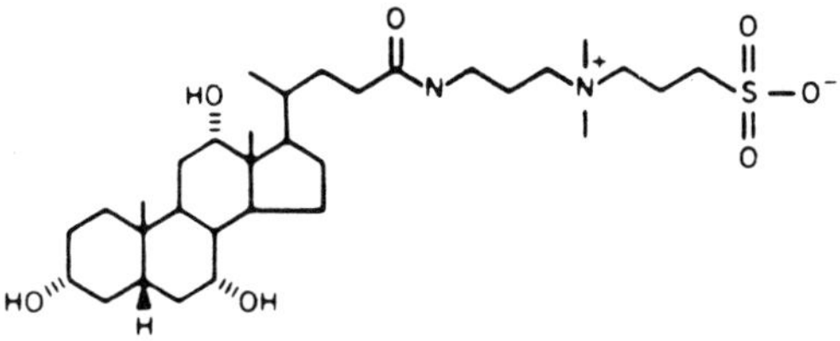

CHAPS

FIGURE 1   Structures of digitonin and CHAPS.

2. The Need for Alternate Detergents

A major disadvantage of digitonin is its low critical micelle concentration (0.02%) and high micelle molecular weight. This not only makes digitonin difficult to remove, but since the concentration needed for solubilization (0.5–1%) is so much higher than its critical micelle concentration, the nature of the resultant protein-digitonin mixed micelle is unknown. It is quite possible that these micelles contain more than one protein per micelle, an undesirable situation when the goal is to obtain a homogeneous enzyme preparation. For these reasons, the use of alternate detergents with higher critical micelle concentration values which are also compatible with callose and cellulose synthases has been investigated.

B.   CHAPS and Other Detergents

Due to the aforementioned problems with digitonin we began to investigate the ability of other detergents to solubilize callose synthase.

One detergent that seemed applicable was CHAPS, a zwitterionic derivative of cholic acid (16,17). The hydrophobic cholic acid portion of CHAPS bears a structural similarity to the digitogenin portion of digitonin (Fig. 1). The major advantages of this detergent over digitonin include a relatively high critical micelle concentration (0.6 vs. 0.02%), its ability to be dialyzed, and a much lower micellar molecular weight (6500 vs. 70,000 kD).

1. Solubilization in CHAPS

CHAPS is now routinely utilized for solubilization of callose synthase from beet (18) and other sources. Optimization of this procedure was based on careful investigation of the role of divalent cations and chelators in solubilization mixtures. It was found that inclusion of the divalent cations Mg^{2+} or Ca^{2+} in either digitonin or CHAPS solubilization mixtures inhibited callose synthase solubilization. On the other hand, the chelators EDTA and EGTA greatly promoted solubilization of red beet callose synthase. Figure 2 illustrates callose synthase release

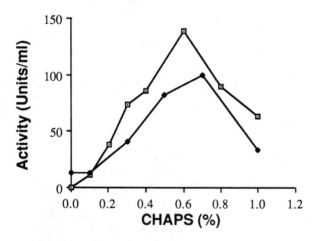

FIGURE 2   Effect of CHAPS concentration on callose synthase solubilization. Upper curve: red beet storage tissue. Lower curve: carrot. Red beet microsomes were solubilized in the presence of 1 mM EDTA, 1 mM EGTA, and 50 mM Tris-HCl, pH 7.5 and the CHAPS-solubilized enzyme was assayed as described (21) with 0.02% digitonin. Carrot microsomes were solubilized in the presence of 50 mM Tris-HCl, pH 7.0. Samples were centrifuged at 80,000 *g* for 30 min. Assays were conducted with 0.02% digitonin. The specific activities of the beet and carrot enzymes solubilized at 0.6 and 0.7% CHAPS were 162 and 78 nmol/min/mg, respectively, assayed at 1 mM UDP-Glc. (From Refs. 18 and 19.)

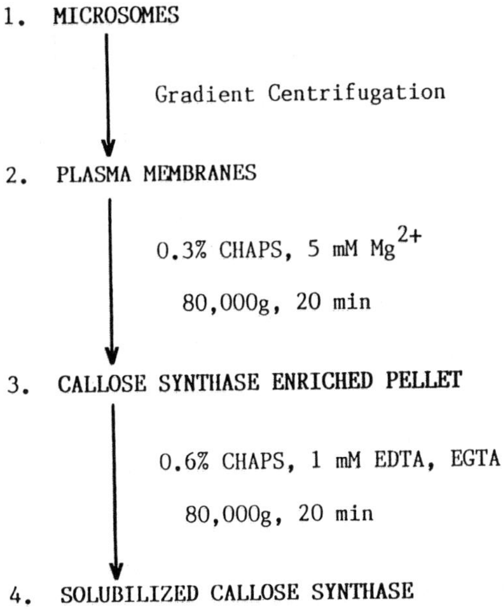

1. MICROSOMES

        Gradient Centrifugation

2. PLASMA MEMBRANES

        0.3% CHAPS, 5 mM Mg^{2+}

        80,000g, 20 min

3. CALLOSE SYNTHASE ENRICHED PELLET

        0.6% CHAPS, 1 mM EDTA, EGTA

        80,000g, 20 min

4. SOLUBILIZED CALLOSE SYNTHASE

FIGURE 3   Schematic for the sequential solubilization of beet callose synthase. (From Ref. 18.)

from red beet and carrot (19) microsomes as a function of CHAPS concentration. Optimal conditions for solubilization varied somewhat between the two sources. Red beet callose synthase was 70% solubilized in the presence of 0.6% CHAPS, 1 mM EDTA, 1 mM EGTA, and 50 mM Tris-HCl, pH 7.5. Carrot callose synthase was similarly solubilized by CHAPS; however, it was necessary to exclude both chelators and divalent cations from the solubilization mixture, since both had inhibitory effects on activity.

The current procedure for solubilization of callose synthase from red beet involves a two-step extraction of an enriched plasma membrane fraction prepared from microsomes by discountinuous sucrose gradient density centrifugation (Fig. 3). The first step consists of exposing plasma membranes to 0.3% CHAPS in the presence of Mg^{2+}. This step solubilizes little activity but approximately 50% of the protein. The presence of Mg^{2+} is critical because it suppresses solubilization. Solubilization is then achieved with 0.6% CHAPS in the presence of 1 mM EDTA and EGTA. It is necessary to conduct assays of the CHAPS-solubilized callose synthase in the presence of 0.01% digitonin. Without digitonin, the solubilized enzyme is essentially inactive. Possible reasons for the digitonin requirement are that it substitutes for steroidal lipids or that it causes a conformational change which exposes the active site of callose synthase to substrate.

CHAPS was also found to be an effective solubilization agent for both the callose and cellulose synthases from the fungus *Saprolegnia monoica* with 0.5% CHAPS (20).

2. Phospholipid Depletion by Triton X-100

Efforts to solubilize red beet callose synthase with Triton X-100 did not succeed. Nonetheless these findings were important in showing the importance of the phospholipid environment of the membrane for activity (21). Extraction of red beet microsomal or plasma membranes with 2% Triton X-100 resulted in the depletion of over 90% of total membrane phospholipid and was accompanied by the reduction of callose synthase activity by 80−90%. Activity in the Triton X-100, $100,000g$ fraction could be reconstituted by the inclusion of various phospholipids in assay mixtures. The most effective phospholipid was phosphatidylethanolamine, which returned 110−114% of the original activity at 0.05%. The specific activity of callose synthase in the Triton X-100-extracted fraction was increased approximately ninefold relative to microsomes. The association of callose synthase activity with phospholipid was independently confirmed in gel filtration experiments with CHAPS (18, discussed below). These and other results (22) provide evidence that callose synthase is modulated by boundary lipid.

3. Other Detergents

Morrow and Lucas (23) reported solubilization of callose synthase from sugar beet petiole with 0.01% zwittergent 3−4. However, low specific activities of the solubilized enzyme suggest that it may be unstable in this detergent. Octylglucoside appears to inactivate callose synthase from plant tissue. However, n-octylglucoside and Triton X-100 were effective in solubilizing cellulose synthase from *A. xylinum* (7).

III. PURIFICATION OF CALLOSE AND CELLULOSE SYNTHASES

A. Glycerol Gradient Centrifugation

Rate-zonal glycerol gradient centrifugation was used to prepare a callose synthase-enriched fraction from digitonin-solubilized membranes (24). A 40-fold increase in specific activity over microsomes was obtained. This approach was chosen based on morphological studies suggesting that cellulose synthase (or callose synthase if it is a deregulated form of the cellulose synthase complex) exists as a large multisubunit complex (25–29). These experiments demonstrated that it is possible to separate many contaminating proteins from callose synthase while retaining enzyme activity and is now used routinely by several laboratories to obtain callose synthase-enriched fractions (30).

B. Product Entrapment

To date two membrane-bound glycosyltransferases responsible for biosynthesis of structural polysaccharides have been purified close to homogeneity, chitin synthase from *Saccharomyces cerevisiae* (31) and cellulose synthase from *A. xylinum* (7). Both purifications took advantage of the fact that the solubilized enzyme forms an insoluble product, in which significant amounts of synthase are entrapped, followed by recovery of purified enzyme.

The cellulose synthase was purified using a repeated product entrapment procedure from trypsinized membranes solubilized with Triton X-100. Pretreatment with trypsin was shown to be important for elimination of contaminating polypeptides during the entrapment steps and had no deleterious effect on cellulose synthase activity. Lithium dodecyl sulfate polyacrylamide gel electrophoresis revealed two prominent polypeptides at 83 and 93 kD. The 83-kD polypeptide was implicated as being a component of cellulose synthase.

This approach has also been utilized for enrichment of callose synthase from mung bean (30).

C. Gel Filtration and Ion Exchange Chromatography

Several attempts to purify digitonin-solubilized callose synthase have been reported. Henry et al. (32) obtained a 10-fold purification on a very small (1.5 × 0.15 cm) DEAE column; however, attempts to recover activity from gel filtration columns with digitonin-containing buffer were unsuccessful. Neither did repeated efforts to purify digitonin-solubilized, gradient-purified beet callose synthase by gel filtration and other column techniques succeed. This problem has now been attributed to the enzyme's requirement for phospholipid. The presence of digitonin in elution buffers appears to cause dissociation of boundary phospholipid from the enzyme complex, resulting in inactivation (18).

The basic assumption for designing a satisfactory elution buffer is to include effectors that provide maximal stability to the enzyme being measured. In the case of many membrane-bound enzymes, detergent is routinely incorporated into elution buffers to maintain the enzyme in a soluble state. However, activity losses under these

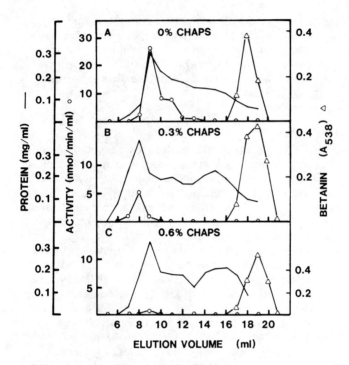

FIGURE 4   Gel filtration profiles of CHAPS-solubilized callose synthase.   Callose synthase solubilized from microsomes by the one-step
method was passed through gel filtration columns equilibrated with
(A) 50 mM Tris, pH 7.5, (B) buffer and 0.3% CHAPS, and (C) buffer and 0.6% CHAPS.   Symbols: (—○—) activity, (———) protein,
(—△—) betanin.   (From Ref. 18.)

conditions led to a careful reexamination of the detergent requirements for conducting gel filtration with CHAPS- and digitonin-
solubilized callose synthase (18).   Only upon the removal of detergent from elution buffers could activity peaks be readily recovered from a 0.9 × 15 cm Sepharose CL-4B.   With both CHAPS- and
digitonin-solubilized enzyme, activity yields were a direct function
of the level of detergent present in the elution buffer.   For example,
with no CHAPS 39% of the activity was recovered (Fig. 4A), whereas
at 0.3% (Fig. 4B) and 0.6% (Fig. 4C) yields less than 10 and 2%,
respectively, were obtained.
     It was demonstrated that when detergent was present in the elu-
tion buffer essential phospholipid was separated from the enzyme

(18). Under conditions whereby phospholipid was removed from the enzyme complex, activity recovery was low. High activity recovery was only obtained under conditions favoring association of phospholipid with solubilized enzyme. Whether using CHAPS- or digitonin-solubilized enzyme, the same overall effect was observed. However, whereas a sharp activity peak was obtained with CHAPS, the digitonin peak was broad and largely superimposable with the peak of protein (18). The large overlap between protein and activity suggests that digitonin-callose synthase micelles are heterodisperse and contain a mixture of proteins. In contrast, CHAPS-callose synthase micelles are likely to contain fewer contaminating proteins.

An active enzyme fraction of cellulose synthase from *A. xylinum* could be recovered after passage through a gel filtration column (7). A Triton X-100-solubilized fraction was applied to a Sephacryl S-300 column and eluted with a Tris buffer (pH 7.5) containing 10 mM $MgCl_2$, 1 mM EDTA, 0.1% Triton X-300, and 20% glycerol. Although recovery of enzyme activity ranged from 60 to 100%, little purification over the initial precolumn solubilized fraction was observed.

D. Affinity or Immunoaffinity Chromatography

Affinity chromatography has been utilized with wide success for purification of a variety of enzymes and receptors. Its application for purifying glucan synthases warrants further exploration. An affinity method utilizing a sperimine-Sepharose 4B column for partial purification of digitonin-solubilized callose synthase from suspension-cultured soybean cells has been reported (33). A broad peak of activity was eluted which is typical of columns run in the presence of digitonin (18,34,35). Although a large increase in specific activity was not obtained, approaches such as this using immobilized ligands to which glucan synthase is known to bind show potential. It would be most desirable to design immobilized ligands that are known to bind specifically with the catalytic site of the enzyme in conjunction with high critical micelle detergents. Several affinity resins such as uridine 5'-diphosphoglucuronic acid-agarose and uridine 5'-diphosphate-agarose are commercially available. Uridine 5'-diphosphoglucuronic acid-agarose has been used in the purification of UDP glucuronosyltransferase (36). Preliminary research from our laboratory indicates that callose synthase binds to this affinity resin but is difficult to elute in an active form.

The availability of partially purified glucan synthase preparations offers the additional possibility of designing immunoaffinity-based purification methods. The key to this approach is to raise antibodies to proteins present in the partially pure preparation and identify those specific for glucan synthase subunits. Nodet et al.

(20) developed a rapid method to screen monoclonal antibodies to callose and cellulose synthases from the fungus *S. monoica*. Several hybridoma clones which precipitate callose synthase have been identified by the laboratories of Fevre (20) and Delmer (37). Although still in the preliminary stages of characterization, one positive clone precipitated 50% of callose synthase activity and reacted with the 56-, 54-, and 48-kD polypeptides on Western blots (37).

E. Isoelectric Focusing, Chromatofocusing, and
   Fast Protein Liquid Chromatography (FPLC)

These are relatively new techniques which have not yet been thoroughly explored for the purification of integral membrane proteins but which show great promise. The success of such procedures is strongly dependent on the use of detergents such as CHAPS, which maintains solubilized proteins in a disaggregated state. A good example of how these techniques may be applied is the purification of lipase by CHAPS solubilization followed by isoelectric focusing in the presence of CHAPS (38). Digitonin-solubilized callose synthase was partially purified through an anion exchange FPLC, but enzyme stability was a problem (39). A similar situation was encountered with CHAPS-solubilized callose synthase from celery petioles (40). Yields and specific activities were both low after separation through an anion exchange column. A possible explanation is removal of essential phospholipids as illustrated with gel filtration chromatography (18).

IV. OTHER APPROACHES FOR IDENTIFYING
    POLYPEPTIDE COMPONENTS OF
    CALLOSE AND CELLULOSE SYNTHASES

Purification of an enzyme may be the most direct way to establish subunit composition, but it is not the only way. Many alternate strategies for identifying polypeptide chains in either membrane-bound or partially purified preparations exist and need to be investigated further with respect to membrane-bound polysaccharide synthases. Use of photoaffinity labels such as 5-azidouridine $5'-[\beta-{}^{32}P]$diphosphate glucose ($5N_3[{}^{32}P]$UDP-Glc) appears to be especially promising (6).

A. Affinity Labeling

1. UDP-Pyridoxal

The affinity label uridine diphosphopyridoxal (Fig. 5), which was used to identify the active site of glycogen synthase (41), was

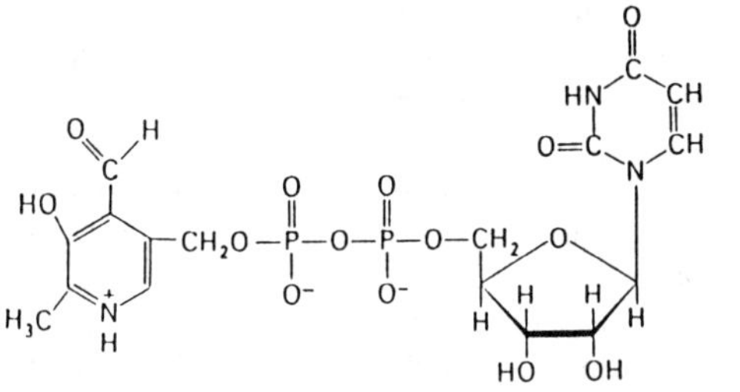

UDP-pyridoxal

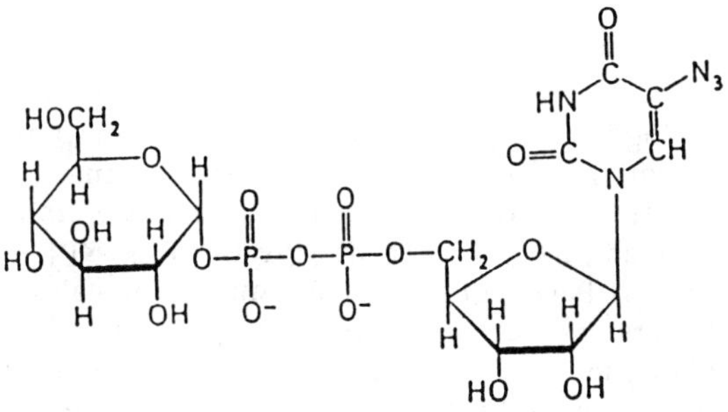

$5N_3UDP-Glc$

FIGURE 5   Structures of UDP-pyridoxal and 5-azidouridine-5'-diphosphate glucose ($5N_3$UDP-Glc).

shown by Read and Delmer (39) to react with callose synthase from mung bean. Callose synthase was inactivated by UDP-pyridoxal with a K_{inact} of approximately 3.8 μM. In addition, protection against inactivation by substrate and effectors was obtained, evidence that the reaction was occurring at the UDP-Glc binding site. However, under callose synthase assay conditions UDP-[^{3}H]pyridoxal labeled numerous polypeptides in a membrane fraction. The reagent was reacting nonspecifically with lysine residues not associated at the enzyme's active site. A 42-kD polypeptide was labeled specifically in the presence of Mg^{2+} but may be a glycosylated protein involved with other polysaccharide synthases such as the Golgi-associated glucan synthase I (42,43).

UDP-pyridoxal is also an effective inhibitor of red beet callose synthase (44). Potential UDP-Glc-binding polypeptides of callose synthase were identified by a two-step labeling procedure. First nonspecific residues were blocked by irreversible modification with formaldehyde or UDP-pyridoxal in the presence of substrate to protect lysines at the UDP-Glc binding sites. In the second step, proteins were recovered, reacted with [^{14}C]formaldehyde or UDP-[^{3}H]pyridoxal in the absence of UDP-Glc, and polypeptide labeling patterns analyzed. Polypeptides that became labeled were those at 69, 57, 54, 25, 22, and 8 kD. From further studies (below) utilizing $5N_3$[^{32}P]UDP-Glc, evidence points to the 57-kD polypeptide as the best candidate for the UDP-Glc-binding polypeptide of callose synthase.

2. Photoaffinity Labeling with 5-Azidouridine 5'-[β-^{32}P]Diphosphate Glucose

Azido-labeled nucleotide analogs have become powerful tools in the identification and characterization of nucleotide-binding proteins (45). Upon UV irradiation, the photoreactive azide covalently links into the protein. We have used the photoaffinity probe $5N_3$[^{32}P]-UDP-Glc (46) to identify the UDP-Glc-binding polypeptide of callose synthase from red beet storage tissue (6). The structure of $5N_3$[^{32}P]UDP-Glc is shown in Fig. 5. Callose synthase was purified from plasma membranes by the two-step (sequential) CHAPS solubilization procedure (described above and in Ref. 47), followed by product entrapment. The 57-kD polypeptide was consistently photolabeled in all callose synthase fractions and was particularly enriched in product-entrapped samples (Fig. 6). Photoinsertion of the 57-kD polypeptide was closely correlated with catalytic properties of the callose synthase such as ion dependence, pH optima, and inactivation by phospholipase treatment (48). These data strongly suggest that the 57-kD polypeptide represents the substrate-binding and cation-regulated component of the callose synthase complex.

In an effort to identify the catalytic subunit of cellulose synthase the herbicide 2,6-dichlorobenzonitrile (DCB), a specific inhibitor of cellulose biosynthesis, was investigated. Delmer et al. (49) synthesized the photoreactive analog 2,6-dichlorophenylazide (DCPA) and used it to probe a DCB receptor. An 18-kD polypeptide whose appearance in cotton fibers parallels in the onset of secondary wall biosynthesis was specifically labeled. Since the 18-kD polypeptide was easily dissociated from membranes, it is unlikely to be the catalytic subunit. However, it may serve as a regulatory subunit and will undoubtedly be the subject of further investigation.

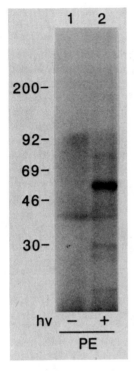

FIGURE 6   Photoaffinity labeling of product-entrapped red beet callose synthase (specific activity 2100 nmol/min/mg). Product-entrapped callose synthase (20 μg) was incubated with 20 μM 5N$_3$[^{32}P]UDP-Glc. Samples exposed to UV illumination are marked (+) whereas dark controls are marked (−).

B.   Chemical Modification with Group-Specific
     Reagents

Chemical modification of amino acid residues at the active site of enzymes has been successfully utilized to identify subunits in a number of enzymes. This approach is similar in principle to affinity labeling but differs in that the reagents are not substrate analogs. Examples of enzymes whose subunits have been identified by this approach include the plasma membrane [H$^+$]-ATPase from *Neurospora* with the cysteine-specific reagent [^{14}C]-N-ethylmaleimide (50) and the Na$^+$/glucose cotransporter with the tyrosine-specific reagent n-acetylimidazole (51).

Successful utilization of this approach generally requires that
(1) chemical modification of an essential residue correlates with a
first-order loss of enzyme activity and (2) in the presence of substrate or substrate and appropriate effectors, protection against inactivation by the group-specific reagent is obtained. When these
conditions are met, it is possible to identify subunits by gel electrophoresis of proteins that have been reacted with radiolabeled modification reagents in the presence and absence of protectants. Subunits that have been protected against reaction with a radiolabeled
modification reagent will be absent on autoradiography of gels.
This approach is generally more laborious than affinity labeling due
to the absolute requirement for protection in impure systems. However, its value may lie in obtaining independent confirmation of results obtained utilizing affinity reagents.

C. Complementation Experiments Using Cellulose-Deficient Mutants

Cloning the cellulose and callose gene will lead to a wealth of information concerning the structure, function, and homology of the
microbial and plant glucan synthase enzymes. For example, bacterial
strains (e.g., *Acetobacter* sp.) deficient in cellulose production have
been reported (52,53). A genomic library can be constructed from
the wild type and mated by conjugation with the cellulose-deficient
mutants. Transformants could then be selected based on the ability
to synthesize cellulose. The wild-type gene will complement the
cellulose deficiencies in the mutant. Proper selection of the mutant
is essential since cellulose-deficient mutants may not be mutations
within the structural genes for subunits of cellulose synthase but for
some other biochemical pathway involved in providing the correct effectors or biochemical precursors for cellulose biosynthesis.

V. CONCLUSION

Enzymes that biosynthesize callose and cellulose have proven difficult
to isolate from higher plants or microorganisms. However, over the
past several years significant advances have been made. The importance of the phospholipid milieu has been recognized. New detergents with superior solubilizing properties are replacing digitonin.
Reagents that specifically label catalytic subunits have been developed. With the diverse tools described in this chapter, the prospects for purification and the ability to produce genetic and immunological probes seem bright. Such advances should set the stage for

investigating the regulation of cell wall glucan biosynthesis at the molecular level and resolving the many questions that currently exist.

ACKNOWLEDGMENTS

This research was supported in part by grants from the National Science Foundation (DMB 85-02523 and DCB 89-07202), the U.S. Department of Agriculture (87-CRCR-1-2414), and the Charles and Johanna Busch Memorial Fund, the Rutgers Research Council, and the New Jersey Agricultural Experiment Station with State and Hatch Act funds. New Jersey Agricultural Experiment Station Publication F-10538-1-87.

REFERENCES

1. D. P. Delmer, *Adv. Carboh. Chem. Biochem.*, *41*, 105 (1983).
2. D. P. Delmer, G. Cooper, D. Alexander, J. Cooper, T. Hayashi, C. Nitsche, and M. Thelen, *J. Cell Sci. Suppl.*, *2*, 33 (1985).
3. B. P. Wasserman, L. L. Eiberger, and K. J. McCarthy, *Food Technology*, *40*, 5, 90 (1986).
4. D. P. Delmer, *Ann. Rev. Plant Physiol.*, *38*, 259 (1987).
5. S. R. Jacob and D. H. Northcote, *J. Cell Sci. Suppl.*, *2*, 1 (1985).
6. D. J. Frost, S. M. Read, R. R. Drake, B. E. Haley, and B. P. Wasserman, *J. Biol. Chem.*, *265*, 2162 (1990).
7. F. C. Lin and R. M. Brown, Jr., in *Cellulose and Wood: Chemistry and Technology* (C. Scheurch, ed.), John Wiley and Sons, New York, 1989, p. 473.
8. D. S. Feingold, E. F. Neufeld, and W. Z. Hassid, *J. Biol. Chem.*, *233*, 783 (1958).
9. H. M. Flowers, K. K. Batra, J. Kemp, and W. Z. Hassid, *Plant Physiol.*, *43*, 1703 (1968).
10. C. Peaud-Lenoel and M. Axelos, *FEBS Lett.*, *8*, 224 (1970).
11. C. M. Tsai and W. Z. Hassid, *Plant Physiol.*, *47*, 740 (1971).
12. R. J. Henry and B. A. Stone, *Biochem. J.*, *203*, 629 (1982).
13. U. Heiniger, *Plant Sci. Lett.*, *32*, 35 (1983).
14. L. L. Eiberger, C. L. Ventola, and B. P. Wasserman, *Plant Sci. Lett.*, *37*, 195 (1985).
15. Y. Aloni, R. Cohen, M. Benziman, and D. Delmer, *J. Biol. Chem.*, *258*, 4419 (1983).

16. L. M. Hjelmeland and A. Chrambach, *Methods in Enzymol.*, *104*, 305 (1984).

17. L. M. Hjelmeland, *Proc. Natl. Acad. Sci. USA*, *77*, 6368 (1980).

18. M. E. Sloan, P. Rodis, and B. P. Wasserman, *Plant Physiol.*, *85*, 516 (1987).

19. S. G. Lawson, T. L. Mason, R. D. Sabin, M. E. Sloan, R. R. Drake, B. E. Haley, and B. P. Wasserman, *Plant Physiol.*, *90*, 101 (1989).

20. P. Nodet, J. Grange, and M. Fevre, *Anal. Biochem.*, *174*, 662 (1988).

21. B. P. Wasserman and K. J. McCarthy, *Plant Physiol.*, *82*, 396 (1986).

22. H. Kauss and W. Jeblick, *Plant Physiol.*, *81*, 171 (1986).

23. D. L. Morrow and W. J. Lucas, *Plant Physiol.*, *81*, 171 (1986).

24. L. L. Eiberger and B. P. Wasserman, *Plant Physiol.*, *83*, 982 (1987).

25. S. C. Mueller and R. M. Brown, Jr., *J. Cell Biol.*, *84*, 315 (1980).

26. S. C. Mueller and R. M. Brown, Jr., *Planta*, *154*, 498 (1982).

27. S. C. Mueller and R. M. Brown, Jr., *Planta*, *154*, 501 (1982).

28. A. M. C. Emons, in *Biosynthesis and Biodegradation of Cellulose and Cellulosic Materials* (C. Haigler and P. Weimer, eds.), Marcel Dekker, New York, 1991, pp. 71-96.

29. H. Quader, in *Biosynthesis and Biodegradation of Cellulose and Cellulosic Materials* (C. Haigler and P. Weimer, eds.), Marcel Dekker, New York, 1991, pp. 51-67.

30. T. Hayashi, S. M. Read, J. Bussell, M. Thelen, F.-C. Lin, R. M. Brown, Jr., and D. P. Delmer, *Plant Physiol.*, *83*, 1054 (1987).

31. M. S. Kang, N. Elango, E. Mattia, J. Au-Young, P. W. Robbins, and E. Cabib, *J. Biol. Chem.*, *259*, 14966 (1984).

32. R. J. Henry, A. Schibeci, and B. A. Stone, *Biochem J.*, *209*, 627 (1983).

33. H. Kauss and W. Jeblick, *Plant Sci.*, *48*, 63 (1987).

34. H. Gorissen, G. Aerts, B. Ilien, and P. Laduron, *Anal. Biochem.*, *111*, 33 (1981).

35. K. E. Langley and E. P. Kennedy, *J. Bact.*, *136*, 85 (1978).

36. O. M. P. Singh, A. B. Graham, and G. C. Wood, *Eur. J. Biochem.*, *116*, 311 (1981).

37. D. P. Delmer, in *Cellulose and Wood: Chemistry and Technology* (C. Scheurch, ed.), John Wiley and Sons, New York, 1989, p. 749.

38. W. Stuer, K. E. Jaeger, and U. K. Winkler, *J. Bacteriol.*, *168*, 1070 (1986).

39. S. M. Read and D. P. Delmer, *Plant Physiol.*, *85*, 1008 (1987).
40. R. M. Slay and A. E. Watada, *Plant Physiol.*, *Suppl. 89*, 57 (1989).
41. M. Tagaya, K. Nakano, and T. Fukui, *J. Biol. Chem.*, *260*, 6670 (1985).
42. S. M. Read, M. Thelen, and D. P. Delmer, *Proc. 4th Cell Wall Meeting*, Groupe Parois, Paris, 1986, p. 308.
43. P. M. Ray, K. S. Dhugga, and S. R. Gallaghar, *Plant Physiol.*, *Suppl. 89*, 102 (1989).
44. T. M. Mason, S. M. Read, D. J. Frost, and B. P. Wasserman, *Physiologia Plantarum*, *79*, in press (1990).
45. R. L. Potter and B. E. Haley, *Meth. Enzymol.*, *91*, 613 (1983).
46. R. R. Drake, R. K. Evans, J. K. Wolf, and B. E. Haley, *J. Biol. Chem.*, *264*, 11928 (1989).
47. B. P. Wasserman, D. J. Frost, S. G. Lawson, T. L. Mason, P. S. Rodis, R. D. Sabin, and M. E. Sloan, in *Modern Methods in Plant Analysis*, Springer-Verlag, New York, 1989, p. 1.
48. M. E. Sloan and B. P. Wasserman, *Plant Physiol.*, *89*, 1341 (1989).
49. D. P. Delmer, S. M. Read, and G. Cooper, *Plant Physiol.*, *84*, 415 (1987).
50. R. J. Brooker and C. W. Slayman, *J. Biol. Chem.*, *258*, 222 (1983).
51. B. E. Peerce and E. M. Wright, *J. Biol. Chem.*, *260*, 6026 (1985).
52. I. M. Saxena and R. M. Brown, Jr., in *Cellulose and Wood: Chemistry and Technology* (C. Scheurch, ed.), John Wiley and Sons, New York, 1989, p. 537.
53. S. Valla, in *Cellulose and Wood: Chemistry and Technology* (C. Scheurch, ed.), John Wiley and Sons, New York, 1989, p. 559.

11

Biochemistry of Cellulose Synthesis
in *Acetobacter xylinum*

PETER ROSS, RAPHAEL MAYER, and MOSHE BENZIMAN   *Institute of Life Sciences, Hebrew University of Jerusalem, Jerusalem, Israel*

I. INTRODUCTION

Acetobacter xylinum has long been considered the most successful prokaryotic model for studies on the mechanism of cellulose biogenesis in the intact cell (1–6). In recent years, since the advent of a highly active cell-free preparation for cellulose synthesis derived from this organism (7–9), we have had the privilege of taking a closer look inside this bacterial cell in examining the biochemistry of this process. The in vitro system we are working with is almost 500 times more active than previous preparations (cf. 6) and approaches 90% of the in vivo rate of cellulose synthesis. In this chapter we shall review developments in the enzymology of cellulose synthesis and its regulation. Upon embarking on a detailed discussion of the various enzymes related to this system and their products, we felt it advantageous to provide first a brief overview of the interrelated nature of these components.

The target enzyme in the regulation of cellulose synthesis is the membrane-bound cellulose synthase. Regulation of the enzyme involves an unusual system of "fine control" based on a novel nucleotide, cyclic diguanylic acid (c–di–GMP) and the regulatory enzymes maintaining its intracellular turnover (10–12). A scheme depicting the current model for this system is shown in Fig. 1. Diguanylate cyclase, presumably a membrane-associated enzyme, converts two

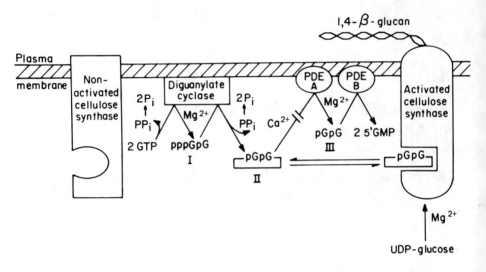

FIGURE 1   Proposed regulatory control pathway of the cellulose
synthase.   (Reproduced by permission from Ref. 12.)

molecules of GTP into a 5'-tetraphosphate dimer (I) which is then
condensed intramolecularly to form cyclic diguanylic acid (II).
Cyclic diguanylic acid is a highly potent, allosteric activator of the
cellulose synthase.   Deactivation of the synthase occurs when a
membrane-bound phosphodiesterase (PDE-A) cleaves the cyclic
structure of the nucleotide to yield the inactive open dimer (III).
A second phosphodiesterase (PDE-B) splits this product into two
molecules of 5'-GMP.   The first of these degradative steps is blocked
by Ca^{2+} ions, thus preserving the nucleotide activator and extending
its availability to the synthase.   For simplicity, the synthesis of a
single β-1,4-linked glucan chain is depicted, although a more complex form of the synthase, polymerizing several chains simultaneously, might be the active enzyme unit in cellulose biogenesis.

II.   THE CELLULOSE SYNTHASE: THE SPINNERET
      OF CELLULOSE BIOGENESIS

The cellulose synthase catalyzes the initial step in cellulose biogenesis, namely, the polymerization of glucose from UDP-glucose
into β-1,4-linked glucan chains.   Considering the structural

complexity of the final product formed in vivo, this enzyme may also play a role, at least indirectly, in the noncovalent steps of crystallization and fibril formation (5). The study of the cellulose synthase in vitro has begun to provide some clues as to the molecular mechanism that mediates this complex process.

A. Nature of the Polymerization Reaction

The alkaline-insoluble product formed from UDP-glucose in vitro by the synthase in both membrane-bound and solubilized forms has been characterized enzymatically and chemically as a β-1,4-linked glucan (7,9). The second product of the reaction, released in amounts equimolar to the glucose incorporated into insoluble material, has now been determined to be UDP (Table 1). This supports the following general equation for the cellulose synthase reaction:

$$\text{UDP-glucose} + [\beta\text{-}1,4\text{-glucose}]_n \longrightarrow \text{UDP} + [\beta\text{-}1,4\text{-glucose}]_{n+1}$$

This formulation excludes a direct transfer of hexose-1-P, which occurs in some of the partial reactions of bacterial mucopeptide (14) or glycoprotein (15) synthesis, where a lipid-linked intermediate is

TABLE 1   Stoichiometry of the Cellulose Synthase Reaction (pmol/min/mg protein)[a]

Exp	β-1,4-Glucan formed	UDP-glucose consumed	UDP formed	UMP formed
1	196	225	210	<5
2	150	165	155	<5

[a]Cellulose synthase activity was assayed under standard conditions (10–12) in the presence of c-di-GMP (10 μM). Parallel incubations were carried out employing as substrate either [14C]UDP-glucose, for quantitation of β-glucan by filter assay (7–12), or [α-32P]UDP-glucose, for determination of uracil phosphates by thin layer ion exchange chromatography (11). [α-32P]UDP-glucose was prepared from [α-32P]UTP and glucose-1-P, essentially as described (12a).

the acceptor of the sugar phosphate. This conclusion is compatible with the failure to identify such an intermediate by in vivo labeling of *A. xylinum* cells with [^{14}C]glucose (16) and with the finding that the enzyme maintains high activity following solubilization and purification stages in which such an endogenous lipid component would most probably be removed (9). Significantly, two other β-linked homopolysaccharide-polymerizing enzymes, the β-1,3-linked glucan synthase (17) and chitin synthase (18) from yeast, similarly do not exhibit a requirement for such a lipid component. Furthermore, none of these polymerases, including the cellulose synthase, depend on exogenously added "primer" materials.

The apparent noninvolvement of covalent intermediates in the cellulose synthase reaction suggests that catalysis from UDP-glucose occurs via a direct substitution mechanism, in which the phospho-ester-activating group at the anomeric carbon of one glucose residue is displaced by the C-4 hydroxyl group of another glucosyl residue, inverting the α configuration to form a β-glucosidic bond. Ascertaining the validity of this mechanism, which involves a minimum of bond cleaving and forming steps, as well as determining the actual direction from which elongation of the polyglucan chain proceeds, will have significant implications on how we perceive later stages in the hierarchy of cellulose biogenesis.

Given such a mechanism, the subsequent vectorial discharge of the product might be through a "cotranslational" type of mode, wherein polymerization and extrusion of the growing chain via the purported "extrusion pore" structures in the LPS layer of the organism (2,3) occur simultaneously. This simple mechanism makes it particularly easy to envision the process of microfibril assembly as being accomplished by a multienzyme complex containing several molecules of cellulose synthase, each catalyzing independently the processive synthesis of solitary polyglucan chains within sufficient proximity to ensure their rapid association and crystallization. Considering the other extreme, several glucose-transferring catalytic sites operating in obligatory conjunction may comprise the active form of the synthase. The synthases might operate on a concerted self-assembly principle in which nascent chains serve as templates for the incoming residues of immediately adjacent chains. Compatible with either of these possibilities is the finding that premicrofibril-type structures have been detected as products of the c-di-GMP-activated solubilized enzyme (19). Considering the foregoing, it would be of great interest to firmly establish the true percentage of enzyme product which actually achieves crystalline form, and ultimately if there exists — as suggested in the former mechanism and excluded by the latter — a single β-glucan chain-synthesizing form of the synthase.

While the energy of crystallization and fibril aggregation seems a good candiate for providing the force required for the motion of newly added glycosyl residues away from the enzyme catalytic site and out through the membrane, there is some evidence which points to the contrary; materials such as Calcofluor, which perturb chain association, actually accelerate the polymerization rate (5; see Chap. 5). Alternatively, the energy source for this process may be generated in the course of β-glucoside bond formation and transduced through conformational shifts in enzyme structure. The transmembrane potentials found to be essential for in vivo cellulose synthesis (4) may account for the force presumably necessary to drive and/or organize this highly complex process. A processive mechanism similar to that discussed here for the cellulose synthase, where the polymeric product remains tightly bound to the enzyme between elongation steps, has also been proposed for yeast chitin synthase (18).

B. Characteristics of the Synthase

The cellulose synthase is most probably an integral membrane protein; synthase activity occurs exclusively in the membrane-associated fraction, as determined in a variety of *A. xylinum* strains. This activity cannot be eluted from the membrane fraction by extensive washing in either low or high ionic strength buffers, by changes in pH, or by metal chelating agent (unpublished results). The enzyme has been effectively solubilized by treatment of membranes with digitonin (9). The catalytic and regulatory properties of the solubilized enzyme are remarkably similar to those observed for the membranous enzyme. Thus, enzyme activity in both forms is Mg^{2+}-dependent, optimal at 30°C in the pH range 7.5–8.5, displays typical Michaelis–Menten kinetics with respect to UDP-glucose (K_m = 0.2 mM) (7–9), and is competitively inhibited by the uridine 5'-phosphates, UTP, UDP (K_i = 0.14 mM), and UMP (K_i = 0.71 mM) (8). Similarly, the soluble enzyme retains full sensitivity to c-di-GMP, displaying the same overall stimulation effect (two orders of magnitude) and K_a value for this activator as observed for the membranous synthase. In contrast to previous reports (8,9), which preceeded the discovery of c-di-GMP and relied on an activation system comprised of GTP and a crude soluble fraction, under current standard assay conditions the reaction is linear with time for at least 30 min at 30°C in the presence of saturating concentrations of c-di-GMP. The enzyme shows enhanced stability when immobilized in polyacrylamide gel assays, following nondenaturing electrophoresis (20). Significantly, following either native gel electrophoresis or size exclusion chromatography, the synthase retains its sensitivity to c-di-GMP activation, suggesting that this nucleotide binds directly to the synthase or to a tightly associated regulatory subunit. In both cases, enzyme activity migrates as a single fraction, possibly reflecting the occurrence of only one basic macromolecular form of the enzyme.

From behavior on size exclusion columns, the molecular mass of the digitonin-solubilized synthase is estimated to be in the range of 350−500 kD indicative of a multipeptide structure. The synthase is apparently an —SH enzyme, based on its sensitivity to prarhydroxymercuribenzoic acid (PHMB), which can be reversed by dithiothreitol. The critical sulfhydryl group(s) affected might be related to the enzyme's interaction with both UDP-glucose and c-di-GMP, since high concentrations of either substrate or activator protect the enzyme against PHMB inactivation. Attempts to purify the cellulose synthase employing conventional chromatographic techniques have not been especially successful, but recently the enzyme has been purified 350-fold by affinity chromatography, and its properties and structure in the purified state are currently under investigation.

C. Cellulose Synthase is the Target
Site for Regulation

Essentially four enzymatic steps have been characterized in cell-free extracts of *A. xylinum* that appear to comprise the complete pathway from glucose to cellulose. These are the phosphorylation of glucose by glucokinase (21), the isomerization of glucose-6-phosphate to glucose-1-phosphate by phosphoglucomutase (22,23), the synthesis of UDP-glucose by UDPG-pyrophosphorylase (23,24), and the cellulose synthase reaction. The validity of this series of reactions has also been borne out by kinetic tracer studies in which the in vivo flow of carbon from glucose through sugar phosphates and nucleotide sugar pools into cellulose was quantitatively evaluated (6). Based on this pathway, the total cost incurred in incorporating one glucose residue into polyglucan is only two high-energy phosphate bonds. Estimating from measurements of respiration and cellulose synthesis rates in resting cells (25−27), which indicate that for each glucose molecule completely oxidized two are converted to cellulose, at least 10% of the cell's energy budget is devoted to cellulose production at any one time. Although the physiological role of cellulose synthesis in this organism is still not clear (cf. 1,28), considering the relatively high cellular energy expenditure of this metabolically irreversible process, it is not surprising that the biosynthetic system is governed by a complex regulatory system. The cellulose synthase, the only enzyme unique to the pathway, performs the "committed" step in the cellulose

formation — a metabolic dead end with regard to carbon utilization —
and hence would logically be the prime candidate for strict regulatory control. Furthermore, as demonstrated in cell-free extracts,
the levels of enzyme activities leading to UDP-glucose are in large
excess relative to that of the cellulose synthase, strongly supporting the proposition that the latter comprises the rate-limiting step
in cellulose biogenesis.

The basis for regulation appears to reside in the nearly absolute
dependence of the enzyme on its allosteric effector c-di-GMP.
Under optimal in vitro conditions, but in the absence of this nucleotide, synthase activity is hardly detectable; addition of micromolar
concentrations of the cyclic dinucleotide activator results in up to a
200-fold stimulation of the enzyme (Fig. 2). The interaction between
the activator and the synthase within the membrane is apparently
reversible since washing of membranes exposed to c-di-GMP results
in loss of all their enhanced synthase activity, which can subsequently be restored upon readdition of the activator. Since enhanced
activity is due to an increase in the V_{max} of the enzyme and not to

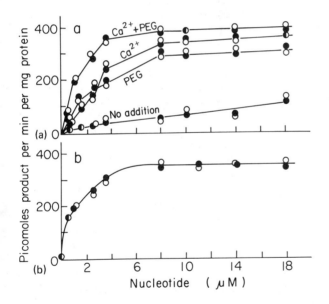

FIGURE 2   Activation of the cellulose synthase by native (○) and
synthetic (○) c-di-GMP in both its (a) membrane-bound and (b)
solubilized forms. Although not shown, Ca^{2+} and PEG do not affect the enzyme solubilized in digitonin, since this detergent de-
stroys c-di-GMP degrading phosphodiesterase. (Reproduced by permission from Ref. 12.).

a reduction in the K_m value for UDP-glucose substrate (8), it
would appear that the synthase only functions as an active catalytic
unit when c-di-GMP is bound to subunit(s) of the enzyme, a condition which bestows an "on-off" quality to the regulation. In this
sense, the cyclic dinucleotide might be thought of as an essential
cofactor of the synthase. Alternatively, c-di-GMP may affect a specific rate-limiting step in the polymerization reaction, perhaps by
promoting the initiation of chain synthesis. Examination of the degree of polymerization of the glucan product synthesized in vitro
might lead to the ability to discern between these possibilities.
There is now some evidence that the chitin synthase produces
chains of homogeneous length (18), although the molecular basis of
such control has not been established.

In addition to c-di-GMP, two additional factors have been found
to promote activation of the cellulose synthase. These are Ca^{2+}
and polyethylene glycol (PEG 4000). The effect of Ca^{2+} is related
to its ability to inhibit c-di-GMP phosphodiesterase, thus preserving
the activator. The effect of PEG, which is to reduce the apparent
K_a for c-di-GMP, was previously attributed to promoting the binding
of c-di-GMP to its target site. However, as with Ca^{2+}, the PEG
effect also appears to be related to preservation of the activator in
crude phosphodiesterase-containing reaction mixtures.

Given that the K_m of the synthase for UDP-glucose corresponds
to physiological concentrations of this guar nucleotide, fluctuations
in the availability of substrate might provide the basis for additional
regulatory control. While in all cellulose-producing strains tested
UDPG-pyrophosphorylase activity is more than adequate to supply
the demand of the synthase for activated precursor, large variations in the level of this enzyme exist between strains of comparable
cellulose-synthesizing capacities (unpublished results). Although
the reasons for such diversity are not clear, it would appear that
some strains function more efficiently with regard to the ability to
utilize UDP-glucose for cellulose synthesis. It is noteworthy that no
other nucleoside diphosphoglucose pyrophosphorylase has been found
in cell-free extracts of this organism (23,24). We have recently characterized a series of Cel⁻ mutants which appeared to be specifically
deficient in this enzyme and which, upon conversion to Cel⁺ phenotype by transconjugation with a plasmid carrying a 2.7-kb DNA
fragment obtained from a library of A. xylinum DNA, were restored
to the higher, wild-type levels of UDPG-pyrophosphorylase (29).
The identification of this DNA fragment as a structural gene for
UDPG pyrophosphorylase demonstrates that the relationship between
cellulose synthesis and the metabolism of UDP-glucose might be
more complex than we previously presumed.

III. CYCLIC DIGUANYLIC ACID, A UNIQUE MESSENGER MOLECULE, MODULATES THE RATE OF CELLULOSE SYNTHESIS

A. Discovery of the Cellulose Synthase Activator: Uncovering the Role of GTP

One of the first exciting moments in studying the membrane-bound cellulose synthase was the finding that the enzyme, when assayed in total cell homogenates prepared in the presence of PEG, was markedly stimulated by GTP (7). Centrifugal fractionation of this crude membrane preparation revealed not that the synthase in the resultant pellet is directly affected by GTP but that the nucleotide's stimulatory effect depended on the presence of the supernatent fraction. The essential supernatant factor proved to be a protein and initially was thought to be a loosely associated regulatory subunit of the synthase (7−9). In the course of experiments designed to test the hypothesis that GTP and the protein factor directly activate the cellulose synthase, the formation of an additional, previously undetected stimulatory factor was observed; preincubating the protein factor with GTP had a dramatic, synergistic effect on their combined ability to stimulate the synthase (10,11). In direct contrast to the properties of the protein factor, the stimulatory factor generated during the preincubation was both heat-stable and of low molecular weight. Formation of the new activator was Mg^{2+}-dependent and highly specific for GTP, which could not be substituted by other nucleotides including the nonhydrolyzable analog guanosine-5'-[β,γ-imino]triphosphate (GppNHp) (7,10). Thus, heat-stable activator formation appeared to be the result of a GTP-utilizing enzymatic reaction. The role of GTP in the formation of the heat-stable activator was clarified by employing either [α-^{32}P]- or [8-^{3}H]-labeled GTP as substrates. Separation of the incubation products by DEAE-cellulose chromatography yielded a distinct UV-absorbing, radiolabeled fraction bearing all of the original heat-stable stimulatory activity, suggesting that the low molecular weight activator arises from GTP. Further purification stages confirmed that the stimulatory activity resides in a single material. Thus it was possible for the first time to demonstrate activation of the synthase not by protein factor and GTP, but by a product of their enzymatic interaction.

B. Elucidation of the Unique Structure of the Activator

Initial attempts to understand the chemical structure of the heat-stable activator immediately indicated that some surprises awaited us

(10,11). The material was obviously a derivative of GTP and was
found by chemical analysis to contain equimolar amounts of ribose,
phosphate, and unmodified guanosine residues; the basic structural
unit of the activator was therefore riboguanosine monophosphate.
Whereas the activator appeared to contain at least four GMP residues, based on its ion exchange and electrophoretic mobility relative to GTP-Mg^{2+}, analysis of partial hydrolyzates did not reveal
the accumulation of intermediate degradation products, as expected
for an oligomeric nucleotide structure. For instance, the final products of mild alkaline hydrolysis, 2'- and 3'-GMP, composed more
than 95% of partial hydrolysis products, and virtually no degradation products, except for the final product 5'-GMP, were detected
upon partial degradation with venom phosphodiesterase. While
these results confirmed the basic GMP unit structure of the molecule
and indicated that these are linked in alkali-labile 2',5'- or 3',5'-
phosphodiester bonds, the absence of degradation intermediates suggested that no more than two GMP units compose its structure.
Another surprising finding was that the compound is resistant to
both alkaline phosphatase treatment and to the periodate β-elimina-
tion reaction, compelling the conclusion that the molecule has neither
monoesterified phosphate groups nor a free terminal ribose residue.
Taken together all of these properties could be attributed to a cyclic
nucleotide structure composed of two or more unmodified GMP residues in 2',5'- or 3',5'-phosphodiester linkage.

   More rigorous structural analysis enabled conclusive identification of the activator compound as bis (3' → 5') cyclic diguanylic
acid (Fig. 3). Chromatography on DEAE-Sephadex in the presence
of 7 M urea, which disrupts nonionic interactions such as hydrogen
bonding, now allowed an estimated net charge of -2 for the activator,
indicating that it contains no more than two GMP residues. The

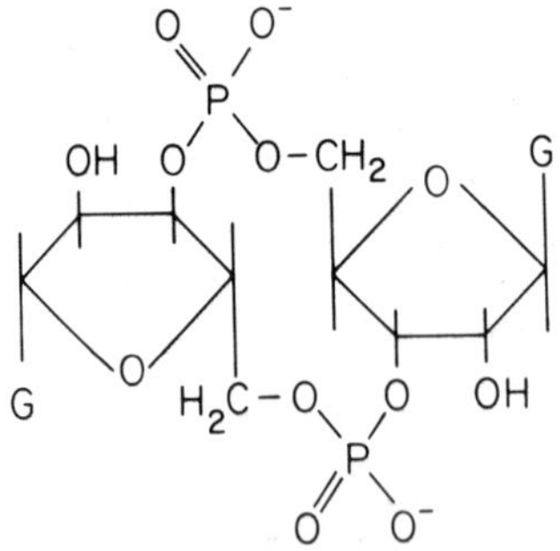

FIGURE 3   Structural formula of bis (3' → 5')-cyclic diguanylic acid
(c-di-GMP). (Reproduced by permission from Ref. 12.)

electrophoretic mobility of the activator relative to that of GTP,
GDP, 5'-GMP, and 3',5'-cGMP, when carried out in the absence of
Mg^{2+}, was also compatible with the dimeric structure. (That the
relative electrophoretic mobility of these compounds is strongly influenced by Mg^{2+} may be of physiological significance, since it indicates that the activator's affinity for this metal probably is far
lower than that of the common guanyl mononucleotides.) Direct
support for a cyclic dimeric structure was obtained by mass spectroscopic analysis, which yielded a relative molecular mass of 690, a
value corresponding to that of cyclic diguanylic acid. The susceptibility of the activator to the 3',5-phosphodiester-specific T1
endonuclease, yielding 3'-GMP as sole final product, strongly supported the assignment of the phosphodiester isomeric bond structure
as $3'-5'$. Conclusive characterization of the activator was finally
established by chemical synthesis of the putative compound. The
chemically synthesized compound and the native activator were
found (1) to be identical in stimulating cellulose synthesis under a
variety of conditions (Fig. 2); (2) to show the same sensitivity and
yield identical products when subjected to a variety of chemical and
enzymatic treatments (Table 2); (3) to yield identical proton and
phosphorous NMR spectra (12); and (4) to be indistinguishable by
HPLC analysis (Fig. 5). These findings represented the first case
of a cyclic dinucleotide of biological origin with a defined physiological function, namely, the regulation of cellulose biogenesis. The
central role that c-di-GMP plays in regulating cellulose synthesis —
an evolutionary early and universal process — raises the prospect
that this or similar compounds may be of general natural occurrence
and significance.

The existence in *A. xylinum* of such a specialized regulatory
system based on a molecule bearing the unusual structure of c-di-
GMP is intriguing, since bacterial cells are equipped with a variety
of regulatory mechanisms based on nucleotides, such as 3',5'-cyclic
AMP (30), diadenosine tetraphosphate (A5'pppp5'A) (31), and magic
spot compounds (ppG3'pp) (32), which might have been adapted to
the cellulose-synthesizing system. It is possible either that the
physiological roles of such compounds are incompatible with their
also serving in the regulation of cellulose synthesis or that the
structure of ci-di-GMP has unique properties which deem it particularly suited as an essential component of active cellulose synthase. Some unusual features of the modifier molecule's structure
are its twofold degree of rotational symmetry and its bivalent arrangement of GMP residues, which presumably reflects a similar
disposition in its binding sites with synthase subunits. [Curiously,
a high-affinity binding site for synthetic cyclic dinucleotides has
been demonstrated for a bacterial RNA polymerase (33).] The
regulatory system is highly specific toward this structure, and of a

TABLE 2   Enzymic and Chemical Susceptibility of c-di-GMP[a]

Treatment	Degradation product
1. Mild alkali	2'-GMP (45%), 3'-GMP (55%)
2. Periodate/β-elimination	Unaffected
3. Bacterial alkaline phosphatase	Unaffected
4. Venom phosphodiesterase	5'-GMP
5. T1 endonuclease	3'-GMP
6. Mung bean endonuclease	5'-GMP
7. Washed membranes A. xylinum:	
Short incubation	pG3'p5'G
Extended incubation	5'-GMP

[a]Samples of native and synthetic c-di-GMP were subjected to the indicated treatments, carried out as in (10–12). Products were identified by their reverse phase HPLC retention times (12).

long list of guanyl nucleotides tested (11), including pG3'p5'G and G3'p5'G, only c-di-GMP activates the enzyme. More particularly, the activation is highly specific toward the dimeric structure; thus neither the lower homolog of c-di-GMP, 3',5'-cGMP, nor the higher homolog 3',5'-cyclic triguanlyic acid, are effective activators of the synthase or inhibitors of c-di-GMP activation. Both of the guanine moieties of the activator also fulfill a crucial structural requirement; substitution of one of these with cytosine, or of both with either adenine, uracil, or thymine, renders the compound inactive in either a stimulatory or inhibitory capacity. Other analogs of c-di-GMP bearing modified phosphate and ribose moieties are currently being tested with the aim of defining a more precise structure–function relationship for the interaction of c-di-GMP with the cellulose synthase and with its degrading phosphodiesterase.

IV.   INTRACELLULAR c-di-GMP LEVELS ARE MAINTAINED
      BY A SYSTEM OF REGULATORY ENZYMES

Considering the unique structure of c-di-GMP and its critical role in maintaining the cellulose synthase in an active state, it could be expected that a highly specialized enzymatic system participates in both its synthesis and degradation. The importance of such enzymes becomes apparent considering that c-di-GMP activation of the synthase is readily reversible, which indicates that alterations in the rate of cellulose synthesis in vivo may be directly dependent on fluctuations in the intracellular concentration of c-di-GMP. Accordingly, the opposing actions of the regulatory enzymes govern the rates of c-di-GMP formation and degradation, and comprise a system for the rapid both positive and negative regulation of cellulose biogenesis.

A. Positive Control: Diguanylate Cyclase

The enzyme responsible for c-di-GMP formation, termed diguanylate cyclase, was originally discovered as the protein factor in the supernatant fraction required for GTP stimulation of the synthase in washed membranes. The enzyme is readily assayed by the methods initially used to elucidate the structure of its unusual product (10,11). Cyclase activity can be determined directly by following the rate of formation of [^{32}P]c-di-GMP from [α-^{32}P]GTP or biologically by the ability of its product to stimulate the cellulose synthase. The enzyme is highly specific for GTP, is optimal at pH 7.0–8.5, and has an absolute requirement for Mg^{2+} ($K_{1/2} \sim 2.5$ mM in the presence of 0.2 mM GTP). The kinetics of the reaction are complex and at 10 mM Mg^{2+} the $K_{1/2}$ for GTP is ~ 90 μM. Diguanylate cyclase has been extensively purified by affinity chromatography on immobilized GTP columns. The purified enzyme contains equivalent amounts of two different peptide subunits of very similar molecular weights (57 and 60 kD); however, based on binding experiments with polyclonal antibodies raised against each of these, the subunits appear to be at least immunologically distinct. Taking a molecular weight of approximately 200 kD estimated for the native enzyme by gel filtration, the simplest model for the quaternary structure of diguanylate cyclase is that of a tetrameric protein composed of two heterogeneous sets of peptide subunits.

The availability of a highly purified preparation of diguanylate cyclase was crucial for examining the c-di-GMP-forming reaction, since crude extracts from A. *xylinum* possess unrelated GTPase activity, exceeding by several orders of magnitude that of the cyclase. The stoichiometry of the c-di-GMP-forming reaction has been studied employing [α-^{32}P]- and [γ-^{32}P]GTP as substrates. As seen in Fig. 4, for each molecule of c-di-GMP produced, two molecules of GTP are consumed and two of inorganic pyrophosphate are released. The reaction proceeds in two distinct steps, first forming the diguanosine tetraphosphate pppG3'p5'G (identified on the basis of radiolabeling experiments and by comparison with chemically synthesized material; Ref. 12), which in the second step is intramolecularly condensed to form c-di-GMP. The intermediary nature in the reaction of the tetraphosphate has been proven by its ability to serve as a direct substrate for the enzyme in forming c-di-GMP (Fig. 5). Each of these two 3',5'-phosphodiester bond-forming steps appears to be catalyzed by the same enzyme since, as observed in the course of purification and antibody inhibition assays, formation of the tetraphosphate is always accompanied by c-di-GMP-forming activity. Within the cell, in contrast to the situation in vitro, this intermediate might be efficiently channeled into c-di-GMP without significantly accumulating as a free, intracellular pool. It is notable that the tetraphosphate intermediate is only a marginal activator of the synthase, an effect most probably due to membrane-associated cyclase. In addition to c-di-GMP-forming activity, highly purified enzyme preparations exhibit inorganic pyrophosphatase activity at levels roughly equivalent to that of diguanylate cyclase; this pyrophosphate cleavage might be an intrinsic activity of the enzyme. Although inorganic pyrophosphate is the second product in the diguanylate cyclase reaction, it does not significantly inhibit the enzyme and preliminary experiments indicate that the cyclase does not readily catalyze the reaction in the reverse direction, i.e., pyrophosphorylysis of c-di-GMP to form tetraphosphate or GTP.

As with its cyclic mononucleotide-forming counterparts, adenylate and guanylate cyclases in both eukaryotic and prokaryotic cells, the diguanylate cyclase of *A. xylinum* occurs in both a cytoplasmic and

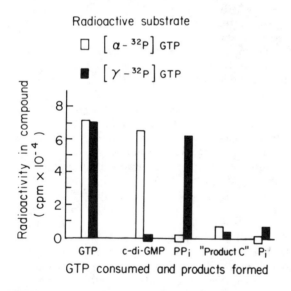

FIGURE 4   Stoichiometry of the diguanylate cyclase reaction. Enzyme incubations were carried out in parallel employing either [α-^{32}P]- or [γ-^{32}P]GTP as substrate and the products analyzed by thin-layer ion exchange chromatography. "Product C" refers to the diguanosine tetraphosphate intermediate in the reaction. (Reproduced by permission from Ref. 12.)

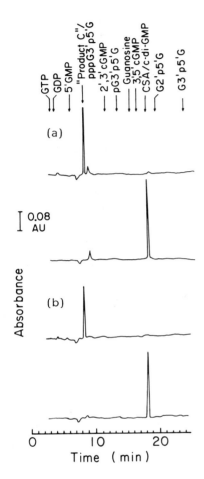

FIGURE 5   Conversion of the tetraphosphate intermediate to c-di-GMP by purified diguanylate cyclase (12). Presented is the HPLC elution profile of synthetic (a) and native (b) pppG3'p5'G prior to (top trace in each part) and following (bottom trace) incubation with purified diguanylate cyclase. (Reproduced by permission from Ref. 12.)

a membrane-associated form. While in the strain employed for most of our studies more than 80% of total diguanylate cyclase activity is present in the high-speed supernatant fraction, the membrane-associated form accounts for more than 50% of total activity in other strains. These variations make it unlikely that the PEG routinely present during cell rupture can explain the enrichment of the enzyme in the membranous fraction. Moreover, membranes prepared in the absence of PEG exhibit comparable levels of cyclase activity to that of PEG membranes. In retrospect, it is now clear that polyethylene glycol is not an essential ingredient of cell rupture medium for obtaining a highly active cellulose synthase preparation provided that the assay is carried out under appropriate conditions, namely, inclusion of $CaCl_2$ and c-di-GMP, which can be added either exogenously or produced in situ by supernatant fraction plus GTP. The membrane-associated state of the cyclase and the regulatory phosphodiesterases raises the possibility that these enzymes are situated in close proximity to the cellulose synthase within the membrane structure of the intact cell. Together these enzymes may compose the discrete proteinaceous complexes implicated by electron microscopy as the presumed cellulose-synthesizing sites within the membrane (2,3). The precise localization studies necessary for clarification of this hypothesis must await the generation of antibodies specific to the cellulose synthase.

B. Negative Control: Cyclic Diguanylate and 5'-Diguanylate Phosphodiesterases (PDE-A and PDE-B)

The unusual structure of the cellulose synthase activator should confer a high degree of stability to the molecule against the action of wide-specificity phosphatases and phosphoryltransferases within the cell. However, in analogy with other cyclic and polyphosphorylated nucleotide regulatory systems, the c-di-GMP activation is not a permanent condition, and a specific enzymatic mechanism for terminating the activation signal exists. This system of negative control was initially revealed when preincubation of synthase-containing membranes with c-di-GMP in the presence of $CaCl_2$ was found to preserve the stimulated synthase activity in comparison to controls lacking Ca^{2+}. Further work showed that a heat-labile component of the membrane preparation was capable of rapidly destroying the activator and, moreover, that $CaCl_2$ prevented this degradation (11). The pathway of c-di-GMP degradation, elucidated as with the diguanylate cyclase reaction by employing ^{32}P-labeled compounds as substrates, was found to be composed of two distinct phosphodiesterases acting in sequence in degrading c-di-GMP to the common metabolite 5'-GMP.

TABLE 3   Characterization of the PDE-A Reaction Product[a]

Treatment	^{32}P-Labeled product and % yield
1. Venom phosphodiesterase	5'-GMP (100%)
2. Bacterial alkaline phosphatase	P_i (50%), G3'p5'G (50%)
3. Periodate/β elimination followed by bacterial alkaline phosphatase	P_i (100%)

[a]The initial product of c-di-GMP degradation was prepared from ^{32}P-labeled c-di-GMP (12) by incubation with washed membranes from A. *xylinum* as described (10,11), and purified by reverse phase HPLC (12). Following the indicated treatments, performed as in (10,11), the products were identified and fractionated for radiochemical quantitation by either thin-layer ion exchange chromatography (11) or reverse phase HPLC (12).

The first of these phosphodiesterases, termed PDE-A, is the most directly involved in regulation, converting the activator to the inactive 5'-phosphorylated linear dimer pG3'p5'G (5'-diguanylate). The structure of this product was actually characterized prior to that of the activator from which it arises, since it proved to be the more amenable to structural analysis by conventional enzymatic and chemical techniques (Table 3). The structure of the product was finally confirmed by comparison with synthetically prepared pG3'p5'G (12). Since the product of the PDE-A reaction is entirely devoid of stimulatory activity, the role of this enzyme is to "turn off" the synthase by inactivating c-di-GMP via the hydrolysis of one of its phosphodiester bonds. The second enzyme in the degradative pathway, termed PDE-B, also hydrolyzes a single 3',5'-phosphodiester bond, converting 5'-diguanylate into two molecules of 5'-GMP. This enzyme evidently serves a catabolic role, regenerating 5'-GMP from one round of c-di-GMP formation and breakdown for de novo GTP synthesis.

The degree of molecular specialization exemplified in this degradative pathway, which incorporates the action of two separate 3',5'-phosphodiester bond-splitting enzymes, is somewhat enigmatic since conceivably a single phosphodiesterase could commit the same sequence of reactions. At least two phosphodiesterases (cf. Table 2), albeit of broad specificity, have been shown to hydrolyze c-di-GMP to 5'-GMP. Possibly by employing a unique enzyme for the initial deactivation step, the cellulose-synthesizing system is provided with an additional locus of regulation. Indeed, PDE-A is significantly and selectively inhibited by Ca^{2+} ions, which is reflected in the potentiating effect of this cation on c-di-GMP activation of the synthase in washed membranes (Fig. 2). At a concentration of 40 μM, Ca^{2+} inhibits the enzyme by 50%; other divalent cations tested (Zn^{2+}, Mn^{2+}, Ni^{2+}, Na^+, K^+) were without effect. The mechanism of Ca^{2+} inhibition has not yet been clarified, but a highly purified form of the enzyme retains its sensitivity to this cation. In contrast, PDE-B is insensitive to comparably low Ca^{2+} concentrations, presumably because this activity is essential for ensuring the rapid breakdown of its inactive substrate and the release of 5'-GMP units.

In addition to their different sensitivities to Ca^{2+}, the two phosphodiesterases appear to differ with respect to the extent of their membrane association; whereas in cell-free extracts, PDE-B activity is distributed approximately evenly between the soluble and membrane fractions, up to 95% of PDE-A activity occurs in washed membranes. Both PDE-A and PDE-B can be released from membranes by extraction in low ionic strength buffers, indicating that neither enzyme is an integral membrane protein but rather attaches to membrane surfaces through ionic interactions. The PDE activities solubilized in this way can be separated by fractionation in polyethylene glycol and by gel filtration. As with the other enzymes of cellulose synthesis (cellulose synthase, diguanylate cyclase), both phosphodiesterases are Mg^{2+}-dependent enzymes ($K_M \sim 0.3$ mM) and exhibit alkaline pH optima (PDE-A, pH 9.8; PDE-B, pH 8.8). The K_m of the phosphodiesterase for their respective substrates falls within the same range as the corresponding K_a value (1 μM) of the cellulose synthase for c-di-GMP. Both enzymes display a high degree of specificity with regard to substrate structure. For example, guanyl nucleotides such as 3',5'-cGMP, cyclic (3',5') triguanylic acid, and G3'p5'G resist degradation when exposed to washed membranes under conditions whereby equivalent amounts of c-di-GMP are completely hydrolyzed to 5'-GMP. Furthermore, PDE-A activity is not significantly affected by the above nucleotides or by 5'-GMP, pG3'p5'G, GDP, ATP, ADP, AMP, or UDP-glucose. The substrate specificity of PDE-B appears to be equally restricted.

V. GENERAL ASSESSMENT

The components and properties of the in vitro cellulose-synthesizing system derived from *A. xylinum* indicate that a highly specialized and complex system exists in the cell for regulation of this process.

The functioning of this regulatory system in relation to the overall
physiology of this organism is still difficult to ascertain, mainly
since *A. xylinum* is an obligative aerobe and appears to maintain a
uniform ratio of cellulose production to oxidation and growth rates
under a variety of conditions (25–27). This phenomenon could be
attributed to the optimized state of the regulatory system, designed
to keep the polymerization process in step with the pace of cellular
metabolism by a mechanism whereby synthase activity is linked to
other GTP-dependent polymerization processes, such as nucleic acid
and protein synthesis. Alternatively, the architectural complexity
of cellulose and the cell-directed nature of its formation (5) imply
that the polymerizing sites must be maintained in rigid coordination;
this could be attained if the elongation and crystallization of glucan
chains arising from these sites were synchronized by an efficient,
localized system for controlling the activity of cellulose synthase
units within the membrane. The c-di-GMP regulatory system might
conceivably function in such a capacity. Another facet of the basic
operation of this regulatory system that needs to be addressed concerns the fact that activity levels of the c-di-GMP-forming and degrading enzymes, as measured in vitro, are roughly equivalent and
fall within the same range as that of the cellulose synthase. On
this basis, it would appear that as much energy is invested in regulating cellulose synthesis as in the overall process of glucose polymerization. If indeed the rapid turnover of c-di-GMP is an integral
feature of the system, this would allow for equally rapid variations
in intracellular concentrations of the activator in response to additional regulatory signals such as Ca^{2+} ions. However, in recognition of the possibility that in vitro rates might not faithfully represent conditions within the intact cell, the estimated turnover rate
of c-di-GMP might actually be much lower.

Calrification of the physiological significance of this novel regulatory system with regard to the overall biochemistry and morphology of cellulose biogenesis will be facilitated through understanding
the genetics of this process. With the current knowledge regarding
the enzyme components of the system, we are now at a stage where
the biochemical characterization of Cel⁻ mutants specifically deficient
in regulatory or glucose-processing enzymes is possible. In this
respect, recent reports on the development of gene transfer techniques for *A. xylinum* and attempts to characterize its complex system of DNA plasmids, some of which may be related to cellulose
synthesis (34), are encouraging. Chemical mutagenesis of *A. xylinum*
cells induces a high frequency of Cel⁻ mutations, the majority of
which apparently do not reside in the structural genes of cellulose
synthesis since production of this polymer can be activated phenotypically in most of the mutants by antibiotics that block RNA or
protein synthesis (28). In this connection, the occurrence in some

strains of *A. xylinum* of an extracellular, water-soluble polysaccharide, containing glucose, mannose, rhamnose, and glucuronic acid (35,36), may warrant further attention. An apparent connection between the synthesis of this polymer and that of cellulose has been proposed based on the observation that synthesis of this heteropolysaccharide is inhibited concurrently with induction of cellulose synthesis by tetracycline (28). While this is unclear, these phenomena may be pointing to the existence of additional regulatory features of cellulose biogenesis.

Considering the complexity and specificity of the enzymes and binding sites in the c-di-GMP regulatory system, it is tempting to speculate that this mechanism did not evolve in isolation in *A. xylinum* but that it may be of much broader distribution in connection with cellulose synthesis in other organisms, with the biogenesis of other extracellular and structural polysaccharides, or even with other physiological activities. There is some evidence that apparent cellulose synthase activity in cell-free preparations from *Agrobacterium tumefaciens* is stimulated by c-di-GMP (20). The GTP-sensitive activity of another β-linked glucan-producing system, the β-1,3-glucan synthase of yeast, was recently shown to require a soluble protein regulatory factor; however, in this case the mechanism of activation is probably not based on c-di-GMP, since this system does not appear to be activated by this nucleotide or to contain an active c-di-GMP-degrading enzyme (E. Cabib, pers. commun.). Attempts in our laboratory, as well as in others, to demonstrate c-di-GMP activation of a plant β-glucan synthase have been unsuccessful (37). It would thus appear that the c-di-GMP regulatory system is not a ubiquitous component to all cellulose-synthesizing systems, although this conclusion may be premature in light of the fact that the in vitro systems currently available from plant sources are far from optimized.

The ability to study the cellulose synthase in a highly active, isolated state has long been considered the Excalibur of the biochemical approach to cellulose biogenesis. The experience with *A. xylinum* demonstrates that the mechanism of synthesis of this polymer can actually be solved when subjected to this approach. It is hoped that the results of these efforts will encourage the continued pursuit of obtaining similar preparations from other biological systems.

VI. ADDENDUM

As part of continuing attempts to comprehend the catalytic and regulatory mechanism of the cellulose synthase from *A. xylinum*, the enzyme has been obtained in a highly purified form and subjected to analysis on the polypeptide level (38).

The cellulose synthase was purified 350-fold from digotonin-solubilized membrane preparations of *A. xylinum*, with an overall yield of 20%, employing the product entrapment technique (18). The high affinity of the enzyme–product complex, indicative of a processive mechanism for cellulose synthesis, is exemplified by the particulate nature of the enzyme in the entrapped state which may be resolubilized by exposure to cellulase.

The entrapped enzyme contains three major peptides — 90, 67, and 57 kD — as revealed by gel electrophoresis in the presence of sodium dodecyl sulfate (SDS-PAGE) and staining in Coomassie blue. As described below, the results of both direct photoaffinity and immunochemical labeling substantiate a role for all three of these peptides as constituents to native cellulose synthase. These results differ significantly from those of other reports (39), in which the cellulose synthase was described as containing a single 83-kD subunit.

Direct photoaffinity labeling of the entrapped synthase has been carried out by UV irradiation in the presence of either $[^{32}P]$c-di-GMP or $[\alpha\text{-}^{32}P]$UDP-glucose followed by SDS-PAGE and autoradiography. The pattern of label incorporation indicates a differential relationship to the structure of the active enzyme with respect to the three peptides. Activator- and substrate-specific binding sites are most closely associated with the 67- and 57-kD subunits, respectively, while no radiolabel incorporation was detected in the 90-kD peptide irrespective of ligand.

The pattern of radiolabel incorporation of each ligand also closely parallels its effect on the level of cellulose synthase activity. In each case the labeling reaction is highly specific in the sense that structural analogs of the ligands are poor competitors for binding sites. The concentrations of either activator or substrate ligand required for half-maximal saturation when binding is measured by photolabeling are roughly equivalent to their respective K_a and K_m values when cellulose synthase activity is assayed directly. Furthermore, the binding reaction and the cellulose synthase reaction share a similar sensitivity to PHMB: preexposure of the purified cellulose synthase to this —SH reagent blocks both the photolabeling and the cellulose-producing reactions while the presence of either c-di-GMP or UDP-glucose during the pre-incubation period effectively prevents the PHMB inactivation.

Immunologically all three of the peptides appear to represent closely related structures. In this analysis a preparation of specific anticellulose synthase antisera, which was generously provided by the Cetus Corporation of Emeryville, California, was used. This rabbit antiserum was produced against a protein translation product derived from the cellulose synthase gene, which was recently isolated and cloned (40). In Western blot analysis of both crude and

highly purified synthase preparations, the antisera strongly label
the 90-, 67-, and 57-kD peptides, but not other peptides, as
present in the total cell extract. Thus while the three major peptides of purified cellulose synthase appear to be functionally distinct,
they are sufficiently related to display immunological cross-reactivity.
One current hypothesis based on these results is that the larger
90-kD band represents an inactive precursor peptide form which
undergoes processing to yield the 67- and 57-kD peptides, both of
which appear to participate directly in the catalytic and regulatory
functioning of the enzyme. It appears, then, that the catalytically
active form of bacterial cellulose synthase (MW = 420 kD) is an
oligomeric protein composed of two different, immunochemically related subunits (67 and 57 kD), derived from a 90-kD precursor.

By the same immunochemical analysis it was possible to demonstrate the presence of related peptides in other cellulose-producing
organisms such as *A. tumefaciens*, while the analysis of other
gram-negative bacteria, including some Cel⁻ strains of *A. xylinum*,
have so far yielded negative results. Thus, *A. tumefaciens* appears to utilize a similar mechanism for cellulose production as described for *A. xylinum*. Indeed, the cellulose-synthesizing enzyme
from this organism is markedly stimulated by c-di-GMP in vitro (20,
41). Furthermore, the results of in vitro assay for c-di-GMP forming and degrading activity and of in vivo occurrence of intracellular
c-di-GMP (41) suggest that the c-di-GMP regulatory system is operative in this organism.

Of great potential importance are recent findings related to the
mechanism of biosynthesis in higher organisms. Employing the same
antisera as above, sets of immunologically related peptides have
been detected in the protein extracts of mung bean, wheat, peas,
and cotton (38). The occurrence of such cellulose synthase-like
structures in plant species suggests that a common enzymatic mechanism for cellulose biogenesis is employed throughout nature.

As part of attempts to gain more insight into the nature of the
c-di-GMP regulatory system, a series of 13 analogous cyclic dimer
and trimer nucleotides were synthesized by the laboratory of
Dr. J. H. van Boom, University of Leiden, and tested for the
ability to mimic c-di-GMP in the regulatory pathway of cellulose synthesis in *A. xylinum* (43). The results of these studies affirm the
role of c-di-GMP as a highly specialized activator of cellulose synthesis as well as the nature of the PDE-A as a c-di-GMP-specific
regulatory enzyme.

REFERENCES

1. M. Schramm and S. Hestrin, *J. Gen. Microbiol.*, *11*, 123 (1954).
2. K. Zaar, *J. Cell. Biol.*, *80*, 773 (1979).
3. R. M. Brown, Jr., in *Structure and Biochemistry of Natural Biological Systems* (W. M. Walk, ed.), Philip Morris, New York, 1979, p. 51.
4. D. P. Delmer, M. Benziman, and E. Padan, *Proc. Natl. Acad. Sci. USA*, *79*, 5282 (1982).
5. M. Benziman, C. H. Haigler, R. M. Brown, Jr., A. R. White, and K. M. Cooper, *Proc. Natl. Acad. Sci. USA*, 77, 6678 (1980).
6. Y. Aloni and M. Benziman, in *Cellulose and Other Natural Polymer Systems: Biogenesis, Structure and Degradation* (R. M. Brown, Jr., ed.), Plenum Press, New York, 1982, p. 341.
7. Y. Aloni, D. P. Delmer, and M. Benziman, *Proc. Natl. Acad. Sci. USA*, *79*, 6448 (1982).
8. M. Benziman, Y. Aloni, and D. P. Delmer, *J. Appl. Polym. Sci.*, *37*, 131 (1983).
9. Y. Aloni, R. Cohen, M. Benziman, and D. P. Delmer, *J. Biol. Chem.*, *258*, 4419 (1983).
10. P. Ross, Y. Aloni, C. Weinhouse, D. Michaeli, P. Weinberger-Ohana, R. Meyer, and M. Benziman, *FEBS Lett.*, *186*, 191 (1985).
11. P. Ross, Y. Aloni, H. Weinhouse, D. Michaeli, P. Weinberger-Ohana, R. Mayer, and M. Benziman, *Carboh. Res.*, *149*, 101 (1986).
12. P. Ross, H. Weinhouse, Y. Aloni, D. Michaeli, P. Weinberger-Ohana, R. Mayer, S. Braun, E. de Vroom, G. A. van der Marel, J. H. van Boom, and M. Benziman, *Nature*, *325*, 279 (1987).
13. J. A. Thomas, K. K. Schlender, and J. Larner, *Anal. Biochem.*, *25*, 486 (1968).
14. J.-M. Ghuysen, J. L. Strominger, and D. J. Tipper, in *Comprehensive Biochemistry*, Vol. 26, Part A (M. Florkin and E. H. Stotz, eds.), Elsevier, Amsterdam, 1968, p. 53.
15. C. J. Waechter and W. J. Lennarz, *Ann. Rev. Biochem.*, *45*, 95 (1976).
16. D. P. Delmer, M. Benziman, A. S. Klein, A. Bacic, B. Mitchell, H. Weinhouse, Y. Aloni, and T. Callaghan, *J. Appl. Polym. Sci., Appl. Polym. Symp.*, *37*, 1 (1983).

17. E. M. Shematek, J. A. Braatz, and E. Cabib, *J. Biol. Chem.*, *255*, 888 (1980).
18. M. S. Kang, N. Elango, E. Mattia, J. Au-Young, P. W. Robbins, and E. Cabib, *J. Biol. Chem.*, *259*, 14966 (1984).
19. F. C. Lin, R. M. Brown, Jr., J. B. Cooper, and D. P. Delmer, *Science*, *230*, 822 (1985).
20. M. P. Thelen and D. P. Delmer, *Plant Physiol.*, *81*, 913 (1986).
21. M. Benziman and B. Rivetz, *J. Bacteriol.*, *111*, 325 (1972).
22. Z. Gromet, M. Schramm, and S. Hestrin, *Biochem. J.*, *67*, 679 (1957).
23. J. Frei-Roitman, M.Sc. thesis, Hebrew University of Jerusalem, Jerusalem, Israel (1974).
24. M. Swissa, Ph.D. thesis, Hebrew University of Jerusalem, Jerusalem, Israel (1978).
25. H. Weinhouse, Ph.D. thesis, Hebrew University of Jerusalem, Jerusalem, Israel (1977).
26. H. Weinhouse and M. Benziman, *Biochem. J.*, *138*, 537 (1974).
27. H. Weinhouse and M. Benziman, *J. Bacteriol.*, *127*, 747 (1976).
28. S. Valla and J. Kjosbakken, *J. Gen. Microbiol.*, *128*, 1401 (1982).
29. S. Valla, D. H. Coucheron, H. Weinhouse, P. Ross, D. Amikam, M. Benziman, and J. Kjosbakken, *Mol. Gen. Genet.*, *217*, 26 (1989).
30. R. L. Perlman and I. Pastan, *J. Biol. Chem.*, *243*, 5420 (1968).
31. P. C. Lee, B. R. Bochner, and B. N. Ames, *J. Biol. Chem.*, *258*, 6827 (1983).
32. M. Cashel and B. Kalbacher, *J. Biol. Chem.*, *245*, 2304 (1970).
33. C. J. Hsu and D. Dennis, *Nucl. Acids Res.*, *10*, 5637 (1982).
34. S. Valla, D. H. Coucheron, and J. Kjosbakken, *Mol. Gen. Genet.*, *208*, 76 (1987).
35. S. Valla and J. Kjosbakken, *Can. J. Microbiol.*, *27*, 599 (1981).
36. R. O. Couso, L. Ielpi, R. C. Garcia, and M. A. Dankert, *Eur. J. Biochem.*, *123*, 617 (1982).
37. D. P. Delmer, *Ann. Rev. Plant Physiol.*, *38*, 259 (1987).
38. R. Mayer, P. Ross, H. Weinhouse, D. Amikam, G. Volman, P. Ohana, and M. Benziman, in *5th Cell Wall Meeting* (S. C. Fry, C. T. Brett, and J. S. G. Reid, eds.), Edinburgh, U.K., August, 1989.
39. F. C. Lin and R. M. Brown, Jr., in *Cellulose and Wood-Chemistry and Technology* (C. Schuerch, ed.), John Wiley, New York, 1989, p. 473.

40. Pers. commun., Cetus Corporation, Emeryville, CA.
41. D. Amikam and M. Benziman, *J. Bacteriol.*, *171*, 6649 (1989).
42. P. Ross, R. Mayer, H. Weinhouse, D. Amikam, Y. Huggirat,
     E. de Vroom, A. Fidder, P. de Paus, L. A. J. M. Sliedregt,
     G. A. van der Marel, J. H. van Boom, and M. Benziman,
     J. Biol. Chem, in press.

12
Cloning of Genes Involved in Cellulose Biosynthesis in *Acetobacter xylinum*

SVEIN VALLA   *University of Trondheim, Trondheim, Norway*

I. INTRODUCTION

Cellulose is a fundamentally important macromolecule in nature (see Chap. 1) that is used by man in many industrial and other applications. Therefore its structure, biosynthesis, and degradation has been studied extensively for many years. Although the studies have resulted in an accumulation of relatively detailed knowledge about some of these aspects, much of the present knowledge on the biochemistry of cellulose synthesis has been achieved in recent years. In addition, the genes controlling these biochemical reactions are still almost completely unknown. In our laboratory we have focused our research on this latter problem.

   As a model organism in these studies, we have used the bacterium *Acetobacter xylinum*. The reason for using a prokaryotic organism as a model was that it was considered probable that it would be technically easier to identify the genes in such a system. Although there exist several other cellulose-producing bacteria in nature, *A. xylinum* has been the most extensively studied organism. This bacterium also has the advantage that it produces large quantities of the polysaccharide, making much of the practical work simpler.

   At the time this project was initiated, none of the enzymes controlling cellulose biosynthesis had been purified. This fact restricted the number of possible approaches to the genetic analysis. We concluded that there were two basic problems that would have to be solved before any gene controlling cellulose biosynthesis could be isolated and characterized. The first of these problems involved the isolation and characterization of mutants that had lost the ability to synthesize cellulose. Second, a gene transfer system would have to be developed in the organism. This chapter briefly reviews the results of these experiments and also describes how it has now been possible to use the results to develop a strategy for isolation of the genes controlling cellulose biosynthesis in *A. xylinum*.

II. EXPERIMENTAL

A. Isolation and Characterization of Cellulose–Negative (Cel⁻) Mutants of *A. xylinum*

Mutants deficient in cellulose synthesis were described in the early work of Schramm and Hestrin (1). We repeated and extended these experiments, and the results of this work are covered in two papers (2,3). In agreement with the results of Schramm and Hestrin, spontaneous Cel⁻ mutants could be isolated easily because of their selective accumulation during growth of the cells under shaking conditions. The selective accumulation of Cel⁻ cells was explained by the selective aggregation of Cel⁺ cells within the accumulating cellulose; a reduced growth rate of cells in the aggregates seemed probable due to restriction of oxygen and nutrients, but this hypothesis was not investigated experimentally. We also found that Cel⁻ mutants could be induced with very high frequencies after treatment of the cells with certain chemical mutagens.

The further characterization of the mutants indicated that there were certain discrepancies between our work and the results reported by Schramm and Hestrin. We found that all our Cel⁻ mutants produced large quantities of a water-soluble extracellular polysaccharide. Schramm and Hestrin did not report a similar observation. We also found that all our spontaneous mutants reverted to the Cel⁺ phenotype during growth under static conditions, whereas Schramm and Hestrin easily found spontaneous nonreverting Cel⁻ mutants. We later found that these discrepancies are almost certainly attributable to the fact that our two groups did not use the same *A. xylinum* strain in the experiments. Since this observation also had several other consequences for our strategy in general, the problem of strain variations will be discussed in more detail below.

The characterization of the Cel⁻ mutants we isolated also resulted in the discovery of genes exerting pleiotropic effects on

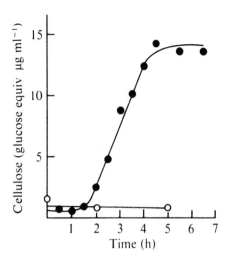

FIGURE 1   Induction of cellulose synthesis in a Cel⁻ mutant of
A. xylinum ATCC 10245 by tetracycline. The mutant used in this
experiment was a spontaneous Cel⁻ mutant called Cel1. Tetracycline
was added to an exponentially growing cell culture at time zero
(filled circles). The open circles represent an aliquot of the culture where no tetracycline was added. For other technical details,
see Ref. 2.

cellulose biosynthesis. This was evident from the observation that
cellulose synthesis could be activated phenotypically in most of the
mutants. As can be seen from Fig. 1, cellulose synthesis can be
induced in the mutants by adding tetracycline to exponentially growing cells. Such induced cells will, however, fall back to the mutant
phenotype when the tetracycline is removed. We later also found
that this activation of cellulose synthesis is quantitatively different
in different mutants, indicating that there is not one unique genetic
event underlying the phenotype of the mutants. It is not yet clear
what the nature of these mutations is, but the experiments clearly
indicate that all the structural proteins necessary for cellulose synthesis are present in the inducible mutants. The mutations thus
somehow interfere negatively with the activity of one or more of
these enzymes. At present it seems most likely that the mutations
are affecting either the level of precursors in the cells (like UDPG)
or the level of the GTP-derived positive regulator of cellulose

synthesis (4; see Chap. 8). It seems in principle possible to envision that many kinds of mutations might affect the level of these
compounds, since they are interacting with many other pathways in
the cell. If the hypothesis described above is correct, one would
predict that the inducible mutants should have a normal cellulose-
synthesizing capacity in vitro. It should also be possible to test
this hypothesis by using the very efficient system for cell-free
cellulose synthesis developed by Aloni et al. (5). To my knowledge
such experiments have not yet been performed.

No matter what explains the properties of the inducible Cel⁻
mutants, it seemed necessary to find a way to discriminate between
this class of mutants and mutants that contain deficiencies in the
structural enzymes controlling cellulose synthesis. This might have
been done by analyzing the properties of the mutants in vitro, as
described above, or simply by using those mutants that were not
inducible for the genetic analysis. However, analysis of the genome
of strain ATCC 10245 wild-type and Cel⁻ mutants indicated that the
genetic events that had taken place in many of the Cel⁻ nonin-
ducible mutants were very complicated. These observations, which
will be described in more detail below, resulted in a search for
another strain of *A. xylinum* that would be easier to use for the
genetic analysis of cellulose biosynthesis.

B. Analysis of the Physical Structure of
the *A. xylinum* Genome and Development
of a Gene Transfer System

Physical analysis of the genome of strain ATCC 10245 showed that
this organism contained a complex system of plasmid DNA molecules
(Figs. 2 and 3). The experiments also showed that the plasmid
profile was drastically altered in many of the nonreverting (and non-
inducible Cel⁻ mutants. These alterations could not be explained as
simply the result of plasmid loss. Instead, new plasmids seemed to
be generated with copy numbers that were sometimes much higher
than those present in the original wild-type strain (6). A more
detailed characterization of some of these new plasmids showed that
the different mutants could be divided in three classes based on
their plasmid profiles (7). When two of these classes were studied
in more detail, it was discovered that all mutants in one class contained a particular 44-kb plasmid (class H), while all the mutants in
the other class contained a unique 49-kb plasmid (class I). Restriction endonuclease analysis of these two plasmids showed that they
contained mostly the same sequences (Fig. 4a, lanes 1 and 2), but
the 44-kb plasmid was apparently not simply a deletion derivative of
the 49-kb plasmid. This latter hypothesis has been confirmed experimentally (D. H. Coucheron, unpublished results).

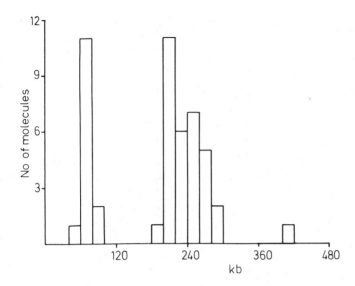

FIGURE 2   Size distribution of plasmids in *A. xylinum* ATCC **10245.**
The plasmid DNA fraction (from the wild type) was purified by CsCl/
ethidium bromide density gradient centrifugation and then inspected
in an electron microscope. Measurement of the contour lengths of
the individual molecules was performed on a digitizing table. For
other details of the procedure, see Ref. 12.

A further understanding of the origin of the *A. xylinum* plasmids became possible when it was discovered that the cells were
diploid for at least some of the sequences in the plasmids. This
conclusion was deduced from the observation that at least two of
the *Bam*H1 fragments generated by digestion of the 49-kb plasmid
were only partly present in the 44-kb plasmid, whereas whole fragments could be generated by digesting total DNA from the class H
strains (Fig. 4b, lanes 1 and 2). Since parts of these fragments
were not present in the 44-kb plasmid, such a part could be used
as a DNA probe for identifying the location of the "missing" sequences in the genome of the class H strains. This analysis clearly
demonstrated that the sequences were located in the chromosomal
DNA fraction of these strains. Together with several other lines of
evidence (7), these experiments indicated that there exists a complex interaction between the plasmids and chromosomal DNA. One
possible interpretation of the results is that the plasmids are actually generated by some rearrangement process in the chromosome.

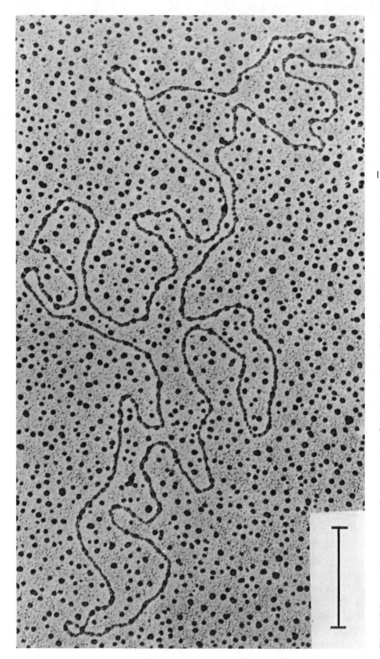

FIGURE 3   Electron micrograph of 44-kb plasmid from a nonreverting Cel⁻ mutant of *A. xylinum* ATCC 10245.   The plasmid fraction was isolated from the nonreverting Cel⁻ mutant Cel 7 (7). Bar = 0. 42 μm.

Even if this should turn out not to be true, the experiments clearly
showed that it would presumably not be possible to understand the
genetic events that had taken place in the Cel⁻ mutants by studying
the plasmids alone. Together with the observed complexity of the
phenotypes of the mutants (see above), these results therefore indicated that a further genetic analysis of this system would probably
be very difficult.

The genetic analysis of *A. xylinum* ATCC 10245 also included
the development of a gene transfer system in this organism. Access
to such a system would be a necessity for any strategy that involved the use of modern recombinant DNA technology in *A. xylinum*.
These studies showed that plasmids of incompatibility group P1
could be transferred conjugatively from *E. coli* to *A. xylinum* with
satisfactory efficiency. All selective markers on the plasmids were
also well expressed in *A. xylinum* (6). The gene transfer system
could be shown to be useful for inserting the transposon Tn1 into
the naturally occurring *A. xylinum* plasmids. The Tn1-labeled
plasmids could also be mobilized from one *A. xylinum* strain to
another, but it was not possible to demonstrate transfer of the
ability to synthesize cellulose in these experiments. In spite of
this the development of a gene transfer system was considered very
important since it opened up the possibility of using recombinant
DNA technology in the further genetic studies of *A. xylinum*.

C.   Strain Variations Within *A. xylinum*

Due to the complexity of the genetics of *A. xylinum* ATCC 10245, it
was decided to study other strains of this species in the hope of
finding a strain whereby genetic analysis of cellulose synthesis could
be performed more easily. The discrepancies between our results
and those of Schramm and Hestrin (described in Sec. II.A) indicated that there existed at least some differences between the
strains, but it was not clear whether these differences were of any
importance in relation to the analysis of cellulose biosynthesis.

We used five different strains of *A. xylinum* (in addition to
ATCC 10245) for studying strain variations. All strains were obtained from ATCC and were numbered 10821, 14851, 23768, 23769,
and 23770, respectively. It was immediately obvious that there existed at least some differences between the strains because the
colony morphologies on agar medium varied significantly. Strain
10821, for instance, formed soft colonies on agar medium, whereas
the other strains formed the rough colonies normally associated
with cellulose production. The colonies obtained from the cells of
strain 23770 turned out to form two easily distinguishable colony

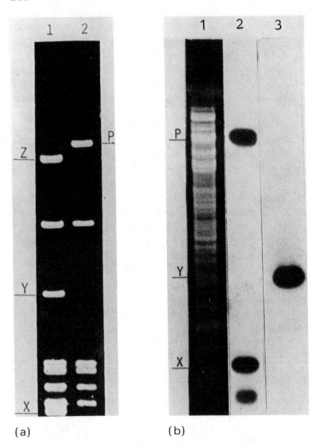

(a)   (b)

FIGURE 4   (a) Restriction endonuclease analysis of the 44- and
49-kb plasmids isolated from Cel⁻ mutants of *A. xylinum* ATCC
10245.   Lane 1: *Bam*H1-digested 49-kb plasmid from a spontaneous
Cel⁺ revertant (Cel1R6) of the spontaneous Cel⁻ mutant Cel1.   This
strain belongs to a class of mutants which lacks the 44-kb plasmid.
Lane 2: *Bam*H1-digested 44-kb plasmid from the nonreverting Cel⁻
mutant Cel7.   This strain belongs to a class of mutants which lacks
the 49-kb plasmid.   Fragments X, Y, Z, and P are those fragments
which distinguish the two plasmids from each other, and the sizes
of these fragments are 2.8, 5.4, 16, and 19 kb, respectively.   DNA/
DNA hybridization analysis had shown that there is strong sequence
homology between X and P and between Z and P, whereas there is
a weak homology between Y and P.   (b) Identification of sequences
from plasmid DNA that are also present in a chromosomal location in
the *A. xylinum* genome.   Lane 1: *Bam*H1-digested total DNA from

morphology variants (called small and large, respectively). These
two variants were present in the original sample from ATCC.
When the strains were grown in liquid media, no major difference between them could be observed as long as the cells were
grown under static growth conditions, although the growth rate
seemed to vary to some extent. All strains gave rise to the characteristic cellulose pellicle, which floats on top of the medium.
Under shaking growth conditions, strain 10821 did not form the
cellulose aggregates characteristic of the other strains. It was also
interesting to observe that the large-colony morphology variant of
strain 23770 (see above) produced large quantities of a soluble
extracellular polysaccharide in addition to the cellulose, while this
was not the case for cultures of the small-colony morphology variant
(D. H. Coucheron, unpublished results). Genetic studies later
demonstrated that the two variants probably have the same origin
such that one is a mutant of the other (D. H. Coucheron, unpublished results). We also found that Cel⁻ mutants of the 10245 strain
obtained directly from ATCC did not produce significant quantities
of the extracellular polysaccharide described previously for this
strain (see above). Our original 10245 strain was obtained from
J. Colvin, and genetic analysis indicated that it and ATCC 10245
derived from the same origin with subsequent mutations occurring
(similar to the development of the two-colony morphology variants
of strain 23770; D. H. Coucheron, unpublished results).

D. Identification of a New Strain Useful for Genetic
   Analysis and Cloning of DNA Fragments Involved
   in Cellulose Biosynthesis in *A. xylinum*

The experiments described above clearly demonstrated that there
exist many strain differences within the bacteria belonging to the
species *A. xylinum*. Due to the difficulties experienced in the
genetic studies of strain 10245, it seemed potentially possible that
some of the other strains might be simpler to use for genetic experiments. One advantage with all the other strains (except 23770) was

---

Cel7. Lane 2: Hybridization of ^{32}P-labeled fragment X against
*Bam*H1-digested total Cel7 DNA. The origin of the hybridization
signal not corresponding to X or P is not known. Lane 3: Hybridization of ^{32}P-labeled fragment Y against *Bam*H1-digested total
Cel7 DNA. The weak homology against fragment P (see above)
needs longer time of exposure for visualization. For other technical
details, see Ref. 7.

that they did not produce significant quantities of water-soluble
extracellular polysaccharides. It seemed possible that the synthesis
of such a polymer might have indirect effects on cellulose synthesis
and could be involved in the determination of the phenotypes of the
Cel⁻ mutants of strain 10245. These considerations thus excluded
the use of strain 23770 as an alternative to 10245. Strain 10821
was also excluded because of its atypical colony morphologies, which
made it difficult to distinguish between Cel⁺ and Cel⁻ cells on agar
medium (see above).

Among the remaining three strains, we also tested another property that we suspected to be related to the pleiotropic effects of
genes on cellulose synthesis, namely, their comparative frequencies
of mutation. In strain 10245 it was observed that unusually high
frequencies of Cel⁻ mutants could be induced after treatment with
certain mutagens (3), possibly because a very large number of
genes were affecting cellulose synthesis indirectly in this strain.
Experiments with mutagenesis of strains 14851, 23768, and 23769 indicated that the latter strain behaved similarly to 10245 in mutagenesis experiments. At comparable frequencies of cell killing by
the mutagen (N-methyl-N'-nitro-N-nitrosoguanidine), about 30-fold
lower frequency of mutation to the Cel⁻ phenotype was observed in
the other two strains. Of these, strain 14851 was more inconvenient
to work with because it formed very irregular and slow-growing
colonies on agar medium. We therefore decided to study strain
23768 in more detail.

One critical test concerning the possible advantage of using
strain 23768 for the genetic analysis of cellulose synthesis was to
study the phenotype of Cel⁻ mutants isolated from this strain. A
large number of such mutants were isolated after mutagenesis of the
wild type with N-methyl-N'-nitro-N-nitrosoguanidine. Further
analysis of seven of these mutants showed that none could be induced phenotypically to produce cellulose in the way observed for
strain 10245. This result thus gave further support to the hypothesis that strain 23768 might be a good candidate for genetic studies
of cellulose synthesis in *A. xylinum*.

Based on the above results, we also tested whether cloning
vectors derived from plasmids of incompatibility group P1 could be
transferred to strain 23768 in the same way described previously
for strain 10245 (see Sec. II.B). These experiments showed that
the two strains behaved similarly in conjugative plasmid transfers,
indicating that it might be possible to use recombinant DNA technology in further studies. We therefore constructed a gene library
of DNA from strain 23768 wild type in plasmid pVK100 (8). The
library was established in *E. coli* , and by mating the library with
14 cel⁻ mutants isolated from strain 23768, we were able to show
that three of the mutants could be complemented (visual inspection

of transconjugant colonies). Further studies showed that one particular *E. coli* clone was responsible for complementation of all three
mutants and we were able to isolate this clone from the library.
The clone contained a pVK100 plasmid [pVK100 (246)] with a 2.8-kb
DNA insert. Biochemical analysis showed that all the three mutants
were deficient in UDPG-pyrophosphorylase, while this enzyme activity was restored in the corresponding pVK100 (246) transconju-
gants. The 2.8-kb DNA fragment was also subcloned into a small
high-copy number *E. coli* vector (pUC19) and then transformed into
an *E. coli* strain known to carry a mutation in the structural gene
(*galU*) for UDPG-pyrophosphorylase. Comparison between the
UDPG-pyrophosphorylase activities in the pUC19 (246) transformant
and the isogenic strain containing the same vector without insert,
showed that the presence of the insert in pUC19 resulted in about
a 330-fold increase in the activity of UDPG-pyrophosphorylase (9).
These data thus clearly demonstrated that the cloned DNA fragment
contains the *A. xylinum* structural gene for UDPG-pyrophosphorylase.

The gene library described above was constructed in such a way
that it could not be expected to represent the entire *A. xylinum*
genome. We therefore extended the experiments by constructing a
new and representative library of *A. xylinum* ATCC 23768 in the
broad host range vector pRK311 (10). This library was then used
for complementation of eight Cel⁻ mutants that were not complemented
by the previous library. Restriction endonuclease analysis of the
complementing clones, which were isolated independently from the
gene library, showed that they all originated from the same region
in the *A. xylinum* genome (11). The work was recently extended
both by isolating new mutant classes and by identifying new clones
that can complement these mutants. The results of this work,
which will soon be published (13,14), has convinced us that the
strategies outlined in this paper can be used for a detailed analysis
of the mechanism of cellulose synthesis in *A. xylinum*.

III. CONCLUSIONS

The purpose of this chapter was to present an experimental strategy
for cloning the genes involved in cellulose synthesis in *A. xylinum*.
Characterization of Cel⁻ mutants of this organism indicated that at
least in certain strains there are probably many genes that affect
cellulose synthesis indirectly. Although it would be interesting to
know the function of these genes, we were interested in first
studying the genes coding for the structural enzymes in the cellulose biosynthetic pathway. Such an analysis was difficult to perform in the strain we have used in most of our experiments (strain

ATCC 10245), both due to the problems described above and because the genetics in general seemed to be complex and unusual in
this strain. By analyzing the properties of other natural isolates
of A. xylinum, we found that several of the strains could be distinguished from each other on the basis of simple phenotypic criteria. One of the strains (ATCC 23768) had properties indicating
that the problems associated with the use of strain 10245 in the
studies of the genetics of cellulose biosynthesis could be avoided.

Strain 23768 could be used for cloning of DNA fragments that
complemented all tested Cel⁻ mutants. One of the cloned fragments
contains a gene that can complement a deficiency in the E. coli
galU gene, which is the structural gene for UDPG-pyrophosphorylase.
It could thus be concluded that the cloned gene represents the corresponding structural gene from A. xylinum. The recent extension
of these experiments have resulted in isolation of both new mutant
classes and new complementing genes, and it therefore seems clear
that a rapid progress in the understanding of both the biochemistry
and the genetics of cellulose synthesis can be expected.

ACKNOWLEDGMENTS

The work performed in our laboratory was supported by the Nor-
wegian Research Council for Science and the Humanities, the Royal
Norwegian Council for Scientific and Industrial Research, and Anders
Jahres fond til vitenskapens fremme.

REFERENCES

1. M. Schramm and S. Hestrin, J. Gen. Microbiol., 11, 123 (1954).
2. S. Valla and J. Kjosbakken, Can. J. Microbiol., 27, 599 (1981).
3. S. Valla and J. Kjosbakken, J. Gen. Microbiol., 128, 1401
   (1982).
4. P. Ross, Y. Aloni, H. Weinhouse, D. Michaeli, P. Weinberger-
   Ohana, R. Mayer, and M. Benziman, Carboh. Res., 149, 101
   (1986).
5. Y. Aloni, D. P. Delmer, and M. Benziman, Proc. Natl. Acad.
   Sci. USA, 79, 6448 (1982).
6. S. Valla, D. H. Coucheron, and J. Kjosbakken, J. Bacteriol.,
   165, 336 (1986).
7. S. Valla, D. H. Coucheron, and J. Kjosbakken, Mol. Gen.
   Genet., 208, 76 (1987).
8. V. C. Knauf and E. W. Nester, Plasmid, 8, 45 (1982).

9. S. Valla, D. H. Coucheron, J. Kjosbakken, H. Weinhouse, P. Ross, D. Amikam, and M. Benziman, *Mol. Gen. Genet.*, *217*, 26 (1989).
10. G. Ditta, T. Schmidhauser, E. Yakobson, P. Lu, X. Liang, D. R. Finlay, D. Guiney, and D. R. Helinski, *Plasmid*, *13*, 149 (1985).
11. S. Valla, in *Cellulose and Wood-Chemistry and Technology* (C. Schuerch, ed.), John Wiley and Sons, New York, 1989, pp. 559–571.
12. S. Valla, D. H. Coucheron, and J. Kjosbakken, *Arch. Microbiol.*, *134*, 9 (1983).
13. E. Fjaervik, K. Frydenlund, S. Valla, Y. Huggirat, and M. Benziman, *FEMS Microb. Lett.* Submitted (1990).
14. E. Fjaervik, J. Blatny, B. Bleivik, K. Frydenlund, and S. Valla, (1990). Manuscript in preparation.

II
Biodegradation

PAUL J. WEIMER

As Part I of this volume demonstrated, there have been considerable advances over the past two decades in our understanding of how cellulose is synthesized in both plants and bacteria. Studies of cellulose biosynthesis began at a relatively gross level (e.g., micro-anatomy) and progressed to the point where the approach has become highly molecular. Cellulose biodegradation, despite a somewhat longer history as a scientific subdiscipline, has traditionally focused on such areas as microbial physiology and chemical engineering, and only recently moved toward molecular approaches to probe its more essential mysteries.

It is instructive to place our understanding of cellulose biodegradation into a historical context. Unlike studies on cellulose biosynthesis, which tend to be unabashedly basic in nature, the focus of cellulose biodegradation was until recently dominated by its potential applications to problems in the food, fuel, chemical, and textile industries. What researcher in this area would not like to have a dollar for every journal article or grant proposal which began by extolling the abundance of cellulose and the nearly limitless potential for putting this king of structural biopolymers to work? As a result, the field has attracted a wide variety of researchers, from microbiologists to engineers, food scientists to chemists, geneticists, and enzymologists.

Although research on cellulose biodegradation began a century ago, it accelerated considerably with the work of the U.S. Army group at Natick, Massachusetts, in the early 1950s. The original

problem — fungal degradation of military tents and clothing — has
long been a dead issue in the age of plastics and other man-made
polymers. But the Natick group laid the groundwork for future
studies by developing a number of analytical methods and by providing the first working model of the process of cellulose biodegradation.

The "energy crises" of the 1970s focused further attention on
cellulose as a potential replacement for fossil hydrocarbons as a
source of fuels and chemicals. In retrospect, some of this work
was driven more by panic than by science, and only peripherally
recognized the costs associated with collecting and processing cellulosic substances prior to their saccharification and/or fermentation.
Nevertheless, this period probably represented a sort of golden age
of cellulose biodegradation research, at least in terms of research
funding. Studies during this period led to the discovery of new
cellulolytic microbial species, and the introduction of such seminal
concepts as the cellulosome organelle and the synergism among different enzymes of the cellulase complex.

Part II of this volume includes a compendium of information on
cellulose degradation in a variety of environments and by a variety
of agents. Much of this information has heretofore been scattered
among the biological, chemical, and engineering literature. In general, this section is organized to move from the grosser levels
toward the finer. It begins with a consideration of analytical problems related to measuring the numbers and activities of the cellulolytic agents (Chap. 13) and the substrate-related factors that affect
degradability (Chap. 14). It then moves to a description of two
animal systems in which cellulose degradation plays a major nutritional role (Chaps. 15 and 16). These are followed by descriptions
of the individual physiological groups of cellulose-degrading bacteria
themselves (Chaps. 17–20). No chapters on the degradation of
cellulose by whole fungi have been included, owing to the fact that
virtually all of the work on fungal cellulose degradation has been
carried out with fungal enzymes or enzyme complexes (Chap. 21).
Likewise, degradation of cellulose by protozoans is treated only peripherally (e.g., in Chap. 16), since these intractable organisms seem
to have succeeded in driving even the most peristent researchers
into other, more approachable areas. The enzymology (Chaps. 22
and 23) of cellulose degradation is considered next, and draws upon
earlier descriptions of methodologies, substrates, and microbial
agents. Owing to a greater understanding of fungal than bacterial
cellulases, the former are considered first in order to provide a
frame of reference for the latter. Finally, the rapid advances in
molecular genetics of cellulose degradation are discussed (Chap. 24),
with particular emphasis on the cloning of cellulolytic genes into
more tractable noncellulolytic microorganisms.

What does the future of the field hold? Undoubtedly, the largest item on the agenda is the understanding of the enzymology of cellulolysis at the molecular level, i.e., the specific chemistry of the binding of celluloytic enzymes and of β-1,4-glucan bond cleavage. The structure of individual enzymes is now being elucidated, principally by sequence analysis of the corresponding genes following their molecular cloning into well-established genetic systems. However, it will be a long stretch to translate this sequence information to an understanding of the complex actions of the many proteins involved in cellulolysis, whether these proteins are physically held together in discrete organelles (such as the cellulosome) or are freely excreted into the environment. Thus multidisciplinary approaches which combine genetics, biochemistry, and microbiology will probably be most fruitful in advancing our understanding of the cellulolytic process. Part of this understanding will no doubt come from molecular modeling studies in which the fit of enzyme to substrate can be visualized and honed by the use of computers. Such approaches have been enormously useful in predicting the structure of cellulose itself and have now been extended to a modeling of the enzyme – substrate complex. This work will also undoubtedly benefit from multidisciplinary collaborations, which should bridge the gulf between biologists (for many of whom "cellulose is cellulose is cellulose") and chemists (for many of whom enzymes are merely a particularly unusual though interesting type of catalyst).

Despite a trend toward the molecular, studies on cellulose utilization at the level of the whole organisms are by no means obsolete or unnecessary. The very recent discovery that the hoatzin (an obligate leaf-eating bird from South America) possesses an active cellulolytic foregut fermentation has extended the role of cellulolytic microorganisms in animal nutrition, and may provide clues to the evolution of pregastric digestive processes, as well an elegant system for comparative microbial ecology.

On the applied side, the major goals for research in the industrial conversion of cellulosic materials will continue to be enzyme overproduction coupled with the development of more efficient engineering processes. As we are currently between what will undoubtedly be periodic shortages of fossil fuels, applied research in cellulose utilization is presently underfunded. In a sense, this affords us the opportunity to reexamine the works of the 1970s and 1980s with a more critical eye and to focus our attentions more carefully. It would seem that the items most in need of further work are overproduction of active cellulase complexes, as well as such ancillary engineering processes as improved pretreatments of cellulosic biomass, or recovery of fermentation products from dilute aqueous solution. The cloning and expression of genes from cellulolytic organisms into noncellulolytic organisms will undoubtedly continue to receive a good deal of effort (and well they should). Even if these engineered organisms turn out to be useless as industrial mono-cultures or as effective cellulolytic agents in complex and highly competitive environments such as the rumen, they will teach us much about the nature of cellulose biodegradation, an intellectual goal once thought to be virtually unattainable.

13

Quantitative and Semiquantitative Measurements of Cellulose Biodegradation

PAUL J. WEIMER[*] E. I. du Pont de Nemours & Company, Incorporated, Wilmington, Delaware

I. INTRODUCTION

The complex structure of cellulose and cellulosic materials introduces considerable problems to the researcher attempting to measure its biodegradation. The rate and extent of cellulose degradation by microorganisms and their enzymes is dependent in part on physical and chemical parameters such as temperature, pH, and composition of the reaction mixture. Moreover, because cellulose is a solid substrate of complex structure, a number of substrate-related factors are major determinants of degradability (Table 1). A general discussion of the effect of cellulose structure on its degradability may be found in Chap. 14 of this volume or in several excellent reviews (23–26).

The field of cellulose biodegradation has attracted the interest of researchers from many different scientific disciplines and has involved the study of a wide variety of different organisms. It is thus little wonder that a large number of methods for measuring cellulose biodegradation have been developed, each seemingly tailored to the unique needs of the individual experimenter. It is the purpose of this chapter to present an overview of these disparate methods, along with a discussion of their particular advantages and drawbacks.

Two general types of substrates are used to measure cellulose biodegradation (Table 2). The first group includes relatively unaltered natural substrates such as pure crystalline cellulose or biomass;

---

*Current affiliation: US Dairy Forage Research Center, USDA/ARS, Madison, Wisconsin.

TABLE 1   Substrate-Related Factors Which Affect Rate
and Extent of Cellulose Biodegradation

Factor	Ref.
Degree of polymerization	1 – 3
Degree of substitution	4 – 6
Crystallinity	2, 3, 7 – 14
Available surface area	3, 7, 8, 9, 12, 15
Pore volume and its distribution	11, 16 – 18
Association with hemicellulose and lignin	9, 18 – 22

TABLE 2   Cellulosic Substrates Classified According to Their
Structure and Biodegradability

Substrate	Reference for method of recovery or preparation
Biomass	
Untreated wood	27 – 29
Pretreated wood	30 – 35
Agricultural residues	36 – 39
Pure cellulose	
Valonia cellulose	40, 41
Cotton linters	42
Filter paper	43
Bacterial cellulose	44, 45
Microcrystalline cellulose	14, 46, 47
Amorphous celluloses	48 – 50
Reprecipitated cellulose	51
Cellodextrins	52 – 55
Modified celluloses	
Covalently dyed celluloses	56 – 60

TABLE 2   (Continued)

Substrate	Reference for method of recovery or preparation
Nonionic-substituted celluloses (e.g., hydroxyethylcellulose)	4
Ionic-substituted celluloses (e.g., carboxymethylcellulose)	4, 61
Covalently dyed cellulose derivatives	62, 63

the second includes modified cellulosic substrates whose degradation occurs more rapidly (e.g., substituted celluloses) or is more easily observed (e.g., dyed celluloses). Within each class there is a continuum of degradability which reflects the structural similarity or dissimilarity of each substrate to native cellulose.

II. ENUMERATION OF CELLULOLYTIC POPULATIONS

A. Liquid Culture Methods

Liquid culture methods involve sequential (usually 10-fold) dilution of source material in specific media which enrich for cellulose-degrading microorganisms. Such media usually contain powdered cellulosic substrates but alternatively contain filter paper strips (64). After an appropriate incubation period, population estimates are determined following examination of cultures to determine the highest dilution which exhibits net cellulose removal. This is essentially a "most probable number" (MPN) method whose statistical accuracy is enhanced by using multiple replicates at each dilution, with three regarded as a minimum and ten being preferred. Statistical tables for the interpretation of MPN data are readily available (e.g., see 65). However, such MPN techniques usually underestimate the true population, since a fraction of the population will not grow in the liquid culture medium.

Cells in liquid culture may also be enumerated by direct microscopic counting using a phase contrast microscope and Petroff – Hausser counting chamber; however, this method is only useful for pure cultures or for defined mixed cultures whose cellulolytic subpopulation has a morphology distinct from that of the associated noncellulolytic microorganisms. Furthermore, this method usually overestimates the viable cell population, as it does not readily distinguish viable cells from nonviable ones.

Relative to the solid media techniques described below, liquid culture methods have numerous advantages, including (1) minimal stress to the cells (cf. the thermal death associated with agar plating; see Ref. 66); and (2) reduced labor and materials costs, particularly for anaerobic cultures.

B. Solid Medium Cultivation Methods

Solid culture media are often employed in situations where it is necessary to differentiate the cellulolytic and noncellulolytic components of a cellulose-degrading population (e.g., for enumeration of the numbers of cellulolytic organisms in a field sample or laboratory culture containing associated noncellulolytic organisms). Because these techniques yield discrete colonies presumably derived from single microbial cells, these methods are also used to isolate pure cultures from their associated contaminants.

Such media usually incorporate agar as solidifying agent at concentrations ranging from 1 to 4%, depending on the subsequent incubation temperature of the solid media. Other solidifying agents, such as Gelrite, a modified bacterial polysaccharide sold by the Kelco Division of Merck, San Diego, (67–69), can also be used for specific applications, e.g., those involving incubation at higher temperatures.

The medium containing the solidifying agent is generally dispensed in molten form into Petri dishes, but may be laid out in other configurations, e.g., within flat-sided bottles (70), or as a thin layer attached to the inside wall of anaerobic roller tubes (71,72). Cultures or natural source material, appropriately diluted, are mixed directly into the molten medium just above its solidification temperature, followed by immediate cooling to reduce losses of cells to thermal damage and to prevent settling of the cellulosic substrate. The choice of cellulosic substrate is dictated more by its physical characteristics (e.g., settling rate from solution) than by its biodegradability, since most truly cellulolytic organisms from natural environments are capable of degrading a wide variety of cellulosic substrates, albeit at different rates. Viable counts of cellulolytic microbes are generally determined by counting the number of colonies which produce zones of clearing on cellulose agar, and correcting for dilution (73,74).

In the case of pure cultures, nonspecific heterotrophic media lacking cellulose may be used to ensure more rapid colony formation and higher plating efficiency (75,76). Such media may also be used for enumerating subpopulations in defined mixed cultures whose members have different colony morphologies (77).

Specialized procedures have also been developed for the use of agar media for cultivating strictly anaerobic bacteria (e.g., see 75,

76). Most of these procedures are derived from the early pioneering work of Hungate (71,72) and Skinner (70).

III. MEASUREMENT OF CELLULOSE DEGRADATION BY MICROBIAL CULTURES

A. Weight Loss Methods

Gravimetric quantitation of cellulose degradation by microbial cultures is complicated by the presence of microbial cells, which interfere with determination of residual substrate by weight loss methods. This problem can generally be circumvented by addition of a lytic agent (such as lysozyme/sodium dodecyl sulfate or 10% formic acid) to the culture prior to recovery of residual substrate. An additional and somewhat thornier problem is introduced by the nature of cellulose catabolism. There are many different potential intermediates and end products in this multistep process. Included among the intermediates are not only soluble carbohydrates (monomers and oligomers) and intermediary metabolites, but also partially degraded polysaccharides which may be too small to recover by centrifugation or filtration, yet too large to be assayed by conventional analytical techniques. Consequently, determination of cellulose degradation by weight loss methods generally underestimates the amount of residual substrate, and thus overestimates cellulose degradation. Measurements of the fraction of substrate lost in this fashion are not straightforward and usually require measurement of known end products followed by calculation of unrecoverable substrate by difference (see below). Despite these difficulties, gravimetric methods remain extremely popular and have been applied to a wide variety of different experimental systems.

On its simplest level, weight loss methods have been used to measure cellulose degradation by pure cultures and defined mixed cultures (e.g., see 78, 79), but they are more commonly used to measure cellulose degradation in complex, relatively undefined mixed cultures.

Weight loss methods have also been modified to incorporate substrates enclosed with nylon mesh bags (80,81). These bag methods are particularly useful in ecological studies, e.g., in the rumen (81) or in sediments (80), where in situ emplacement and recovery are required. The same caveats apply as above, with an added consideration being the possibility that the microflora entering the bag may not represent the whole microbial population.

The in vitro dry matter digestibility (IVDMD) assay (31,42,82 – 87) is a specialized weight loss technique used to measure the degradation of complex cellulosic substrates by the complex mixture of microorganisms in rumen fluid. The method involves anaerobic incubation of a known amount of cellulosic substrate in a salts buffer with rumen fluid obtained from a ruminant animal equipped with a fistula or rumen cannula. After a fixed period of incubation time (e.g., 48 hr often followed by an incubation with pepsin), residual substrate is recovered either by low-speed centrifugation, decantation, filtration following treatment with a detergent (85) or drying. IVDMD data correlate reasonably well with in vivo degestibility obtained by chemical analysis of input (feed) and output (feces) (88,89), although there is some tendency for the in vitro method to underestimate in vivo digestibility for animals fed low-nitrogen diets (81,83). Some loss of ultrafine residual fiber or dry matter is inevitable during recovery. The IVDMD method is often used not only to measure digestibility after prolonged incubation but also to obtain kinetic data (rate constants and lag times; 87). Comparisons among kinetic experiments are complicated by differences in the microbial activity of the inocula due to variations in animal feeding schedule and in the collection and processing of the inocula.

A more artificial measurement of substrate degradability can be obtained by sequentially incubating the substrates in the presence of cellulase and pepsin (90−93) instead of rumen fluid. The incubation time in the presence of cellulase is usually about 48 hr, while the incubation time with pepsin is adjusted (e.g., 0, 24, or 48 hr) depending on the protein content of the substrate.

These digestibility assays are most widely used to determine differential utilization of natural substrates (16,81,93−98) or to evaluate the effectiveness of physical and chemical pretreatments on enhancing the biodegradability of woody (93,94,97,99−104) or nonwoody (84,93,95,98,100,105,106) biomass. They may also be used to determine the effects of feed additives (81,107) or other nutrients (95) on biomass utilization.

Differences in cellulose and hemicellulose digestibilities may be determined by combining weight loss data with analysis of individual sugars in the residual substrates (e.g., by GC of acid hydrolysis products before and after exposure to microorganisms; see 108). Digestibility has also been determined by averaging weight loss data with data on cell yield and sugar production (109,110).

While gravimetric methods are widely used, other methods of measuring residual substrate may be utilized. For example, residual substrate may be determined by total hydrolysis (usually with sulfuric acid) and subsequent determination of individual sugars by HPLC (54) or of total reducing sugars using colorimetric procedures involving phenol (58,111) or anthrone (112) in sulfuric acid.

The reduction in height of a column of wet-milled filter paper by growing cultures of anaerobic bacteria has also been used to measure cellulolytic activity (113).

Covalently dyed celluloses may also be measured by residual substrate methods, such as measurement of remaining dye following its extraction from the residual substrate (114).

B. Measurements of Product Formation

Measurement of product formation is usually performed as an ancillary technique with other procedures (e.g., weight loss measurements) for the purpose of answering specific questions regarding cellulose degradation. These include determination of pathways of carbohydrate catabolism (78,115) or evaluation of bioconversion schemes for producing specific chemicals such as ethanol (78,116, 117) or single-cell protein (118,119).

These measurements may be carried out to various levels of completeness, depending on what is known of the microbial population and its metabolic characteristics. For pure cultures and defined mixed cultures, this usually takes the form of a complete fermentation balance (78,115,120). For relatively undefined systems (e.g., soils or sediments), the most important measurements appear to be methane and carbon dioxide (which arise from complete mineralization of cellulose and hemicellulose, and which can be quantitated by gas chromatography). These types of studies are more easily performed in closed systems where all components, including gases, may be retained for analysis (98,99). This is relatively easy for anaerobic systems but is more problematic for aerobes because of their considerable oxygen demand, which necessitates an open system with air sparging and trapping of products such as CO_2.

C. Radioisotopic Methods

Methods employing radioactivity labeled cellulose are usually used to follow the formation of specific catabolic products or to measure the disappearance of cellulose in the presence of other substrates (e.g., hemicellulose or lignin) which are degraded to similar products. Carbon-14 has been the most widely used isotope (78,121– 125), although at least one use of [³H]-cellulose has been reported (103).

The [¹⁴C]-cellulose used in biodegradation studies is generally a mixture of unlabeled substrate and a smaller amount of labeled substrate. The latter may either be a commercial product (e.g., *Canna indica* cellulose from ICN Radiopharmaceuticals, Irvine, CA; see 78,125) or may be prepared by growing cuttings of plants in the presence of [¹⁴C]-glucose (122,123). Mixing of labeled and unlabeled substrates allows the use of optimal concentrations at reasonable specific radioactivities but introduces the possibility that the two substrates may be differentially utilized (i.e., that they

may not form a common substrate pool) as a result of differences
in particle size, crystallinity, etc.

The degradation of lignocelluloses specifically labeled in the cellulose or lignin moiety has been studied in a number of microbial systems, including thermophilic (78,125) and mesophilic soil microflora
(121), sediment microflora (121), and defined mixed cultures of bacteria (78).

Degradation of labeled substrates is usually determined by measuring radioactive gases (CH_4 and CO_2) by liquid scintillation counting (LSC; see 122,123,125) or gas proportional counting (78), and/
or measuring soluble products (usually without fractionation into
separate components) by LSC.

D. Miscellaneous Methods

Microcalorimetry has been used to demonstrate the differences in the
rate of degradation of celluloses of different crystallinities (127).
Calorimetry has the additional advantage of providing thermodynamic information associated with microbial growth, such as growth
yield and heat evolution. Obviously, this technique is limited to
whole cells (rather than enzymes) and is more useful in aerobic systems than in anaerobic ones owing to the more complete utilization of
substrate (and thus greater heat evolution) in the former. Calorimetry can also be used to measure heat of dehydration of cellulose,
which has been shown to correlate very well with crystallinity index (128).

Biodegradation of cellulose may also be reflected in changes in
the characteristics (rather than just the amount) of residual substrate. Thus, an increase in cellulose crystallinity has been taken
as evidence for a preferential attack of the less resistant amorphous
regions of the cellulose microfibrils (25). Also, a decrease in the
tensile strength of cellulose thread has been used as a crude measure of cellulose biodegradation (129).

Osmometry may be used to measure the molecular weight of cellulose before and after its biodegradation. Soluble contaminating
materials generally do not interfere with the determinations, since
they are readily removed by the membrane of specifically constructed
osmometers (130,131).

Transmission electron microscopy (TEM) has been used to determine the length of cellulose fibers before and after biodegradation (132). Both TEM and scanning electron microscopy (SEM) have
found utility in assessing the extent of attachment of microorganisms
to pure cellulose (133-136) or to plant cell wall components (137-
140). Both SEM and TEM have been widely used to detect morphological changes in plant cell wall structure during biodegradation
(137-143). An excellent, detailed review of the use of electron

microscopy to visualize cellulose degradation was presented by White (144).

IV. IDENTIFICATION OF ORGANISMS WITH ENHANCED CELLULOLYTIC ACTIVITIES

The desire to develop technologies for the conversion of biomass to fuels, industrial chemicals, and single-cell protein has stimulated a search for strains with altered cellulolytic capabilities, such as cellulase hyperproduction (145,146), or resistance to catabolite repression or fermentation end products (145). In these cases, the goal is to simultaneously identify the altered strain and permit its isolation from the bulk population.

Early strategies focused on the development of methods for rapidly testing individual isolates for cellulolytic acitivity. Following the initial preparation by Fernley (56) of cellulose-containing covalently linked dye molecules, a number of methods were developed based on the release of soluble dye mono- or oligosaccharides from these substrates (57,147,148). By conducting assays in separate tubes (each inoculated with a single isolate), reasonably quantitative data on the capabilities of each isolate could be ascertained (147). Dyed celluloses have a number of advantages in assaying cellulolytic activity. Because of their relatively low DS, they retain most of the physical and chemical properties of their parent substrates (149). With careful preparation, substrates release dye only upon depolymerization to soluble saccharides (57,147). Rates of dye release have been shown to be proportional to other widely accepted measurements of cellulose degradation such as release of reducing sugars from filter paper (147). Because of the usefulness of dyed substrates in a cellulase assay systems in vitro (see Sec. V.C below), there has been a considerable effort to synthesize and characterize different dyed celluloses and hemicelluloses (56,57,147,149,150 – 153).

Other methods, such as liquefaction of hydroxyethylcellulose (154), have also been developed for measuring cellulolytic activities of individual isolates in vitro, but in general these are not as sensitive as the methods employing dyed substrates.

More useful perhaps has been the development of agar plate assays which permit the simultaneous identification of the desired trait and isolation of the organism from a population of nonidentical cells. The utility of agar plate assays have been enhanced enormously (particularly in the case of fungi) by the use of compounds which restrict the growth of individual colonies (155 – 160) to maximize the number of colonizes which may be screened on an individual plate.

The first detection procedures to be developed involved measurement of the diameter of a zone of clearing associated with conversion of opaque, insoluble cellulose to soluble monomers or oligomers. Substrates used for this purpose have included macerated cellulose powder (70), microcrystalline cellulose (161), or phosphoric acid-swollen cellulose (162-163). The chief advantage of such methods is their relative simplicity (i.e., no additional visualization procedure is required). However, these methods are slow and usually lack the sensitivity necessary for detecting strains with only slightly altered capabilities (145). Some of these methods are summarized in Table 3.

More recently, advances have been made in the detection of cellulolytic activity by employing reagents which enhance visualization of residual substrate. These procedures generally employ soluble, substituted celluloses (e.g., carboxymethylcellulose) to permit rapid substrate degradation followed by precipitating or staining the excess substrate to yield clearing zones. Examples of these methods are provided in Table 4. Such overlay methods have

TABLE 3   Examples of Direct Methods for Visualization of Cellulolytic Activity on Agar Plates

Substrate	Indication of cellulolysis	Remarks	Ref.
Microcrystalline cellulose	Clearing	Requires prolonged incubation	161, 165
Amorphous cellulose	Clearing	Requires prolonged incubation	162-164, 166
Dyed amorphous cellulose	Dye release	Relatively rapid	151, 153
Dyed hydroxy-ethyl cellulose	Clearing	Clearing zones very pale unless postprecipitation employed	152
Cellobiose/ 2-deoxyglucose	Growth on cellobiose	Enhanced β-glucosidase activity yields larger colonies; cellobiase mutants killed upon utilizing 2-DOG	145

TABLE 4 Examples of Indirect Visualization of Cellulolytic Activity on Agar Plates

Substrate	Precipitation or reaction reagent	Remarks	Ref.
Carboxymethylcellulose	Congo red/NaCl	Method adaptable to other glucans; Post-trt w/HCl stabilizes zones	167
	Cetylpyridium chloride	Clearing indicates hydrolysis	145
	Hexadecyltrimethyl-ammonium chloride	Qualitative assay; conditions optimized	5
Dyed hydroxyethyl-cellulose	Ethanol/acetone/Na acetate	Weak clearing noted in absence of precipitating reagent	152, 168
Various	Cellobiose + Glucostat reagent	Enhanced β-glucosidase activity yields red spots	145
Esculin	Ferric ammonium citrate[a]	Esculetin, produced by action of aryl-β-glucosidase, yields black precipitate	145

[a]Incorporated directly into the agar.

expanded utility in that they may be used to detect not only strains which produce more hydrolytic enzyme but also mutants which produce altered enzymes or which circumvent regulatory controls of metabolism. Obviously, recovery of the cells by patching or replica plating onto fresh media is required to preserve the culture prior to treatment with the visualization reagents.

Mutants resistant to catabolite repression may be identified as those which synthesize detectable enzyme components (using the screening techniques described in Tables 3 and 4) when grown in the presence of high concentrations of such known catabolite repressors as glucose, 2-deoxyglucose, or glycerol). Constitutive mutants may be detected by plating cultures onto media containing a simple growth substrate which does not also serve to induce the synthesis of cellulolytic enzymes. Visualization of cellulolytic activity is usually carried out in an agar overlay (see Table 4). Mutants which produce enzymes resistant to end-product inhibition may be detected by incorporating high concentrations of end-product inhibitor (e.g., glucose or cellobiose) into the overlay agar. Note that several of these compounds (e.g., glucose) can serve multiple functions; as a result, some care is necessary in designing the screening strategy, and careful testing of the phenotypes of the isolates is required.

V. ASSAY OF CELLULOLYTIC ENZYMES

A. Total Cellulase Activity vs. Activity of Individual Components of the Cellulase Complex

Assays of the activity of cellulolytic enzymes include the determination of specific activities, kinetic parameters, and the location of individual cleavage sites within the polymer or oligomer chain. The optimum measurement of specific activity is determination of the number of $1,4-\beta$-glucan bonds hydrolyzed per unit enzyme per unit time. Such measurements are complicated by the fact that fragments of different lengths display differences in chemical reactivity and in hydrodynamic behavior, and thus respond differently to measurement techniques. No analytical method circumvents this problem entirely, although some do a better job than others. As might be expected, the results and their interpretation vary tremendously and are largely dependent on the substrate used.

In many cases (e.g., for evaluating strains for biomass conversion potential) it is not necessary to perform a detailed analysis of the individual cellulolytic enzymes, but only to assay the net activity of the mixture of individual enzymes which together make up what might loosely be called the "cellulase complex." However, attempts to develop a widely accepted standard assay for total cellulase activity have met with considerable resistance. Among these, the filter paper assay of Mandels et al. (43,169) has enjoyed the widest use. This method involves incubation of enzyme with standard 50 mg (1 × 6 cm) strips of filter paper under defined conditions (50 mM citrate buffer, pH 4.8, 50°C for 1 hr), followed by measurement of reducing sugars released. Cellulase activity is reported in filter paper units (FPUs), with 1 FPU equaling the release of 1 mg of reducing sugar (see Sec. V.B below) per hr per ml of culture filtrate. Since the assay is substrate-limited at high activities (>3 FPUs), it has been suggested (170) that the assay may be improved by doubling the amount of filter paper and cutting the incubation time in half. Because of its historical popularity, the filter paper method has been used to gather a great many data on cellulose biodegradation. Nevertheless, the rather wide acceptance of the method has given it a longevity that is not justified by its inconvenience and insensitivity. The method is useful only for assaying exoglucanase-containing cellulase complexes of sufficient activity to overcome the kinetic constraints imposed by the very low surface area of the substrate. In recent years there has been a pronounced trend toward the use of microcrystalline cellulose powder as a substrate for assaying the activity of the cellulase complex. Regardless of the substrate used, the preferred method of measuring activity involves measurement of the release of reducing sugars (see Sec. V.B below).

The historical development of separate assays for different cellulase components may be traced to the observation by Reese et al. (3) that the capability of hydrolyzing cellulose derivatives such as CMC is much more widely distributed among microbes than the ability to degrade natural or unmodified celluloses. Specific assays for these separate activities were thus developed, and were integrated into the evolving the $C_1 - C_x$ hypothesis of cellulose biodegradation (4).

In the intervening decades, CMC has come to be used as a substrate for endoglucanases (1,4-β-glucan glucanohydrolase, EC 3.2.1.4), which randomly hydrolyze cellulose chains; microcrystalline celluloses (such as Avicel or MN300) have come to be used as substrates for exoglucanases (cellobiohydrolases, EC 3.2.1.91), which sequentially liberate cellobiose from the nonreducing end of the polymeric chain. However, in recent years it has become apparent that at least some purified endoglucanases may display considerable activity toward Avicel and some exoglucanases have been reported to hydrolyze CMC. Furthermore, the possibility of mechanistic differences within each class of cellulase components has thrown the whole matter of cellulase assays (and results derived from these assays) into a state of considerable confusion. A more complete discussion of this overlap is presented in Chap. 21. The discussion which follows lays out some of the principles for assaying the "classical" cellulase components.

Exoglucanase activity may be more properly determined by spectrophotometric measurement of the release of p-nitrophenol from the artificial substrates p-nitrophenyl-β-D-cellobioside (pNP(Glc)$_2$) or p-nitrophenyl-β-D-lactoside (pNPL) in the presence of D-glucono-1, 5-β-lactone, a powerful inhibitor of β-glucosidases (171). This method requires correction for the low level of activity of endoglucanases toward these substrates, a correction which requires knowledge of the ratio of hologlucosidic and aglucosidic bond cleavage rates. Since this ratio must be determined experimentally, purification of the contaminating endoglucanase(s) is required. This makes the assay useful for purified exoglucanases but impractical for mixtures of cellulolytic enzymes.

Assays for endoglucanase activity in culture filtrates and crude enzyme mixtures is also problematic, as none completely eliminates the contribution of other cellulolytic enzymes. However, viscometric methods (see Sec. V.C below) do appear to minimize such contributions. The activity of isolated endoglucanases can be measured by release of reducing sugars from CMC, release of dyed oligomers from dyed CMC (63), or can be determined viscometrically. Quantitation of individual oligosaccharides is occasionally employed (e.g., for mechanistic studies, see Sec. V.B below).

β-glucosidase (EC 3.2.1.21) in cell-free systems (146,173,174) or in whole or permeabilized cells (175) is generally determined by measuring the release of p-nitrophenol from p-nitrophenyl-β-glucoside (pNPGlc). Activity of β-glucosidase or cellobiase toward cellobiose may be determined by specific measurement of glucose formation (see Sec. V.B below).

In some bacteria, transport of hydrolytic products of cellulose into the cell is performed by cellobiose phosphorylase (EC 2.4.1.20) and cellodextrin phosphorylase (EC 2.4.1.49), which respectively catalyze the following reactions:

$$\text{Cellobiose} + P_i \longrightarrow \text{glucose} + \alpha\text{-glucose-1-P}$$

and

$$(1,4\text{-}\beta\text{Glucosyl})_n + P_i \longrightarrow (1,4\text{-}\beta\text{-glucosyl})_{n-1} + \alpha\text{-glucose-1-P}$$

The latter enzyme is normally assayed by determining the rate of P_i release in the back reaction, using cellobiose as glucosyl acceptor (176):

$$\alpha\text{-Glucose-1-P} + \text{cellobiose} \longrightarrow \text{cellotriose} + P_i$$

The former enzyme is assayed similarly, except that xylose is used as glucosyl acceptor (177).

B. Reducing Sugar Methods

With minor exceptions, the component monosaccharides of cellulosic materials are classified as reducing sugars in that they are all capable of reducing Tollen's reagent [which contains $Ag(NH_3)_2$] or Fehling's solution (which contains alkaline Cu^{2+}). The chemistry of detecting free reducing end groups of saccharides was first worked out many decades ago and is in fact one of the cornerstones of carbohydrate chemistry. The application of total reducing sugar methods for quantitative measurement of cellulose degradation has evolved in the direction of assay simplicity, versatility, and reproducibility. Reducing sugar methods involve removal of subsamples from reaction mixtures, followed by separation of the soluble saccharides from residual insoluble substrates by filtration or centrifugation, followed by destructive conversion of the soluble saccharides to colored reaction products whose concentration may be measured colorimetrically. These methods have found wide acceptance due to their simplicity and the fact that minimal skills and equipment are required.

The most commonly used detection agent for reducing sugars formed from cellulose is 3,5-dinitrosalicylic acid (DNS). Although its use for this purpose was first described by Sumner and Sisler (178,179), the most widely used modification of the method is that of Miller (180,181). Samples are boiled in the presence of DNS and buffer for 15 min, and after cooling the colored reaction product is measured at 640 nm. Advantages of the assay include relative simplicity, a reasonably linear relationship between absorbance and concentration, and good stability of the colored product. The chief disadvantages of the assay are (1) variable color formation with variations in boiling time, (2) relatively poor sensitivity (lower detection limit of ~0.025 mg glucose/ml), and (3) differential response of glucose, cellodextrins, xylose, and xylodextrins.

The Nelson modification (182,183) of the Somogyi method (184) is also widely employed for measurement of reducing sugars, particularly glucose. In this method, deproteinized sample is boiled with alkaline cupric ions; the resulting cuprous ions are reacted with arsenomolybdate reagent to yield molybdenum blue, whose absorbance is measured at 500−520 nm. The gradual development of this assay (see 182−185) provides an interesting example of stepwise improvements in the sensitivity and utility of a wet chemical method. The sensitivity of the method is only slightly greater than that of the DNS assay (186). The chief advantages of this procedure are its linearity and the stability of the molybdenum blue complex. Disadvantages include the requirement for deproteinizing the sample and the instability of the alkaline cupric working solution. In addition, the widely varying response of different sugars makes the assay impractical for analysis of hydrolytic products of complex natural biomass materials.

The ferricyanide method (51,187−190), despite its age, is still
occasionally used (e.g., see 172,191). In this method, sugars are
reacted with an alkaline ferricyanide solution in a boiling bath, and
the ferrocyanide produced is reacted with ferric iron to produce
Prussian blue, whose absorbance is measured at 690 nm. The method's chief disadvantage is the requirement that samples be freed of
interfering materials such as proteins.

An assay (181) based on the reaction of reducing sugars with
I_2/KI followed by treatment with Na_2CO_3 and H_3PO_4 and measurement of color formation at 480 nm is relatively simple as no boiling
step is required. The method is specific for aldehyde groups and
has been touted as particularly useful for kinetic studies of cellulose hydrolysis. However, the method gives poor results for hydrolytic products of cellulose derivatives such as CMC.

Another method (192, based on 193,194) uses an alkaline aqueous solution of tetrazolium blue dye. The reaction mixture is
steamed for short period (~30 sec) and the colored formazan product is extracted with toluene from the cooled reaction mix prior
to measurement at 570 nm. The primary disadvantages of the assay is the strong dependence of color formation on heating time
and the inconvenience of the extraction step. However, the method appears to be particularly useful for assaying low levels of cellulolytic activity, since the extraction procedure concentrates the
reaction product.

Specific sugars released by biodegradation of cellulosic materials have been separated by paper chromatography (55,173) or
thin-layer chromatography (172,195,196), and subsequently estimated by triphenyltetrazolium chloride (173), silver nitrate (55,
197), or sulfuric acid (172,198). However, these methods have
for the most part been supplanted by HPLC (53,54,77,199,200) owing to its simplicity and accuracy.

Glucose can also be assayed enzymatically using hexokinase
coupled to glucose-6-phosphate dehydrogenase (201−203) or glucose oxidase coupled to peroxidase (204). Glucose can also be
measured in the presence of cellobiose based on differences in the
extinction coefficients of these sugars following their reaction with
the Nelson−Somogyi reagent (205). These assays are particularly
useful for assays of cellobiase or β-glucosidase.

C. Dye Release Methods

The use of chromogenic substrates to measure enzymatic activity
toward soluble substrates has a long and distinguished history.
Only recently, however, has this technique been adapted to measurement of cellulase activity. If the dyed substrates are derived
from microcrystalline or amorphous celluloses, the activity measured

is analogous to the "total cellulase" activity described above; dyed substrates derived from substituted celluloses or cellodextrins can be used to estimate the activity of specific enzymatic components (63).

Considerable differences have been observed among different celluloses and cellulose derivatives with respect to their reactivity with various dyes (149). Avicel SF microcrystalline cellulose has been found to be particularly amendable to dyeing with Remalzol Brilliant Blue R (RBB) to yield a material having a very low DS (0.025) and strongly resembling the parent substrate (149). Thus this material has been employed most frequently in enzymatic assays of dyed celluloses. Other dyes appear to be more useful than RBB for cellodextrins (56). The structures of some of the dyes and dyed substrates are shown in Fig. 1. The sensitivity of assays based on dyed substrates is a complex function of numerous factors, such as the extinction coefficients (or fluorescent yield) of the dye, its degree of substitution in the substrate, and the susceptibility of the dyed substrate to hydrolysis. The most sensitive assays use fluorescent dyes attached to soluble cellulose derivatives (62,63).

A continuous spectrophotometric assay has been developed for measuring cellulolytic activity in vitro using RBB-Avicel (150,206). The enzymatic reaction is run in a stirred, temperature-controlled cell and the soluble reaction products are passed through a nylon filter into a spectrophotometer's flow cell prior to their return to the reaction mixture. The assay has been used to obtain specific activity as well as kinetic and activation energy data for *Trichoderma* cellulase.

Enzymatic degradation of RBB-Avicel SF has also been coupled to a commercial autoanalyzer instrument for rapid analysis of many samples in a fixed-time (namely, 11.5 min) assay (207). This technique is especially useful for screening fractions for enzymatic purification procedures, or for testing of compounds for stimulation or inhibition of cellulase activity. The primary advantages of automated assays are speed and reproducibility, while the chief disadvantage is the requirement for capital investment and operator training.

Chromogenic substrates have also found utility in mechanistic studies of cellulase components. For example, a homologous series of 4-methylumbelliferylglycosides of cellooligosaccharides has been used to study the chain cleavage by the cellobiohydrolase I (CBH I) and endoglucanase of *T. reesei* (208). The free 4-methylumbelliferone (= 4-methyl-7-hydroxycoumarin) and the various glycosides [4-MeUmb-β-(D-Glc)$_n$, where n = 1−6] were separated by HPLC and their concentrations estimated by their absorbances at 313 nm. This study demonstrated that the CBH I does not act as a true

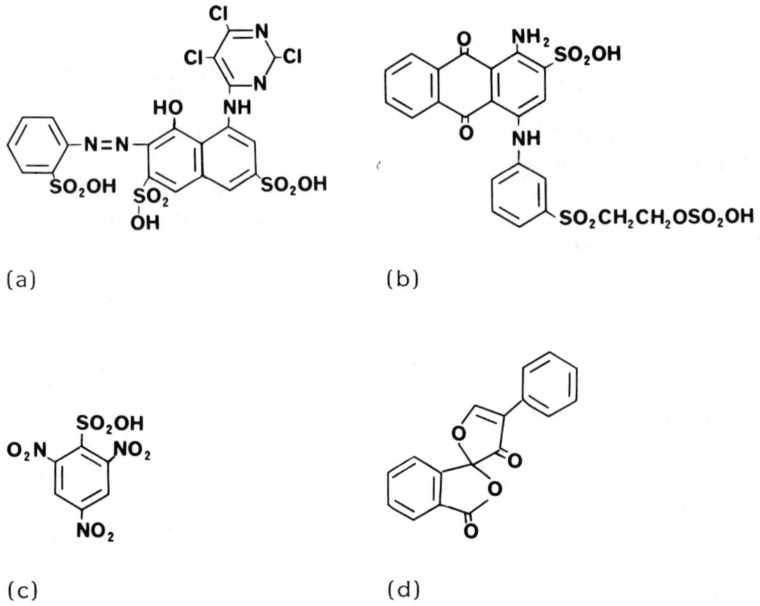

(a)   (b)

(c)   (d)

FIGURE 1   Examples of dyes used for covalent attachment to cellulose: (a) Reactone Red 2B, one of the most promising dye candidates reported in the pioneering work of Fernley (56); (b) Remalzol Brilliant Blue R, the most commonly used dye for preparation of dyed cellulose; (c) TNBS (2,4,6-trinitrobenzylsulfonic acid), used to dye diaminoethyl-CMC (62); (d) fluorescamine, a fluorophore used to dye animoethyl-CMC (62).

cellobiohydrolase, and that the endoglucanase displayed a certain specificity with respect to cleavage site (i.e., a nonrandom attack of substrate).

D. Viscometric Methods

Viscometric methods for measuring cellulose degradation involve measurement of inherent, intrinsic, or specific viscosities of reaction mixtures under precisely controlled physical and chemical conditions. While they can be applied to ionic-substituted celluloses (e.g., CMC, see 174,209) complications are introduced by the difficulty in preparing reaction mixtures of reproducible initial viscosities and by the strong dependence of viscosity of ionic substrates on pH, ionic strength, and concentration of divalent cations

(210). These difficulties can be overcome by use of nonionic cellulose derivatives, particularly hydroxyethylcellulose (HEC) (210-212). The primary advantage of viscometric methods is that they can provide absolute measurement of enzyme activity (the so-called activity number, A_n, or the number of bonds broken per unit time). In addition, viscometric methods are extremely sensitive for endoglucanases since relatively few bond scissions yield large decreases in viscosity; by contrast, viscosity changes are minimal during attack by exoglucanases, particularly when high-DP celluloses are used as substrates.

Detailed discussions of theoretical aspects of viscometry and their relationship to measurement of cellulose hydrolysis are available (211,213-217) and will not be discussed in detail here. However, two specific examples of the use of viscometry will be provided.

Viscometric assays have been applied to measurement of cellulase activity under different growth conditions known to regulate cellulose biosynthesis (218). The advantage here is that as long as the same substrate is used, one can express cellulolytic activity in arbitrary units, e.g., proportional to the slope of $(\eta_{sp})_{t_o}/$ $(\eta_{sp})_t$ vs. time. This obviates some of the mathematical derivations of viscometric constants and should thus be of particular interest to biologists whose reluctance to embrace viscometry probably reflects their collective discomfort with mathematics beyond the level of simple algebra.

Because viscometry can also be used to determine the molecular weights of isolated cellulose (213,215), this technique should be applicable to assessing changes in the degree of polymerization of substrate resulting from biodegradation, provided that the substrate is sufficiently cleaned up prior to analysis.

ADDENDUM

After over 10 years of consideration, a committee under the auspices of the International Union of Pure and Applied Chemistry (IUPAC) (219) recently recommended a set of standard procedures for the measurement of cellulase activities. Included within the recommendations are specific, step-by-step protocols for measurement of endoglucanase activity (using both CMC and HEC as substrates), filter paper activity, and cellobiase, as well as procedures for protein (by the Folin method) and reducing sugars (by the DNS method). The committee admits that the procedures are more designed for the application-oriented cellulose worker (e.g., one involved in biomass conversion) than the enzymologist. Furthermore, the committee notes that the procedures are heavily

weighted toward fungal (particularly *Trichoderma*) cellulase systems
and that modification of the assays would enhance their utility in
bacterial cellulase systems. Despite these shortcomings, the docu-
ment is noteworthy and laudable in that it attempts to bring order
to a field that has lacked standard assay methodologies, and it presents the recommendations in a convenient format immediately usable
in the research laboratory. Whether these recommendations will be
completely embraced by cellulase researchers remains to be seen. It
should be noted that Chan et al. (220) reported that even slight
changes in the IUPAC protocols lead to significant changes in the
measured cellulase activities.

It is somewhat ironic that prior to publication of the IUPAC recommendations the DNS method for determining reducing sugars was
shown by Breuil and Saddler (221) to be inferior to the Nelson –
Somogyi procedure in several important aspects. The Nelson – Somogyi
method gives a closer estimate of the actual number of hemiacetal linkages cleaved, is less influenced by variations in incubation conditions,
and is less subject to interference by materials present in crude lignocellulosic substrates. These facts, coupled with the inherently
greater sensitivity of the assay, led the authors to conclude that the
Nelson – Somogyi procedure, although less convenient, was a superior
assay method.

Schwald et al. (222) directly compared both of the above colorimetric methods to a method based on HPLC analysis of sugars released
during hydrolysis. These authors conclude that the HPLC technique
is superior in cases where the amount of β-glucosidase in the assay
mixture is insufficient to prevent the buildup of cellobiose or cellodextrins. In this case, the colorimetric methods underestimate the amount
of hydrolysis because of their reduced sensitivity to cellobiose and
oligomers. Although enzyme mixtures are often tested to ensure adequate amounts of β-glucosidase, these tests must be interpreted cau-
tiously; Chan et al. (220) demonstrated a differential hydrolysis of
cellobiose and salicin (p-nitrophenyl-β-D-glucoside, a substrate analog
sometimes used in the measurement of β-glucosidase activity. Thus
crude enzyme mixtures thought to have sufficient β-glucosidase activity
to overcome end-product inhibition of cellobiose may in fact be β-
glucosidase-deficient.

REFERENCES

1. B. Focher, A. Marzetti, M. Cattaneo, P. L. Beltrame, and
    P. Carniti, *J. Appl. Polym. Sci.*, *26*, 1989 (1981).
2. V. P. Puri, *Biotechnol. Bioeng.*, *26*, 1219 (1984).
3. D. D. Y. Ryu, S. B. Lee, T. Tassinari, and C. Macy,
    Biotechnol. Bioeng., *24*, 1047 (1982).
4. E. T. Reese, R. G. H. Siu, and H. S. Levinson, *J. Bacteriol.*,
    59, 485 (1950).

5. L. Hankin and S. L. Anagnostakis, *J. Gen. Microbiol.*, *98*, 109 (1977).
6. R. F. Boyer and M. A. Redmond, *Biotechnol. Bioeng.*, *25*, 1311 (1983).
7. L. T. Fan, Y.-H. Lee, and D. H. Beardmore, *Biotechnol. Bioeng.*, *22*, 177 (1980).
8. Y.-H. Lee and L. T. Fan, *Biotechnol. Bioeng.*, *24*, 2383 (1982).
9. M. M. Gharpuray, Y.-H. Lee, and L. T. Fan, *Biotechnol. Bioeng.*, *25*, 157 (1983).
10. C. E. Dunlap, J. Thomson, and L. C. Chiang, *AIChE Symp. Ser.*, *72*, 58 (1976).
11. P. J. Weimer and W. M. Weston, *Biotechnol. Bioeng.*, *27*, 1540 (1985).
12. L. T. Fan, Y.-H. Lee, and D. H. Beardmore, *Biotechnol. Bioeng.*, *23*, 419 (1981).
13. T. Sasaki, T. Tanaka, N. Nanbu, Y. Sato, and F. Kainuma, *Biotechnol. Bioeng.*, *21*, 1031 (1979).
14. B. Henrissat, H. Driguez, C. Viet, and M. Schulein, *Biotechnology*, *3*, 722 (1985).
15. D. F. Caulfield and W. E. Moore, *Wood Sci.*, *6*, 375 (1974).
16. J. E. Stone, A. M. Scallan, E. Donefer, and E. Ahlgren, *Adv. Chem. Ser.*, *95*, 219 (1969).
17. H. E. Grethlein, *Biotechnology*, *3*, 155 (1985).
18. K. W. Lin, M. R. Ladisch, M. Voloch, J. A. Patterson, and C. A. Noller, *Biotechnol. Bioeng.*, *27*, 1427 (1985).
19. H. E. Woodman and J. Stewart, *J. Agr. Sci.*, *22*, 527 (1932).
20. M. A. Millett, A. J. Baker, and L. D. Satter, *Biotechnol. Bioeng. Symp. Ser.*, *5*, 193 (1975).
21. M. J. Latham, B. E. Brooker, G. L. Pettipher, and P. J. Harris, *Appl. Env. Microbiol.*, *35*, 1166 (1978).
22. M. J. Latham, D. G. Hobbs, and P. J. Harris, *Ann. Rech. Vet.*, *10*, 244 (1979).
23. E. B. Cowling, *Biotechnol. Bioeng. Symp. Ser.*, *5*, 163 (1975).
24. S. P. Rowland, *Biotechnol. Bioeng. Symp. Ser.*, *5*, 183 (1975).
25. L. T. Fan, Y.-H. Lee, and D. H. Beardmore, *Adv. Biochem. Eng.*, *14*, 102 (1980).
26. I. G. Gilbert and G. T. Tsao, *Ann. Rep. Ferm. Proc.*, *6*, 323 (1983).
27. B. L. Browning, *Methods of Wood Chemistry*, Interscience, New York, 1967.
28. H. F. J. Wenzl, *The Chemical Technology of Wood*, Academic Press, New York, 1970.
29. E. Sjostrom, *Wood Chemistry: Fundamentals and Applications*, Academic Press, New York, 1981.
30. M. M. Chang, Y.-C. T. Chou, and G. T. Tsao, *Adv. Biochem. Eng.*, *20*, 15 (1981).

31. M. A. Millett, A. J. Baker, W. C. Feist, R. W. Mellenberger, and L. D. Satter, *J. Anim. Sci.*, *31*, 781 (1970).
32. E. A. DeLong, Can. Patent 1,096,374 (Feb. 24, 1981).
33. M. Kumakura and I. Kaetsu, *Biomass*, *3*, 199 (1982).
34. R. H. Marchessault, S. L. Mahotra, A. Y. Jones, and A. Perovic, In *Wood and Agricultural Residues* (J. Solts, ed.), Academic Press, New York, 1983, p. 401.
35. Y.-C. T. Chou, *Biotechnol. Bioeng. Symp. Ser.*, *17*, 19 (1986).
36. M. G. Jackson, *Anim. Feed Sci. Technol.*, *2*, 105 (1977).
37. L. Jurasek, *Dev. Ind. Microbiol.*, *20*, 177 (1979).
38. L. T. Fan, M. M. Gharpuray, and Y.-H. Lee, *Biotechnol. Bioeng. Symp. Ser.*, *11*, 29 (1969).
39. J. M. Gould, *Biotechnol. Bioeng.*, *26*, 46 (1984).
40. H. Chazny and B. Henrissat, *Carboh. Polym.*, *3*, 161 (1983).
41. A. Kulshreshtha and N. E. Durletz, *J. Polym. Sci. Phys. Ed.*, *11*, 487 (1973).
42. G. Halliwell, *J. Gen. Microbiol.*, *17*, 153 (1957).
43. M. Mandels, R. E. Andreotti, and C. Roche, *Biotechnol. Bioeng. Symp. Ser.*, *6*, 21 (1976).
44. K. H. Gardner and J. Blackwell, *Biopolymers*, *13*, 1975 (1974).
45. S. Hestrin, *Methods Carboh. Chem.*, *3*, 4 (1963).
46. Y. Tomita, H. Suzuki, and K. Nisizawa, *J. Ferment. Technol.*, *52*, 233 (1974).
47. M. Gritzaldi and R. D. Brown, *Adv. Chem. Ser.*, *181*, 237 (1979).
48. C. S. Walseth, *TAPPI*, *35*, 288 (1952).
49. M. Nummi, P. C. Fox, M.-L. Niku-Paavola, and T.-M. Enari, *Anal. Biochem.*, *116*, 133 (1981).
50. M. Nummi, M.-L. Niku-Paavola, T.-M. Enari, and V. Raunio, *Biochem. J.*, *215*, 677 (1983).
51. G. Halliwell, *Biochem. J.*, *95*, 270 (1965).
52. G. L. Miller, *Anal. Biochem.*, *2*, 133 (1960).
53. S. N. Freer and R. W. Detroy, *Biotechnol. Lett.*, *4*, 453 (1982).
54. J. B. Russell, *Appl. Env. Microbiol.*, *49*, 572 (1985).
55. A. Hedges and R. S. Wolfe, *J. Bacteriol.*, *120*, 844 (1974).
56. H. N. Fernley, *Biochem. J.*, *87*, 90 (1963).
57. P. Hagen, E. T. Reese, and O. A. Stamm, *Helv. Chim. Acta*, *49*, 2278 (1966).
58. R. Montgomery, *Biochem. Biophys. Acta*, *48*, 591 (1961).
59. R. L. Moore, B. Basset, and M. J. Swift, *Soil Biochem. Biophys.*, *11*, 311 (1979).
60. M. J. Swift, In *Sourcebook of Experiments for the Teaching of Microbiology* (S. B. Primrose and A. C. Wardlaw, eds.), Academic Press, New York, 1982, p. 603.
61. G. Halliwell, *Biochem J.*, *85*, 67 (1962).

62. J. S. Huang and J. Tang, *Anal. Biochem.*, *73*, 369 (1976).
63. B. V. Cleary, *Carboh. Res.*, *86*, 97 (1980).
64. S. O. Mann, *J. Appl. Bacteriol.*, *31*, 241 (1968).
65. H. O. Halvorson and N. R. Ziegler, *J. Bacteriol.*, *25*, 101 (1933).
66. J. A. Z. Leedle and R. B. Hespell, *Appl. Env. Microbiol.*, *39*, 709 (1980).
67. K. S. Kang, G. T. Veeder, P. J. Mirrasoul, T. Kaneto, and I. W. Cottrell, *Appl. Env. Microbiol.*, *43*, 1086 (1982).
68. R. Moorhouse, G. T. Colegrove, P. A. Sandord, J. Baird, and K. S. Kang, In *Solution Properties of Polysaccharides* (D. A. Brand, ed.), American Chemical Society, New York, 1981, p. 111.
69. D. Shungu, M. Valiant, V. Tutlane, E. Weinberg, B. Weissberger, L. Koupal, H. Gadenbusch, and E. Stapley, *Appl. Env. Microbiol.*, *46*, 840 (1983).
70. F. A. Skinner, *J. Gen. Microbiol.*, *22*, 539 (1960).
71. R. E. Hungate, *Bacteriol. Rev.*, *14*, 1 (1949).
72. R. E. Hungate, *The Rumen and Its Microbes*, Academic Press, New York, 1966.
73. A. Kistner, *J. Gen. Microbiol.*, *23*, 565 (1960).
74. N. O. van Gylswyk, *J. Gen. Microbiol.*, *60*, 191 (1970).
75. H. W. Scott and B. A. Dehority, *J. Bacteriol.*, *89*, 1169 (1965).
76. C. S. Stewart, C. Paniagua, D. Dinsdale, K.-J. Cheng, and S. H. Garrow, *Appl. Env. Microbiol.*, *41*, 504 (1981).
77. S. N. Freer and R. E. Wing, *Biotechnol. Bioeng.*, *27*, 1085 (1985).
78. P. J. Weimer and J. G. Zeikus, *Appl. Env. Microbiol.*, *33*, 289 (1977).
79. C. David, R. Fornaser, C. Greindl-Fallon, and N. Vanlauten, *Biotechnol. Bioeng.*, *27*, 1591 (1985).
80. B. van Hofsten and N. Edberg, *OIKOS*, *23*, 29 (1972).
81. R. W. Van Keuren and W. W. Heinemann, *J. Anim. Sci.*, *21*, 340 (1962).
82. J. M. A. Tilley and R. A. Terry, *J. Br. Grassld, Soc.*, *18*, 104 (1963).
83. D. J. Minson and M. N. McLoed, *CSIRO Aust. Div. Trop. Pastures Tech. Paper No. 8* (1972).
84. R. R. Johnson, *J. Anim. Sci.*, *25*, 855 (1966).
85. H. K. Goering and P. J. Van Soest, *Forage Fiber Analysis*, Agr. Res. Service/U.S. Dept. Agr. Handbook No. 379, U.S. Department of Agriculture, Washington, D.C., 1970.
86. R. F. Barnes, *J. Anim. Sci.*, *26*, 1120 (1967).

87. P. J. Weimer, J. M. Lopez-Guisa, and A. D. French, *Appl. Environ. Microbiol.*, *56*, 2421 (1990).

88. R. J. Wilkins and R. C. Grimes, *Aust. J. Anim. Prod.*, *6*, 334, (1966).

89. M. N. McLeod and D. J. Minson, *J. Br. Grassld. Soc.*, *24*, 244 (1969).

90. M. J. Playne, *Biotechnol. Bioeng.*, *26*, 426 (1984).

91. M. N. McLeod and D. J. Minson, *Anim. Feed Sci. Technol.*, *3*, 277 (1978).

92. M. N. McLeod and D. J. Minson, *Anim. Feed Sci. Technol.*, *5*, 347 (1980).

93. V. P. Puri and H. Mamers, *Biotechnol. Bioeng.*, *25*, 3149 (1983).

94. R. W. Mellenberger, L. D. Satter, M. A. Millett, and A. J. Baker, *J. Anim. Sci.*, *32*, 756.

95. R. F. H. Dekker and G. N. Richards, *Aust. J. Biol. Sci.*, *25*, 1377 (1972).

96. W. J. Pigden and D. P. Heaney, *Adv. Chem. Ser.*, *95*, 245 (1969).

97. W. D. Kitts, C. R. Krishnamurti, J. A. Shelford, and J. G. Huffman, *Adv. Chem. Ser.*, *95*, 279 (1969).

98. W. H. Pfander, S. Grebing, G. Hajny, and W. Tyree, *Adv. Chem. Ser.*, *95*, 298 (1969).

99. H. Tarkow and W. C. Feist, *Adv. Chem. Ser.*, *95*, 197 (1969).

100. R. K. Wilson and W. J. Pigden, *Can. J. Anim. Sci.*, *44*, 122 (1964).

101. A. I. Virtanen, *Nature* (London), *158*, 795 (1946).

102. E. J. Lawton, W. D. Bellamy, R. E. Hungate, M. P. Bryant, and E. Hall, *TAPPI*, *34*, 113A (1951).

103. T. J. Klopfenstein, V. E. Krause, M. J. Jones, and W. Woods, *J. Anim. Sci.*, *35*, 418 (1972).

104. P. J. Weimer and Y.-C. T. Chou, *Appl. Env. Microbiol.*, *52*, 733, (1986).

105. T. J. Klopfenstein, R. R. Bartling, and W. R. Woods, *J. Anim. Sci.*, *26*, 1492 (1967).

106. B. A. Dehority and R. R. Johnson, *J. Dairy Sci.*, *44*, 2242 (1961).

107. E. Donefer, I. O. A. Adeleye, and T. A. O. C. Jones, *Adv. Cem. Ser.*, *95*, 328 (1969).

108. G. F. Collings and M. T. Yokoyama, *Appl. Env. Microbiol.*, *39*, 566 (1980).

109. Y. W. Han and C. D. Callihan, *Appl. Microbiol.*, *27*, 159 (1974).

110. J. R. Guggolz, R. M. Saunders, G. O. Kohler, and T. Klopfenstein, *J. Anim. Sci.*, *33*, 151 (1971).

111. M. Dubois, K. Gilles, J. K. Hamilton, P. A. Rebers, and F. Smith, *Anal. Biochem.*, *28*, 350 (1956).
112. R. G. Spiro, *Meth. Enzymol.*, *8*, 4 (1966).
113. D. G. Brandon, *Eur. J. Appl. Microb. Technol.*, *7*, 281 (1979).
114. J. F. M. Hoeniger, *Appl. Env. Microbiol.*, *50*, 315 (1985).
115. N. O. Van Gylswyk and C. E. G. Roche, *J. Gen. Microbiol.*, *64*, 111 (1970).
116. P. J. Blotkamp, M. Takagi, M. S. Pemberton, and G. H. Emert, *AIChE Symp. Ser.*, *74*, 85 (1978).
117. D. I. C. Wang, C. L. Cooney, A. L. Demain, R. F. Gomez, and A. J. Sinskey, U.S. Department of Energy Report COO-4198-4, U.S. Department of Energy, Washington, D.C., 1977.
118. C. D. Callihan and C. E. Dunlap, *Compost Sci.*, *10*, 6 (1969).
119. D. L. Crawford, E. McCoy, J. M. Harkin, and P. Jones, *Biotechnol. Bioeng.*, *15*, 833 (1973).
120. G. B. Patel, A. W. Khan, B. J. Agnew, and J. R. Colvin, *Int. J. Syst. Bacteriol.*, *30*, 179 (1980).
121. D. L. Crawford, R. L. Crawford, and A. Pometto III, *Appl. Env. Microbiol.*, *33*, 1247 (1977).
122. A. E. Maccubbin and R. E. Hodson, *Appl. Env. Microbiol.*, *40*, 735 (1980).
123. R. Benner, A. E. Maccubbin, and R. E. Hodson, *Appl. Env. Microbiol.*, *47*, 381 (1984).
124. R. Benner, A. E. Maccubbin, and R. E. Hodson, *Appl. Env. Microbiol.*, *47*, 998 (1984).
125. R. Benner and R. E. Hodson, *Appl. Env. Microbiol.*, *50*, 971 (1985).
126. J. C. Stewart, C. S. Stewart, and J. Heppinstall, *Biotechnol. Lett.*, *4*, 459 (1982).
127. Z. Dermoun and J. P. Belaich, *Biotechnol. Bioeng.*, *27*, 1005 (1985).
128. M. S. Bertran and B. E. Dale, *J. Appl. Polym. Sci.*, *32*, 4241 (1986).
129. M. Laurent, *Adv. Water Pollut. Res.*, *4*, 917 (1969).
130. P. M. Doty and H. M. Spurlin, In *Cellulose and Cellulose Derivatives* 2nd ed., Part III (E. Ott and H. M. Spurlin, eds.), Interscience, New York, 1955, p. 1133.
131. H. Vink, In *Cellulose and Cellulose Derivatives* Part IV, (N. M. Bikales and L. Segal, eds.), Wiley-Interscience, New York, 1971, p. 453
132. K. M. Paralikar and S. P. Bhatawdekar, *J. Appl. Polym. Sci.*, *29*, 2573 (1984).
133. J. R. Colvin, L. C. Sowden, G. B. Patel, and A. W. Khan, *Curr. Microbiol.*, *7*, 13 (1982).

134. G. Gaudet and B. Gaillard, *Arch. Microbiol.*, *148*, 150 (1987).
135. B. Berg, B. van Hofsten, and G. Petterson, *J. Appl. Bacteriol.*, *35*, 215 (1972).
136. T. K. Ng, P. J. Weimer, and J. G. Zeikus, *Arch. Microbiol.*, *114*, 1 (1977).
137. D. E. Akin and H. E. Amos, *Appl. Microbiol.*, *29*, 692 (1975).
138. D. E. Akin, *Appl. Env. Microbiol.*, *40*, 808 (1980).
139. K.-J. Cheng, D. E. Akin, and J. W. Costerton, *Fed. Proc.*, *36*, 193 (1977).
140. K.-J. Cheng, J. P. Fay, R. E. Howarth, and J. W. Costerton, *Appl. Env. Microbiol.*, *40*, 613 (1980).
141. D. E. Akin, *J. Bacteriol.*, *125*, 1156 (1976).
142. D. E. Akin, D. Burdick, and G. E. Michaels, *Appl. Env. Microbiol.*, *27*, 1149 (1982).
143. D. E. Akin and L. L. Rigsby, *Appl. Env. Microbiol.*, *50*, 825 (1985).
144. A. R. White, In *Cellulose and Other Natural Polymer Systems* (R. M. Brown, Jr., ed)., Plenum Press, New York, 1982, p. 489.
145. B. S. Montenecourt and D. E. Eveleigh, *Adv. Chem. Ser.*, *181*, 289 (1979).
146. J. A. Brown, S. A. Collin, and T. M. Wood, *Enzyme Microb. Technol.*, *9*, 355 (1987).
147. R. P. Poincelot and P. R. Day, *Appl. Microbiol.*, *23*, 875 (1972).
148. R. E. Smith, *Appl. Env. Microbiol.*, *33*, 980 (1977).
149. M. Leisola and M. Linko, *Anal. Biochem.*, *70*, 592 (1976).
150. T. K. Ng and J. G. Zeikus, *Anal. Biochem.*, *103*, 42 (1980).
151. D. A. J. Wase and A. K. Vaid, *Proc. Biochem.*, *18* (Dec), 35 (1983).
152. V. Farkas, M. Liskova, and P. Beily, *FEMS Microbiol. Lett.*, *28*, 137 (1985).
153. A. M. Deschamps and J. M. Lebault, *Ann. Microbiol.* (Paris) (A), *131*, 77 (1980).
154. J. Zemek, L. Kuniak, and J. Augustin, *Abh. Akad. Wiss. DDR*, *3*, 47 (1981).
155. J. P. Martin, *Soil Sci.*, *69*, 215 (1950).
156. G. W. Steiner and R. D. Watson, *Phytopathology*, *55*, 1009 (1969).
157. G. C. Papovizas, *Phytopathology*, *57*, 848 (1967).
158. E. A. Curl, *Can. J. Microbiol.*, *14*, 182 (1969).
159. P. Ander and K.-E. Ericksson, *Sven. Papperstidn.*, *78*, 643 (1975).
160. P. Ander and K.-E. Ericksson, *Arch. Microbiol.*, *109*, 1 (1976).

161. A. Kaufmann, T. Fegan, P. Doleac, C. Gainier, D. Wittich, and A. Glann, *J. Gen. Microbiol.*, *94*, 405 (1976).
162. M. R. Tansey, *Arch. Mikrobiol.*, *77*, 1 (1976).
163. G. S. Rautela and E. B. Cowling, *Appl. Microbiol.*, *14*, 892 (1966).
164. B. S. Montenecourt and D. E. Eveleigh, *Appl. Env. Microbiol.*, *33*, 78 (1977).
165. K. M. H. Nevalainen, E. T. Palma, and M. J. Bailey, *Enzyme Microb. Technol.*, *2*, 59 (1980).
166. R. Mullings and J. H. Parish, In *The Flavobacterium-Cytophaga Group* (H. Reichenbach and O. B. Weeks, eds.), Verlag Chemie, Weinheim, 1981.
167. R. M. Teather and P. J. Wood, *Appl. Env. Microbiol.*, *43*, 777 (1982).
168. P. Beily, D. Mislovicova, and R. Toman, *Anal. Biochem.*, *144*, 142 (1985).
169. M. Mandels and J. Weber, *Adv. Chem. Ser.*, *95*, 391 (1969).
170. H. L. Griffin, *Anal. Biochem.*, *56*, 621 (1973).
171. M. V. Deshpande, K.-E. Ericksson, and L. G. Petterson, *Anal. Biochem.*, *138*, 481 (1984).
172. R. M. Gardner, K. C. Doerner, and B. A. White, *J. Bacteriol.*, *169*, 4581 (1987).
173. G. A. Somkuti, *J. Gen. Microbiol.*, *81*, 1 (1974).
174. M. Hrmova, P. Beily, M. Vrsanka, and E. Petrakova, *Arch. Microbiol.*, *138*, 371 (1984).
175. K. D. Haggett, W. Y. Choi, and N. W. Dunn, *Eur. J. Appl. Microbiol. Biotechnol.*, *6*, 189 (1978).
176. J. K. Alexander, *Meth. Enzymol.*, *28*, 944 (1972).
177. J. K. Alexander, *Meth. Enzymol.*, *28*, 948 (1972).
178. J. B. Sumner, *J. Biol. Chem.*, *65*, 393 (1925).
179. J. B. Sumner and E. B. Sisler, *Arch. Biochem.*, *4*, 333 (1944).
180. G. L. Miller, *Anal. Chem.*, *31*, 426 (1959).
181. G. L. Miller, R. Blum, W. E. Glennon, and A. L. Burton, *Anal. Biochem.*, *2*, 127 (1960).
182. N. Nelson, *J. Biol. Chem.*, *153*, 375 (1944).
183. M. Somogyi, *J. Biol. Chem.*, *160*, 69 (1945).
184. P. A. Schaffer and M. Somogyi, *J. Biol. Chem.*, *100*, 695 (1933).
185. M. Somogyi, *J. Biol. Chem.*, *160*, 62 (1945).
186. J. E. Hodge and B. T. Honfreiter, *Meth. Carboh. Chem.*, *1*, 17 (1962).
187. O. Folin and M. Malmros, *J. Biol. Chem.*, *83*, 115 (1929).
188. M. Somogyi, *J. Biol. Chem.*, *117*, 771 (1937).
189. J. T. Park and M. J. Johnson, *J. Biol. Chem.*, *181*, 149 (1949).

190. G. Halliwell and M. Riaz, Biochem. J., 116, 35 (1970).
191. R. M. Hoffman and T. M. Wood, Biotechnol. Bioeng., 27, 81 (1985).
192. R. Mullings and J. H. Parish, Enzyme Microb. Technol., 6, 491 (1984).
193. N. D. Cheronis, Mikrokhim. Acta, 16, 925 (1956).
194. K. Mopper and E. T. Degens, Anal. Biochem., 45, 147 (1972).
195. C. E. Weill and P. Hanke, Anal. Chem., 34, 1736 (1962).
196. E. Stahl and U. Kaltenbach, In Thin-layer Chromatography, A Laboratory Handbook (E. Stahl, ed.), Academic Press, New York, 1965, p. 461.
197. W. E. Trevelyen, D. P. Procter, and S. S. Harrison, Nature, 166, 444 (1960).
198. J. C. Touchstone and M. F. Dobbins, In Practice of Thin-layer Chromatography (J. C. Touchstone and M. F. Dobbins, eds.), John Wiley and Sons, New York, 1978, p. 161.
199. E. A. Johnson, M. Sakajoh, G. Halliwell, A. Madia, and A. L. Demain, Appl. Env. Microbiol., 43, 1125 (1982).
200. W. Schwald and J. N. Saddler, Enzyme Microb. Technol., 10, 37 (1988).
201. R. Lachenicht and E. Bernt, In Methods of Enzymatic Analysis, Vol. 3 (H. U. Bergmeyer, ed.), Verlag Chemie, Weinheim, FRG, 1981, p. 1201.
202. H. U. Bergmeyer and E. Bernt, In Methods of Enzymatic Analysis, Vol. 13 (H. U. Bergmeyer, ed.), Verlag Chemie, Weinheim, 1974, p. 1205.
203. M. D. Joshi and V. Jagannathan, Meth. Enzymol., 9, 371 (1973)
204. J. B. Lloyd and W. J. Whelan, Anal. Biochem., 30, 467 (1969).
205. G. Peiji, Biotechnol. Bioeng., 29, 903 (1987).
206. T. K. Ng and J. G. Zeikus, U.S. Patent 4,430,032 (Sept. 6, 1983).
207. M. Leisola and V. Kauppinen, Biotechnol. Bioeng., 20, 837 (1978).
208. H. van Tilbeurgh, M. Claeyssens, and C. K. de Bruyne, FEBS Lett., 149, 152 (1982).
209. S. Nemec, Mycopathol. Mycol. Appl., 52, 283 (1974).
210. J. J. Child, D. E. Eveleigh, and A. S. Sieben, Can. J. Biochem., 51, 39 (1973).
211. K.-E. Ericksson and G. Petterson, Arch. Biochem. Biophys., 124, 160 (1968).
212. W. Klop and P. Kooiman, Biochem. Biophys. Acta, 99, 102 (1965).
213. M. L. Huggins, In Cellulose and Cellulose Derivatives, 2nd ed., Part III (E. Ott and H. M. Spurlin eds.), Interscience, New York, 1955, p. 1189.

214. K. E. Almin and K.-E. Ericksson, *Biochem. Biophys. Acta*, *139*, 238 (1967).
215. H. Vink, In *Cellulose and Cellulose Derivatives*, Part IV (N. M. Bikales and L. Segal, eds.), Wiley-Interscience, New York, 1971, p. 468.
216. K. E. Almin, K.-E. Ericksson, and B. Petterson, *Eur. J. Biochem.*, *51*, 207 (1975).
217. K. Manning, *J. Biochem. Biophys. Meth.*, *5*, 189 (1981).
218. G. Canevascini, M.-R. Coudray, J.-P. Rey, R. J. G. Southgate, and H. Meier, *J. Gen. Microbiol.*, *110*, 291 (1979).
219. IUPAC Commision on Biotechnology, *Pure Appl. Chem.*, *59*, 258 (1987).
220. M. Chan, C. Breuil, W. Schwald, and J. N. Saddler, *Appl. Microbiol. Biotechnol.*, *31*, 413 (1989).
221. C. Breuil and J. N. Saddler, *Enzyme Microb. Technol.*, *7*, 327 (1985).
222. W. Schwald, M. Chan, C. Breuil, and J. N. Saddler, *Appl. Microbiol. Biotechnol.*, *28*, 398 (1988).

14

Structural Features of Cellulose and Cellulose Derivatives, and Their Effects on Enzymatic Hydrolysis

BONAVENTURA FOCHER and ANNAMARIA MARZETTI *Stazione Sperimentale per la Cellulosa, Carta e Fibre Tessili Vegetali ed Artificiali, Milan, Italy*

PIER LUIGI BELTRAME and PAOLO CARNITI *Università di Milano, Milan, Italy*

I. INTRODUCTION

The structural properties of cellulosic materials are among the most important factors governing their enzymatic hydrolysis. These properties are essentially related to the origin and history of each cellulosic material and can be modified by adequate pretreatment in order to enhance the yield and rate of the hydrolysis reaction.

Useful relationships between structural features and enzymatic degradation of cellulose are required not only for practical purposes but also for a better understanding of the reaction mechanism.

II. PRETREATMENTS OF CELLULOSIC MATERIALS AND CHANGES OF THEIR STRUCTURAL FEATURES

Pretreatments of cellulosic materials to facilitate their enzymatic hydrolysis can essentially be of a physical and chemical type; they are widely reviewed up to 1982 by Chang et al. (1) and Fan et al. (2). (See also Addendum.)

A. Physical Methods

1. Dry and Wet Milling

Celluloses submitted to dry milling undergo a transition from crystalline to amorphous state at different rates depending on their initial structure (3). X-ray diffraction profiles indicate that native cellulose is degraded mainly in the direction of 004, 101, and 10$\bar{1}$ crystallographic planes during the initial phases of dry milling, while the 002 plane is affected only when milling time is increased (4). Milled native cellulose, when wetted or exposed to high relative humidity, recrystallizes as either cellulose I or cellulose II depending on the extent of ball milling. By short milling it retains the memory of the initial order, whereas by a prolonged treatment it converts to cellulose II.

The CP/MAS ^{13}C-NMR spectrum of ball-milled cellulose I is clearly different from that of the parent cellulose (5). All sharper features typical of C-4 and C-6 carbons of cellulose I have vanished and the crystallinity of the sample is not detectable.

The Raman spectrum (6) shows the broadening of all the bands and the occurrence of few and unresolved features. More in particular, a linearly disordered state of the chains is identifiable (7) in which the chain conformations differ from those present in cellulose I and II.

Microscopic observations indicate the complete disintegration of the fiber structure even if some agglomeration of particles is still evident. This "fusion" of the ball-milled cellulose as well as the collapsing of the capillary structure during the treatment could be a possible explanation of the low value of the surface area obtained (8). However, the drying of the sample under vacuum or by solvent exchange can influence the measurement of the surface area (9).

Cellulose II is more resistant to dry milling because the orientation of the 101 planes on the surface of the microfibrils leads to a higher cohesion energy (10).

The wet-milling treatment is found to provoke fibrillation and delamination of cellulose but its crystallinity and degree of polymerization (DP) are practically unaffected owing to the plasticizing action of water.

All the other milling treatments (two-roll, hammer, colloid, vibro energy milling) are more or less effective depending on the experimental conditions (time, temperature, etc.).

2. High-Pressure Treatment

In steam explosion pretreatment, high-pressure saturated steam heats and permeates the cellulosic substrate, causing the hydrolysis of more accessible glycosidic links and hemicellulose acetate groups. The hydrolysis is autocatalyzed by the acetic acid produced. After contacting with steam the substrate is quickly flashed to atmospheric pressure and the water inside the substrate vaporizes and expands rapidly, disintegrating the cellulosic material (11). Fiber fragmentation and the increase of specific surface area lead to a material highly accessible to chemical and enzymatic reagents.

X-ray diffraction studies show that the native cellulose crystals are preserved with an increase of the lateral size/perfection as results from the sharpening of the 002 plane signal (12).

The CP/MAS ^{13}C-NMR spectra (13) indicate the reduction of the shoulders of C-4 and C-6 carbon signals and a general sharpening of resonance peaks. When *Valonia* cellulose is exploded after treatment at 230°C for 4 min, no dramatic changes in the spectrum are observed, whereas when the treatment is carried out at 260°C, a remarkable splitting of C-1, C-4, and C-6 signals occurs due to a modification in the environment of the microfibrils, probably at the level of the chain conformation (14).

Increasing the treatment time also results in an increased pore size of the exploded substrate, owing to fiber fractures and disrupting of cell walls into microfibrils; as a consequence, an enhancement of the accessibility of the cellulosic material to cellulase components is obtained (15).

3. High-Energy Radiation

When fibrous cellulose is subjected to γ irradiation, cellulosic chain depolymerization (16) and loss in fiber breaking strength are observed. The radiation acts perpendicularly to the orientation of cellulose fibrils and affects only the intramolecular hydrogen bonds; the intermolecular hydrogen bonds appear unaltered even by 100-Mrad doses. X-ray diffraction indicates that the supermolecular structure of cellulosic materials is not affected by the treatment (17).

The apparent increase of the surface area is due to the formation of new spaces between microfibrils as well as to an extensive depolymerization. The cellulose irradiated by 100-Mrad doses is soluble in dilute sodium hydroxide solutions, and its subsequent regeneration leads to formation of cellulose II with high values of moisture regain and water retention (16).

B. Chemical Pretreatments

The intracrystalline swelling of cellulosic materials is achieved by such chemical reagents as NaOH solutions (18), liquid NH_3 (19–21), and aliphatic amines (22). They are capable of disrupting the inter- and intramolecular hydrogen bonds and converting the crystal lattice of native cellulose into celluloses II and III. The cellulose

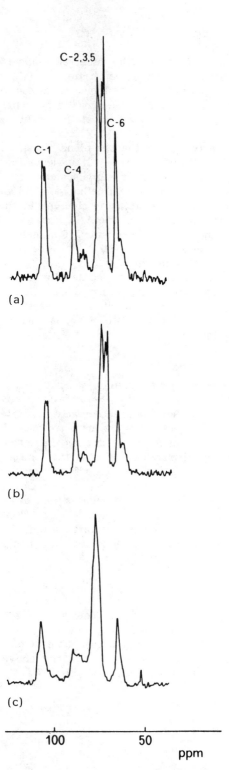

(a)

(b)

(c)

allomorphs differ with respect to chain conformation (6), three-dimensional packing, and morphology (23).

Agents such as aqueous $ZnCl_2$ solutions, which decrystallizes native cellulose, largely modify the specific surface area and the order/disorder ratio of the material (24). During the regeneration step, the reorganization of cellulosic chains is not allowed. The X-ray diffractograms, as well as the CP/MAS ^{13}C-NMR spectra (Fig. 1), are very similar to those of amorphous material, providing direct evidence of the effectiveness of the pretreatment.

However, the most significant modifications of morphology and supermolecular structure of native cellulose are achieved by dissolving it in both traditional (25) and new organic solvents of cellulose (26). The regeneration of cellulose can lead either to cellulose II or to a three-dimensional amorphous network depending on experimental conditions. In the latter case, X-ray diffractograms and CP/MAS ^{13}C-NMR spectra indicate the disappearance of the ordered region, and specific surface area measurements show an increase in the permeability of the material.

III. THE MODE OF ENZYMATIC DEGRADATION OF CELLULOSE: KINETICS AND MATHEMATICAL MODELS

The enzymatic hydrolysis of cellulose by cellulase is rather complex owing to (1) the heterogeneity of the system, (2) the dual nature of the insoluble substrate (crystalline and amorphous), and (3) the component multiplicity of the enzymatic system. This complexity is increased by the fact that the structure of the substrate changes during the process and that the different enzyme components act in a synergistic fashion.

Four types of enzymes are mainly involved in the enzymatic cleavage of cellulose: endo-1,4-β-D-glucanases (EC 3.2.1.4) (E_1), exocellobiohydrolases (EC 3.2.1.91) (E_2), β-D-glucosidases (EC 3.2.1.21) (E_3), and exo-1,4-β-D-glucosidases (EC 3.2.1.74) (E_4) (27).

The kinetics of hydrolysis of insoluble cellulose by cellulase was reviewed in detail up to 1979 by Lee et al. (28), and subsequently the state of the art on this topic was given in some reports (29–31). Here only recent kinetic models will be briefly illustrated.

FIGURE 1　CP-MAS ^{13}C-NMR spectra of (a) untreated cotton cellulose; (b) $ZnCl_2$-treated cotton cellulose; (c) cellulose pulp regenerated from DMSO-PF solution. (From Ref. 26.)

298   Focher et al.

Surface of Macropore

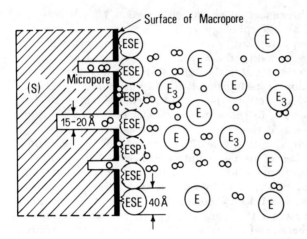

FIGURE 2   Representation of the mechanism of enzyme hydrolysis
of cellulose proposed by Fan and Lee (29). E = cellulase, (S) =
cellulose, E_3 = β-glucosidase, ESE = effective portion of adsorbed
enzyme, ESP = dead-end complexes, o = glucose, ∞ = cellulose.

    Three factors mainly affect the enzymatic hydrolysis of insoluble
cellulose: the properties and mode of action of cellulase, the structural properties of cellulose, and the mode of interaction between
the enzymes and the cellulose molecules. All of these have been
taken into account by Fan and Lee (29) in their development of a
comprehensive kinetic model based on the mechanism depicted in
Fig. 2. It was derived according to the following major assumptions: (1) the insoluble cellulose is hydrolyzed first to cellobiose
by the synergistic action of E_1 and E_2, these enzymes being considered in the mechanism as a whole (E); (2) the reaction to give
cellobiose is controlled by the extent of soluble protein (enzyme)
adsorption and occurs at the surface of solid cellulose, being inhibited by cellobiose (C) and glucose (G); (3) the decomposition of
cellobiose to give glucose in the liquid phase is competitively inhibited by glucose. The dual nature of cellulose, crystalline and
amorphous, was ignored but the conversion of cellulose (S) into a
less digestible form (S') was considered. The influence of mass
transfer on the overall rate was assumed to be negligible. The
mechanism results in the following steps:

Heterogeneous and solid phase reactions:

$$(E) + (S) \rightleftharpoons (ES) \qquad \text{Adsorption}$$

$$\left. \begin{array}{l} (ES) + (C) \rightleftharpoons (ESC) \\[2mm] (ES) + (G) \rightleftharpoons (ESG) \end{array} \right\} \quad \text{Product inhibition}$$

$$(ESE) \longrightarrow (E) + (C) \qquad \text{Hydrolysis to cellobiose}$$

$$(S) \longrightarrow (S)' \qquad \text{Transformation of cellulose}$$

Homogeneous reactions:

$$(E_3) + (C) \rightleftharpoons (E_3C) \longrightarrow (E) + G) \qquad \text{Hydrolysis to glucose}$$

$$(E_3) + (G) \rightleftharpoons (E_3G) \qquad \text{Product inhibition}$$

where $(ESE) = (ES) - (ESC) - (ESG)$ is the effective portion of the adsorption complex. This comprehensive model combines specific models for several key aspects of the enzymatic hydrolysis of insoluble cellulose which were previously derived independently of each other.

A kinetic model similar to that of Fan and Lee was presented by Wald et al. (30), who discriminated between amorphous (S_a) and crystalline (S_c) cellulose and considered only cellobiose inhibition on the adsorption complex (ES_a). In addition, cellobiose hydrolysis was assumed to involve a mixed-type product inhibition.

The difference between amorphous and crystalline cellulose was accounted for also by Ryu et al. (32). They considered the adsorption step as $(E) \longrightarrow (E^*)$ and the subsequent reactions occurring on the surface of cellulose as

$$(E^*) + (S_a) \rightleftharpoons (E^*S_a) \xrightarrow{k_a} (E^*) + (P)$$

$$(E^*) + (S_c) \rightleftharpoons (E^*S_c) \xrightarrow{k_c} (E^*) + (P)$$

Similar adsorption and reaction steps have been suggested by Asenjo (33) and Holtzapple et al. (34), but they did not differentiate between crystalline and amorphous cellulose.

The mechanism of cellulase adsorption on cellulosic materials, which is a prerequisite step for the subsequent hydrolysis reaction, is still not completely understood, although it has been investigated by several authors (27,35−40). Ryu et al. (37) reported some results useful to elucidate the synergism between E_1 and E_2 on the basis of a competitive adsorption model. On the other hand, the adsorption process can hardly be kinetically distinguished from the formation of the Michaelis−Menten substrate−enzyme complex and the equations involving adsorption can be rewritten as apparent Michaelis−Menten expressions (33). Moreover, in the case of $ZnCl_2$-treated (decrystallized) cotton cellulose the value of the Michaelis constant K_m of the enzymatic hydrolysis was found to be well correlated with Langmuir adsorption parameters (36).

Assuming that the adsorption process can be implicitly accounted for by the Michaelis−Menten equation, Beltrame et al. (41) developed a mathematical model according to most of the suggestions of Okasaki and Moo−Young (42). It involves enzymes E_1, E_2, and E_3. The targets of E_1 are all the glucosidic bonds except those of cellobiose; they are randomly hydrolyzed. The targets of E_2 are the nonreducing ends of all the molecules with degree of polymerization equal or greater than 3; cellobiose is the product of this hydrolysis. Eventually, cellobiose is hydrolyzed to glucose by enzyme E_3 (Fig. 3).

The general scheme for the action of either E_1 or E_2, involving a mixed-type inhibition by both cellobiose and glucose, is shown in Fig. 4 where G_i is a polysaccharide molecule ($i \geqslant 3$); EG, EC, EG_iG, and EG_iC are dead-end complexes, and EG_i is an intermediate complex. The action of E_3 was considered inhibited by glucose competitively and by its substrate, as independently determined (43). The differential equations that represent the whole mechanism of the hydrolytic process are as follows:

$$\frac{dC_{RS}}{dt} = \frac{V_{max1}C_B}{D_1} + \frac{V_{max2}C_M}{D_2} + \frac{V_{max3}C_C}{D_3}$$

$$\frac{dC_C}{dt} = \frac{2V_{max1}C_M}{D_1} + \frac{V_{max2}(C_M + C_{tetra})}{D_2} - \frac{V_{max3}C_C}{D_3}$$

$$\frac{dC_G}{dt} = \frac{2V_{max1}C_M}{D_1} + \frac{V_{max2}C_{tri}}{D_2} + \frac{2V_{max3}C_C}{D_3}$$

where C_{RS} is the concentration of reducing sugars; C_B and C_M are the concentrations of bonds and reducing ends, targets of E_1 and E_2,

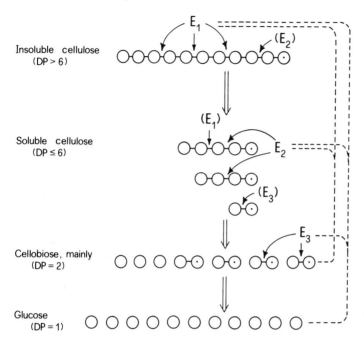

FIGURE 3 Action of a three-enzyme cellulase complex proposed by Okasaki and Moo–Young (42). (○-) glucose unit; (-◉) non-reducing end of cellulose or cellodextrin; (– – –) inhibition by end-products.

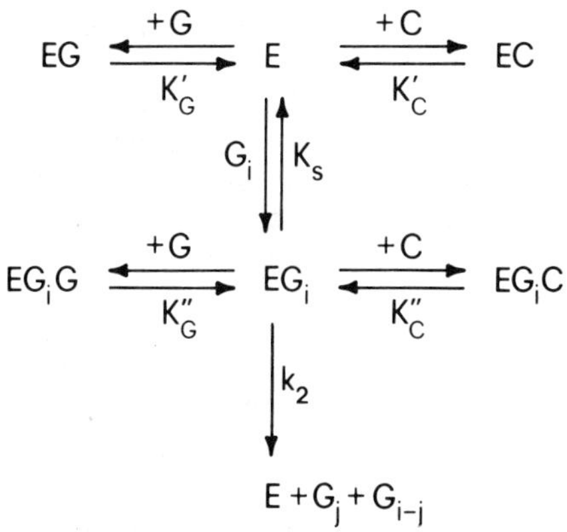

$$EG \underset{K'_G}{\overset{+G}{\rightleftharpoons}} E \underset{K'_C}{\overset{+C}{\rightleftharpoons}} EC$$

$$G_i \updownarrow K_s$$

$$EG_iG \underset{K''_G}{\overset{+G}{\rightleftharpoons}} EG_i \underset{K''_C}{\overset{+C}{\rightleftharpoons}} EG_iC$$

$$\downarrow k_2$$

$$E + G_j + G_{i-j}$$

FIGURE 4 Mechanistic scheme for the action of enzyme E on poly-saccharide G_i involving inhibition by cellobiose (C) and glucose (G).

302   *Focher et al.*

respectively; C_{tri} and C_{tetra}, the concentrations of cellotriose and cellotetraose, respectively; and

$$D_1 = K_{m1}(1 + C_G/K'_{1G} + C_C/K'_{1C}) + C_B(1 + C_G/K''_{1G} + C_C/K''_{1C})$$

$$D_2 = K_{m2}(1 + C_G/K'_{2G} + C_C/K'_{2C}) + C_M(1 + C_G/K''_{2G} + C_C/K''_{2C})$$

$$D_3 = K_{m3}(1 + C_G/K'_{3G}) + C_C + C_C^2/K''_{3C}$$

V_{maxi}, K_{mi}, K'_{iG}, K''_{iG}, K'_{iC}, and K''_{iC} refer to enzyme E_i

(i = 1, 2, 3)

Significant dependencies of the above parameters on temperature were accounted for according to linear Arrhenius – van't Hoff-type relationships.

The model focuses on the mode of action of each enzyme and, in particular, on the synergistic effects of E_1 and E_2. It does not consider the multiplicity of the substrate. The use of such a model allows a comparison between differently structured substrates on the basis of different kinetic parameter values. Thus, the structural features of cellulose can be related to the kinetics of its hydrolysis. Such an approach was followed when studying various pretreated cotton celluloses (26,41).

Gusakov et al. (44) formulated a mathematical model for the enzymatic hydrolysis of cellulose in a batch reactor. This model considers the four types of enzymes E_1-E_4. E_1, E_2, and E_4 have been assumed to attack both amorphous and crystalline cellulose to give cellobiose and glucose, the latter product being formed at a higher rate from the amorphous regions. Cellulases are competitively inhibited by the relevant products, and during the hydrolysis cellobiose-producing enzymes are inactivated according to a first-order kinetic equation. Each enzymatic reaction, including the hydrolysis of cellobiose by E_3, has been interpreted within the framework of Michaelis –Menten kinetics. The maximum rates were determined from empirical dependencies not directly proportional to the enzyme concentration. The above model has been included in a more comprehensive mathematical model designed for the enzymatic hydrolysis of cellulose in a plug-flow column reactor (45). In this case, the kinetics of the adsorption of the enzymes on cellulose was considered in order to take into account the distribution of cellulases along the reactor.

In other interesting reports on the modeling of enzymatic hydrolysis of cellulose, thermal enzyme deactivation has also been

considered (46) and simple empirical rate expressions for practical
purposes have been developed (47,48). (See also Addendum.)

IV. RELATIONSHIPS BETWEEN STRUCTURAL FEATURES AND ENZYMATIC HYDROLYSIS

Despite the fact that enzymatic hydrolysis of cellulose has been the
subject of a large number of studies in recent years, accurate relationships between structural features and hydrolysis rate have not
yet been found (49). Most of the investigators regard the accessibility and the crystallinity of cellulosic substrates as the major
factors affecting the kinetics of hydrolysis.

The surface area of wood and cotton fibers includes the external
surface (great as 1 m^2/g) and the internal surface. The latter at-
tains the maximum extent when the fiber is saturated with water
($300-600$ m^2/g). Without an effective pretreatment, however, most
of the internal area of the porous cellulose fibers is not accessible
to the cellulase system. According to Cowling et al. (50), the capillary internal structure affects the adsorption of the enzyme, which
is a prerequisite step for subsequent catalytic reaction. Stone et al.
(51) found a linear relationship between the initial reaction rate and
the surface area accessible to molecules of the same size of the cellulase system. On the other hand, Fan et al. (8) suggested that
surface area is not a limiting factor for hydrolysis, this being mainly dependent on the fine structural order well represented by crystallinity. Lee et al. (9), while confirming the occurrence of linear
relationships between surface area and hydrolysis rate, attributed
the discrepancies of the results to different experimental conditions.

The minor role of crystallinity in explaining the hydrolysis of
cellulose is outlined by several authors. Caufield and Moore (52)
indicated that the enzyme digestibility of crystalline regions is enhanced by grinding to a greater extent than is the digestibility of
amorphous regions. Thus they proposed that the overall increase in
the extent of hydrolysis is apparently a result of decreased size
rather than reduced crystallinity. Weimer and Weston (53) found a
weak negative correlation between degradation rate and crystallinity,
and concluded that the crystallinity index is less than a limiting
factor in cellulose degradation. A modest increase in the crystallinity index during hydrolysis is observed by several authors (54,
55), whereas the hydrolysis rate drops rapidly. On the other
hand, crystallinity is often reported as one of the critical structural
features that govern cellulose degradation (56,57).

Focher et al. (26) examined variously pretreated celluloses with
different crystallinities and surface areas, and reported the influence of these and other structural features on the kinetics of the

enzymatic hydrolysis. Such an influence was measured by the extent of the variations induced in the values of the kinetic coefficients that best described the trend of the reaction course for the various substrates (maximum rates, Michaelis constants, inhibition constants). In particular, for the E_1 component of the enzymatic complex, the differences among the substrates seemed to affect more the K_{m1} and the term $1/K_{1G}''$ than the V_{max1} values. For the E_2 component, the V_{max2} and $1/K_{2G}'$ values were found to be quite close to each other for all the substrates, likely due to the endwise attack of cellobiohydrolase. Interestingly, however, the K_{m2} values revealed rather significant differences among the substrates.

Another factor that might affect the enzymatic hydrolysis of cellulose is represented by the molecular dimensions of the cellulose chains, as expressed by the degree of polymerization. However, it has been reported that it does not influence either the rate or the yield of the hydrolysis reaction (16).

As a result of all the above observations to date, the search for a preferential structural parameter that essentially governs the rate of enzymatic hydrolysis has been unsuccessful. Difficulties in this search arise from the fact that it is virtually impossible to vary a single structural feature without affecting the others.

A fundamental topic that deserves more attention in future investigations on enzymatic hydrolysis of cellulose is represented by the conformational aspects of the interactions between enzyme and substrate that promote the formation of the adsorptive as well as the Michaelis complex. Mere physical contact between enzyme and substrate is not sufficient for the formation of such complexes. It is necessary that cellulose macromolecules have a peculiar conformation that does not create steric hindrance to the attack of glycosidic linkage (58) and at the same time allows a successful interaction with the enzyme molecules. For instance, the nonnative forms of cellulose with different chain conformations are more easily degraded than the parent polymer. To gain a deeper insight into the enzyme – substrate interaction according to the above suggestion, Henrissat and Chanzy (59) showed a preferential topology in the adsorption of a cellobiohydrolase (CBHI) on *Valonia* microcrystals. In particular, it was observed that while all equatorial electron diffraction lines vanished upon enzyme adsorption, the meridional reflections are still present and are apparently unaffected.

Pretreatments of cellulosic materials, besides affecting some selected crystallographic planes, can modify the conformation of the macromolecules of the parent polymer (10). As a result, the interaction between enzyme and substrate, as well as the subsequent hydrolysis reaction, is dramatically influenced. (See also Addendum.)

V. ENZYMATIC HYDROLYSIS OF CELLULOSE DERIVATIVES

By exchanging hydrogen atoms of anhydroglucopyranoside rings with various functional groups, the number of the intra- and intermolecular hydrogen bonds of cellulose chains are reduced in proportion to the degree of substitution (DS) and to the pattern of their occurrence along the molecular chains. Depending on the solvating capacity of the substituents as well as on the pattern of substitution, the DS value at which complete solubility is attained ranges from 0.4 to 0.7. Thus, the compound loses its initial ordered structure and becomes susceptible to the attack by the cellulase system. However, when DS approaches one substituent per anhydroglucose, the enzymatic hydrolysis rate decreases, and at higher DS values a complete immunity to cellulolytic action results. Cellulase systems can in fact break glycosidic linkages only where there are two or more contiguous unsubstituted anhydroglucose units (60−64).

Carboxymethylcellulose (CMC), like other cellulose derivatives, contains a nonuniform distribution of substituted residues. Hence, cellulase hydrolysis yields large fragments arising from the more highly substituted regions, whereas small fragments are obtained from those regions of the parent polymer less accessible to the derivatization agents. Generally, for high-DS cellulose derivatives, the cellulase system produces low glucose yields; only for insoluble (low-DS) CMC are high glucose yields reported (65).

The susceptibility of cellulose derivatives to enzymatic attack varies with the type of substituent group. Perlin et al. (66) suggested that the difference may be due either to the nature of substituents themselves or to the variations in the pattern of substitution. Boyer et al. (67) indicated that chemical derivatization can also modify the electrostatic properties of the cellulose chains and thus the interaction with enzyme. An increase in the negative charge on the macromolecule results in a reduction of the hydrolytic rate, whereas the presence of positive charges markedly increases cellulolytic action. The negative charges on the amino acid residues of cellulase involved in the substrate binding and catalytic action are responsible for this behavior.

The mechanism of enzymatic hydrolysis of soluble cellulose derivatives is simpler with respect to insoluble cellulose owing to the homogeneity of the reaction, but the substitution pattern of the substrate must be adequately taken into account. The hydrolysis begins with the rapid cleavage of a few of the longest and the most accessible unsubstituted parts of the substrate by the action of E_1. The resulting shorter polymer chains are the substrates for other cellulase components.

The various components of the cellulase system interact syner-gistically at an early stage of the reaction to enhance the reaction rate (68,69). This synergism is described in more detail in Chap. 21.

VI. ADDENDUM

(Sec. II) Most of the recent studies on the pretreatment of cellu-losic materials deal with the optimization of the operative conditions of autohydrolysis steam treatments, both with or without explosion, which have been resulted the most promising ones in terms of ef-fectiveness, suitability for different substrates and products, energy consumption, and environmental impact (70−73).

(Sec. III) The synergism by different enzymes in cellulose hydroly-sis has been considered and partially explained in light of the ad-sorption behavior of the cellulases by Beldman et al. (74). In their previous report (75), the adsorption parameters of purified endo-glucanases and exoglucanases were compared with the Michaelis con-stant of cellulose hydrolysis, confirming the observed (36) strict correlation between these two kinds of parameters.

Adsorption of cellulase on cellulosic materials was considered in other recent reports (76−79). In particular, Converse et al. (76) employed mathematical models involving an adsorption step that ac-count for hydrolysis inhibition. The role of pore size in the sub-strate in enzyme diffusion and the resulting consequences on syner-gism by the components of cellulase in cellulose hydrolysis have been examined by Tanaka et al. (80).

A comparison between kinetic models for enzymatic cellulose degradation was made considering the reaction carried out in aque-ous two-phase systems (81). Several interesting topics relating to the mode of action of enzymes in cellulose degradation were discussed in a recent symposium (82).

(Sec. IV) Referring to the influence of structural features of cel-lulosic materials on their bioconversion, the more recent results emphasize the importance of the surface area available to the enzymes and of the pore size and distribution inside these materials with re-spect to such parameters as crystallinity index and degree of polymer-ization of cellulose (80,83−85).

REFERENCES

1. M. M. Chang, Y.-C. T. Chou, and G. T. Tsao, *Adv. Biochem. Eng.*, *20*, 15 (1981).

2. L. T. Fan, Y.-H. Lee, and M. M. Gharpuray, *Adv. Biochem. Eng.*, *23*, 157 (1982).
3. G. Centola and D. Borruso, *Ind. Carta*, *16*(3), 87 (1962).
4. R. Butnarn and G. R. Simionescu, *Cellulose Chem. Technol.*, *7*, 641 (1973).
5. D. L. VanderHart and R. H. Atalla, *Macromolecules*, *17*, 1465 (1984).
6. R. H. Atalla, in *Wood and Agricultural Residues* (J. Soltes, ed.), Academic Press, New York, 1983, p. 59.
7. R. H. Atalla, *J. Appl. Polym. Sci.*, *Appl. Polym. Symp.*, *37*, 295 (1983).
8. L. T. Fan, Y.-H. Lee, and D. H. Beardmore, *Biotechnol. Bioeng.*, *22*, 177 (1980).
9. S. B. Lee, I. H. Kim, D. D. Y. Ryu, and H. Taguchi, *Biotechnol. Bioeng.*, *25*, 33 (1983).
10. G. Centola, *Teintex*, *7*, 513 (1963).
11. S. J. Gracheck, D. B. Rivers, L. C. Woodford, K. E. Gidding, and G. H. Emert, *Biotechnol. Bioeng. Symp. Ser. 11*, 47 (1981).
12. R. H. Marchessault, S. L. Malhotra, A. Y. Jones, and A. Perovic, in *Wood and Agricultural Residues* (J. Soltes, ed.), Academic Press, New York, 1983, p. 401.
13. M. G. Taylor, Y. Deslandes, T. Blum, R. H. Marchessault, M. Vincendon, and J. Saint Germain, *Tappi*, *66*(6), 92 (1983).
14. T. Tanahashi, M. Karina, and T. Higuchi, in *Proc. 4th ISWPC Symposium*, Paris, April 27–30, 1987, p. 343.
15. W. R. Grous, A. O. Converse, and H. E. Grethlein, *Enzyme Microbiol. Technol.*, *8*, 274 (1986).
16. B. Focher, A. Marzetti, C. Santoro, V. Sarto, and L. D'Angiuro, *Die Angew, Makromol. Chemie*, *102*, 187 (1982).
17. B. Focher, A. Marzetti, M. Cattaneo, P. L. Beltrame, and P. Carniti, *J. Appl. Polym. Sci.*, *26*, 1989 (1981).
18. J. O. Warwicker and A. C. Wright, *J. Appl. Polym. Sci.*, *11*, (1967).
19. M. Lewin and L. G. Roldan, *J. Polym. Sci.*, *Part C*, *36*, 213, 659 (1971).
20. J. O. Warwicker, *Cellulose Chem. Technol.*, *6*, 85 (1972).
21. H. Schleicher, C. Daniels, and B. Philipp, *J. Polym. Sci.*, *Part C*, *47*, 251 (1974).
22. P. K. Chidambareswaran, S. Sreenivasan, N. B. Patil, H. T. Lokhande, and S. R. Shukla, *J. Appl. Polym. Sci.*, *22*, 3089 (1978).
23. A. Sarko, in *New Industrial Polysaccharides* (V. Crescenzi, I. C. M. Dea, and S. S. Stivala, eds.), Gordon and Breach, New York, 1985, p. 87.
24. S. M. Bertrabet, E. H. Daruwalla, H. T. Lokhande, and M. R. Padhye, *Cellulose Chem. Technol.*, *3*, 309 (1969).

25. M. R. Ladish and C. R. Tsao, *Science, 201*, 743 (1978).
26. B. Focher, A. Marzetti, V. Sarto, P. L. Beltrame, and P. Carniti, *J. Appl. Polym. Sci., 29*, 3329 (1984).
27. A. A. Klyosov and L. Rabinowitch, in *Enzyme Engineering Future Directions* (L. B. Wingard, Jr., I. V. Berezin, and A. A. Klyosov, eds.), Plenum Press, New York, 1980, p. 83.
28. Y.-H. Lee, L. T. Fan, and L. S. Fan, *Adv. Biochem. Eng., 17*, 131 (1980).
29. L. T. Fan and Y.-H. Lee, *Biotechnol. Bioeng., 15*, 2707 (1983).
30. S. Wald, C. R. Wilke, and H. W. Blanch, *Biotechnol. Bioeng., 26*, 221 (1984).
31. W. D. Eigner, A. Huber, and J. Schurz, *Cellulose Chem. Technol., 19*, 579 (1985).
32. D. D. Y. Ryu, S. B. Lee, T. Tassinari, and C. Macy, *Biotechnol. Bioeng., 24*, 1047 (1982).
33. J. A. Asenjo, *Biotechnol. Bioeng., 25*, 3185 (1983).
34. M. T. Holtzapple, H. S. Caram, and A. E. Humphrey, *Biotechnol. Bioeng., 26*, 775 (1984).
35. A. A. Huang, *Biotechnol. Bioeng., 17*, 1421 (1975).
36. P. L. Beltrame, P. Carniti, B. Focher, A. Marzetti, and M. Cattaneo, *J. Appl. Polym. Sci., 27*, 3493 (1982).
37. D. D. Ryu, C. Kim, and M. Mandels, *Biotechnol. Bioeng., 26*, 488 (1984).
38. J. Y. Stuart and D. L. Ristroph, *Biotechnol. Bioeng., 27*, 1056 (1985).
39. M. Tanaka and R. Matsuno, *Enzyme Microb. Technol., 7*, 197 (1985).
40. M. Tanaka, H. Nakamura, M. Taniguchi, T. Morita, R. Matsuno, and T. Kamikubo, *Appl. Microbiol. Biotechnol., 23*, 263 (1986).
41. P. L. Beltrame, P. Carniti, B. Focher, A. Marzetti, and V. Sarto, *Biotechnol. Bioeng., 26*, 1233 (1984).
42. M. Okazaki and M. Moo-Young, *Biotechnol. Bioeng., 20*, 637 (1978).
43. P. L. Beltrame, P. Carniti, B. Focher, A. Marzetti, and V. Sarto, *Chim. Ind.* (Milan), *65*, 398 (1983).
44. A. V. Gusakov, A. P. Sinitsyn, and A. A. Klyosov, *Enzyme Microb. Technol., 7*, 346 (1985).
45. A. V. Gusakov, A. P. Sinitsyn, and A. A. Klyosov, *Enzyme Microb. Technol., 7*, 383 (1985).
46. G. Caminal, J. Lopez-Santin, and C. Solà, *Biotechnol. Bioeng., 27*, 1282 (1985).
47. K. Ohmine, H. Ooshima, and Y. Harano, *Biotechnol. Bioeng., 25*, 204 (1983).

48. M. T. Holtzapple, H. S. Caram, and A. E. Humphrey, *Biotechnol. Bioeng.*, *26*, 936 (1984).
49. L. T. Fan, Y.-H. Lee, and D. H. Beardmore, *Adv. Biochem. Eng.*, *14*, 101 (1980).
50. E. B. Cowling and T. K. Kirk, *Biotechnol. Bioeng. Symp. Ser.* *6*, 95 (1976).
51. J. E. Stone, A. M. Scallan, E. Donefer, and E. Ahlgreen, *Adv. Chem. Ser.*, *95*, 219 (1969).
52. D. F. Caufield and W. E. Moore, *Wood Sci.*, *6*, 375 (1974).
53. P. J. Weimer and W. M. Weston, *Biotechnol. Bioeng.*, *27*, 1540 (1985).
54. L. Segal, J. J. Creely, A. E. Martin, and C. M. Conrad, *Text. Res. J.*, *29*, 786 (1959).
55. K. Nisizawa, *J. Ferment. Technol.*, *51*, 267 (1973).
56. B. Norkrans, *Phys. Plantarum*, *3*, 75 (1950).
57. C. S. Walseth, *Tappi*, *35*, 233 (1952).
58. R. H. Atalla, *Adv. Chem. Ser.*, *181*, 55 (1979).
59. B. Henrissat and H. Chanzy, in *Cellulose: Structure, Modification and Hydrolysis* (R. A. Young and R. H. Rowell, eds.), John Wiley and Sons, New York, 1986, p. 337.
60. E. T. Reese, *Ind. Eng. Chem.*, *49*, 89 (1957).
61. M. G. Wirick, *J. Polym. Sci.*, *A-1, 6*, 1705 (1968).
62. M. G. Wirick, *J. Polym. Sci.*, *A-1, 6*, 1965 (1968).
63. K. E. Eriksson and B. H. Hollmark, *Arch. Biochem. Biophys.*, *133*, 233 (1969).
64. L. F. McBurney, in *Cellulose and Cellulose Derivatives* (E. Ott, H. H. Spurlin, and M. W. Grafflin, eds.), Interscience, New York, 1954, p. 105.
65. G. C. Dominguez, C. R. Engler, and E. J. Soltes, *Biotechnol. Bioeng. Symp. Ser.*, *15*, 98 (1985).
66. S. S. Battacharjee and A. S. Perlin, *J. Polym. Sci., Part C*, *36*, 509 (1971).
67. R. F. Boyer and H. A. Redmond, *Biotechnol. Bioeng.*, *25*, 1311 (1983).
68. M. Fujii and M. Shimizu, *Biotechnol. Bioeng.*, *28*, 878 (1986).
69. T. M. Wood and S. J. McCrae, *Adv. Chem. Ser.*, *181*, 181 (1979).
70. R. Overend and E. Chornet, *Phil. Trans. Roy. Soc. Lond. A*, *321*, 523 (1987).
71. T. M. Wood and J. N. Saddler, *Meth. Enzymol.*, *160*, 3 (1988).
72. E. Chornet and R. Overend, in *Research on Thermochemical Biomass Conversion* (A. Bridgwater and J. Knester, eds.), Elsevier, New York, 1989.

73. B. V. Kokta, R. Chen, H. Y. Zhan, D. Barrette, and
    R. Vit, *Pulp and Paper Canada*, *89*, 3, T91 (1988).
74. G. Beldman, A. G. J. Voragen, F. M. Rombouts, and
    W. Pilnik, *Biotechnol. Bioeng.*, *31*, 173 (1988).
75. G. Beldman, A. G. J. Voragen, F. M. Rombouts, M. F.
    Searle-van-Leeuwen, and W. Pilnik, *Biotechnol. Bioeng.*, *30*,
    251 (1987).
76. A. O. Converse, R. Matsuno, M. Tanaka, and M. Taniguchi,
    Biotechnol. Bioeng., *32*, 38 (1988).
77. W. Steiner, W. Sattler, and H. Esterbauer, *Biotechnol. Bioeng.*,
    32, 853 (1988).
78. G. Gonzàlez, G. Caminal, C. de Mass, and J. Lòpez-Santin,
    Biotechnol. Bioeng., *34*, 242 (1989).
79. A. Kyriacou, R. J. Neufeld, and C. R. MacKenzie, *Biotechnol.
    Bioeng.*, *33*, 631 (1989).
80. M. Tanaka, M. Ikesaka, R. Matsuno, and A. O. Converse,
    Biotcchnol. Bioeng., *32*, 698 (1988).
81. C. F. Mandenius, B. Nilsson, I. Persson, and F. Tjerneld,
    Biotechnol. Bioeng., *31*, 203 (1988).
82. J. P. Aubert, P. Beguin, and J. Millet, *Biochemistry and
    Genetics of Cellulose Degradation*, Academic Press, New York,
    1988.
83. D. E. Eveleigh, *Phil. Trans. Roy. Soc. Lond. A*, *321*, 435
    (1987).
84. Z. Dermoun and J. P. Belaich, *Appl. Microbiol. Biotechnol.*,
    27, 399 (1988).
85. K. K. Y. Wong, K. F. Deverell, K. L. Mackie, T. A. Clark,
    and L. A. Donaldson, *Biotechnol. Bioeng.*, *31*, 447 (1988).

15
Role of Microbial Cellulose Degradation in Reptile Nutrition

KATHERINE TROYER   *U.S. National Museum of Natural History,*
Smithsonian Institution, Washington, D.C.

I. INTRODUCTION

The image of magnificent plant-eating dinosaurs peacefully defoliating a prehistoric landscape is a universally familiar one; of all animals, only their predatory contemporaries possibly retain a greater imaginative appeal to the general public. Consequently, it is common knowledge that the largest and, in terms of biomass and evolutionary longevity, most successful vertebrate herbivores in the history of the earth were reptiles. Today, in contrast, relatively few members of this class subsist primarily on plants; therefore, the utilization of cellulose is restricted to a small subset of modern reptiles

Because cellulosic materials are derived from structural portions of plants, addressing the topic of cellulose degradation in metazoans implicitly necessitates a discussion of herbivory. Two aspects of reptilian herbivory have been consistently troublesome. First, it has not been clear exactly what is meant by the description of a given species as "herbivorous." The term "herbivory" has generally implied the consumption of any type of plant material — fruit and other highly digestible structures as well as the fibrous and less digestible plant parts such as leaves, stems, and buds. The importance of fiber content of the diet and microbial fermentation of cell wall material has been long recognized by animal scientists (reviewed in Ref. 1) and the technology for measuring fiber content of food and feces has been available for some time (2). However, field biologists have been slow to make use of this knowledge in interpreting the role of anatomy and nutrition in the ecology and behavior of non-domestic herbivores (cf. 3–10).

Another issue that routinely arises in considerations of reptile herbivory concerns the relationship between body size, metabolic requirements, and type of diet — insectivorous or carnivorous vs. herbivorous — observed and expected in reptiles. In an enduringly provocative paper, Pough (11) calculated metabolic requirements for lizards of different body sizes, along with the rate of energy acquisition predicted from animal vs. plant foods, based on gross energy contents of sample materials from the literature. His results indicated that lizards below a certain body size (roughly 100 g) could not maintain a positive energy balance on a diet of plants. Pough concluded that lizards with an adult body size below the 100-g threshhold must be insectivorous, and the young of herbivorous species must begin life as insectivores, and could only shift to a vegetarian diet as body size reached the 100-g limit.

These conclusions appeared to be substantiated by the information on reptile diets that was available at the time, and in a number of smaller lizard species the proportion of plant material (all types) in the diet does increase with body size, e.g., *Anolis aeneus*, *A. richardi*, and *A. roquet* (12); *Liolaemus nitidus* (13); *Sceloporus poinsetti* (14); *Tropidurus pacificus* (15); *Uma scoparia* (16). However, additional data on lizard diets have complicated the picture. For example, in the family Iguanidae, the young of all but one large herbivorous lizard species are herbivorous from the time of hatching, even though body sizes at this stage are an order of magnitude below Pough's "threshold for herbivory" (17,18). In addition, several small species (5–20 g) within the genus *Liolaemus* have diets based primarily on leafy plant materials (see Sec. II.B). The question remains, therefore, to what extent small (<100 g) reptiles can utilize diets of fibrous plant materials, and whether intrinsic constraints of metabolism, ecology, or life history restrict evolution of adaptations for cellulose utilization in these reptiles.

II. SQUAMATA: SAURIA (LIZARDS)

A. Large Herbivores

Most of the larger herbivorous lizards (body size approximately 100 g and larger) display adaptations for herbivory that are analogous to those found in nonruminant mammals, with characteristic features of the digestive tract, structure, dentition, and, in many cases, behavior, associated with processing high-fiber food and maintaining a symbiotic fermentative microflora. The hindgut is relatively enlarged and usually elaborated in structure: it may be divided into two or more compartments, separated by circular valves. In addition, the posterior region or chamber of the hindgut may possess a series of infoldings, or semilunar valves, that partially occlude the lumen and may serve to retard the passage of digesta in addition to increasing the absorptive surface of the organ. Similar modifications of hindgut structure are found in several different families of lizards, demonstrating that this set of adaptations for utilizing a fibrous diet has evolved independently (17,18).

An example of a relatively elaborate hindgut structure, that of the neotropical green iguana (*Iguana iguana*), is illustrated in cutaway view in Fig. 1. In this species, the large intestine is divided

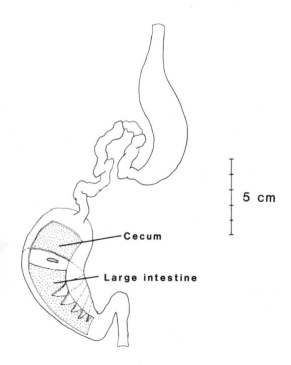

5 cm

Cecum

Large intestine

FIGURE 1   Cutaway view of the digestive tract of the green iguana *Iguana iguana*.

by a circular valve into two separate chambers; the anterior chamber in this type of arrangement has traditionally been named the "cecum," by analogy to the mammalian structure located at this site (e.g., 19). The posterior chamber (hereafter referred to as the "large intestine" proper) contains a number of semilunar valves that varies geographically, but is most commonly six (17).

Although relatively little direct evidence exists to date, it is generally believed that the cecum and/or large intestine of large herbivorous lizards houses a microbial fermentation system that degrades plant fiber. The only published measurements of fermentation rate in this group are those of McBee and McBee (20), using the green iguana (*I. iguana*). Their data are reported for the entire hindgut contents rather than on a per gram dry weight basis, and therefore cannot be compared directly with literature values for mammals. However, estimates of the proportion of daily energetic requirement satisfied by VFA (volatile fatty acid) production averaged 34%. This is ample evidence that the hindgut fermentation of fiber is a significant energy source in these animals, but because the absolute daily energy needs of reptiles are roughly an order of magnitude lower than those of comparably sized mammals (e.g., 21), the absolute rate of fermentation in the iguanas must be considerably slower than in mammals.

Even less information is available on the microbial species responsible for the hindgut fermentation. Direct counts of bacterial cells in hindgut contents from *I. iguana* showed concentrations around 10^{10} g^{-1}, similar to concentrations found in the rumen (20,32); ciliate protozoans were also observed in variable numbers (32). A limited number of roll tube cultures based on a glucose-cellobiose-starch medium demonstrated *Clostridium* and *Leuconostoc* as the dominant anaerobic genera; no *Bacteroides* or *Ruminococcus* were identified (20). Much more comprehensive culturing of hindgut contents from *Iguana* and other herbivorous lizards must be performed in order to elucidate the identities and biochemical roles of fermentative bacteria in these animals, and to determine how much the hindgut fermentation systems of reptiles differ from those of mammals.

Because of the difficulty of carrying out digestion trials with herbivorous reptiles (9,22), most studies of digestion in these animals have been based on artificial, low-fiber diets (reviewed in 23). Data are available for fiber utilization in the green iguana fed a natural, high-fiber food: leaves of the tropical tree *Lonchocarpus pentaphyllus* (Leguminosae), which has an average fiber content [about 50% NDF (neutral detergent fiber)] but a high index of lignification (0.18). In iguanas of all age groups combined, *Lonchocarpus* leaf NDF had a digestibility of 54%, and cellulose 42%. The NDF digestibility in iguanas compares favorably with values for mammalian herbivores (ruminants 60%, nonruminants 20–30%). Cellulose digestibility in iguanas was somewhat low compared to ruminant (90%) and nonruminant (40–60% mammals; the relatively low cellulose utilization may reflect the higher lignification of the iguana diet (0.18 vs. about 0.09 for the mammalian forages; comparative values from Refs. 24 and 25).

The only other published account of cellulose utilization in large herbivorous lizards (26) reports a mean digestive efficiency of 39% for the Galapagos land iguana (*Conolophus pallidus*) on a relatively low-fiber food, i.e., fruits of *Opuntia echios* (mean "cellulose" content 22.8%). Unfortunately, the authors used an older (1938) method for cellulose analysis which they failed to adequately describe (27, in 26). Although the reported value is similar to that of *I. iguana* on *Lonchocarpus* leaves, it cannot be assumed that their results are comparable with those obtained by the detergent method of Van Soest (2).

The dentition of large herbivorous lizards is usually modified for shearing fibrous material rather than grasping and piercing prey (28,29). The teeth tend to be laterally compressed and bladelike, and are often multicuspate (29), whereas the teeth of insectivorous lizards are primarily simple, conical structures (28). None of these animals displays adaptations for grinding food to reduce particle size, which is an important step in food processing by mammalian herbivores (25); it has been argued that the jaw structure of lizards cannot be readily modified to permit a grinding action (30). Even without grinding their food, however, green iguanas digested cell contents, notably protein, highly effectively: protein true digestibility was 88%, a value comparable to that seen in mammals (9).

Neonates of large herbivorous species display coprophagic behavior that functions to inoculate the hindgut with the fermentative microflora from conspecifics (31,32; Iverson, Werner, Swingland, pers. commun.). In the green iguana, this process entails leaving the low, open habitat normally preferred by hatchlings and remaining in the forest canopy, near one or more older conspecifics, for a period of days (31,32). Recently hatched iguanas appear actively attracted to iguana fecal material: when offered choice tests between various materials, experimental hatchlings consumed feces even in preference to food (Troyer and Rand, unpublished results).

Other behavior patterns are also involved in the herbivorous lifestyle of this group. Although nearly all species of large herbivorous lizards are herbivorous throughout life, the immature growing animals presumably require a relatively higher intake of protein and energy than that needed by adults. This difference in requirements may be reflected in differential diet selection among the same array of available foods. Juvenile green iguanas feeding on the same array of plants as adults selected proportionately more of the younger foliage, achieving diets significantly higher in digestible protein (10).

B. Small Herbivores

A number of smaller lizard species, i.e., those with body mass less than 100 g, have been described as herbivorous, but in those reports where the type of plant material is actually identified, the primary plant food in the diet is often low-fiber material such as fruit.

However, many species do consume leafy material, and leaves may even compose the major portion of the diet of certain species [e.g., *Liolaemus magellanicus* (33); *L. nitidus* (13); *L. fitzingeri*, Troyer unpublished].

I have found no description of digestive tract structure or function in these species that deals with the possibility of microbial fermentation. My preliminary (unpublished) observations within one group, species of the South American genus *Liolaemus*, suggest that some level of microbial fermentation may be functioning even in very small species (5–20 g). Species with a substantial proportion (e.g., roughly one-third) of leafy material in the diet tend to have enlarged hindguts compared to entirely or primarily insectivorous species, and the plant remains in the hindgut contents of the more herbivorous groups often appear relatively more degraded than similar material in the hindguts of primarily insectivorous species. The hindgut structure of two *Liolaemus* species are compared in Fig. 2: the species *L. fitzingeri* has a diet consisting of about 81% leafy materials, while the similarly sized *L. elongatus* is primarily insectivorous (diet about 17% leaves). It is obvious that the hindgut of the herbivorous species is relatively larger, and this may be demonstrated statistically by comparing the ratio of large intestine/head–body length in

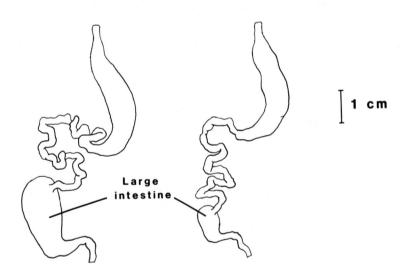

Large intestine

1 cm

Liolaemus fitzingeri   *Liolaemus elongatus*

FIGURE 2   Digestive tracts of small lizards *Liolaemus fitzingeri* (primarily herbivorous) and *Liolaemus elongatus* (primarily insectivorous).

specimens of the two species (*fitzingeri* mean = 0.313, n = 10; *elongatus* mean = 0.135, n = 12; Mann–Whitney U test, U = 0, $p < 0.001$).
Even the most herbivorous *Liolaemus* species, however, show no signs of elaboration in hindgut structure — valves, compartments — such as those found in the larger herbivores. If the function of the semilunar valves is to increase surface-to-volume ratio in the hindgut for optimal absorption rate of fermentation products, then possibly in small herbivores with intrinsically larger surface-to-volume ratios, the gut surface areas are sufficient for maximal absorption without valves. This interpretation is supported by the observation (18) that the modal number of hindgut semilunar valves in large herbivorous lizard species is positively correlated with body size, and the smallest herbivore in the series (*Dipsosaurus dorsalis*) has no semilunar valves in the large intestine.

Small herbivorous lizards may display quantitative modification of the dentition as well as digestive tract structure. I have noted a tendency for teeth to be somewhat more flattened and more cuspate in the herbivorous *Liolaemus* species, suggesting a shift in function from grasping live insect prey to shearing tough plant material (unpublished). Again, these observations are from work in progress; much more information of this kind is needed from a variety of species before the extent of adaptation for herbivory in small lizards can be assessed and compared with that seen in their larger counterparts.

In summary, some small lizards may utilize fibrous plant materials as a significant portion of their diet; however, the extent to which cellulosic materials are degraded in these animals is presently unknown. The evidence to date does not appear to support the original premise of a 100-g body size "threshold for herbivory" (11); in fact, recent observations, such as the discovery of a 5-g lizard species with a diet consisting entirely of leaves (33), suggest that there may be no minimum body size at which a lizard could satisfy day-to-day metabolic requirements on a diet based partly or even primarily on plant structural materials. However, there still remains the question of why so few small lizards are truly herbivorous, and why the young of at least some of these species rely primarily on animal matter for food (e.g., 13,14,34).

It may be that energetic constraints do exist but operate at a level different from that of maintenance metabolism. For example, the factor limiting rate of digestion of fibrous material is rate of microbial fermentation rather than digestion/absorption of soluble nutrients (10), and this rate appears to be independent of body size in herbivorous reptiles (6,7,9). Therefore, the rate at which a lizard can extract energy and nutrients from plant food might be sufficient for adult maintenance but insufficient for growth and/or reproduction, regardless of absolute adult body size. The sparse life

history information available for small herbivorous lizards suggests
that species with herbivorous diets may show reduced reproductive
rates compared with closely related insectivorous species (35). In
fact, it has been argued that in lizards, reduced reproductive potential is a consequence of low-quality diet that limits evolution of herbivory to species that live on islands and are hence subject to reduced predation (36-38). It is clear than in order to assess the extent of, and constraints on, cellulose utilization in small reptiles,
much more information is needed on diet, digestive tract structure
and function, and reproductive pattern for a much larger number of
species.

III. TESTUDINES (TURTLES AND TORTOISES)

Studies of diet and digestion in herbivorous testudinians are, as for
other reptiles, quite scarce. Both direct and indirect evidence suggest that these animals also house microbial fermentation systems
that degrade fibrous materials. Testudinians in general display much
less modification of the hindgut structure than the large herbivorous
lizards. In the green sea turtle, the "cecum," site of the highest
fermentative activity, is simply an expanded region at the anterior
extremity of the large intestine (5), and this appears to be the pattern in other herbivorous turtles as well as tortoises (e.g., 39).
The digestive tract structure of an immature specimen of the Bolson
desert tortoise (Gopherus flavomarginatus) is illustrated in Fig. 3.
        By far the most comprehensive information on testudinian cellu-
lose utilization is found in the state-of-the-art work of Bjorndal on
the green sea turtle, Chelonia mydas (5,6,40,41). This species feeds
on a variety of marine plants throughout its range, from seagrasses
to "seaweeds" (marine algae). The microbial fermentation system in
green sea turtles degrades cellulosic materials with a high degree of
efficiency. In Bjorndal's (5,6) study area for C. mydas, the turtles
fed primarily on the seagrass Thalassia testudinum; this material has
an NDF or cell wall content (mean of 58.9%) and index of lignification
(0.09) similar to that of typical domestic cattle forages in the tem-
perate zone (25). Protein digestibility in this species, however, was
inexplicably low compared to that of iguanas as well as mammalian herbivores on natural diets [38.2%, (Ref. 6); Bjordnal's explanation that
cell contents might be nutritionally unavailable prior to fermentation
of cell walls is not the case in mammals (25) and iguanas (9), and is
probably incorrect].
        Only two published studies describe hindgut fermentation in this
species. The cecum is reported to contain concentrations of bacteria
similar to levels found in the rumen 10^{10} ml^{-1}, and protozoans are
also abundant in this organ (42). Indirect evidence also suggests

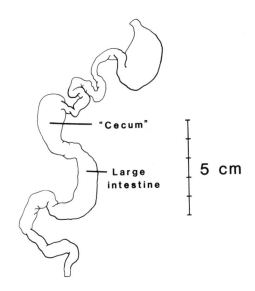

FIGURE 3   Digestive tract of Bolson tortoise *Gopherus flavomargi-natus.*

that the fermentative microflora in *C. mydas* may be specific to the type of food normally consumed: algae vs. seagrasses (6). Bjorndal (6) reported that when turtles accustomed to feeding only on seagrass were fed marine algae in captivity, they digested it markedly more poorly than conspecifics that normally fed on this food. Clearly, more information is needed on the nature and activity of the hindgut microflora in these animals.

In the one case where fermentation rate was measured (5), results were reported on a wet weight basis of the fluid fraction only and therefore cannot be compared with literature values for mammalian herbivores. However, based on this figure (12 mM liter^{-1} hr^{-1}), green sea turtles obtain about 15% of their daily energy needs from their hindgut fermentation (5), compared with values of about 34% for iguanas (20), and $70-80$% vs. $10-20$% for ruminant and nonruminant mammals, respectively (24). As discussed in the case of iguanas, because the energetic requirements of the reptiles are roughly one-tenth those of mammals, and the proportion of daily energy obtained from fermentation is similar, it follows that the absolute fermentation

rate must be approximately an order of magnitude slower in the reptiles. Digestibility of the cellulose fraction was estimated at about
90% in both studies (5,42); this very high efficiency of cellulose utilization is similar to that of ruminants on forages of similar fiber content and lignification, and higher than nonruminant values, e.g.,
20–30% (25).

A particularly fascinating aspect of herbivory in green sea turtles is their habit of routinely recropping the same beds of seagrass,
thus maintaining areas of "pasture" with a higher organic matter content and lower lignification index than ungrazed areas (6,41). This
behavior pattern clearly enhances digestibility of the available diet,
and resembles the intensive grazing pattern of the African ungulates,
which results in a similar improvement of the food supply (43). This
is an especially striking example of convergence in vertebrate herbivore ecology as well as digestive function.

Little published information is available on fiber utilization by terrestrial testudinians, all of which are primarily or strictly herbivorous (44). Bjorndal (7) found high digestive efficiencies in the go-
pher tortoise (Gopherus polyphemus) on a natural food, leaves of
the legume Aeschynomene americana. Digestive efficiency of cell wall
(NDF) averaged 73%, and protein digestibility was also high, with a
mean of 71%. The Aldalbran giant tortoise (Geochelone gigantea) is
reported to digest the holocellulose (cellulose + hemicellulose) content
of "tortoise turf" with about 42% efficiency (45, in 41) and a highly
lignified NDF was digested by an unidentified species with about 22%
efficiency (Hintz, unpublished, in 24). Clearly, much more information
is needed on diet and digestion in the herbivorous Testudinia, most
of which are threatened or endangered species (40,44).

IV. OTHER REPTILES

A. Rhynchocephalia (Tuatara)

The relict species Sphenodon, common name tuatara, is reported to
consume primarily animal material such as arthropods, snails, and
bird eggs and hatchlings (46). I have found no descriptions of gut
morphology in this species. The dentition is unusual in that the
degree of tooth occlusion is low and the jaw mechanics produce a
sawing motion during mastication (47). The teeth themselves are
relatively unspecialized, being generally conical in form like those
of insectivorous lizards (28,48). The limited anatomical and ecological evidence suggests that this species does not make significant
use of fibrous plant materials.

B. Squamata: Serpentes (Snakes)

No herbivores are known among the snakes, which are specialized
predators that consume living animals whole (49).

C. Crocodylia and Other Archosaurs

Modern crocodiles, alligators, and caiman are all aquatic predators that are generally insectivorous as hatchlings and increasingly carnivorous as the body size increases (49). Many extinct archosaurs, however, were highly specialized terrestrial herbivores (50). Various workers have speculated that the gastroliths or "stomach stones" found in various herbivorous species aided in the physical breakdown of fibrous food (e.g., 51), although gastroliths are also found in nonherbivorous dinosaurs and modern crocodilians (50). Several groups possessed dentition that was modified for grinding, notably the hadrosaurs ("duck-billed" dinosaurs) (52−54) and the ceratopsians (horned dinosaurs) (54,55). In contrast, the sauropods (brontosaur-type dinosaurs) had very unspecialized dentition in spite of their presumably fibrous diet (50). However, the high digestive efficiencies of modern reptiles without grinding dentition (e.g., 5,7,9) may indicate that the importance of grinding food has been overrated for these animals.

Fossilized digesta and feces ("coprolites") have been reported occasionally, although identifications of such materials are not always reliable (e.g., 50). At least one description of indisputable fossilized stomach contents has been published for a sauropod dinosaur (56): the bolus was composed primarily of leaves and stems. Evidence of microbial fermentation in dinosaurs must, of course, be interpreted indirectly, but the scanty information currently available, along with comparative data from living reptilian herbivores, suggests that cellulose and hemicellulose utilization was probably commonplace in these animals.

V. CONCLUSIONS

Based on the very few species that have been studied, herbivorous reptiles are capable of degrading cellulose and other cell wall materials extremely effectively. In contrast to mammalian herbivores, reptiles display very little physical processing of food: the teeth function primarily to snip off large pieces of leaves rather than to grind (reviewed in 48). Most of the breakdown of plant material is accomplished enzymatically, whether by endogenous or microbial systems. Digestible fractions, notably protein, may be extracted very efficiently through the action of an extremely acidic stomach and a (presumably) effective small intestine [*I. iguana*; (9)], although protein digestibility is low in some groups [*Chelonia mydas*; (6)]. Cell wall fractions, including cellulose, are degraded by a microbial fermentation system housed in a modified hindgut (5,6,9,20), and the proportion of daily energetic requirements that is satisfied by fermentation products may be substantial [15−38% (5,20)].

Available data on the utilization of cellulose and/or other cell wall materials by reptiles are summarized in Table 1. It is obvious

TABLE 1 Utilization of Cellulose by Herbivorous Reptiles

Species	Food	Apparent digestibility NDF	Apparent digestibility cellulose	Daily energy from hindgut fermentation (%)
Sea turtle Chelonia mydas (5,6)	Seagrass Thalassia	—	0.82	15
Sea turtle Chelonia mydas (42)	Seagrass Syringodium	—	0.90	—
Gopher tortoise Gopherus polyphemus (7)	Leaves Aeschynomene	0.73	—	—
Aldabran tortoise Geochelone gigantea (45, in 41)	"Tortoise turf"	(0.40)[a]	—	—
Green iguana Iguana iguana (9,20) and Troyer unpublished	Leaves Lonchocarpus	0.54	0.42	34
Land iguana Conolophus pallidus (26)	Fruit Opuntia	—	(0.39)[b]	—

[a]Holocellulose = hemicellulose + cellulose.
[b]Cellulose analysis of uncertain comparability.

that the study of reptile herbivory at the biochemical level is in a primordial state. We need much more empirical information on digestive system function, especially the species composition, activity, and contribution to metabolic requirements of microbial fermentation systems in these animals. Iverson's (17) survey of digestive tract anatomy in large lizard herbivores, and Bjorndal's (5−7,41) and Troyer's (9,10,32) analyses of digestive system structure and function in large herbivorous reptiles, demonstrate that microbial fermentation of cellulose is significant in these groups. However, only a modicum of information is available on what bacterial species are responsible for the fermentation and how the fermentation in reptilian herbivores compares with that of their mammalian counterparts. In addition, almost nothing is currently known about fiber utilization in the majority of large lizards, small lizards, freshwater turtles, tortoises, or, most unfortunately, dinosaurs.

At this point it appears that there are no immediate energetic limitations on the body size at which a reptile can rely on a high-fiber diet, in contrast to early assertions to the contrary (11) and subsequent reports claiming to lower some explicit or implicit body size threshold for herbivory, (e.g., 34). Reptilian herbivores that are known to degrade cellulosic materials appear to do so as extensively as mammals (5,9,41); however, fermentation rate and food turnover rate appear to be substantially slower in the reptiles (5,9,10,20). This difference may be due in part to the more variable body temperatures of reptiles under natural conditions (57) and the relatively smaller absorptive area of their digestive tracts (58). In any event, it suggests that herbivory may impose limitations on energy transformation, and therefore growth, reproduction, and behavior, in reptiles (22,40,41). Because so many reptilian herbivores are also endangered species (22,40,44), the nature and extent of life history constraints associated with adaptations for cellulose utilization are compelling areas for future research, both pure and applied.

REFERENCES

1. R. E. Hungate, *The Rumen and its Microbes*, Academic Press, New York, 1966.
2. P. J. Van Soest, *J. Anim. Sci.*, *26*, 119 (1967).
3. K. Milton, *Am. Nat.*, *114*, 362 (1979).
4. K. Milton, P. J. Van Soest, and J. Robertson, *Physiol. Zool.*, *53*, 402 (1981).
5. K. A. Bjorndal, *Comp. Biochem. Physiol.*, *63A*, 127 (1979).
6. K. A. Bjorndal, *Marine Biol.*, *56*, 147 (1980).
7. K. A. Bjorndal, *Copeia*, *1987*, 714 (1987).

8. E. S. Dierenfield, H. F. Hintz, J. B. Robertson, P. J. Van
    Soest, and O. T. Oftedahl, *J. Nutr.*, *112*, 636 (1982).
9. K. Troyer, *Physiol. Zool.*, *57*, 1 (1984).
10. K. Troyer, *Oecologia*, *61*, 201 (1984).
11. F. H. Pough, *Ecology*, *54*, 837 (1973).
12. T. W. Schoener and G. C. Gorman, *Ecology*, *49*, 819 (1968).
13. F. M. Jaksic and E. R. Fuentes, *Stud. Neotrop. Fauna Env.*,
    15, 109 (1980).
14. R. E. Ballinger, M. E. Newlin, and S. J. Newlin, *Am. Midl.
    Nat.*, *97*, 482 (1977).
15. D. Schluter, *Oikos*, *43*, 291 (1984).
16. J. E. Minnich and V. H. Shoemaker, *Copeia*, *1972*, 650 (1972).
17. J. B. Iverson, *J. Morphol.*, *163*, 79 (1980).
18. J. B. Iverson, In *Iguanas of the World: Their Behavior,
    Ecology, and Conservation* (G. M. Burghardt and A. S. Rand,
    eds.), Noyes, Park Ridge, NJ, 1982, p. 60.
19. E. Lonnberg, *Bihang. Till. K. Svenska Vet.-Akad. Handlingar*,
    28, P. A. Norstedt and Sons, Stockholm, 1902.
20. R. H. McBee and V. H. McBee, In *Iguanas of the World: Their
    Behavior, Ecology, and Conservation* (G. M. Burghardt and
    A. S. Rand, eds.), Noyes, Park Ridge, NJ, 1982, p. 77.
21. F. H. Pough, *Am. Nat.*, *115*, 92 (1980).
22. K. Troyer, *Herpetologica*, *39*, 317 (1983).
23. K. Troyer, *Comp. Biochem. Physiol. A.*, *87*, 623 (1987).
24. R. Parra, In *The Ecology of Arboreal Folivores* (G. G. Mont-
    gomery, ed.), Smithsonian Institution, Washington, D.C.,
    1978, p. 205.
25. P. J. Van Soest, *Nutritional Ecology of the Ruminant*, O & B
    Books, Corvallis, OR, 1982.
26. K. A. Cristian, C. R. Tracy, and W. P. Porter, *Herpetologica*,
    40, 205 (1984).
27. E. W. Crampton and L. A. Maynard, *J. Nutr.*, *15*, 383 (1938).
28. N. Hotton III, *Am. Midl. Nat.*, *53*, 88 (1955).
29. R. R. Montanucci, *Herpetologica*, *24*, 305 (1968).
30. J. H. Ostrum, *Evolution*, *17*, 368 (1963).
31. K. Troyer, *Science*, *216*, 540 (1982).
32. K. Troyer, *Behav. Ecol. Sociobiol.*, *14*, 189 (1984).
33. F. M. Jaksic and K. Schwenk, *Herpetologica*, *39*, 457 (1983).
34. A. Burques, O. Flores-Villela, and A. Hernandez, *J. Herpetol.*,
    20, 262 (1986).
35. J. J. Schall, *J. Herpetol.*, *17*, 406 (1983).
36. H. Szarski, *Evolution*, *16*, 529 (1962).
37. D. H. Janzen, *Ecology*, *54*, 687 (1973).
38. T. J. Case, *Acta Biotheoretica*, *28*, 54 (1979).

39. C. L. Guard, In *Comparative Physiology: Primitive Mammals* (K. Schmidt-Nielson, L. Bolis, and C. R. Taylor, eds.), Cambridge University Press, New York, 1980, p. 43.
40. K. A. Bjorndal, In *Biology and Conservation of Sea Turtles* (K. A. Bjorndal, ed.), Smithsonian Institution, Washington, D.C., 1982, p. 111.
41. K. A. Bjorndal, *Copeia*, *1985*, 736 (1985).
42. T. M. Fenchel, C. P. McRoy, J. C. Ogden, P. Parker, and W. E. Rainey, *Appl. Env. Microbiol.*, *37*, 348 (1979).
43. S. J. MacNaughton, *Science*, *191*, 92 (1976).
44. R. B. Brury, In *North American Tortoises: Conservation and Ecology* (R. B. Brury, ed.), USFWS Wildlife Research Report 12, U.S. Dept. Interior, Washington, D.C., 1982.
45. J. Hamilton and M. Coe, *J. Arid Env.*, *5*, 127 (1982).
46. W. H. Dawbin, *Endeavour*, *21*, 16 (1962).
47. P. L. Robinson, In *Morphology and Biology of Reptiles* (A. d'A. Bellairs and C. B. Cox, eds.), Linnean Soc. Symp. Ser. 3, Academic Press, New York, 1976, p. 43.
48. A. G. Edmund, In *Biology of the Reptilia*, Vol. 1, *Morphology*, Part A (C. Gans, A. d'A. Bellairs, and T. S. Parsons, eds.), Academic Press, New York, 1969, p. 117.
49. A. d'A Bellairs, *The Life of Reptiles*, Weidenfield and Nicholson, London, 1969.
50. J. O. Farlow, *Paleobiology*, *13*, 60 (1987).
51. R. T. Bakker, In *A Cold Look at the Warm-Blooded Dinosaurs* (R. D. K. Thomas and E. C. Olson, eds.), AAAS Sel. Symp. 28, Westview, Boulder, CO, 1980, p. 351.
52. J. H. Ostrum, *Bull. Am. Mus. Nat. Hist.*, *122*, 33 (1961).
53. D. B. Weishampel, *Acta Palaeontol. Polon.*, *28*, 271 (1983).
54. D. B. Norman, *Symp. Zool. Soc. Lond.*, *52*, 521 (1984).
55. J. H. Ostrum, *Evolution*, *20*, 290 (1966).
56. W. L. Stokes, *Science*, *143*, 576 (1964).
57. A. S. Rand, In *The Ecology of Arboreal Folivores* (G. G. Montgomery, ed.), Smithsonian Institution, Washington, D.C., 1978, p. 115.
58. W. H. Karasov, E. Petrossian, L. Rosenberg, and J. M. Diamond, *J. Comp. Physiol. B*, *156*, 599 (1986).

16

Cellulose Degradation in Ruminants

BURK A. DEHORITY   *Ohio Agricultural Research and Development Center, Ohio State University, Wooster, Ohio*

I. INTRODUCTION

The ruminant animal occupies a unique position among mammals in that it can obtain a majority of its energy requirements from cellulose. This is possible because of a large-scale microbial fermentation which takes place in the first two compartments of the ruminant stomach. Cellulolytic species of anaerobic bacteria, ciliate protozoa, and chytridiomycete fungi are all present in the rumen and can degrade cellulose to end products which serve as energy sources for the host animal. Volatile fatty acids are the principal end products produced, along with lesser amounts of lactic acid, carbon dioxide, and hydrogen. Additional products of the fermentation which can be used by the host are vitamins and microbial protein.

Cellulose comprises about 20–40% of the dry matter in forages consumed by ruminants. Availability of this energy to the animal, however, is dependent on its degradation in the rumen, which can vary between 30 and 90% (1). Type of forage and maturity appear to be the major factors affecting digestibility among the forage plants. The relationship between these two variables and digestibility is generally assumed to be the result of differences in lignification.

The present chapter will focus on the microorganisms which are active in the degradation of cellulose in the rumen and those factors which influence the rate and extent of degradation. These would include the rumen environment, animal and plant differences, and

interactions between the different microorganisms. Some information
will also be presented on cellulose degradation in the hindgut of both
ruminant and nonruminant herbivorous mammals.

II. THE RUMEN ENVIRONMENT

With the evolutionary development of a pregastric microbial fermentation, multichambered stomachs have evolved. This provided a way
to slow down ingesta passage rate and allowed physical separation of
the ingesta from the acid-secreting regions of the stomach. Adequate
production of buffered saliva was also needed to maintain pH in the
physiological range. A number of ruminant-like herbivores have developed, ranging from the hampster rat with a two-chambered stomach (2), tree sloths with three-chambered stomachs, and colobid monkeys whose stomachs have four parts (3). The hippopotamus stomach also contains four main chambers; however, the stomach anatomy
is unique to this animal species (2,3).
    Old World camels and New World camelids are sometimes called
pseudoruminants. This is based on the fact that they ruminate like
the true ruminants, i.e., forestomach contents are regurgitated, rechewed, and reswallowed. However, their stomach contains only three
chambers, the first of which contains glandular sacs (4).

A. Anatomy of the Rumen

In the true ruminant, the stomach has four compartments: the rumen,
reticulum, omasum, and abomasum (5,6). The entire inner surface of
the rumen, reticulum, and omasum is lined with squamous epithelium,
while the abomasum or pyloric region contains glandular cells. A
schematic diagram of the ruminant stomach is shown in Fig. 1. The
rumen and reticulum are separated by the reticuloruminal fold and
the anterior of the rumen itself is divided into sacs by various pillars. These pillars aid in contraction of the sacs and concomitant
mixing of the digesta (5).
    Capacity of the reticulorumen varies with size of the animal and
type of diet consumed, ranging from 3 to 15 liters in sheep and from
35 to 100 liters in cattle. In general, about 80−90% of the total stomach contents are found in the reticulorumen, where an extensive microbial fermentation takes place (5). After fermentation, ingesta
passes through the omasum and into the abomasum. The larger food
particles can be selectively retained at the reticuloomasal orifice and
fermented further. Rumination, in which the reticulorumen contents
are rechewed, provides a very thorough mechanical breakdown of
fibrous plant particles. Factors such as particle size, specific

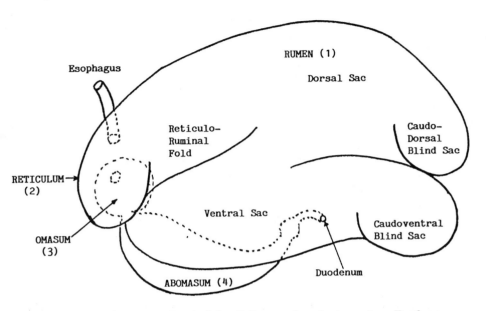

FIGURE 1   Diagrammatic sketch of the ruminant stomach. Feed en-
ters through the esophagus and is fermented by the microbial population in the rumen (1) and reticulum (2). As particle size is reduced, the particulate matter passes through the omasum (3) down
into the true stomach or abomasum (4).

gravity, and concentration of solids all appear to influence the rate
at which ingesta moves out of the reticulorumen. After the abomasum, the digestive tract of the ruminant is quite similar to that of
the nonruminant.

B.   Physical Characteristics

The rumen has been likened to a large fermentation vat in which the
animal maintains a fairly constant environment through various physiological mechanisms. A steady supply of food and continuous removal of fermentation products and undigested residues provides
ideal conditions for development and maintenance of a rather large
microbial population.

Temperature in the rumen is relatively constant, generally ranging between 38 and 40°C (5,7). When an active fermentation is taking place, the temperature may temporarily rise to as high as 41°C

immediately after feeding. Ingestion of large volumes of cold water
or frozen forage can cause rumen temperature to decrease $5-10°C$
and $1-2$ hr may be required for the temperature to return to normal. However, this appears to have little effect on overall food utilization.

Rumen pH is probably the most variable factor in the fermentation environment $(5-7)$. Fluctuations occur with time after feeding,
nature of the feed, and frequency of feeding. The normal range of
rumen pH is approximately between 5.5 and 7.0, with outer limits of
4.5 and 7.5. In general, with animals fed once a day, minimum pH
values are generally attained about $2-6$ hr after feeding which coincides with maximum production of fermentation acids. Concentrate-
type feeds contain readily available carbohydrates and generally
cause a greater reduction in rumen pH. As would be expected, less
fluctuation in pH occurs when animals consume small quantities of
feed at numerous times throughout the day. Large amounts of paro-
tid saliva are secreted by the ruminant animal, especially during the
process of eating (5,6). Ruminant saliva contains large quantities
of bicarbonate and phosphate, and has a pH of 8 or above. Thus,
the volatile fatty acids produced by fermentation are partially neu-
tralized and, along with bicarbonate, provide buffering capacity to
maintain rumen pH between 5 and 7 (8).

Prior to feeding, osmotic pressure of rumen fluid is generally hy-
potonic with respect to plasma (250 mosm/kg compared to 300 mosm/
kg). Rumen fluid osmolality rises to hypertonic levels (400 mosm/
kg) immediately after feeding but returns to normal within $8-10$ hr.
(9).

Although dry matter content might be expected to vary considerably, fluctuating with the type of feed, sampling time, drinking
behavior, etc., most values for rumen contents fall into the range
of $10-15\%$ (5). Fairly rapid exchange of water ($30-70$ ml/min) can
occur between rumen contents and blood (10).

Since most of the rumen microorganisms are obligate anaerobes,
the oxidation-reduction potential or Eh of rumen contents is a very
important parameter. Values of rumen contents range from about
-150 to -350 mV (7). Most of the reducing substances in rumen contents are associated with feed particles and colloidal material, not
the bacterial cells (11).

Principal gases in the rumen are CO_2 (65%), CH_4 (27%), N_2 (7%),
and traces of H_2, H_2S, CO, and O_2 (6). Although surface tension
can markedly affect microbial physiology, little information is available on possible beneficial or inhibitory effects in the rumen. Reported values for rumen fluid vary from $45-59$ dynes/cm (6).

III. RUMEN MICROORGANISMS

Within the limits of the above described environment, a microbial
population of bacteria, ciliate protozoa, flagellate protozoa, and

fungi live under symbiotic conditions with the host. As mentioned earlier, these microorganisms carry out a pregastric fermentation of feedstuffs, particularly the cellulose portion, providing energy to the animal from an otherwise unavailable source.

In general, bacteria and protozoa first appear in young ruminants between 1 and 3 weeks of age (12–18). However, the populations differ from those found in mature animals, e.g., total aerobic–facultative anaerobic bacterial numbers in calves were 90 × 10^7 per gram after 5 weeks (14). Microbial populations similar to those found in mature animals appear to become established at about 3–4 months of age (13,14,16). Flagellate protozoa and/or fungal zoospores have been observed in rumen contents of calves and lambs as early as 6 days of age (17).

For domestic ruminants (cattle, sheep, and goats), total protozoan numbers have been reported to range from 0.06 to 4 × 10^6 per ml (6). As an overall estimate, 1 × 10^6 protozoa per ml rumen contents appears to be "normal" across most rations. Total bacterial numbers vary with counting procedure, direct or viable, as well as diet. For example, Maki and Foster (19) found that the viable count was 3–12% of the direct count in cows fed high roughage and 57–73% in cows fed no roughage (high concentrate). The viable counts ranged from 0.3 to 1.8 × 10^9 per g with high roughage and from 12.7 to 22.9 × 10^9 per g with no roughage (19). Similar numbers have been obtained in more recent studies with cattle, using improved media (20) and anaerobic plating techniques (21). In a study where sheep were abruptly switched from an all-roughage to high-concentrate diet, bacterial numbers ranged from 3.3 to 55.3 × 10^9 per g of rumen contents (22).

Total numbers of cellulolytic bacteria are difficult to assess because not all of the cellulolytic organisms are included with a single counting procedure (23; B. A. Dehority, unpublished results). However, van Gylswyk (24) compared numbers obtained by counting zones of clearing in cellulose agar roll tube and testing cellulolytic activity of bacterial strains randomly isolated from total count medium. The two methods gave similar values for sheep fed teff hay diets, ranging from 23 to 38 × 10^6 cellulolytic bacteria per g of rumen contents. The cellulolytic bacteria ranged from 0.8 to 7.2%. Cellulolytic numbers in cellulose agar roll tubes ranged from 8.8 to 582 × 10^6 per g of rumen contents from sheep which were abruptly switched from hay to high concentrate (22). A slightly wider range of values was reported from cattle on all-hay and high concentrate diets, i.e., 10–900 × 10^6 cellulolytic bacteria per g of rumen contents (21).

IV. CELLULOLYTIC MICROORGANISMS IN THE RUMEN

Hungate (6) presents an excellent historical review on the early scientific observations and experiments which suggested the involvement of bacteria and protozoa in the degradative activities which occur in the rumen. By the middle 1800s cellulose was identified as a major carbohydrate constituent of plants and was shown to be utilized by the ruminant animal. The rumen was subsequently shown to be the site of this digestion. Knowledge was rather limited in this area until 1940, when both chemical and microbiological methods were developed which allowed a more thorough investigation of the rumen fermentation. Interest in rumen microbial activity has expanded markedly during the last 40 years and considerable progress has been made in our understanding of the microorganisms involved and their mode of action.

A. Ciliate Protozoa

Microscopic observations of the rumen protozoa revealed that the larger entodiniomorph protozoa could ingest plant particles, which led early workers to speculate about their possible cellulolytic activity. Becker et al. (25) compared cellulose digestion between faunated (protozoa present) and defaunated (protozoa removed) goats, and concluded that the protozoa were not responsible for, or did not appreciably contribute to, the digestion of cellulose in the rumen. However, their observations did not preclude the possibility that the protozoa were cellulolytic, but just that they were not essential for cellulose degradation in the rumen.

Veira (26) summarized a number of studies which compare apparent digestibility of organic matter, dry matter, energy, acid detergent fiber, crude fiber or cellulose in faunated and ciliate-free ruminants. In almost all cases apparent digestibility was slightly higher in the faunated ruminants, but the differences were generally less than five digestibility units. In a recent study by Orpin and Letcher (27), no differences in organic matter digestibility of two timothy hay samples (66.6 and 82.3% digestible) were found following defaunation.

Using a nylon bag technique, Kurihara et al. (28) observed that cellulose digestion was reduced by about 50% when the bags were suspended in the rumen of a defaunated animal. In conventional digestion trials, Jouany and Senaud (29) found a slight decrease in lignocellulose digestion (3 – 10%) with defaunated animals; however, cellulose digestion was also depressed in the defaunated animals when estimated with nylon bags which would not permit entry of protozoa. Obviously the increased nylon bag cellulose digestion in the faunated animals cannot be attributed to cellulolytic activity of the protozoa.

When rumen digestibility is compared to total tract digestibility, defaunation appears to primarily decrease rumen digestibility. Data compiled by both Veira (26) and Demeyer (30) suggest that the decrease in ruminal digestion can be largely, but not completely, compensated for by increased digestion in the cecum and large intestine.

In 1942, Hungate (31) reported the successful cultivation of the rumen ciliate *Eudiplodinium neglectum*, using grass and cellulose as energy sources. After thorough washing of the protozoan cells, a cell-free extract was prepared which digested cellulose as determined by the appearance of reducing sugars and was inactivated by boiling. Similarly prepared extracts from the culture debris, which should have contained most of the bacteria and cellulose, did not produce reducing sugars when incubated with cellulose. In a subsequent study (32), similar culture growth and cell-free cellulolytic activity were demonstrated for *Eudiplodinium maggii* and *Polyplastron multivesiculatum*. *Diplodinium dentatum*, but not *Entodinium caudatum*, could be grown with cellulose and grass as substrates. Hungate's conclusions have been criticized on the basis that cellulase from intracellular bacteria could be responsible for the cellulose degradation and that actual disappearance of the insoluble cellulose was not measured (33). Sugden (34) used Hungate's procedures to culture *P. multivesiculatum* (33); however, when streptomycin was added to suppress bacterial growth the cultures died out within 3 or 4 days. Since streptomycin was not toxic to the protozoa themselves, she concluded that cellulolytic bacteria within the cells were responsible for the ability of *P. multivesiculatum* to utilize cellulose. The above results could have been confounded in view of the fact that none of the protozoa have subsequently been grown for any period of time in the absence of living bacteria (35).

Using a ciliate-free sheep, Abou Akkada et al. (36) established a rumen population of small entodinia and *P. multivesiculatum* as the only protozoal species. The small entodinia were readily removed by washing and decantation so that fairly large, clean preparations of *P. multivesiculatum* could be obatined. Cell-free extracts degraded cellulose powder or cotton wool with good agreement of two assay procedures, i.e., formation of soluble carbohydrates and loss of insoluble substrate. Negligible cellulolytic activity was found when an extract was prepared with rumen liquor from a defaunated sheep. Similar experiments with *Ophryoscolex tricoronatus* showed that this species also possessed an active cellulase (33). Workers in New Zealand studied cellulase activity in cell-free extracts prepared from *Epidinium ecaudatum*, mixed *Entodinium* species, and *Eremoplastron bovis*, none of which could degrade insoluble cellulose (37−40). However, cell-free extracts from all of these species did contain cellodextrinase activity. Other workers reported very limited activity in *Entodinium* against soluble or chemically modified cellulose derivatives (35,41,42).

Using powdered dried grass as the sole substrate for in vitro cultures, Coleman et al. (43) hoped to select for cellulolytic species

of protozoa. A total of six species, all in the subfamily Diplodiniinae
(44), were maintained under these conditions for periods ranging
from 3 months to well over a year. One species (*Diplodinium mono-
canthum*) died out 13 months after isolation. Cell-free extracts from
the five remaining species (see Table 1) were able to release soluble
14C-labeled compounds into the medium when incubated with [14C]-
cellulose. However, as pointed out by the authors, although extracts
from all species of rumen protozoa that can live solely on dried grass
will digest cellulose, the origin of the cellulase is uncertain. Even
though extracts of the culture medium and particulate matter would
not degrade cellulose, the presence of cellulolytic bacteria within the
protozoan cell must be considered a possibility. Coleman (45) later
used antibiotics to control bacterial contribution to cellulase activity
in *E. maggii*. He estimated that 70% of cellulase activity inside the
protozoa is soluble and not derived from bacterial sources.

   Ostracodinium gracile was cultured in the author's laboratory for
periods of up to 3 months using microcrystalline cellulose as the only
added substrate (B. A. Dehority, unpublished results). Cellulose,
either purified or in dried grass, was required for growth of *O. gracile*. No visible clear zones were seen when 1:10 dilutions of the protozoan cultures were blended and used to inoculate cellulose agar roll
tubes, indicating that cellulolytic bacteria, if present, were intracellular and not freed by the blending treatment. Generation time for
this species was estimated to be 48 – 72 hr.

   In a recent report, Coleman (46) measured cellulase activity in
15 species of rumen protozoa using six different assay procedures.
Substrates varied from soluble carboxymethylcellulose to microcrystalline cellulose. In general, those species which degraded appreciable amounts of the soluble cellulose derivatives also degraded insoluble cellulose. The rumen protozoa identified in this study plus
several species identified in previous studies are listed in Table 1.
Coleman (46) attempted to fractionate the total cellulase activity in
rumen contents by also preparing extracts from the particulate matter and free bacteria. In rumen samples taken just prior to feeding,
62% of the carboxymethylcellulase (CMCase) activity was associated
with the protozoa, 27% with the plant debris, and 10.6% with the
washed free bacteria. In a subsequent study (47), these observations were extended to include sampling times of 1.5, 6.5, and 24
hr after feeding. Sheep were used which contained either no ciliates,
a single species of one of three different cellulolytic protozoa, an
amylolytic protozoan (*Entodinium caudatum*), or a mixed population.
In general, total rumen CMCase activity was lowest in defaunated
animals, next highest in the animal with *E. caudatum* or the mixed
population, and highest in those containing single species of cellulolytic protozoa. Except for the defaunated animals and the animal with
E. caudatum, distribution of the CMCase activity among protozoa,
debris, and bacteria was similar to that of the previous study.

TABLE 1   Cellulolytic Rumen Protozoa

Species	Activity[a]	Ref.
Eudiplodinium maggii	Strong	31, 32, 43, 45, 46
Epidinium caudatum[b]	Strong	43, 46
Ostracodinium dilobum[c]	Strong	46
Metadinium affine[d]	Moderate	43, 46
Eudiplodinium bovis[e]	Moderate	31, 32, 46
Ophryoscolex caudatus	Moderate	46
Polyplastron multivesiculatum	Moderate	36, 46
Diplodinium pentacanthum	Weak	43, 46
Enoploplastron triloricatum	?	43
Ophryoscolex tricoronatus	?	33
Entodinium caudatum	Trace	41, 46
Ostracodinium gracile	?	f

[a]Activities are based on a relative scale, from (46). No information is available on those species listed as "?"

[b]*Epidinium ecaudatum caudatum* Dogiel 1927.

[c]*Ostracodinium obtusum dilobum* Dogiel 1927.

[d]*Diploplastron affine* Kofoid and MacLennan 1932 (*Diplodinium affine* Dogiel and Fedorowa 1925).

[e]*Eremoplastron bovis* Kofoid and MacLennan 1932 (*Eudiplodinium neglectum* Dogiel 1927).

[f]B. A. Dehority (unpublished).

In summary, the quantitative contribution of protozoa to rumen cellulose degradation is not entirely clear. Digestion trials would suggest that cellulose digestion is similar or slightly less in ciliate free animals (26,27). However, additional studies indicate that cellulose digestion is decreased in the rumen and increased in the hindgut of ciliate-free animals, resulting in similar values for total tract cellulose digestion (30,48,49). The actual extent of cellulolytic activity by the protozoa in vivo could be affected by several factors, i.e., numbers, rate of cellulose digestion, and rate of growth of the individual species. For example, reported generation times for several cellulolytic species determined from growth curves in vitro, range from 8.5 hr for *Diplodinium pentacanthum* to 26 hr for *Epidinium caudatum* (35). In addition, questions concerning the possible contribution of intracellular bacteria to ciliate degradation of cellulose remains unanswered because of our inability to grow the rumen protozoa in axenic culture.

B. Bacteria

Based on numbers and their ability to hydrolyze purified and intact forage cellulose, *Fibrobacter succinogenes*, *Ruminococcus flavefaciens*, and *Ruminococcus albus* appear to be the predominant cellulolytic bacteria in the rumen of domestic cattle and sheep (50–60). Although many strains of *Butyrivibrio fibrisolvens*, which occur in high numbers on a variety of rations, are weakly cellulolytic, this species probably contributes little to overall cellulose digestion in the rumen (51,59,61,62). Other cellulolytic bacteria which occasionally occur in moderate numbers in the rumen are *Eubacterium cellulosolvens* and several species of *Clostridium*; however, they are considered to be of doubtful significance in rumen cellulose digestion (55,56,59,63,64).

1. Fibrobacter succinogenes

Fibrobacter succinogenes is an anaerobic, non-spore-forming, gram-negative, pleomorphic rod, usually varying in diameter from 0.3 to 0.8 μm and in length from 1 to 2 μm. Smaller forms and rounded bodies are observed as cultures age. With few exceptions, *F. succinogenes* ferments only glucose, cellobiose, and cellulose. Primary end products of *F. succinogenes* are acetic and succinic acids.

Although *F. succinogenes* is one of the more active rumen cellulolytics, particularly in its ability to degrade highly crystalline cellulose (65), it has rarely been isolated from cellulose agar roll tubes. The cellulase enzyme of *F. succinogenes* appears to be cell-bound, requiring an intimate association between the cell envelope and the substrate, for cellulose digestion to occur (66). Hungate (50,51) had proposed earlier that some strains were able to migrate through the agar to cellulose particles whereas other strains could not. If the agar concentrations in cellulose roll tubes is reduced from 2.0 to 0.5% *F. succinogenes* is able to form clear zones (67), which would support the suggested mode of cellulase activity.

2. Ruminococci

The *Ruminococcus flavefaciens* cell is almost spherical, about 0.8–0.9 μm in diameter, and generally occurs in chains. It is nonmotile, non-spore-forming, anaerobic, gram-positive or negative or

variable, and ferments cellulose and cellobiose. Considerable strain
variation occurs in other sugars utilized. Principal end products are
acetic, formic, and succinic acids. *Ruminococcus albus* differs in that
it usually occurs singly or as a diplococcus; cells are 0.8 to 2.0 μm
in diameter; hydrogen, ethanol, acetic acid, formic acid, and lactic
acid are produced in various combinations and proportions; little if
any succinate is produced. Other general characteristics of *Ruminococcus* are an obligate requirement for NH_3 and accumulation of reducing sugars in the medium when grown on an excess of cellulose.

Some of the cellulolytic cocci isolated from rumen contents of
reindeer differed from the two described *Ruminococcus* species (68),
producing approximately equivalent amounts of ethanol, formic acid,
and lactic acid, with lesser amounts of acetic acid and only a trace
of hydrogen. Aside from cellobiose, these seven strains utilized only
polysaccharides for growth, i.e., cellulose, xylan, and pectin. Since
the magnitude of difference with the previously described strains is
low, it might suggest that they are simply a biotype of *R. flavefaciens* or *R. albus*.

Five strains of anaerobic, gram-variable cellulolytic cocci were
isolated from the cecal contents of guinea pig and classified as belonging to the genus *Ruminococcus*. They differed from most previously described strains of the two recognized cellulolytic species
of this genus in that lactate was the major fermentation product
along with lesser amounts of formate, ethanol, and a trace of succinate. No growth occurred at 30°C; however, good growth was observed at 38 and 45°C. Xylose, arabinose, sucrose, and lactose were
fermented by all strains and rumen fluid was required for growth in
a complete medium containing all nutrients previously found to be required by species in this genus (69).

3. Butyrivibrio fibrisolvens

The first isolation of *Butyrivibrio fibrisolvens* was reported by Hungate in 1950 (51). Although he isolated the species in cellulose agar
medium, he noted an incomplete digestion of cellulose in the vicinity
of the colony. Bryant and Small (61) established the species on the
basis of a study of 48 strains which had been isolated in 40% rumen
fluid-glucose-cellobiose-agar medium; however, only three of their
strains were cellulolytic. Hungate (6) suggested that in general
those strains isolated in cellulose medium are cellulolytic, whereas
only a few cellulolytic strains are isolated on soluble sugar medium.
Dehority (62) isolated five strains of *B. fibrisolvens* on a xylan medium in 1966. Quantitative measurement of cellulose digestion for
these isolates, with 0.75% purified cellulose substrate, gave digestibilities of 0, 5.6, 10.0, 9.8, and 23.2%. Under similar conditions,
F. succinogenes, R. albus, and *R. flavefaciens* will digest about
90-100% of the cellulose.

Butyrivibrio fibrisolvens is an anaerobic, non-spore-forming,
gram-negative, motile, slightly curved rod. Dimensions usually vary
between 0.4 and 0.8 μm in width and 1.0 and 3.0 μm in length. In
general, CO_2, rumen fluid, and NH_3 are not required for growth.
Starch is hydrolyzed. A fairly large number of carbohydrates are
fermented, with considerable differences occurring between strains.
End products of fermentation are hydrogen, formic, butyric, and
lactic acids, and ethanol. Slight production or an uptake of acetic
acid is generally also observed (61,62,70). Although *Butyrivibrio*
consistently stains gram-negative, the ultrastructure of *Butyrivibrio*, as seen by electron microscopy, indicates that its cell wall is
characteristic of a gram-positive organism (71). However, the cell
wall is extremely thin which may account for the gram-negative staining reaction.

4. Eubacterium cellulosolvens

In 1958, Bryant et al. (56) isolated a single strain of a new species
of cellulose digesting bacteria from the bovine rumen. It was an anaerobic, gram-positive, peritrichous rod with pointed ends that produced primarily lactic acid in rumen fluid-cellobiose medium. They
classified the organism as *Cillobacterium cellulosolvens* and, based
on its low numbers, judged it to be relatively unimportant in the
rumen.
    Van Gylswyk and Hoffman (63) isolated and characterized nine
strains of cellulolytic *Cillobacterium* from the ovine rumen in 1970.
They estimated that the organism occurred at numbers of at least
10^6 per gram rumen contents from animals fed diets containing teff
hay. Although end products of their strains differed from that of
the type strain, they broadened the species description and classified the organisms as *C. cellulosolvens*.
    More recently, seven strains of cellulolytic, gram-positive, motile
rods were isolated by Prins et al. (64). They were identified as
C. cellulosolvens, now named *Eubacterium cellulosolvens*. The organisms were isolated from 10^{-8} dilutions of rumen contents. The
authors concluded that despite the high numbers observed and the
high rate of cellulose digestion observed in vitro, their importance
as cellulolytic bacteria in the rumen was doubtful since so many soluble carbohydrates are fermented. They suggested its importance
in cellulose digestion in the rumen to possibly be similar to that of
B. fibrisolvens.

5. Other Less Numerous Cellulolytic Bacteria

Hungate isolated and described two species of spore-forming cellulolytic rods, *Clostridium lockheadii* and *Clostridium longisporum* (55).
In later studies from South Africa, two strains of cellulolytic

Clostridium were isolated which differed from previously described species (59). Based on their sporadic occurrence in high dilutions of rumen contents, species of *Clostridium* are not considered to be of much importance in rumen cellulose digestion.

Two additional gram-negative, non-spore-forming rods have been isolated from the rumen on single occasions. Both produce butyric acid but differ both morphologically and physiologically from *B. fibrisolvens*. One species isolated by Leatherwood and Sharma is unnamed (72), while the other was described as *Fusobacterium polysaccharolyticum* by van Gylswyk (73).

6. Cellulolytic Activity

The ability of the cellulolytic rumen bacteria to digest or solubilize purified cellulose in shown in Table 2. Conditions varied considerably between the studies, but in general the predominant species, i.e., *F. succinogenes, R. albus,* and *R. flavefaciens*, can all solubilize a considerable amount of powdered cellulose. In contract, only *F. succinogenes* could solubilize appreciable quantities of undegraded cotton fibers (65). Considerable variation was observed between *Ruminococcus* strains in cotton fiber solubilization.

Table 3 presents data on solubilization of cellulose from intact forages, which would be the normal substrate for these organisms. Although the data are limited, *F. succinogenes* appears able to solubilize more cellulose than the ruminococci, which in turn are more cellulolytic than *B. fibrisolvens*. As shown with the purified cellulose, there is considerable strain variation with *Ruminococcus*. Since most strains of *B. fibrisolvens* are also hemicellulolytic, pectinolytic, and amylolytic (62,76−80), their contribution to cellulose digestion in the rumen is probably minimal.

C. Fungi

Recognition of anaerobic fungi as contributors to cellulose digestion in the rumen occurred just recently. Warner (81) observed in 1966 that numbers of the flagellate protozoa *Neocallimastix frontalis* increased dramatically just after feeding. Numbers then decreased and reached a low point about 10 hr after feeding and remained at that level until the next feeding. Rapid multiplication and subsequent cell lysis did not seem probable to Warner, who suggested that the cells possibly sequestered on the rumen wall after feeding and migrated back into the rumen contents in response to feeding. Later investigations by Orpin (82) indicated that a constituent of the feed was responsible for or triggered the increased concentration of *Neocallimastix*. Using centrifugation techniques, he determined that *Neocallimastix* was associated with the larger feed particles and suggested that the organisms sequestered there or that

TABLE 2 Solubilization of Purified Cellulose by the Principal Cellulolytic Bacteria in the Rumen

Organism	% solubilized[a]			% solubilized[b]		% solubilized[c]	
	No. of strains	Cellulose powder	Cotton fibers	No. of strains	Cellulose powder	No. of strains	Cellulose powder
F. succinogenes	1	88	97			1	69
R. albus	2	88,88	10,40	2	78,79	1	64
R. flavefaciens	2	72,90	0,55	2	38,39	1	63
B. fibrisolvens				2	15,20		
E. cellulosolvens				2	0,0		

[a]Data from (65).
[b]Data estimated from graphs in (74).
[c]Data from (75).

TABLE 3 Solubilization of Cellulose from Intact Forages by Cellulolytic Rumen Bacteria

Organism	No. of strains	% solubilized[a]		No. of strains	% solubilized[b]
		Bromegrass	Alfalfa		Teff hay
F. succinogenes	2	79,81	51,52		
R. albus	1	57	43	2	43,56
R. flavefaciens	4	49–65	20–48	2	39,66
B. fibrisolvens	1	15	6	10	10–37
Clostridium sp.				2	10,11

[a]Data from (76).
[b]Data from (77).

its life cycle involved a multiple reproduction phase which was associated with the particulate matter. Subsequent studies indicated
that feeding stimulated a reproductive body on a vegetative phase
of the organism to differentiate and liberate the flagellates (83). The
liberated flagellates lost motility within an hour and developed into
the vegetative phase, explaining the observed decrease in numbers.
The morphology and life cycle of these organisms resembed the phy-
comycete fungi, and Orpin was able to confirm their identity as fungi
based on the occurrence of chitin in their vegetative cell wall (84).
Thus the flagellates released from the vegetative stage (sporangia)
would be correctly termed as zoospores.

Using scanning electron microscopy, Bauchop (85) examined plant
fragments obtained from the rumen of cattle and sheep at various
times after feeding. He found that large numbers of the fungal zoospores rapidly attached to plant fragments in the rumen. After attachment, the small zoospores (6 – 10 μm) lost their flagella and thalli
were formed, with the development of rhizoids and sporangia. Because of their rapid attachment to plant materials, zoospore numbers
are probably not a reliable indication of the importance of fungi in
the rumen fermentation. Bauchop (85) suggests that the quantity
of fungal mass (vegetative rhizoid tissues and its enzymic activities)
would be the important factor in assessing the importance of the
fungi.

Zoospores of N. frontalis show a site preference for attachment
to plant material, i.e., stomata, cut ends, and damaged tissues (86).
This preference appears to be a chemotactic response to soluble carbohydrates diffusing from the plant tissues (87). Orpin and Hart
(88) studied the digestion of several forage plants using pure cultures of the rumen fungi N. frontalis, Piromonas communis, and
Sphaeromonas communis. Dry matter digestibility ranged from 30 –
45%, varying with the different fungal and plant species (perennial
rye grass, mixed hay, and wheat straw leaves). Of the structural
carbohydrates, cellulose and hemicellulose digestion were greatest,
ranging from 40 to 58%. Concentration of lignin was decreased in
plant particles after fermentation by the fungi (approximately 20%
lignin loss with wheat straw leaves); however, there was no indication that the fungi utilized lignin as an energy source.

Windham and Akin (89) attempted to evaluate the contribution of
fungi to forage (alfalfa and bermuda grass) digestion in vitro by
using penicillin and streptomycin to inhibit bacterial growth and cycloheximide to inhibit fungal growth. Dry matter, neutral detergent
fiber, acid detergent fiber, and cellulose digestion by the bacteria
alone was equal to that of the total population. Digestion of all fractions occurred but was less with only the anaerobic fungi. No lignin
digestion was observed. In general, the authors concluded that bacteria were the most active fiber degraders in the rumen.

Two rumen species of *Neocallimastix* have now been identified. *N. frontalis* and *N. patriciarum* (90). They are classified as chytridromycete fungi (91). *Neocallimastix* species are polyflagellated (7 – 17 flagella) and differ from *P. communis* and *S. communis*, which are uniflagellate. Classification of these latter species is uncertain, and Heath et al. (91) even suggested they may be an alternate life cycle stage for the genus *Neocallimastix*.

V. FACTORS INFLUENCING FORAGE CELLULOSE DEGRADATION

Much of our information on forage cellulose digestion has been obtained in vitro using mixed-culture fermentation procedures. Systems have varied from a simple static fermentation to an automated continuous culture-type system using semipermeable vessels. These procedures are described in several reviews (92,93) and articles (94 – 96). In general, the problem to be investigated will determine the procedures which are suitable; however, the use of natural substrates, relatively short fermentation periods, and the employment of whole-rumen contents as inoculum are desirable conditions in that these would most closely simulate the normal in vivo environment.

Another procedure which has been used to study forage cellulose digestion is the nylon or Dacron bag technique (93,97,98). Small nylon or Dacron bags are filled with a forage, tied tightly, and placed in the rumen for varying periods of time. The bag material must be of fine enough mesh to prevent the forage particles from washing out and the bags must be suspended in such a manner that they do not become lodged in the bottom of the rumen. This procedure was recently studied in detail by Meyer and Mackie (99), who found that the microbial population inside the bag was markedly influenced by bag pore size and generally differed from the population in the rumen contents. They concluded that values for rate and extent of feed or forage dry matter disappearance obtained in synthetic fiber bags suspended in the rumen should be interpreted with caution.

Table 4 presents some data collected in our laboratory with mixed-culture fermentations which illustrate the effects of different forage species as well as plant maturity on the availability of forage cellulose (100). Cellulose digestion varies among grasses as well as between grasses and legumes, decreasing as the plant matures. These data and other studies (101) further suggest that cellulose digestion decreases to a greater extent with maturity in grasses than in legumes. Cellulose digestion can be markedly increased, generally to a value of 80% or more, by physical reduction of forage particle size (ball milling) or chemical isolation of the cellulose (100,102). Presumably lignin, which increases in concentration as plants mature, is a major contributory factor to decreased cellulose digestion.

TABLE 4 Effect of Forage Species and Maturity Stage on Extent of Cellulose Digestion[a]

Forage		Date cut	Description	Cellulose digested (%)
Grasses	Timothy, 1st stage	5 – 24	Vegetative stage	83.0
	Timothy, 2nd stage	6 – 7	Heading	75.7
	Timothy, 3rd stage	6 – 20	Headed	62.4
	Timothy, 4th stage	7 – 16	Seed ripe	45.6
	Bromegrass, 1st stage	6 – 10	Early bloom	53.1
	Bromegrass, 2nd stage	6 – 26	Seed stage	42.7
	Orchardgrass, 1st stage	6 – 6	Early bloom	64.6
	Orchardgrass, 2nd stage	6 – 18	Early seed	53.7
Legumes	Alfalfa, 1st stage	5 – 18	Early bloom	67.8
	Alfalfa, 2nd stage	6 – 26	Late bloom	58.1
	Alfalfa, 3rd stage	7 – 17	Seed stage	54.8
	Red clover	9 – 29	Full bloom	55.6

[a]Data from (100).

Physical form of the forage can be an important factor with regard to cellulose digestion in the animal. Working with sheep and goats, Quick and Dehority (103) noted a significant increase in dry matter and fiber digestion when animals were fed long hay as compared to chopped or pelleted hay. This might be partially explained on the basis that dry matter intake and turnover time were significantly reduced for animals fed long forage as compared to pellets. Animals fed chopped forage had intermediate values. Thus the physical form of the forage, in addition to plant type and maturity stage, can influence the rate and extent of cellulose degradation in vivo.

VI. SYNERGISM BETWEEN BACTERIAL SPECIES

As the various rumen bacteria were isolated and studied in detail, it became obvious that they differed in their abilities to degrade the structural carbohydrates in forages (6). This led to speculation that ruminal digestion of forages was probably the result of synergism between the various bacterial species. Although this concept was difficult to test with mixed cultures, pure cultures provided a simple method to investigate this possibility.

Dehority and Scott (76) were able to show that by combining two organisms, a cellulolytic and a hemicellulolytic, intact forage cellulose digestion was significantly increased. In other studies (104,105) it was found that the cellulolytic species were capable of degrading forage hemicelluloses, i.e., converting the hemicellulose from a form insoluble in 80% acidified ethanol to a form which was soluble. However, most of the cellulolytic strains could not utilize the hemicelluloses as an energy source. A subsequent study (106) suggested that solubilization of the hemicellulose might be a nonspecific action of the cellulase enzyme. Many of the so-called hemicellulolytic organisms were unable to utilize hemicellulose from intact forages. However, when the hemicellulose was chemically isolated from the forage or a hemicellulose-degrading strain was also included in the fermentation, marked utilization occurred (79). Thus for hemicellulose utilization, the synergistic effect between bacteria is the result of one bacterial species altering (degrading) the forage substrate so that it is available to the second species. This can be seen quite readily in the data shown in Table 5. Similar results were obtained with respect to pectin utilization from intact forages (80).

In general, many interrelationships exist between the different rumen microorganisms. These relationships can be beneficial to one or both partners (commensalism and mutualism) or competetive, where one partner suffers (parasitism). These concepts and examples of their occurrence in the rumen are discussed in detail by Prins (107).

TABLE 5 Degradation and Utilization of Hemicellulose from Intact Fescue Grass or Isolated Fescue Grass Hemicellulose by Pure Cultures of Rumen Bacteria[a]

Strain[b]	Degradation(%)		Utilization(%)	
	Forage	Isolated hemicellulose	Forage	Isolated hemicellulose
B34b	66.6	88.5	3.0	0
H10b	44.8	87.5	38.0	83.8
B34b + H10b	67.3	91.3	64.8	87.8
H8a	2.7	82.0	2.0	80.4
B34b + H8a	69.0	93.9	67.7	87.0

[a]Data from (79).
[b]Ruminococcus flavefaciens B34b; Butyrivibrio fibrisolvens H10b; Bacteroides ruminicola H8a.

VII. OTHER HERBIVOROUS MAMMALS

In addition to the true ruminants, there are other mammalian families which by means of microbial fermentation can utilize cellulose as an energy source (3,108,109). Pregastric fermentations occur in the foregut of tree sloths, colobid monkeys, kangaroos, camelids, hippopotami, and peccaries. Hindgut or postgastric fermentation has been observed in a wide range of mammals, i.e., rat, mouse, vole, beaver, porcupine, rabbit, pika, pig, man, monkey, ruminants, horse, zebra, ass, tapir, rhinoceros, elephant, wart hog, guinea pig, capybara, gorilla, and chimpanzee. Unless those animals with a hindgut fermentation practice coprophagy (reingestion of feces), the postgastric fermentation tends to be much less efficient than the pregastric fermentation. Primarily this would be the result of limited absorption of fermentation end products and loss of synthesized microbial cells.

Vander Noot and Gilbreath (110) compared the cellulose digestibility of four forages (alfalfa, timothy, orchard grass, and bromegrass) between steers and geldings, and found it to be significantly higher in the steers (63.8 vs. 48.1). However, in another study the concentration of cellulolytic bacteria was found to be similar in steer rumen ingesta and pony cecal contents (111). Even in the ruminant animal, some cellulose digestion occurs in the hindgut. Lewis and Dehority (112) determined that 11.1, 13.9, and 18.9% of the cellulose in diets containing 100, 40, and 20% orchard grass hay (remainder corn) was digested in the cecum. Rumen digestibility of cellulose on these diets was 66.6, 45.5, and 32.4%, respectively. Presumably, the 20% hay – 80% corn diet resulted in a low-rumen pH which slowed down or inhibited cellulose digestion (75), thus passing more available cellulose into the lower digestive tract.

VIII. CONCLUSIONS

Microbial degradation of plant cellulose in the gastrointestinal tract is a large and important source of energy for herbivorous mammals. The herbivores are divided into two types, those with a pregastric or foregut fermentation and those with a postgastric or hindgut fermentation. Ruminants, with their pregastric fermentation, probably comprise the largest numbers of herbivores in the world today. One of the main reasons for this is the widespread domestication of cattle, sheep, and goats. Anaerobic bacteria, ciliate protozoa, and fungi are present in the rumen and have the ability to digest cellulose from a variety of plants. End products of microbial digestion as well as the microbial cells themselves are then available as energy sources to the animal. Although all three types of microorganisms are probably

functional in the rumen fermentation, their individual importance
may vary with diet and animal species.

Information on the extent of cellulose digestion in nonruminant
foregut fermentors is extremely limited; however, values for digestibility in the hippopotamus and llama appear to be similar to those obtained in ruminants (113). Considerable variation between studies
has been noted for fiber digestion in the different hindgut fermenting herbivores (113,114). In general, when compared to the rumen,
cellulose and fiber digestion are considerably lower in the smaller
hindgut fermenting herbivores (rabbits, guinea pigs, capybara, voles,
etc.) and more closely approximate ruminant values in the larger species (horse, zebra, rhinoceros, elephant, etc.). The small herbivores are primarily cecum fermentors, having an enlarged cecum
which retains solutes and small particles, while the less digestible
fibrous materials pass rapidly on through the digestive tract. This
allows utilization of a high-fiber diet without the development of a
large hindgut, which is advantageous because of the relationship between body size and food requirements (115). In contract, fermentation takes place in the cecum and enlarged proximal colon of the
larger herbivores. Contents of these areas mix freely and act as
one large fermentation site.

IX. ADDENDUM

Since this chapter was written, the bacterial species *Bacteroides
succinogenes* has been placed in a new genus, *Fibrobacter* (116).
The name has been changed throughout this chapter, but the reader
should be aware of this name change when referring back to the literature.

An updated reference or source book on rumen microbiology, *The
Rumen Microbial Ecosystem*, edited by P. N. Hobson, was just published (117). The purpose of this book was to summarize the information on rumen microbiology and biochemistry which has appeared
in the literature since 1966 when Hungate published his classic book,
The Rumen and Its Microbes (6). The reader is referred to these
two books as a source of more in-depth information about those areas
of this chapter which were quite brief because of space constraints.

Rumen fungi have received a considerable amount of attention in
the last few years. Lowe et al. (118) were able to isolate two species of fungi from saliva and feces of sheep, suggesting two possible
routes of transfer between animals. Single strains of fungi from each
of the three genera which occur in the rumen were recently examined
for their ability to degrade purified cellulose and cellulose from intact
wheat straw (119). The *Sphaeromonas* strain could not utilize purified cellulose and only 38% of cellulose from wheat straw. Both the

Neocallimastix and *Piromonas* strains digested the purified cellulose and about 54% of wheat straw cellulose. These data confirm previous observations by Orpin and Hart (88).

Three pure cultures of rumen bacteria were used to study synergism in the utilization of cellulose, hemicellulose, and pectin from intact forage (120). Each strain was characterized as using only a single forage polysaccharide, i.e., *Fibrobacter succinogenes*, cellulolytic; *Bacteroides ruminicola*, hemicellulolytic; *Lachnospira multiparus*, pectinolytic. The organisms were used separately and in all possible combinations to inoculate fermentation media containing intact orchard grass as the sole substrate. In general, cellulose digestion was not improved by the addition of either of the other strains. Hemicellulose utilization was markedly increased by combining the cellulolytic and hemicellulolytic strains, similar to previous results (Table 5). However results of pectin fermentation from the intact forage were quite unexpected. Degradation (solubilization) and utilization of intact forage pectin by the pectinolytic strain were minimal, whereas both the cellulolytic and hemicellulolytic strains degraded considerable amounts. By itself the hemicellulolytic strain utilized most of the degraded pectin and both degradation and utilization were increased when the cellulolytic strain was added to the fermentation. For all parameters studied, the combination of three strains was no better than the best combination of any two. In addition to further demonstrating the synergism which can occur between organisms, data from this study suggest that isolation and characterization of rumen bacteria on purified substrates could be misleading with regard to their importance or role in the rumen itself.

REFERENCES

1. P. J. Van Soest, *Nutritional Ecology of the Ruminant*, O & B, Corvallis, Oregon, 1982.
2. R. J. Moir, The comparative physiology of ruminant-like animals, In *Physiology of Digestion in the Ruminant* (R. W. Dougherty, ed.), Butterworths, Washington, D.C., 1965, p. 1.
3. T. Bauchop, In *Microbial Ecology of the Gut* (R. T. J. Clarke and T. Bauchop, eds.), Academic Press, New York, 1977, p. 223.
4. A. Vallenas, J. F. Cummings, and J. F. Munnell, *J. Morphol.*, *134*, 399 (1971).
5. D. C. Church, *Digestive Physiology and Nutrition of Ruminants*, Vol. 1, 2nd ed., Oregon State Univ. Bookstores, Corvallis, Oregon, 1975.
6. R. E. Hungate, *The Rumen and its Microbes*, Academic Press, New York, 1966.

7. R. T. J. Clarke, In *Microbial Ecology of the Gut* (R. T. J. Clarke and T. Bauchop, eds.), Academic Press, New York, 1977, p. 36.

8. B. Emmanuel, M. J. Lawlor, and D. M. McAleese, *Br. J. Nutr.*, *23*, 805 (1969).

9. A. C. I. Warner and B. D. Stacy, *Quart. J. Exp. Physiol.*, *50*, 169 (1965).

10. R. F. Willes, V. E. Mendel, and A. R. Robblee, *J. Anim. Sci.*, *31*, 85 (1970).

11. R. L. Baldwin and R. S. Emery, *J. Dairy Sci.*, *43*, 506 (1960).

12. Ph. Gouet, G. Fonty, J. P. Jouany, and J. M. Nebout, *Can. J. Anim. Sci.*, *64* (Suppl.), 163 (1984).

13. G. Fonty, J. P. Jouany, J. Senaud, pH Gouet, and J. Grain, *Can. J. Anim Sci.*, *64* (Suppl.), 165 (1984).

14. P. O. Williams and W. E. Dinusson, *J. Anim. Sci.*, *34*, 469 (1972).

15. A. Ziolecki and C. A. E. Briggs, *J. App. Bacteriol.*, *24*, 148 (1961).

16. M. P. Bryant, N. Small, C. Bouma, and I. Robinson, *J. Dairy Sci.*, *41*, 1747 (1958).

17. J. M. Eadie, *J. Gen. Microbiol.*, *29*, 563 (1962).

18. M. P. Bryant and N. Small, *J. Dairy Sci.*, *43*, 654 (1960).

19. L. R. Maki and E. M. Foster, *J. Dairy Sci.*, *40*, 905 (1957).

20. D. R. Caldwell and M. P. Bryant, *Appl. Microbiol.*, *13*, 794 (1966).

21. J. A. Z. Leedle, M. P. Bryant, and R. B. Hespell, *Appl. Env. Microbiol.*, *44*, 402 (1982).

22. J. A. Grubb and B. A. Dehority, *Appl. Microbiol.*, *30*, 404 (1975).

23. C. S. Stewart, C. Paniagua, D. Dinsdale, K.-J. Cheng, and S. H. Garrow, *Appl. Env. Microbiol.*, *41*, 504 (1981).

24. N. O. van Gylswyk, *J. Gen. Microbiol.*, *60*, 191 (1970).

25. E. R. Becker, J. A. Schulz, and M. A. Emmerson, *Iowa St. J. Sci.*, *4*, 215 (1930).

26. D. M. Veira, *J. Anim. Sci.*, *63*, 1547 (1986).

27. C. G. Orpin and A. J. Letcher, *Anim. Feed Sci. Technol.*, *10*, 145 (1984).

28. Y. Kurihara, T. Takechi, and F. Shibata, *J. Agr. Sci.*, *90*, 373 (1978).

29. J. P. Jouany and J. Senaud, *Ann. Rech. Vet.*, *10*, 261 (1979).

30. D. I. Demeyer, *Agr. Env.*, *6*, 295 (1981).

31. R. E. Hungate, *Biol. Bull.*, *83*, 303 (1942).

32. R. E. Hungate, *Biol. Bull.*, *84*, 157 (1953).

33. A. R. Abou Akkada, The metabolism of ciliate protozoa in relation to rumen function, In *Physiology of Digestion in the Ruminant* (R. W. Dougherty, ed.), Butterworths, Washington, D.C., 1965, p. 335.

34. B. Sugden, *J. Gen. Microbiol.*, *9*, 44 (1953).

35. G. S. Coleman, In *Biochemistry and Physiology of Protozoa* (M. Levandowsky and S. H. Hunter, eds.), Academic Press, New York, 1979, p. 381.

36. A. R. Abou Akkada, J. M. Eadie, and B. H. Howard, *Biochem. J.*, *89*, 268 (1963).

37. R. W. Bailey, *N. Z. J. Agr. Res.*, *1*, 825 (1958).

38. R. W. Bailey and R. T. J. Clarke, *Nature*, *198*, 787 (1963).

39. R. W. Bailey and B. D. E. Gaillard, *Biochem. J.*, *95*, 758 (1965).

40. R. W. Bailey and R. T. J. Clarke, *Nature*, *199*, 1291 (1963).

41. A. Bonhomme-Florentin, *J. Protozool.*, *22*, 447 (1975).

42. J. Delfosse-Debusscher, D. Thines-Sempoux, M. Vanbelle, and B. Latteur, *Ann. Rech. Vet.*, *10*, 255 (1979).

43. G. S. Coleman, J. I. Laurie, J. E. Bailey, and S. A. Holdgate, *J. Gen. Microbiol.*, *95*, 144 (1976).

44. G. Lubinsky, *Can. J. Zool.*, *35*, 141 (1957).

45. G. S. Coleman, *J. Gen. Microbiol.*, *107*, 359 (1978).

46. G. S. Colemen, *J. Agr. Sci.*, *104*, 349 (1985).

47. G. S. Coleman, *J. Agr. Sci.*, *106*, 121 (1986).

48. R. Knight, J. Sutton, A. McAllan, and R. H. Smith, *Proc. Nutr. Soc.*, *37*, 14A (1978).

49. J. R. Lindsay and J. P. Hogan, *Aust. J. Agr. Res.*, *23*, 321 (1972).

50. R. E. Hungate, *J. Bacteriol.*, *53*, 631 (1947).

51. R. E. Hungate, *Bacteriol. Rev.*, *14*, 1(1950).

52. M. P. Bryant and L. A. Burkey, *J. Diary Sci.*, *36*, 205 (1953).

53. M. P. Bryant and R. N. Doetsch, *J. Dairy Sci.*, *37*, 1176 (1954).

54. B. A. Dehority, *J. Dairy Sci.*, *46*, 217 (1963).

55. R. E. Hungate, *Can. J. Microbiol.*, *3*, 289 (1957).

56. M. P. Bryant, N. Small, C. Bouma, and I. M. Robinson, *J. Bacteriol.*, *76*, 529 (1958).

57. B. D. W. Jarvis and E. F. Annison, *J. Gen. Microbiol.*, *47*, 295 (1967).

58. A. Kistner and L. Gouws, *J. Gen. Microbiol.*, *34*, 447 (1964).

59. B. S. Shane, L. Gouws, and A. Kistner, *J. Gen. Microbiol.*, *55*, 445 (1969).

60. N. O. van Gylswyk and C. E. G. Roche, *J. Gen. Microbiol.*, *64*, 11 (1970).

61. M. P. Bryant and N. Small, *J. Bacteriol.*, *72*, 16 (1956).
62. B. A. Dehority, *J. Bacteriol.*, *91*, 1724 (1966).
63. N. O. van Gylswyk and J. P. L. Hoffman, *J. Gen. Microbiol.*, *60*, 381 (1970).
64. R. A. Prins, F. van Vugt, R. E. Hungate, and C. J. A. H. V. van Vorstenbosch, *Antonie van Leuwenhoek*, *38*, 153 (1972).
65. G. Halliwell and M. P. Bryant, *J. Gen. Microbiol.*, *32*, 441 (1963).
66. D. Groleau and C. W. Forsberg, *Can. J. Microbiol.*, *27*, 517 (1981).
67. J. M. Macy, J. R. Farrand, and L. Montgomery, *Appl. Env. Microbiol.*, *44*, 1428 (1982).
68. B. A. Dehority, Microbes in the foregut of arctic ruminants, In *Control of Digestion and Metabolism in Ruminants* (L. P. Milligan, W. L. Grovum and A. Dobson, eds.), Prentice-Hall, Englewood Cliffs, N. J., 1986, p. 307.
69. B. A. Dehority, *Appl. Env. Microbiol.*, *33*, 1278 (1977).
70. W. E. C. Moore, J. L. Johnson, and L. V. Holdeman, *Int. J. Syst. Bacteriol.*, *26*, 238 (1976).
71. K.-J. Cheng and J. W. Costerton, *J. Bacteriol.*, *129*, 1506 (1977).
72. J. M. Leatherwood and M. P. Sharma, *J. Bacteriol.*, *110*, 751 (1972).
73. N. O. van Gylswyk, *J. Gen. Microbiol.*, *116*, 157 (1980).
74. N. O. van Gylswyk and J. P. L. Labuschagne, *J. Gen. Microbiol.*, *66*, 109 (1971).
75. P. Hiltner and B. A. Dehority, *Appl. Env. Microbiol.*, *46*, 642 (1983).
76. B. A. Dehority and H. W. Scott, *J. Dairy Sci.*, *50*, 1136 (1967).
77. S. G. Kock and A. Kistner, *J. Gen. Microbiol.*, *55*, 459 (1969).
78. B. A. Dehority, *J. Bacteriol.*, *99*, 189 (1969).
79. J. A. Coen and B. A. Dehority, *App. Microbiol.*, *20*, 362 (1970).
80. C. M. Gradel and B. A. Dehority, *Appl. Microbiol.*, *23*, 332 (1972).
81. A. C. I. Warner, *J. Gen. Microbiol.*, *45*, 213 (1966).
82. C. G. Orpin, *J. Gen. Microbiol.*, *84*, 395 (1974).
83. C. G. Orpin, *J. Gen. Microbiol.*, *91*, 249 (1975).
84. C. G. Orpin, *J. Gen. Microbiol.*, *99*, 215 (1977).
85. T. Bauchop, *Appl. Env. Microbiol.*, *38*, 148 (1979).
86. C. G. Orpin, *J. Gen. Microbiol.*, *98*, 423 (1977).
87. C. G. Orpin and L. Bountiff, *J. Gen. Microbiol.*, *104*, 113 (1978).
88. C. G. Orpin and Y. Hart, *J. Appl. Bacteriol.*, *49*, x (1980).
89. W. R. Windham and D. E. Akin, *Appl. Env. Microbiol.*, *48*, 473 (1984).

90. C. G. Orpin and E. A. Munn, *Trans. Br. Mycol. Soc.*, *86*, 178 (1986).
91. I. B. Heath, T. Bauchop, and R. A. Skipp, *Can. J. Bot.*, *61*, 295 (1983).
92. R. R. Johnson, *J. Anim. Sci.*, *22*, 792 (1963).
93. R. R. Johnson, *J. Anim. Sci.*, *25*, 855 (1966).
94. R. A. Weller and A. F. Pilgrim, *Br. J. Nutr.*, *32*, 341 (1974).
95. M. Abe and T. Iriki, *Br. J. Nutr.*, *39*, 255 (1978).
96. F. Nakamura and Y. Kurihara, *Appl. Env. Microbiol.*, *36*, 500 (1978).
97. J. D. Hopson, R. R. Johnson, and B. A. Dehority, *J. Anim. Sci.*, *22*, 448 (1963).
98. E. R. Orskov, F. D. DeB Hovell, and F. Mould, *Trop. Anim. Prod.*, *5*, 195 (1980).
99. J. H. F. Meyer and R. I. Mackie, *App. Env. Microbiol.*, *51* 622 (1986).
100. B. A. Dehority and R. R. Johnson, *J. Dairy Sci.*, *44*, 2242 (1961).
101. D. C. Tomlin, R. R. Johnson, and B. A. Dehority, *J. Anim. Sci.*, *24*, 161 (1965).
102. L. D. Kamstra, A. L. Moxon, and O. G. Bentley, *J. Anin. Sci.*, *17*, 199 (1958).
103. T. C. Quick and B. A. Dehority, *J. Anim. Sci.*, *63*, 1516 (1986).
104. B. A. Dehority, *J. Bacteriol.*, *89*, 1515 (1965).
105. B. A. Dehority, *Appl. Microbiol.*, *15*, 987 (1967).
106. B. A. Dehority, *Appl. Microbiol.*, *16*, 781 (1968).
107. R. A. Prins, In *Microbial Ecology of the Gut* (R. T. J. Clarke and T. Bauchop, eds.), Academic Press, New York, 1977, p. 73.
108. B. A. Dehority, *Insect Sci. Appl.*, *7*, 279 (1986).
109. R. H. McBee, In *Microbial Ecology of the Gut* (R. T. J. Clarke and T. Bauchop, eds.), Academic Press, New York, 1977, p. 185.
110. G. W. Vander Noot and E. B. Gilbreath, *J. Anim. Sci.*, *31*, 351 (1970).
111. D. L. Kern, L. L. Slyter, J. M. Weaver, E. C. Leffel, and G. Samuelson, *J. Anim. Sci.*, *37*, 463 (1973).
112. S. M. Lewis and B. A. Dehority, *Appl. Env. Microbiol.*, *50*, 356 (1985).
113. D. L. Frape, In *Straw and Other Fibrous By-products as Feed* (F. Sundstol and E. Owen, eds.), Elsevier, New York, 1984, p. 487.
114. R. Parra, Comparison of foregut and hindgut fermentation in herbivores, In *The Ecology of Arboreal Folivores* (G. G. Montgomery, ed.), Smithsonian Institution Press, Washington, D.C., 1978, p. 205.

115. I. D. Hume and A. C. I. Warner, Evolution of microbial diges-
     tion in Mammals, In *Digestive Physiology and Metabolism in
     Ruminants* (Y. Ruckebusch and P. Thivend, eds.), MTP Press,
     Lancaster, 1980, p. 665.
116. L. Montgomery, B. Flesher, and D. Stahl, *Int. J. Syst. Bac-
     teriol.*, *38*, 430 (1988).
117. P. N. Hobson (ed.), *The Rumen Microbial Ecosystem*, Elsevier,
     London, 1988.
118. S. E. Lowe, M. K. Theodorou, and A. P. J. Trinci, *J. Gen.
     Microbiol.*, *133*, 1829 (1987).
119. G. L. R. Gordon and M. W. Phillips, *Appl. Env. Microbiol.*,
     55, 1703 (1989).
120. J. M. Osborne and B. A. Dehority, *Appl. Env. Microbiol.*, *55*,
     2247 (1989).

17
Cellulose Degradation by Mesophilic Anaerobic Bacteria

A. W. KHAN and G. B. PATEL   *National Research Council of Canada, Ottawa, Ontario, Canada*

I. INTRODUCTION

Among the cellulolytic microorganisms, mesophilic cellulolytic anaerobes play an important role in the digestion of cellulosic biomass in the rumen, sewage sludge, soil, and other anaerobic habitats containing cellulose. Work on the understanding of the rumen cellulolytic anaerobes has been ongoing for a long time (1,2) to gain knowledge on problems applicable to ruminant nutrition. Among the most common cellulolytic agents isolated from the rumen and studied under laboratory conditions are bacteria belonging to *Fibrobacter*, *Butyrivibrio*, *Eubacterium*, and *Ruminococcus* genera. More recently attempts have been made to obtain mutants of some of these anaerobes in order to improve cellulose degradation in vitro (3), and for molecular cloning of cellulase genes (4; see Chap. 2A).

On the other hand, work on the understanding of mesophilic anaerobes present in sewage sludge, soil, and other habitats containing cellulosic waste has been scanty (5). The main reason for this has been a lack of interest. In the early 1970s, the awareness of environmental pollution and the energy shortage stimulated a systematic study of anaerobes from these environments. The development of enrichment culture techniques, especially the adaptation of

---

*Issued as NRCC Nol 28447

methanogenic cellulolytic microbial ecosystems in the laboratory, on
synthetic media containing cellulose as the sole carbon source (6) led
to the isolation of a mesophilic anaerobe, *Acetivibrio cellulolyticus*,
from sewage sludge (7,8). Subsequently, more cellulolytic anaerobes
were isolated from this enrichment culture (9,10).

During the last 10 years, a number of spore-forming, mesophilic
cellulolytic anaerobes belonging to the genus *Clostridium* have been
isolated from anaerobic reactors utilizing woody biomass (11,12), from
river sediment containing paper mill waste (13,14), and from decaying grasses (15). However, little research has been done to understand enzyme formation and cellulose degradation by these anaerobes.
During this period the controversy over the ability of the protozoa
to degrade cellulose has been resolved (16,17), by confirming the
earlier findings on the participation of ciliates in the digestions of
cellulose in ruminants (18). The existence of anaerobic fungi capable of degrading cellulose has also been shown (19,20).

This chapter covers work carried out during the last 15 years
on the isolation, physiology, and biochemistry of mesophilic-celluloly-
tic anaerobes available in pure culture. Only obligate anaerobes that
require prereduced media (having initial redox potentials lower than
-115 mV) for growth and that have the ability to utilize fibrous or
crystalline cellulose as the sole carbon source for growth are discussed. Anaerobes that produce endoglucanase or β-glucosidase but
are unable to grow on cellulose (21) are not included. The potential
of these anaerobes in biotechnological applications is also discussed.

II. HABITATS AND ISOLATION

In nature, there are numerous habitats where environmental conditions are amenable to the proliferation of mesophilic, cellulolytic
microorganisms. These habitats include the rumen and the intestinal
tracts of ruminant animals, the gastrointestinal tracts of nonruminant
mammals including man, organic sediments of natural waters, anaerobic sewage sludge (22,23), and the hindguts of cellulose ingest-
ing insects such as termites (24) and cockroaches (25). Anoxic conditions prevail in these environments due to the consumption of oxygen by the facultative anaerobic microorganisms.

The E_h of the rumen contents is highly negative (-250 mV and
lower), the temperature is between 38 and 42°C, and the pH is buffered between 6 and 7 due to the high concentration of HCO_3^- and
PO_4^{3-} contained in the influx of large quantities of saliva (22). Until recently the rumen environment had been the most extensively
studied ecosystem where mesophilic anaerobes are present (1,2,22,

26,27). In the rumen, a great variety of non-spore-forming bacteria, ciliate protozoa (16,22), anaerobic fungi (20,28), and spore-forming bacteria are present. Although cellulose digestion is a common property of many aerobic fungi, the presence of anaerobic cellulolytic fungi was only recently shown (19,29,30). The environment in the anaerobic sewage digester is also highly negative (E_h well below -150 mV), the operation temperature is around 35°C and the pH is maintained at 6.8 − 7.2. Bacteria that are identical or functionally similar to the species found in the rumen are also present in anaerobic digesters and other natural habitats (22). However, anaerobic fungi and ciliate protozoa have not yet been observed in anaerobic digesters.

The microorganisms present in these anaerobic environments can be broadly grouped as the fermentative bacteria that solubilize complex polymers such as cellulose, the hydrogen-producing acetogenic bacteria that break down long-chain fatty acids to acetic acid, and the methanogenic bacteria that produce methane from acetate, formate, H_2 and CO_2. The synergistic activities of these microorganisms allow the ruminants to efficiently harness the carbon and energy stored in the cellulose- and hemicellulose-rich forage and feed, and to digest cellulosic materials in sewage to methane and carbon dioxide. The relationship between the microorganisms present in the rumen or in sewage sludge is complex. These microorganisms are dependent on one another for the carbon source, growth factors, removal of toxic metabolic end products, etc. (22,31−33). Although the role of protozoa in cellulose digestion in the rumen is probably small (22), it has been demonstrated that the ciliate-free population in the rumen is unstable (34). The protozoa store large amounts of reserve polysaccharides which when metabolized produce essential amino acids and maintain a favorable proportion and composition of volatile fatty acids (34).

Ecosystems such as the rumen, anaerobic sewage sludge digesters, and anoxic organic sediments of natural waters, where cellulose-rich substrates are being actively digested, are the obvious sources of inocula to attempt the isolation of cellulolytic anaerobes. Generally, it may be preferable to isolate these anaerobes directly from the natural source (35). However, often enrichment techniques in laboratory media are necessary if the organisms of interest are not predominant or are being suppressed in the natural environment. Enrichment techniques are generally carried out under excess substrate conditions and therefore select microorganisms on the basis of maximum growth rate on the substrate provided (35). The anaerobic technique of Hungate (1) and later modifications using roll tubes (2) and serum bottles (36) have greatly facilitated the cultivation of cellulolytic anaerobes.

The composition of the laboratory medium should provide adequate nutrients, but isolation will likely be more successful if the medium is similar to the natural habitat of the organism. The medium of Hungate (1), which contains mineral salts, vitamins, bicarbonate, rumen fluid, and cysteine and sodium sulfide as a reducing solution for the isolation of rumen anaerobes, and that of Khan et al. (37) for the isolation of sewage sludge anaerobes have been successful. For cellulolytic anaerobes, the degree of anaerobiosis of the growth medium as well as the reagents used to achieve it can have a bearing on the success of isolation. For example, the oxidized products of certain reducing agents are highly inhibitory to some cellulolytic anaerobes (38). The nature of the cellulosic substrate incorporated into the medium also influences the species isolated because of the varying capabilities of cellulolytic organisms to digest different kinds of cellulosics. Generally, the cellulosic material used for the enrichment or present in natural habitats is preferable. For most work, ball-milled cellulose, filter paper, and tissue paper have been used successfully (6,39). The nature of the gas phase used to exclude air from the culture vessel also influences the isolation, e.g., the absence of HCO_3^- in the medium and CO_2 in the gas phase makes it difficult to isolate such cellulolytic anaerobes as *Butyrivibrio* (2), *Fibrobacter succinogenes* (40), and *Ruminococcus flavefaciens* (41), which require CO_2 for growth.

For isolation purposes, cellulose-agar plates in conjunction with an anaerobic chamber or anaerobic jars and the roll tube techniques are most commonly used (8,12,42–44). These techniques enable the identification of cellulolytic colonies which have a clearing zone around them as a result of hydrolysis of cellulose by extracellular cellulase. In the case of microorganisms that produce little or no extracellular cellulase, the clearing zone may not be easily visible or may be visible only beneath the colonies. For example, some strains of *R. albus* excrete cellulase of very high molecular weight and thus produce an elliptical zone of cellulose hydrolysis underneath the colonies in vertically incubated roll tubes (45). The nature of the cellulose used in the agar medium also determines if an organism will grow and produce a zone of cellulose hydrolysis. For example, none of the 11 strains of *F. succinogenes* which degraded dewaxed cotton fibers formed a clearing zone with HCl-treated ball-milled filter paper as the substrate since they were unable to degrade that source of cellulose (1). Also, some anaerobes do not grow well in agar, probably because the high concentration of agar (1.5%, w/v) prevents a direct contact between bacteria and cellulose (43). This problem has been alleviated by premixing inoculum in 0.5% cellulose-agar before layering on the plate. In some organisms, a zone of cellulose hydrolysis is observed, but the colonies are almost invisible (46). For the isolation of spore formers such as *Clostridium* species, heating at 80°C for 10 min, before serial dilution and plating is a useful technique to eliminate non-spore-forming bacteria. For the isolation of protozoa and anaerobic fungi, the use of antibiotics to suppress the growth of bacteria which are closely associated with these microorganisms and the use of micromanipulation techniques to pick and transfer single protozoan cells have been used for axenic isolation (47).

Maintenance and preservation of the pure culture isolates is very important. Some cellulolytic anaerobes isolated earlier, namely, *C. lochheadii* and *C. longisporum* (48), are no longer available. Many cellulolytic bacteria have been shown to lose their cellulolytic ability if not routinely transferred in cellulose-containing media (49). For long-term preservation, stock cultures of cellulolytic anaerobes are stored by liquid nitrogen or freeze-drying techniques.

III. PHYSIOLOGY

A list of well-characterized mesophilic cellulolytic anaerobes that are currently available in pure culture for research purposes is given in Table 1. This list does not include *C. lochheadii* and *C. longisporum* [characterized earlier but not available at present (48)], *Bacteroides rumicola* [which has only a limited ability to grow on cellulose (64)], and *Coprococcus* strains [not yet fully characterized (65)].

To date most of the non-spore-forming anaerobic bacteria and anaerobic fungi have been isolated from the rumen (Table 1). All the spore formers listed in Table 1 belong to the genus *Clostridium*, and most have been isolated from environments other than the rumen. The ciliate protozoa were isolated from termites. Most of the cellulolytic bacteria are gram-negative, except for *Eubacterium (Cillobacterium) cellulosolvens* and *C. cellulolyticum*, which are gram-positive. The *Ruminococcus* species are variable in the Gram stain reaction. All these anaerobes grow optimally at 30–40°C and under neutral pH conditions. Although the microorganisms listed are obligate anaerobes, some such as *A. cellulolyticus* can tolerate brief exposure to air without loss of cell viability (66). The taxonomical considerations and classification of anaerobic fungi have been discussed elsewhere (20,28).

Anaerobes such as *A. cellulolyticus* (50), *R. flavefaciens* (27, 61), and *C. populeti* (11) have been shown to produce a yellow pigment which is released in the medium and coats the cellulose particles, whereas *C. cellulovorans* (12) and *R. albus* (27) do not produce such a pigment. While formation of a high molecular weight yellow pigmented polysaccharide by a thermophilic anaerobe, *Clostridium thermocellum*, has been implicated in cellulose digestion (67), the role of yellow pigment in cellulose degradation by the mesophilic

TABLE 1   Characterized Mesophilic Cellulolytic Anaerobes Available
in Pure Culture

Microorganisms	Source	Ref.
(a) Bacteria		
Acetivibrio cellulolyticus	Sewage	8, 50
Acetivibrio cellulosolvens[a]	Sewage	9
Bacteroides cellulosolvens	Sewage	10
Fibrobacter succinogenes[b]	Rumen	51, 52
Butyrivibrio fibrisolvens	Rumen	53
Clostridium cellobioparum	Rumen	26, 54
Clostridium cellulolyticum	Decayed grass	15
Clostridium cellulovorans	Digester on wood	12
Clostridium chartatabidium	Rumen	55
Clostridium lentocellum	Estuarine mud	14
Clostridium paprysolvens	Estuarine mud	13
Clostridium polysaccharolyticum	Rumen	54, 56, 57
Clostridium populeti	Digester on wood	11
Eubacterium cellulosolvens	Rumen	58, 59, 60
Ruminococcus albus	Rumen	48, 61
Ruminococcus flavefaciens	Rumen	48, 61
Ruminococcus flavefaciens var *lacticus*	Rumen	27, 61
(b) Fungi		
Neocallimastix frontalis PN1	Rumen	20
Neocallismastix patriciarum H8	Rumen	29
Piromonas communis	Rumen	30
(c) Protozoa		
Trichomitopsis (Trichomonas) termopsidis	Termite	24, 62
Trichonympha sphaerica	Termite	47

[a]Described as nonmotile, unable to use xylose (9); describes as
motile, able to use xylose and as synonymous with *A. cellulolyticus*
(63).
[b]Recently renamed as *Fibrobacter succinogenes*.

anaerobes has not been established. *Acetivibrio cellulolyticus* (68)
and *F. succinogenes* (44) have also been shown to accumulate large
quantities of glycogen-like polysaccharides in the cells during growth,
when cultivated in media containing high concentrations of cellobiose.
This may serve as a carbohydrate reserve in times of starvation.

Non-spore-forming isolates from sewage sludge are able to grow
in synthetic media containing inorganic salts and vitamins (6,7) and
do not require yeast extract or other complex additives for growth.
By contrast, isolates from rumen and spore-forming anaerobes require rumen fluid or factors present in yeast extract for growth
(Table 2). More specific nutritional requirements, such as vitamins
and sulfur sources for the sewage sludge (38,69), and rumen isolates (64) have also been studied. The nutritional requirements of
bacteria isolated from the rumen are complex and many require rumen
fluid for growth in laboratory media. Many strains of *B. succinogenes, R. albus,* and *R. flavefaciens* require one or more of the
following acids for growth: isobutryic, isovaleric, 2-methylbutyric,
and n-valeric (40,41,70). Some species also require CO_2 for growth
(40,41,44), although it is produced during growth. Inorganic nitrogen and reduced sulfur compounds (38,71) such as sulfide are ac-
ceptable nutrients for rumen anaerobes. The anaerobic fungus
N. particiarum requires hemin, thiamine, and biotin (71); *N. frontalis* requires unidentified components present in cell-free rumen
fluid, yeast extract, and peptone (20) for growth. The protozoan
T. termopsidis needs fetal bovine serum and $NaHCO_3$ for growth
(72). Adequate nutrition has been shown to substantially increase
cellulose digestion by pure cultures grown in laboratory media (38,
69).

The non-spore-forming cellulolytic isolates from sewage sludge
readily utilize a limited number of substrates such as cellulose, cellobiose, and salicin as carbon sources (8-10). Unlike the rumen
isolates, the non-spore-forming sewage isolates do not normally utilize glucose or xylose. *Acetivibrio cellulolyticus* can be adapted for
growth on glucose but at a very low growth rate (73). Glucose-
adapted cultures of *A. cellulolyticus* do not lose their ability to degrade cellulose (50). In contrast to sewage isolates, many of the
cellulolytic bacteria, fungi, and protozoa isolated from the rumen
and other habitats can grow on a wide range of carbohydrates in
addition to cellulose and cellobiose. These are starch, xylan, pectin, glucose, maltose, galactose, sucrose, and xylose (11,60,74).
The fungus *N. frontalis* preferentially utilizes soluble carbohydrates,
glucose being the most preferred sugar (75). This organism does
not degrade cellulose until all of the glucose in the culture medium
has been utilized (74). Many of these cultures lose their ability to
degrade cellulose when they are continuously subcultured on soluble
sugars.

Depending on the particular species and growth conditions, the
major metabolic end products of the mesophilic cellulolytic anaerobes
contain one or more organic acids in varying proportions. These

TABLE 2 Major Cellulosic Substrates and Growth Materials
Reported for the Cultivation of Mesophilic Cellulolytic Anaerobes

Bacteria	Cellulosic materials	Growth materials	Ref.
A. cellulolyticus[a]	AV, CF, CP, DC, FP	None	7, 8
A. cellulosolvens[a]	CP, DC, FP	None	9, 81
B. cellulosolvens[a]	AV, CF, CP, DC, FP	None	10, 81
F. succinogenes	AF, CF, CP, FP	IsoBA, IsoVA PABA, FA[b]	51, 52, 98
B. fibrisolvens	AF, FP	AA, Amino acids, PA, FA[c]	27, 53, 117
C. cellobioparum	CF	Biotin, CO_2, RF	26, 54
C. cellulolyticum	CP	YE	15
C. cellulovorans	AV, FP	YE	12
C. chartatabidium	FP	RF	55
C. lentocellum	AV, CF, DC FP	YE	14
C. paprysolvens	CP	YE	13
C. polysaccharolyticum	FP	AA, CO_2 peptone, RF	54, 56, 57
C. populeti	AV, FP	YE	11
E. cellulosolvens	FP	RF or YE and trypticase	58, 59
R. albus[a]	CF. FP	IsoBA, methylBA, CO_2, PPA, RF	48, 61, 70, 98
R. flavefaciens[a]	AF, CF, FP	Same as R. albus	61, 94, 98

[a]Utilize cellobiose, but not glucose or xylose.

[b]Fatty acids, straight chain > 5 carbon.

[c]Fatty acids, branched chain.
Abbreviations: AF, agricultural fibers obtained from grasses, straw,
silage, or vegetables. AV, "Avicel" or any other crystalline-cellulose
preparation. CF, cotton fibers, dewaxed or mercerized, cheese cloth.
CP, cellulose powder, Whatman CF11 or MN 300. DC, Delignified,
ball-milled, wood cellulose such as Solka-floc or Alpha floc. FP,
filter paper, ball-milled, powder or strips. AA, acetic acid. BA,
butyric acid. PABA, p-aminobenzoic acid. PPA, phenylpropionic
acid. RF, rumen fluid. VA, valeric acid. YE, yeast extract.

acids include acetic, propionic, butyric, lactic, and succinic (8–12, 20,29,47,50,58,59,72,76). In addition, they all produce hydrogen and/or carbon dioxide. Some of these anaerobes also produce ethanol (8–10). The rumen protozoan *Eudiplodinium maggii*, although not isolated in pure culture, has been shown to produce acetic, propionic, and butyric acids, CO_2, and H_2 (17). The anerobic fungus *N. frontalis* also carries out a mixed acid-type fermentation, and the end products, including H_2, are similar to those produced by bacteria (20). Attempts are being made to determine the metabolic pathways for the production for formate, lactate, acetate, and ethanol in anaerobic fungi and to understand whether or not these pathways are similar to those in bacteria (77).

In laboratory cultures, the accumulations of acetic acid in the medium and H_2 in the head space have been shown to inhibit the growth of and, consequently, cellulose degradation by *A. cellulolyticus* (66,78) and *C. cellobioparum* (76). In natural habitats these and other metabolites produced during the growth are disposed of through microbial interaction with other species of anaerobes (e.g., both H_2 and acetic acid are removed by methanogens). In vitro the use of *Methanosarcina barkeri* adapted to utilize acetic acid in co-culture with *A. cellulolyticus* improved cellulose degradation (78,79). Alternately, the removal of acetic acid by the use of $CaCO_3$ or a cation exchange resin also improved cellulose degradation by *A. cellulolyticus*, *A. cellulosolvens*, and *B. cellulosolvens* (80,81). Also, the interaction between cellulolytic anaerobes and saccharolytic bacteria (82), or between cellulolytic protozoa and the host methanogens (72), or between cellulolytic bacteria and the anaerobic spirochete *Treponema bryantii* (83), have been shown to improve cellulose degradation. These beneficial effects may be a result of the removal of inhibitory metabolites or due to the supply of desired growth nutrients as a result of symbiotic interactions. In addition to these factors, the cellulosic materials in natural habitats are broken down by the concerted effort of more than one cellulolytic anaerobe, and consequently, cellulose degradation in natural habitats is generally more extensive than that observed in pure cultures.

IV. LIGNOCELLULOSICS AS SUBSTRATE

Cellulose and hemicellulose present in plants and woody materials are closely associated with lignin. The proportion of these components and the complexity of their association with lignin varies between species, and between woods, branches and foliage of the same tree, and with age and site (84). This association is more complex in the woody plants than in agricultural residues and grasses, and prevents the majority of known microorganisms from attacking it. Virtually all of the cellulolytic mesophilic anaerobes so far isolated are unable to attack native lignocellulosic materials (85–87). For making lignocellulolic materials suitable for the growth of these

anaerobes in vitro, a pretreatment is required to disrupt this association and to remove lignin, lignin breakdown products, and other
inhibitory substances (88,89). Even the soluble lignin–carbohydrate
complexes are not significantly digested in the rumen in spite of the
concerted effort of a number anaerobes (86). Also the bound pheno-
lic materials left after the removal of lignin impose a hindrance to its
utilization by the anaerobes (87). Steam- and explosion-decompres-
sion pretreatment of aspenwood followed by washing with NaOH to
remove lignin and other inhibitory materials has been found suitable
for the growth of isolates from the sewage sludge (80,85), and
H_3PO_4 pretreatment and delignification of newsprint for the growth
of rumen anaerobes (90). Pretreatments for the use of straw (90,91),
jute straw powder (92), exfermented sugar cane chips (93), Kentucky
blue grass, alfalfa, and corn silage (94,95) as substrates for the
growth of rumen anaerobes in vitro have been described.

    Native cellulose has an orderly crystalline structure which offers
a strong resistance to microbial attack (see Chap. 14). Often for
the growth of pure cultures, grinding, ball milling or other pretreatment is required to increase the available surface area and to
reduce the crystallinity for enhancing its susceptibility to microbial
attack. Non-spore-forming isolates from sewage sludge have the
ability to degrade crystalline cellulose (e.g., Avicel, FMC Corp.,
Philadelphia), delignified ball-milled pulp (e.g., Solka or Alpha floc,
Brown & Co., Berlin, New Hampshire), cotton battings, cheese cloth,
Whatman cellulose powder and filter paper, cardboard and tissue
paper (7,80,85). On the other hand, many rumen bacteria capable
of digesting various forms of cellulose present in grasses and plants
are unable to digest cellulose powder. For example, only two strains
of R. albus, two strains of R. flavefaciens, and three out of 48
strains of Butyrivibrio were able to solubilize powdered cellulose
(96). The strains of R. albus and R. flavefaciens that digested
powdered cellulose were unable to degrade cotton (97). It appears
that only F. succinogenes has the ability to digest both types of
cellulose (98). However, there is a great discrepancy between the
reported cellulolytic ability of various strains of rumen anaerobes.
Some strains attack only partially digested cellulose whereas, others
can digest crystalline cellulose (99). Usually, for growing rumen
bacteria in vitro, ball-milled filter paper (49,100) or ball-milled cellulose is used (101). The major cellulolytic anaerobes found in the
rumen, namely, F. succinogenes, R. albus, R. flavefaciens, and
B. fibrisolvens, also have the ability to degrade hemicellulose, the
second major polymer present in plants and woody materials (94,
102–104).

V. CELLULASE ENZYME SYSTEM

The enzymes associated with cellulose degradation by mesophilic anaerobic bacteria are described in detail in Chap. 22. Several aspects of the enzymes will be discussed here with regard to their effects on the physiology and ecology of the organisms. Most of the mesophilic anaerobes described in this chapter have been shown to produce constitutive cellulase enzymes (105–109). The formation, properties, and action of cellulase enzymes produced by *F. succinogenes* (100, 110-113), *R. albus*, (70,97,114-116), *R. flavefaciens* (108,117,118), *A. cellulolyticus* (119–124), and *B. cellulosolvens* (106) have been studied. In *A. cellulolyticus* and *B. cellulosolvens* the cellulase formation appears regulated by catabolite repression (124,125). Hemicellulases produced by sewage isolates have also been studied (126) and those produced by rumen anaerobes have been reviewed (127). The enzyme system produced by these anaerobes has the ability to degrade filter paper, crystalline cellulose (exoglucanase activity), carboxymethylcellulose (endoglucanase activity), xylan (xylanase activity), p-nitrophenyl-β-D-xylopyranoside (β-xylosidase activity), and p-nitrophenyl acetate (esterase activity). The enzyme system produced by *A. cellulolyticus* also has the ability to decrease the turbidity of crystalline cellulose suspensions (121). These activities have been detected both in the growth medium and in association with the cell. However, in *A. cellulolyticus* and *F. succinogenes*, β-glucosidase activity is largely cell-associated (100,124), while *B. cellulosolvens* and *R. flavefaciens* possess little or no β-glucosidase activity (106,128).

Although isolates from both the rumen and the sewage sludge rapidly degrade cellulose during growth, they do not produce an extracellular cellulase system that is highly active in vitro. It appears that in these anaerobes the dissolution of cellulose is performed by the cell-bound enzymes (129,130), and their effectiveness is due to the combined effect of high concentration of enzyme molecules and favorable alignment on the cell surface (131). These parameters are not attained when either the cell extracts or the cellular supernatant liquid are used. The adherence of cellulolytic bacteria to cellulose particles (39,100) may explain the dependence of these anaerobes on cell-associated enzymes. The closer the association of the bacterium with the substrate, the greater the benefit it can derive from cell-associated enzymes. In natural habitats the further the bacterium is from the cellulose, the less chance it has to recover the products of cellulose hydrolysis. Cellulose is completely insoluble and therefore cellulolytic bacteria must produce cellulases located on the outer surface of the cells, and must cling to the cellulose fibers undergoing disintegration, to receive maximum return to the cell from its hydrolytic enzymes (132). *Ruminococcus albus* and other rumen bacteria have frequently been reported to adhere to cellulosic material by means of capsular material (133,134). In *A. cellulolyticus*, the amorphous outer cell wall layer has been implicated in such an attachment (135).

The extensive work carried out on rumen bacteria (100,108,112, 114,136,137) has indicated that the extracellular cellulase activities are found associated with (1) sedimentable membrane fractions, (2) nonsedimentable high molecular weight materials, and (3) low molecular weight (45,000) materials. *Acetivibrio cellulolyticus*, an isolate from sewage sludge, has also been shown to produce multiple endoglucanases differing in both size and molecular weight (123). The endoglucanase of *F. succinogenes* can be released from the membrane fractions by treatment with bovine pancreatic trypsin (112) and Triton X-100 (113). This work also indicates that extracellular cellulase enzymes in these bacteria are associated with sedimentable subcellular vesicles. These vesicles appear to be released from the intact cell by bleb or an outer membrane formation and have been associated with culture aging (139). Several other bacteria are known to release blebs during normal growth without detectable cell lysis (138,140,141). This may represent a general mechanism for the release of these enzymes in the broth and may not be essential for cellulose degradation (139) as cellulose degradation is mainly carried out by the action of cell-associated enzymes (129,130).

VI. CONCLUSIONS

Non-spore-forming cellulolytic anaerobes, especially those isolated from sewage sludge, possess the unique ability to hydrolyze both cellulosic and hemicellulosic materials and to accumulate cellobiose, glucose, and xylose in the broth (81). Most of these anaerobes readily utilize cellobiose but not glucose and xylose. The mechanism of assimilation of cellulose by these organisms is obscure. However, cellobiose metabolism is better understood in *R. flavefaciens* and *Clostridium thermocellum* (128,142,143). It seems probable that cellulose digestion first proceeds by the release of cellobiose units by the action of cellulolytic enzymes as proposed in the case of *R. flavefaciens* (144). Following this, there is the possibility that cellobiose is metabolized by the direct mechanism of phosphorolytic cleavage to glucose-1-phosphate and glucose by cellobiose phosphorylase, and results in the release of free glucose which intact cells do not metabolize (145). A second possibility is that cells have a specific permease or carrier system for cellobiose, but not for glucose. In *R. flavefaciens*, both the phosphorylation and the internal mechanism for glucose metabolism exist (128). *Bacterioides cellulosolvens* appears to possess little or no β-glucosidase activity and appears to utilize cellobiose through cellobiose phosphorylation (106), and consequently glucose is released. On the other hand, *A. cellulolyticus* possesses cellobiase activity which is mainly cell-associated (123) and the possibility of cellobiose being hydrolyzed to glucose and utilized internally cannot be ruled out in the case of this anaerobe. Extracellular glucose accumulation can take place as a result of released cellobiase activity. It should be pointed out that *A. cellulolyticus* has been adapted to grow on glucose, although the growth of the adapted strain on glucose was poorer than on cellobiose (73). However, further work is required to establish the mechanism of utilization of cellobiose and glucose.

The property of some of the mesophilic cellulolytic anaerobes to accumulate sugars in the medium may be useful for industrial application. For example, these anaerobes in pure cultures can be used for the single-step conversion of cellulose to sugars or to other useful products by means of coculture techniques. The accumulation of sugars in significant amounts has been shown to take place in a medium containing high cellulose content (5% w/v or more) (81) and during the stationary phase of growth (146). In order to exploit this property, cells of *B. cellulosolvens* harvested in the stationary phase of growth have been incubated with cellulose to produce sugars. This technique has all the benefits of an immobilized cell system and its potential for further development is worthy of consideration. Although the yield of ethanol from cellulosic materials by *A. cellulolyticus* (147) and other anaerobes is too low for industrial production, the sugars produced by *A. cellulolyticus* and *B. cellulosolvens* have been converted to ethanol in reasonable amounts by using an ethanologenic microorganism in coculture (85,148,149). The coculture technique can be further developed for the production of solvents or by the use of triculture to convert ethanol to high dollar value products such as esters. Most of the mesophilic cellulolytic anaerobes produce acetic acid, carbon dioxide, and hydrogen as end products. The production of these and other products using pretreated aspen wood (85) and the use of cellulolytic anaerobes in coculture with other species of anaerobes (78) have also been studied.

Little or no work on the use of spore-forming anaerobes and non-spore-forming rumen anaerobes has been carried out relative to their potential use for biotechnological purposes. Perhaps the reasons have been the initial interest in rumen nutrition rather than for industrial application, and the complexity of the growth medium required for rumen isolates. The isolates from sewage sludge can be cultivated in synthetic medium (6,37), but the isolates from the rumen require rumen fluid (1,150,151) and/or other supplements.

Even the medium developed without rumen fluid contained trypticase and yeast extract (118,152), which are not strongly desirable for large-scale fermentation processes. Recently, a synthetic medium comparable to the one used for the cultivation of isolates from sewage sludge has been developed (153) for the cultivation of R. albus, and this may help in further studies on this aspect. More recently, the ability to fix N_2 has been demonstrated in some new mesophilic anaerobic isolates from estuarine mud and soil, and in Clostridium paprysolvens (154). The ability of anaerobes to fix nitrogen using cellulose as an energy and carbon source not only has a great implication in the carbon and nitrogen cycling in nature but these organisms would be more economical to grow as they would not require combined nitrogen supplements.

There are a number of inherent advantages in using cellulolytic mesophilic anaerobes in biotechnological processes. These include low cell yield, high product-to-substrate ratio, no stringent gas transfer requirement (as there are in case of aerobes), and a lower temperature requirement than for thermophilic anaerobes. Biotechnological processes such as the conversion of cellulose to sugars, or cellulose to ethanol, based on the use of mesophilic anaerobes also require fewer fermentation steps as compared to those based on the use of aerobic microorganisms and cellulose as a substrate. However, there are a number of problems including a lack of information on the use of these anaerobes in large-scale fermentations. Most of the work so far has been carried out in vials, flasks, or small fermentors, and knowledge about the physiology and biochemistry of these anaerobes required for scale-up work is limited. It should also be recognized that the inherent recalcitrant nature of the cellulose-lignin matrix and the complexity and insolubility of cellulosic materials impose formidable problems with respect to economical bioconversion of this resource to chemicals and gaseous and liquid fuels. It is difficult to increase the initial cellulose concentration to more than 50 g/liter in view of the fluidity of the broth and the low specific rates of cellulose digestion (153). In addition, it should also be recognized that research on cellulolytic anaerobes other than for strictly academic purposes is less than a decade old. The concept of their use in biotechnological processes, like all other new concepts, is being viewed with skepticism and apprehension. Further information through research and development would help in their evaluation for industrial application.

ACKNOWLEDGMENT

The authors wish to express their appreciation to Karen Lamb for her assistance in preparing the manuscript.

REFERENCES

1. R. E. Hungate, *Bacteriol. Rev.*, *14*, 1 (1950).
2. R. E. Hungate, M. P. Bryant, and R. A. Mah, *Ann. Rev. Microbiol.*, *18*, 131 (1964).
3. M. Taya, K. Ohmiya, T. Kobayashi, and S. Shimizu, *J. Ferment. Technol.*, *61*, 197 (1983).
4. S. Kawai, H. Honda, T. Tanase, M. Taya, S. Iijima, and T. Kobayashi, *Agr. Biol. Chem.*, *51*, 59 (1987).
5. L. R. Maki, *Antonie van Leeuwenhoek*, *20*, 185 (1954).
6. A. W. Khan, *Can. J. Microbiol.*, *23*, 1700 (1977).
7. A. W. Khan, J. N. Saddler, G. B. Patel, J. R. Colvin, and S. M. Martin, *FEMS Microbiol. Lett.*, *7*, 47 (1980).
8. G. B. Patel, A. W. Khan, B. J. Agnew, and J. R. Colvin, *Int. J. Syst. Bacteriol.*, *30*, 179 (1980).
9. A. W. Khan, E. Meek, L. C. Sowden, and J. R. Colvin, *Int. J. Syst. Bacteriol.*, *34*, 419 (1984).
10. W. D. Murray, L. C. Sowden, and J. R. Colvin, *Int. J. Syst. Bacteriol.*, *34*, 185 (1984).
11. R. Sleat and R. A. Mah, *Int. J. Syst. Bacteriol.*, *35*, 160 (1985).
12. R. Sleat, R. A. Mah, and R. Robinson, *Appl. Env. Microbiol.*, *48*, 88 (1984).
13. R. H. Madden, M. J. Bryder, and N. J. Poole, *Int. J. Syst. Bacteriol.*, *32*, 87 (1982).
14. W. D. Murray, L. Hofmann, N. L. Campbell, and R. H. Madden, *Syst. Appl. Microbiol.*, *8*, 181 (1986).
15. E. Petitdemange, F. Caillet, J. Giallo, and C. Gaudin, *Int. J. Syst. Bacteriol.*, *34*, 155 (1984).
16. R. T. J. Clarke, *Microbial Ecology of the Gut* (R. T. J. Clarke and T. Bauchop, eds.), Academic Press Inc., New York, 1977, p. 251.
17. G. S. Coleman, *J. Gen. Microbiol.*, *107*, 359 (1978).
18. R. E. Hungate, *Biol. Bull.*, *83*, 303 (1942).
19. C. G. Orpin, *J. Gen. Microbiol.*, *91*, 249 (1975).
20. T. Bauchop and D. O. Mountfort, *Appl. Env. Microbiol.*, *42*, 1103 (1981).
21. S. F. Lee, C. W. Forsberg, and L. N. Gibbons, *Appl. Env. Microbiol.*, *50*, 220 (1985).
22. M. P. Bryant, In *Duke's Physiology of Domestic Animals* , 9th ed. (M. J. Swenson, ed.), Cornell University, Press, Ithaca, New York, 1977, p. 287.
23. R. E. Hungate, *Ann. Rev. Ecol. Syst.*, *6*, 39 (1975).
24. M. A. Yamin, *Appl. Env. Microbiol.*, *39*, 859 (1980).
25. D. L. Cruden and A. J. Markovetz, *Appl. Env. Microbiol.*, *38*, 369 (1979).

26. R. E. Hungate, *J. Bacteriol.*, *48*, 499 (1944).
27. N. O. van Glyswyk and C. E. G. Roche, *J. Gen. Microbiol.*, *64*, 11 (1970).
28. I. B. Heath, T. Bauchop, and R. A. Skipp, *Can. J. Bot.*, *61*, 295 (1983).
29. C. G. Orpin and E. A. Munn, *Trans. Br. Mycol. Soc.*, *86*, 178 (1986).
30. C. G. Orpin, *J. Gen. Microbiol.*, *99*, 107 (1977).
31. A. G. Fredrickson and G. Stephanopoulos, *Science*, *213*, 972 (1981).
32. A. W. Khan and W. D. Murray, *J. Appl. Bacteriol.*, *53*, 379 (1982).
33. R. S. Wolfe, *Adv. Microb. Physiol.*, *6*, 107 (1971).
34. J. M. Eadie and S. Mann, In *Physiology of Digestion and Metabolism in the Ruminant* (A. T. Phillipson, ed.), Oriel Press, New Castle-upon-Tyne, 1970, p. 335.
35. H. W. Doelle, In *Trends in Scientific Research*, Vol. 2 (H. W. Doelle and C.-G. Heden, eds.), Reidel–UNESCO, Paris, 1986, p. 38.
36. T. L. Miller and M. J. Wolin, *Appl. Microbiol.*, *27*, 985 (1974).
37. A. W. Khan, T. M. Trottier, G. B. Patel, and S. M. Martin, *J. Gen. Microbiol.*, *112*, 365 (1979).
38. G. B. Patel, C. Breuil, and B. J. Agnew, *Can. J. Microbiol.*, *28*, 772 (1982).
39. R. E. Hungate, *J. Bacteriol.*, *53*, 631 (1947).
40. J. M. Macy and I. Probst, *Ann. Rev. Microbiol.*, *33*, 561 (1979).
41. L. L. Slyter and J. M. Weaver, *Appl. Env. Microbiol.*, *33*, 363 (1977).
42. S. E. Lowe, M. K. Theodorou, and A. P. J. Trinci, *J. Appl. Bacteriol.*, *59*, XV (1985).
43. J. M. Macy, J. R. Farrand, and L. Montgomery, *Appl. Env. Microbiol.*, *44*, 1428 (1982).
44. C. S. Stewart, C. Paniagua, D. D. Dinsdale, K.-J. Cheng, and S. A. Garrow, *Appl. Env. Microbiol.*, *41*, 504 (1981).
45. T. Bauchop and S. R. Elsden, *J. Gen. Microbiol.*, *23*, 457 (1960).
46. V. H. Varel, S. J. Fryda, and I. M. Robinson, *Appl. Env. Microbiol.*, *47*, 219 (1984).
47. M. A. Yamin, *Science*, *211*, 58 (1981).
48. R. E. Hungate, *Can. J. Microbiol.*, *3*, 289 (1957).
49. N. O. van Glyswyk and J. P. L. Labuschagne, *J. Gen. Microbiol.*, *66*, 109 (1971).
50. G. B. Patel, In *Bergey's Manual of Systematic Bacteriology*, Vol. 1 (N. R. Krieg, ed.), Williams and Wilkins, Baltimore, 1984, p. 658.

51. E. P. Cato, W. E. C. Moore, and M. P. Bryant, *Int. J. Syst. Bacteriol.*, *28*, 491 (1978).
52. L. V. Holdeman, R. W. Kelley, and W. E. C. Moore, In *Bergey's Manual of Systematic Bacteriology*, Vol. I (N. R. Krieg, ed.), Williams and Wilkins, Baltimore, 1984, p. 604.
53. M. P. Bryant, In *Bergey's Manual of Systematic Bacteriology*, Vol. 2 (P. H. A. Sneath, ed.), Williams and Wilkins, Baltimore, 1986, p. 1376.
54. E. P. Cato, W. L. George, and S. M. Finegold, In *Bergey's Manual of Systematic Bacteriology*, Vol. 2 (P. H. A. Sneath, ed.), Williams and Wilkins, Baltimore, 1986, p. 1141.
55. W. J. Kelly, R. V. Amundson, and D. H. Hopcroft, *Arch. Microbiol.*, *147*, 169 (1987).
56. N. O. van Glyswyk, *J. Gen. Microbiol.*, *116*, 157 (1980).
57. N. O. van Glyswyk, E. J. Morris, and H. J. Els, *J. Gen. Microbiol.*, *121*, 491 (1980).
58. R. A Prins, F. van Vugt, R. E. Hungate, and C. J. A. H. V. van Vorstenbosch, *Antonie van Leeuwenhoek*, *38*, 153 (1972).
59. N. O. van Glyswyk and J. J. T. K. van der Toorn, *Int. J. Syst. Bacteriol.*, *36*, 275 (1986).
60. M. P. Bryant, N. Small, C. Bouma, and I. M. Robinson, *J. Bacteriol.*, *76*, 529 (1958).
61. M. P. Bryant, In *Bergey's Manual of Systematic Bacteriology*, Vol. 2 (P. H. A. Sneath, ed.), Williams and Wilkins, Baltimore, 1986, p. 1093.
62. M. A. Yamin and W. Trager, *J. Gen. Microbiol.*, *113*, 417 (1979).
63. W. D. Murray, *Int. J. Syst. Bacteriol.*, *36*, 314 (1986).
64. S. Shimizu, K. Ohmiya, M. Taya, and T. Kobayashi, In *Advances in Biotechnology*, Vol. 2 (M. Moo-Young and C. W. Robinson, eds.), Pergamon Press, Toronto, 1981, p. 149.
65. J. Sukhumavasi, K. Ohmiya, M. Suwana-Adth, and S. Shimizu, *J. Ferment. Technol.*, *62*, 545 (1984).
66. C. Breuil and G. B. Patel, *J. Gen. Microbiol.*, *125*, 41 (1981).
67. L. G. Ljungdahl, B. Pettersson, K.-E. Eriksson, and T. Kliegel, *Curr. Microbiol.*, *9*, 195 (1983).
68. G. B. Patel and C. Breuil, *Arch. Microbiol.*, *129*, 265 (1981).
69. W. D. Murray, *J. Biotechnol.*, *3*, 131 (1985).
70. R. E. Hungate and R. J. Stack, *Appl. Env. Microbiol.*, *44*, 79 (1982).
71. C. G. Orpin and Y. Greenwood, *Trans. Br. Mycol. Soc.*, *86*, 103 (1986).
72. D. A. Odelson and J. A. Breznak, *Appl. Env. Microbiol.*, *49*, 614 (1985).

73. G. B. Patel and C. R. MacKenzie, *Eur. J. Appl. Microbiol. Biotech.*, *16*, 212 (1982).

74. C. G. Orpin and A. J. Letcher, *Curr. Microbiol.*, *3*, 121 (1979).

75. D. O. Mountfort and R. A. Asher, *Appl. Env. Microbiol.*, *49*, 1314 (1985).

76. K.-T. Chung, *Appl. Env. Microbiol.*, *31*, 342 (1976).

77. N. Yarlett, C. G. Orpin, E. A. Munn, N. C. Yarlett, and C. A. Greenwood, *Biochem. J.*, *236*, 729 (1986).

78. A. W. Khan, *FEMS Microbiol. Lett.*, *9*, 233 (1980).

79. V. M. Laube and S. M. Martin, *Appl. Env. Microbiol.*, *42*, 413 (1981).

80. W. D. Murray, *Biomass*, *10*, 47 (1986).

81. C. Giuliano, M. Asther, and A. W. Khan, *Biotechnol. Lett.*, *5*, 395 (1983).

82. W. D. Murray, *Appl. Env. Microbiol.*, *51*, 710 (1986).

83. T. B. Stanton and E. Canale-Parola, *Arch. Microbiol.*, *127*, 145 (1980).

84. G. M. Barton, *Biomass*, *4*, 311 (1984).

85. A. W. Khan, M. Asther, and C. Giuliano, *J. Ferment. Technol.*, *62*, 335 (1984).

86. M. J. Neilson and G. N. Richards, *J. Sci. Fd. Agr.*, *29*, 513 (1978).

87. R. D. Hartley, *J. Sci. Fd. Agr.*, *23*, 1347 (1972).

88. A. Chesson, C. S. Stewart, and R. J. Wallace, *Appl. Env. Microbiol.*, *44*, 597 (1982).

89. A. W. Khan, *Can. Res.*, *17*, 21 (1984).

90. M. Taya, K. Honma, K. Ohmiya, T. Kobayashi, and S. Shimizu, *J. Chem. Eng. Japan*, *14*, 330 (1981).

91. N. Kolankaya, C. S. Stewart, S. H. Duncan, K.-J. Cheng, and J. W. Costerton, *J. Appl. Bacteriol.*, *58*, 371 (1985).

92. M. Rahmatullah, A. Rahman, A. A. Chowdhury, and M. A. Rashid, *J. Ferment. Technol.*, *57*, 117 (1979).

93. C. Rolz, S. de Cabrera, M. J. Valdez, M. del Carmen de Arriola, and J. Valladares, *Biotechnol. Progr.*, *2*, 120 (1986).

94. G. F. Collings and M. T. Yokoyama, *Appl. Env. Microbiol.*, *39*, 566 (1980).

95. L. C. Greve, J. M. Labavitch, and R. E. Hungate, *Appl. Env. Microbiol.*, *47*, 1135 (1984).

96. M. P. Bryant and N. Small, *J. Bacteriol.*, *72*, 16 (1956).

97. T. M. Wood, C. A. Wilson, and C. S. Stewart, *Biochem. J.*, *205*, 129 (1982).

98. G. Halliwell and M. P. Bryant, *J. Gen. Microbiol.*, *32*, 441 (1963).

99. M. P. Bryant, *Fed. Proc.*, *32*, 1809 (1973).

100. D. Groleau and C. W. Forsberg, *Can. J. Microbiol.*, *27*, 517 (1981).
101. K. Ohmiya, K. Nokura, and S. Shimizu, *J. Ferment. Technol.*, *61*, 25 (1983).
102. B. A. Dehority, *Fed. Proc.*, *32*, 1819 (1973).
103. S. G. Kock and A. Kistner, *J. Gen. Microbiol.*, *55*, 459 (1969).
104. J. J. T. K. van der Toorn and N. O. van Gylswyk, *J. Gen. Microbiol.*, *131*, 2601 (1985).
105. R. E. Hungate, *The Rumen and Its Microbes*, Academic Press, New York, 1966.
106. C. Giuliano and A. W. Khan, *Appl. Env. Microbiol.*, *48*, 446 (1984).
107. P. Hintner and B. A. Dehority, *Appl. Env. Microbiol.*, *46*, 642 (1983).
108. G. L. Pettipher and M. J. Latham, *J. Gen. Microbiol.*, *110*, 29 (1979).
109. J. N. Saddler and A. W. Khan, *Can. J. Microbiol.*, *26*, 760 (1980).
110. C. W. Forsberg, T. J. Beveridge, and A. Hellstrom, *Appl. Env. Microbiol.*, *42*, 886 (1981).
111. C. W. Forsberg and D. Groleau, *Can. J. Microbiol.*, *28*, 144 (1982).
112. D. Groleau and C. W. Forsberg, *Can. J. Microbiol.*, *29*, 710 (1983).
113. D. Groleau and C. W. Forsberg, *Can. J. Microbiol.*, *29*, 504 (1983).
114. A. Chesson, C. S. Stewart, K. Dalgarno, and T. P. King, *J. Appl. Bacteriol.*, *60*, 327 (1986).
115. T. M. Wood and C. A. Wilson, *Can. J. Microbiol.*, *30*, 316 (1984).
116. R. J. Stack and R. E. Hungate, *Appl. Env. Microbiol.*, *48*, 218 (1984).
117. D. E. Akin and L. L. Rigsby, *Appl. Env. Microbiol.*, *50*, 825 (1985).
118. M. J. Latham, B. E. Brooker, G. L. Pettipher, and P. J. Harris, *Appl. Env. Microbiol.*, *35*, 1166 (1978).
119. A. W. Khan, *J. Gen. Microbiol.*, *121*, 499 (1980).
120. C. R. MacKenzie and D. Bilous, *Can. J. Microbiol.*, *28*, 1158 (1982).
121. C. R. MacKenzie, D. Bilous, and G. B. Patel, *Appl. Env. Microbiol.*, *50*, 243 (1985).
122. C. R. MacKenzie, G. B. Patel, and D. Bilous, *Appl. Env. Microbiol.*, *53*, 304 (1987).
123. J. N. Saddler and A. W. Khan, *Can. J. Microbiol.*, *27*, 288 (1981).

124. J. N. Saddler, A. W. Khan, and S. M. Martin, *Microbios*, *28*, 97 (1980).
125. W. D. Murray, *Biotechnol. Bioeng.*, *29*, 1151 (1987).
126. A. W. Khan, K. A. Lamb, and M. A. Forgie, *Biomass*, *13*, 135 (1987).
127. R. F. H. Dekker and G. N. Richards, *Adv. Carboh. Chem. Biochem.*, *32*, 277 (1976).
128. W. A. Ayers, *J. Bacteriol.*, *76*, 515 (1958).
129. A. W. Khan, In *Biotechnology and Renewable Energy* (M. Moo-Young, S. Hasnain, and L. Lamptey, eds.), Elsevier, New York, 1986, p. 70.
130. R. Lamed, J. Naimark, E. Morgenstern, and E. A. Bayer, *J. Bacteriol.*, *169*, 3792 (1987).
131. B. v. Hofsten, Topological effects in enzymatic and microbial degradation of highly ordered polysaccharides, In Symposium on Enzymatic Hydrolysis of Cellulose (M. Bailey, T. M. Enari, and M. Linko, eds.), SITRA, Aulanko, Finland, 1975, p. 281.
132. M. R. Pollock, In *The Bacteria*, Vol. 4 (I. C. Gunsalus and R. Y. Stanier, eds.), Academic Press, New York, 1962, p. 19.
133. J. M. Leatherwood, *Fed. Proc.*, *32*, 1814 (1973).
134. H. Patterson, R. Irvin, J. W. Costerton, and K. J. Cheng, *J. Bacteriol.*, *122*, 278 (1975).
135. J. R. Colvin, L. C. Sowden, G. B. Patel, and A. W. Khan, *Curr. Microbiol.*, *7*, 13 (1982).
136. J. M. Gawthorn, *Ann. Rech. Vet.*, *10*, 249 (1979).
137. H. E. Schellhorn and C. W. Forsberg, *Can. J. Microbiol.*, *30*, 930 (1984).
138. D. Hoekstra, J. W. van der Laan, L. de Leij, and B. Witholt, *Biochim. Biophys. Acta*, *455*, 889 (1976).
139. G. Gaudet and B. Gaillard, *Arch. Microbiol.*, *148*, 150 (1987).
140. R. S. Munford, C. L. Hall, and P. D. Rick, *J. Bacteriol.*, *144*, 630 (1980).
141. S. MacIntyre, T. J. Trust, and J. T. Buckley, *Can. J. Biochem.*, *58*, 1018 (1980).
142. T. K. Ng and J. G. Zeikus, *J. Bacteriol.*, *150*, 1391 (1982).
143. C. J. Sih and R. H. McBee, *Proc. Mont. Acad. Sci.*, *15*, 21 (1966).
144. W. A. Ayers, *J. Bacteriol.*, *76*, 504 (1958).
145. E. J. Swisher, W. D. Storvick, and K. W. King, *J. Bacteriol.*, *88*, 817 (1964).
146. C. Giuliano and A. W. Khan, *Biotechnol. Bioeng.*, *27*, 980 (1985).
147. D. W. Armstrong and S. M. Martin, *Biotechnol. Bioeng.*, *25*, 2567 (1983).
148. A. W. Khan and W. D. Murray, *Biotechnol. Lett.*, *4*, 177 (1982).

149. J. N. Saddler, M. K.-H. Chan, and G. Louis-Seize, *Biotechnol. Lett.*, *3*, 321 (1981).
150. M. P. Bryant and I. M. Robinson, *J. Dairy Sci.*, *44*, 1446 (1961).
151. M. P. Bryant and I. M. Robinson, *J. Bacteriol.*, *84*, 605 (1962).
152. D. R. Caldwell and M. P. Bryant, *Appl. Microbiol.*, *14*, 794 (1966).
153. M. Taya, T. Kobayashi, and S. Shimizu, *Agr. Biol. Chem.*, *44*, 2225 (1980).
154. S. B. Leschine, K. Holwell, and E. Canale-Parola, *Science*, *242*, 1157 (1988).

18
Cellulose Degradation by Thermophilic Anaerobic Bacteria

RAPHAEL LAMED   *George S. Wise Faculty of Life Sciences, Tel Aviv University, Ramat Aviv, Israel*

EDWARD A. BAYER   *Weizmann Institute of Science, Rehovot, Israel*

I. INTRODUCTION

For many reasons, the potential use of anaerobic thermophilic bacteria in the eventual industrial degradation of cellulose has served as an attractive and stimulating incentive for application-oriented research in the last decade. On the one hand, anaerobic fermentations provide a relatively inexpensive and efficient means for bioconversion processes, in that problematic aeration is precluded. In addition, relatively high product levels are produced per bacterial cell mass. On the other hand, thermophilic microbial processes are advantageous in that expensive refrigeration procedures usually required for large-scale mesophilic fermentations are avoided. Moreover, thermophilic procedures facilitate product recovery, potential contamination by pathogenic microorganisms is usually less problematic, and the resultant enzymes are frequently structurally and functionally stable. These aspects have been subject to extensive review in recent years by numerous authors (1–5) and the interested reader is directed to these sources for additional information.

In spite of the broad interest in this area and the widespread research devoted to the prospect of industrial utilization of anaerobic thermophilic bacteria for cellulose degradation, all engineering attempts to establish commercially viable processes have failed. Part of the reason for this lack of success may be an insufficient understanding of the biochemistry and physiology of the cellulose

___
This work was submitted in July of 1987.

degradation process and the intricate interspecies interactions which characterize this microbial-induced process in nature.

The purpose of this review is to describe the present status of our knowledge concerning thermophilic anaerbic cellulose fermentation. The chapter is designed to provide a concise survey which (1) traces the historical development of the discovery and characterization of thermophilic anaerobic strains, (2) describes the physiological features and intermediary metabolism of the principal representative of this restricted group of microorganisms, and (3) discusses the interactions in coculture fermentations with other microbes.

II. HISTORICAL PERSPECTIVES

The origin and development of thermophilic bacteria is still a mystery. For example, the question of whether thermophiles have evolved from mesophiles or vice versa has yet to be ascertained. It is clear, however, that multiple mutations would have been required for such conversions.

Today thermophilic anaerobic cellulolytic bacteria have been identified in a wide variety of ecological niches, including autothermal environments, solar-heated systems, as well as geothermal habitats. Thus, such strains are essentially ubiquitous and have been isolated from various commonly occurring ecosystems, such as manure, compost, hay, straw, anaerobic digesters, and decaying debris. Other strains have been found in the various hot springs worldwide. The considerable fluctuations which characterize such ecosystems have resulted in a preponderance of spore-forming microorganisms.

Despite the large ecological diversity, *specific* representatives of the above group of bacteria are restricted to only one confirmed genus, *Clostridium*. Moreover, of the two confirmed species, i.e., *Clostridium thermocellum* and *Clostridium stercorarium*, the latter is a relatively new isolate (6) which has yet to be extensively characterized. Other newly isolated *Clostridium* spp. (7) differ from *C. thermocellum* but their properties appear to be similar to those of *C. stercorarium*.

Of particular interest is the recent report (8) of an as yet taxonomically unidentified, extremely thermophilic (optimal growth temperature of 75°C), cellulolytic anaerobe (isolated from hot springs located in Japan) which is apparently distinguished from the genus *Clostridium* in that the new species is *non-spore-forming*. In a very recent sequel to the latter communication (9), a heat-stable β-glucanglucanase from this bacterium was cloned in *Escherichia coli*; interestingly, the strain was referred to as *Thermoanaerobacter cellulolyticus*, although the basis for this classification (i.e., its relatedness to the type species *T. ethanolicus*) was not mentioned or cited. To our knowledge, additional information concerning this interesting new organism is still needed.

An even more recent report (10) came to our attention just prior to submission of the present manuscript. A cellulolytic, anaerobic, extremely thermophilic (75°C temperature optimum) non-spore-forming microorganism was isolated from thermal pool sites in New Zealand. The isolate, which appears to be very similar to the Japanese strain, was provisionally designated *Caldocellum saccharolyticum*. Although the latter bacterium emulates *C. stercorarium* in various properties, its DNA composition appears to be distinct.

Thus, despite recent activity in this area concerning the isolation of new strains, the information accumulated throughout the years on *C. thermocellum* essentially composes the body of knowledge currently available relating to the topic of thermophilic anaerobic cellulolytic fermentation. Consequently, with few exceptions (and unless otherwise stated henceforth in the text), the scope of the present chapter is confined to a single species.

The first example of thermophilic fermentation of cellulose was reported in 1896 by MacFayden and Blaxall (11). In another early work (12), the presence of an extracellular cellulase was demonstrated in anaerobic thermophilic culture media. A decade later, various applicative aspects of thermophilic fermentation of cellulose were first patented (13).

The fermentation of cellulose to ethanol by *C. thermocellum* was first recorded by Viljoen et al. in 1926 (14). However, facile maintenance of stable and pure thermophilic cellulolytic cultures awaited the advent of improved anaerobic procedures and equipment, which were first utilized for this purpose by McBee (15). Throughout the years, a variety of other investigators isolated strains similar to the original *C. thermocellum* strain described by McBee, although their cellulolytic properties were not consistently characterized in detail.

A second related strain (16,17) which differed from *C. thermocellum* in its substrate utilization pattern, was originally classified as *C. thermocellulaseum*, but unfortunately the strain was lost. In retrospect, this strain is presently considered to be a derivative of *C. thermocellum* (18).

In general, the observed differential substrate utilization capacities noted in different "isolates" of *C. thermocellum* have led to the conclusion that the different strains of this species may vary widely in this context. For example, Lee and Blackburn (19) claimed that *C. thermocellum* strain M7 could utilize a wide variety of hexoses, pentoses, di- and trisaccharides as well as the standard substrates,

cellulose and cellobiose. One of the customary questions which has
been raised in many studies involves the possible utilization of xylose
by *C. thermocellum*, since this pentose is the major monomeric unit
of hemicellulose, which often constitutes a large fraction of the plant
cell wall. For example, one strain of *C. thermocellum* purportedly
capable of growing on xylose — in addition to glucose, cellobiose,
and cellulose — was eventually determined to be a stable, mixed culture of *C. thermocellum* and *C. thermosaccharolyticum* (20). Likewise, the latter noncellulolytic pentose-utilizing bacterial species
could be "derived" from other *C. thermocellum* strains. In at least
one of the latter cases (21), a pure isolate of *C. thermocellum* (designated strain LQRI) was obtained which failed to grow on xylose.
Other authors (22–24) also showed that a variety of *C. thermocellum*
strains are unable to utilize xylose, indicating that this might be a
distinguishing feature of the organism. In fact, one of the above
groups (22) employed *C. thermocellum* ATCC 27405 (McBee's original
isolate), which, in contrast to McBee's original findings, was shown
not to utilize xylose. Thus, the xylose controversy may be explained by the discovery that *C. thermocellum* tends to form stable,
mixed cocultures with related pentose-utilizing strains capable of
surviving simultaneous, repeated cotransfer. The apparent inability
of *C. thermocellum* to catabolize xylose is particularly interesting in
view of the unequivocal demonstration of xylanase activity in this
organism (25).

Another controversy pertains to the potential utilization of glucose by *C. thermocellum* (26–28). It has currently been established
that *C. thermocellum* can grow on glucose, but an adaptation time of
several generations is usually required (22,28–32). The adaptation
is facilitated by the addition of yeast extract. This behavior may
be typical for several other cellulolytic bacteria, e.g., *Acetivibrio
cellulolyticus* (Lamed and Naimark, unpublished results) and *Cellvibrio gilvus* (33), which prefer cellobiose as substrate and require
periods of adaptation for growth on glucose.

Johnson and coworkers (34) studied the effects of carbon nutrition on the formation of cellulases in *C. thermocellum*. A direct relationship between repression of true cellulase production and growth
rate was found. Transfer from a medium containing cellobiose to
either sorbitol- or fructose-containing medium was accompanied by a
transient increase in the specific cellulase production rate. Growth
on cellulose is slow and thus also derepresses cellulase production.
Adaptation on glucose was also characterized by very slow growth,
but, in contrast to the observations described above for sorbitol
and fructose, transfer to glucose-containing medium failed to dere-
press cellulase production.

Unlike *C. thermocellum*, the substrate utilization patterns observed for *C. stercorarium* (6), *Caldocellum saccharolyticum* (10,35),

and the new *Clostridium* spp. isolates (7) are remarkably broad. In view of the above discussion, it remains to be proven unequivocally whether a new anaerobic, thermophilic, cellulolytic isolate is indeed a stable, pure monoculture.

In addition to the types of substrates utilized by a given bacterium, no less important industrially is the pattern of products which the bacterium produces. In this context, some products generated by bacteria are relatively useless in a commercial sense, whereas others are economically desirable. For example, the separation and purification of organic acids (such as lactic and acetic acids) are usually costly processes. On the other hand, desired products would include ethanol, butanol, acetone, glucose, amino acids, vitamins, etc. Thus, in *C. thermocellum*, where the direct production of ethanol on an industrial scale was seriously considered in the past, large accumulations of organic acids would be highly disadvantageous. Another problem which has been encountered during attempts to obtain ethanol from cellulosic biomass using *C. thermocellum* is the relative sensitivity of the bacterium to even low levels of accumulated products (36,37), with particular emphasis on the desired product ethanol.

Another applicative approach which has been considered in the past has been to produce large quantities of extracellular cellulases to be employed secondarily for cellulose degradation. In this case, the production of cellulase systems which would act efficiently on natural cellulosic materials would be preferable. In this regard, *C. thermocellum* belongs to a highly select group of microorganisms which produce "true" cellulases (15,27,38), which can effect total solubilization of "crystalline" cellulose. In view of the above, efforts were made to select strains of cellulolytic microorganisms which exhibit improved characteristics regarding substrate utilization, product pattern, ethanol tolerance, cellulase production, etc. Indeed, various improved strains have been described in the literature in which the ethanol to organic acid ratios have reportedly been enhanced under certain conditions (23,39,40). Likewise, numerous reports of strains tolerant to relatively high concentrations of ethanol have been described (41–44).

Although anaerobic thermophilic fermentation of crude biomass materials has justified the large number of research efforts on *C. thermocellum*, only a handful of reports have been published which compare the degradation of purified vs. crude cellulosics by this organism. Several of the published reports used cocultures or compared monoculture with coculture fermentations of these substrates. Some of the work done in this area is the product of industrial research and may therefore not be readily available.

Ng et al. (45) found that pretreatment of wood was essential for obtaining reasonable degradation rates by *C. thermocellum*,

although these results appeared to contradict other related works
(46,47). In another study, a fed-batch, mixed-culture fermentation
on solka floc was compared to that on cornstover (41,48). Although
85% of theoretical ethanol yields were obtained on solka floc, highly
reduced yields were determined for the crude cellulosic substrate.
The latter phenomenon was partly attributed to the enzymatic de-
acetylation of hemicellulose and to the potential toxicity of lignaceous
components in cornstover. Similarly, decreased rates of ethanol production were observed for raw bagasse; these were improved somewhat upon pretreatment with either mild alkali or steam (36).

In a recent report, Lynd et al. (49) compared the continuous
fermentation of Avicel with that of pretreated (220°C, 1% H_2SO_4,
9 sec) mixed hardwood by a monoculture of *C. thermocellum*. The
bacterium grew readily in continuous culture on both substrates.
Quantification of culture parameters revealed that steady-state conditions were obtainable and that both substrates could be degraded
to about 80% of the initial amount, yielding similar ethanol acetate
ratios of about 1:1. Such systems may eventually be useful in establishing limiting factors of continuous culture of crude substrates
by *C. thermocellum*.

In the following sections we will outline our views on the specialized structure of the *C. thermocellum* surface and its correlation
to the cellulose degradation process. We will also discuss the various enzyme pathways and other biochemical phenomena which characterize this bacterium and may participate in the regulation of the
observed product pattern.

III. PHYSIOLOGY AND INTERMEDIARY METABOLISM

A. Surface Ultrastructure of *C. thermocellum*

We recently carried out in-depth studies on the cell surface architecture of *C. thermocellum* employing both transmission and scanning electron microscopy (50–52). In these studies, histocytochem-
ical and immunocytochemical methods were used in order to visualize
cell surface exostructures and receptors. The cell envelope was
found to be organized as follows: the cell membrane is covered by
an extensive cell wall which includes the peptidoglycan layer; on
the exterior face of the cell wall is an electron-dense layer which
is surrounded by a relatively thick layer of electron-transparent,
anionic material (perhaps consisting of complex carbohydrate); on
the surface of the exocellular anionic material are a multitude of
protuberant structures which appear in near-periodic fashion. An example of a typical bacterial cell, stained histocytochemically in order
to visualize the protuberance-like structures, is shown in Fig. 1.

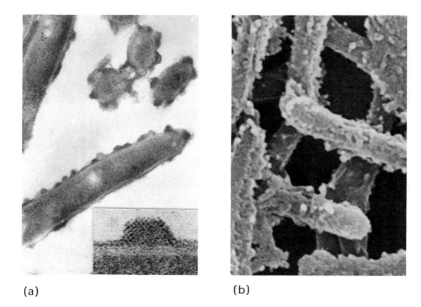

(a)   (b)

FIGURE 1   Transmission electron micrograph of a thin section
(a) and scanning electron micrograph (b) of *Clostridium thermo-cellum*. Cells were stained using cationized ferritin prior to
processing for electron microscopy. Note the protuberance-like
structures which decorate the cell; these structures are visible only
after staining. Insert shows a high magnification of a single protu-berance. Magnifications: 24,000× (a and b) or 100,000× (insert).

to be constant during adaptation and growth on various carbon
sources (e.g., cellobiose, glucose, and fructose), indicating the
constitutive nature of these exostructures (unpublished results).
As shown in the next few sections, these exocellular protuberances
play an important role in the dynamics of the cellulose degradation
process.

B. The Cellulosome and the Cellulase System

Early investigations on the cellulase system of *C. thermocellum* led
to the purification of two endoglucanases (53,54). However, the
relative contribution of the purified components to the total cellu-lase system was not clarified in these studies.

Thus far, direct proof of cellobiohydrolase activity associated
with a given protein component has yet to be reported. It is in-teresting to note, however, that one of the *C. thermocellum*

endoglucanases which has been cloned into E. coli (55) was initially
thought to exhibit cellobiohydrolase activity on the basis of its hydrolysis of the agluconic bond using p-nitrophenyl cellobioside as
substrate; further evidence supported its classification as an endoglucanase. In addition, a cell-bound β-glucosidase has been described (56,57), although we have not been able to demonstrate such
activity associated with *C. thermocellum* strain LQRI (unpublished
results).

In *C. stercorarium*, both endoglucanase and cellobiohydrolase
activities have been partially characterized (58,59). These enzymes
were shown to act synergistically in cellulose degradation by this
organism. The endoglucanases of *C. stercorarium* are purportedly
more thermostable than those of *C. thermocellum*. Similarly, the
thermostability properties of the 1,4-β-glucanase of *Thermoanaerobacter cellulolyticus* (9) and of the endoglucanase from *Caldocellum
saccharolyticum* (10) are improved over those of *C. thermocellum*
cellulases.

Aggregation of cellulolytic enzymes had been noticed previously
(53,60), but the physiological significance of this phenomenon was
only recognized later (61,62). More recent studies (63–65) employing genetic engineering techniques have led to the cloning of about
10 endoglucanases and the purification of several of the cloned gene
products; due to the limitations of the present chapter, however,
details of these studies will not be described here.

One of the major breakthroughs in the initial stages of our work
concerns the observations which led to the development of the cellulosome concept (66). Relatively early in our studies with *C. thermocellum* strains YS, we discovered through a combined immunochemical-
genetic-biochemical approach that the majority of cellulases in both
the extracellular form and the exocellular (cell-associated) form were
organized into a high molecular weight, multiple-enzyme, multifunc-
tional complex which we termed the cellulosome.

The purified cellulosome (whether from cell-bound or cell-free
source) was found to display "true" cellulase activity (67). Like
the crude enzyme system (68,69), the observed cellulolytic activity
was activated by calcium and thiols. Unlike earlier reports relating
to purified endoglucanases from this organism (53,54), the enzymatic
action of the purified cellulosome was subject to end-product inhibition by cellobiose (67), thereby resembling the crude cellulase system in this (68) and other cellulolytic microorganisms (70). Although
different substrates were used in the earlier investigations which involved isolated endoglucanases, the observed cellobiose-induced inhibition may be a defined characteristic of the *intact* cellulosome (or
one or more of its components) and its capacity to solubilize the cellulosic substrate. Interestingly, cellobiose not only fails to inhibit
the capacity of the cellulosome to bind to cellulose (67) but also has

no affect on the in vitro cellular adhesion of the *C. thermocellum*
strain YS to the solid substrate (71). In contrast to our results,
Wiegel and Dykstra (72) reported that the adherence to cellulose of a
different strain of *C. thermocellum* (strain JW1) was subject to inhibition by cellobiose, glucose, and xylose during growth experiments.

An isolated form of the extracellular cellulosome was extensively
characterized (61,66). At least 14 different polypeptide chains were
shown to exhibit endoglucanase activity. Further experiments revealed that some of these activities apparently differ in their specificities (66). One subunit, the 210-kD S1 subunit (the largest
component of the cellulosome) was unique in that it failed to display
cellulolytic activity but proved highly antigenic. Furthermore, an
estimated 25−40% of its content appeared to represent covalently
bound carbohydrate. By far the dominant saccharide was found to
be galactose (73,74). The fact that *Griffonia simplicifolia* isolectin
B_4 (the only lectin out of a panel of over 20) interacted strongly
with the S1 subunit suggested that a terminal α-galactosyl residue
characterized the glycoconjugate. Preliminary evidence indicates
that galactose components of the S1 subunit are attached in 1 → 2
glycosidic linkages. Extensive proteolysis of the cellulosome inferred that the average size of the oligosaccharide chain(s) is quite
defined within the range of 5−10 kD. Additional data suggest that
the oligosaccharides are attached to the protein via O-glycosidic
linkages. It is thought that the S1 subunit may serve either as an
anchoring site of the cellulosome to the cell surface or in an orga-
nizing capacity, perhaps by linking all or part of the various cellulases in the complex to each other, thereby forming and maintaining
the cellulosome complex.

The isolated cellulosome from the growth medium of strain YS
was found to consist of two major types which differed in their respective molecular weights and in their respective content of the S1
subunit (50,61). The major (and larger) type had an average molecular weight estimated at 2.1×10^6 D. Electron microscopic evidence
(negative-stained and dark-field specimens) revealed a heterogeneous
mixture of particles which showed a multiplicity of forms. One of
the dominant structures observed appeared to be composed of multiple subunits and exhibited a relatively uniform size of approximately 18 nm.

Cellulosomes isolated from the cell surface (either by sonication
or by a stripping procedure using buffers) were similar in content
to the extracellular form (66). The two major types which characterized the extracellular form were present, along with relatively
large amounts of a third, very high molecular weight type which
appeared at the void volume on Sepharose 2B. Since the polypeptide pattern of this latter type was similar to that of the isolated
extracellular form, it is presumed to represent multiples of the other
type(s), i.e., polycellulosomal in structure.

More recently, other laboratories have confirmed much of our
original work about the presence and structure of the cellulosome
or cellulosome-like organelles using other strains of C. thermocellum.
Strain-specific variations have been noted in cellulosome structure
and/or content. For example, in strain ATCC 27405, the cellulosome
equivalent apparently bears a molecular weight of about 6×10^6 D
(75). In this strain, our own studies (unpublished results) have
revealed that the S1 subunit is larger ($\sim$250 kD) than the YS analog.
In addition, the two strains (grown under similar conditions) expressed differing (but reproducible) levels of the other cellulosomal
subunits (62). In yet another example, strain JW20 was shown to
comprise cellulosome and polycellulosome equivalents which ranged
in apparent molecular weights from about 5 to 100×10^6 D (76).
From 15 to 20 polypeptides were associated with the isolated material. The S1 equivalent was of similar molecular weight and was
shown to contain most of the cellulosome-associated carbohydrate
(77). The purified cellulosomes were studied by electron microscopy in greater detail (78). Two forms of the cellulosome, designated loose and tight, were detected, and these observations were
considered to reflect mechanistic relevance regarding the capacity
for multiple cleavage of cellulose microfibrils.

The cellulosome thus appears to be an integral part of this organism with strain-specific features regarding relative size, disposition of the subunits within the complex, and arrangement on the
cell surface. The major organizational role of this complex could be
twofold: to effectively deliver the appropriate enzymes to the substrate and to bring into proximity the various complementary enzymes (e.g., exo- and endoglucanases), which then act synergistically in the degradation of cellulose. Moreover, the cellulosome
may be structured in such a way as to enable the protection of various product intermediates and to facilitate their transfer to other
components of the cellulosome for further hydrolysis.

Another finding which may be pertinent to the above mechanistic considerations is that diffusion mobility studies involving laser
bleaching indicated that the cellulosome, once adsorbed onto the
cellulosic surface, is essentially immobile (unpublished results). In
view of the fact that the cellulosome exhibits "true" cellulolytic activity, the data are suggestive of an unusual mechanism of enzyme
action. These findings, together with the confirmed high molecular
weight and multisubunit nature of the cellulosome, may appear (in
superficial terms) to be at variance with a recent report (79) concerning the relatively small dimensions (43 Å) estimated for the
rate-limiting catalytic component(s) of the cellulase system in
C. thermocellum. The latter value was calculated using pore size
distribution data and other complementary physical and kinetic measurements. In reconciling this apparent discrepancy, one may

speculate that the cellulosome subunits are attached to each other by a flexible link (such as the S1 subunit) and that the individual components are sufficiently mobile to enable penetration of cellulosic pores.

C. Interaction with Cellulose

It is quite logical that in nature microorganisms which hydrolyze an insoluble substrate would initially form a physical association in some manner with their substrate. The adherence of a bacterium to its substrate would in theory impart spatial advantages in the subsequent degradation of the substrate. Further ecological precedence would result if the required enzyme mechanism were conveniently accessible. This could be achieved in many ways, e.g., by the arrangement of the enzyme system on the cell surface or by the regulated secretion of the enzyme(s) during or following contact of the cell with the substrate. It would seem to be self-defeating, or at best wasteful, if the hydrolytic enzymes were indiscriminately produced and distributed into the environment. This is especially true for anaerobes, the metabolic yields (moles ATP per hexose catabolized) of which are low compared to that of aerobes. It would seem even more absurd if cells interacted with their insoluble substrate at a distance and depended on diffusion processes for their major nutritional source.

In retrospect, therefore, it may not be considered surprising to have discovered that cells of *C. thermocellum* adhere strongly to cellulose before hydrolysis (Fig. 2). Moreover, the exocellular cellulosome is the molecular factor which mediates the adherence phenomenon, apparently by virtue of its own inherent affinity for cellulose. The fact that, as shown in the last section, the cellulosome also comprises the molecular apparatus which effects saccharification of crystalline cellulose attests to the efficiency and refinement of the system.

The adhesion of *C. thermocellum* strain YS to cellulose was a relatively early observation which stemmed from fermentation studies involving this organism (27,39,40). Following this lead, we eventually isolated an adherence-defective mutant and prepared antibodies specific for a putative cellulose-binding factor (CBF). The mutant was isolated by a procedure which comprised sequential enrichment cycles whereby cellulose was added to a cell suspension and the nonadherent cell fraction was reinoculated into cellobiose medium; the mutant strain was then selected by single-colony isolation (71). The characteristics of this mutant have remained stable for more than 5 years.

Antibodies specific for the CBF were also prepared (71) on a functional basis by first eliciting polyclonal antibodies against

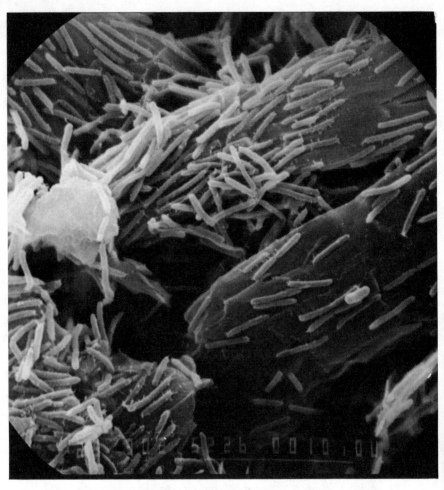

FIGURE 2   Scanning electron micrograph of *C. thermocellum* growing on cellulose.   Cells were stained with cationized ferritin.   Note
the protracted protuberances which connect the cells to the cellulose.   Magnification: 43,000×.

wild-type cells of *C. thermocellum* YS (which adhere to cellulose):
these antibodies were then adsorbed onto mutant cells (which fail
to adhere to cellulose). Thus, the unadsorbed antibody fraction
should be specific for a given surface component(s) which is present
in the wild type but absent in the mutant, i.e., the remaining antibody fraction should be related in some way to the quality of adherence of *C. thermocellum* to cellulose.

The adherence-defective mutant together with the CBF-specific
antibody preparation provided the means by which to further investigate the CBF on the molecular level. The CBF was found to exhibit high levels of cellulolytic activity in addition to its adherence
qualities, and the term "cellulosome" was chosen and characterized
as discussed in the previous section (66).

The adhesion phenomenon was studied further for *C. thermocellum* YS and found to be rather specific for cellulosic substrates;
the cells failed to adhere to a wide variety of other insoluble un-
charged polysaccharides (62,71). Salts, detergents, extremes in
temperature (between 4 and 60°C) and pH (between 4 and 9.5),
mono- and disaccharides (including cellulose degradation products
such as glucose and cellobiose), as well as soluble derivatives of
cellulose all failed to inhibit the adhesion of *C. thermocellum* to its
preferred substrate. Thus far in our studies with *C. thermocellum*
YS, only three materials (or conditions) have been demonstrated to
interfere with the adherence to cellulose: polyethylene imine (the
action of which has yet to be elucidated), water (low ionic strength),
and the isolated cellulosome, which, at relatively low concentrations
(0.1 mg/ml), apparently serves to saturate potential binding sites
on the substrate.

More recently, Wiegel and Dykstra (72) studied the adhesion of
another strain of *C. thermocellum* (JW1). The adherence characteristics of this strain appeared to be very similar to those of strain
YS except that the adherence of JW1 to cellulose appeared to be
strongly inhibited by relevant saccharides, including cellobiose, glucose, and xylose. This latter discrepancy may simply represent a
strain-specific variation in *C. thermocellum*. The fact that in strain
YS cellobiose inhibits the catalytic action without affecting the adhesion properties may provide an elegant tool for biochemical differentiation of the two major functional activities of the cellulosome.

In our studies with strain YS, it was found that the cellulosomes are attached generally at the exterior of the exocellular protuberances shown in Fig. 1 (52). Upon growth of the mutant on
cellobiose rather than cellulose, the protuberances (and cellulosomes)
are lacking on the mutant cell surface (50,71). The interaction of
the wild-type cell with cellulose causes the protuberances to protract to form contact zones which mediate physically between the
cell and the insoluble substrate (51,52). Cellulosome molecules are

attached profusely at the cellulose surface and are connected to the cell by fibrous or amorphous structures. Similar fibrous material was also observed independently by Wiegel and Dykstra (72) who suggested that this material is composed of acid mucopolysaccharides and that the attachment to cellulose occurs via this fibrous material. In contrast, our work supports the contention that the cellulosome is the cellulose-binding factor and that the fibrous material serves in a structural capacity such that the physical connection to the cell is maintained. The composition of this fibrous material and its interaction with the cellulosome on the molecular level are of major importance. In this context we have isolated a 130-kD proteinaceous glycoconjugate which may play a role in the construction of the surface polycellulosome.

In addition, our studies indicate that the levels of exocellular cellulosome are much lower in the stationary phase of growth than those observed in the exponential phase (71). Conversely, the concentration of extracellular cellulosome is much higher in the stationary phase than in the exponential phase. This fact correlates nicely with the observation that in the final stages of growth on cellulose, the cells detach from the substrate.

Several properties of the exocellular moieties have been observed in vitro which may or may not relate to the physiological events of attachment and detachment of the cells. They may be washed off cells by low ionic strength buffers and, once attached to cellulose, the cellulosome may be eluted by water (or similar buffers containing less than 5 mM salt). Wiegel and Dykstra (72) presumed that the detachment of cells is caused by the degradation of cellulose fiber through the action of extracellular cellulase in the medium. Our data, however, indicate that the detachment may also occur by the disjunction of the fibrous material from the cellulose-adsorbed cellulosomes within the protracted protuberances.

In consideration of the above described evidence, our current view of the interaction of *C. thermocellum* with cellulose is shown schematically in Fig. 3. In the figure, cell (a) (e.g., cellobiose-grown), bearing compact multicellulosome-containing protuberances, comes into contact with the cellulose surface. The adherence of the cell to cellulose is mediated via the cellulosome particles which line the exterior of the protuberances. The protuberances protract to form extended "contact corridors" which comprise fibrillar or amorphous material (see cell b). In this state, the cellulosic substrate is degraded exocellularly by the cellulosome primarily to cellobiose, and the latter is assimilated by the cell. At this point and as the process proceeds, some of the cellulose-adsorbed cellulosome complexes become disengaged from the cell surface (e.g., in cell b) via detachment from the fibrous material. Eventually, cells desorb from the cellulose (cell c) leaving cellulose-adsorbed cellulosome and

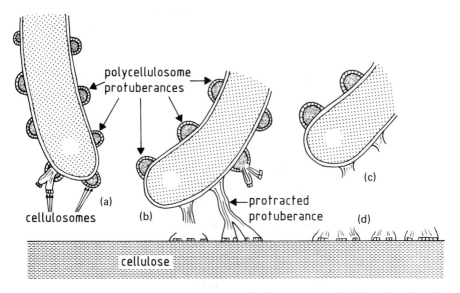

FIGURE 3   Schematic description depicting proposed ultrastructural events which accompany the interaction of *C. thermocellum* with insoluble cellulose (see text for further details).

associated cell-derived material (d). In the absence of the cell, the cellulose-bound cellulosomes continue the degradative process in converting cellulose to cellobiose.

The evolution of such an economical mechanism would be crucial for the survival of cellulose-degrading anaerobes in nature. The greater hydrolytic efficiency at such an early premetabolic stage would presumably compensate for the low (relative to aerobes) ATP yields per mole hexose. In this context, it is interesting to note that cellulosome-like complexes of cellulase-containing aggregates have been identified in several anaerobes, e.g., *Fibrobacter succinogenes* (80), *Ruminococcus albus* (81), *C. cellulosolvens* (82), and *Ruminococcus flavefaciens* (83), and that many cellulolytic bacteria are known to adhere to cellulose. Moreover, we have demonstrated the presence of protuberant-like structures for a variety of evolutionarily divergent strains of cellulolytic bacteria by employing a new staining procedure for scanning electron microscopy (84). This mechanism of interaction with the insoluble substrate may also promote the establishment of stable cocultures with complementary noncellulolytic bacteria as will be discussed in Sec. IV.

D. Transport and Initial Catabolic Steps

We have shown in the last two sections that the attachment of cells
of C. thermocellum to and the enzymatic degradation of the insoluble
cellulosic substrate are connected by a common denominator: the
exocellular cellulosome. It has been demonstrated (66,67) that the
cellulosome in both exo- and extracellular states is capable of solubilizing crystalline cellulose directly to cellobiose and glucose as the
major (>90%) and minor (<5%) sugar products, respectively. It is
clear that this process takes place to apparent completion in the cell
exterior as shown diagrammatically in Fig. 4. In the following sections, we will deal with the cellular processing of these saccharides
into final catabolic products. The scheme in Fig. 4 can serve as a
guide for understanding the relevant enzymatic pathways and key
regulatory steps in C. thermocellum. The reader is, however, re-
minded that significant differences may exist among different strains
of this organism.

     Following degradation of cellulose to cellobiose (and glucose),
the latter must be taken up by the cells. Several reports of uptake
systems in C. thermocellum have appeared in the literature, but
firm evidence as to the mechanism(s) of assimilation of these sugars
is still lacking. Cellular uptake of cellobiose may account for the
fact that the cell-associated cellulase apparatus is far more active
(regarding specific "true" cellulase activity) than purified enzyme
preparations, since clearance of cellobiose from the cell exterior
would obviate its strong inhibitory effect on cellulose hydrolysis
(67,68).

     Regarding cellobiose uptake, both glucose moieties apparently
become phosphorylated in a process which may be related to the
transport of the disaccharide intracellularly (85). It is clear that
cellobiose phosphorylase (86) would be responsible for conversion
of the nonreducing glucosyl moiety to glucose-1-phosphate. The
question is, what happens to the second glucose moiety? Another
riddle concerns the role of the reported presence of cell-associated
cellodextrin phosphorylase (87,88), considering the fact that saccharification proceeds almost quantitatively to cellobiose in this
organism.

     The present status of initial glucose metabolism is sketchy.
Transport of glucose has been demonstrated to be accomplished in
glucose-adapted cells by an ATP-dependent permease (89). A phos-
phoenolpyruvate phosphotransferase system is apparently lacking in
C. thermocellum. Free glucose accumulates in three possible ways:
(1) through the action of an exo(extra)cellular (cellulosomal) cellobiohydrolase on odd-numbered oligocellodextrin chains, (2) by cellobiose phosphorylase as described above (which accounts for the
majority of free glucose release by this organism), and (3) by a
putative β-glucosidase (56). It is not clear whether glucokinase is

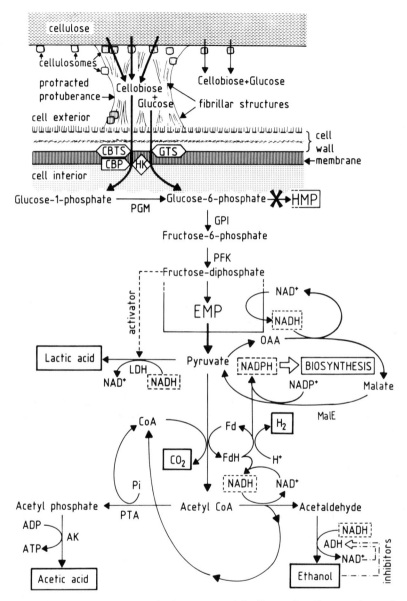

FIGURE 4   Scheme depicting the catabolism of cellulose to major cellular end products in *C. thermocellum*. Abbreviations: ADH = alcohol dehydrogenase; AK = acetate kinase; CBP = cellobiose phosphorylase; CBTS = cellobiose transport system; EMP = Embden −Meyerhof −
Parnas pathway; Fd = oxidized ferredoxin; FdH = reduced ferredoxin;
GPI = glucose phosphate isomerase; GTS = glucose transport system;
HK = hexokinase (glucokinase); HMP = hexose monophosphate pathway; LDH = lactate dehydrogenase; MalE = malic enzyme; OAA = oxaloacetate; PFK = phosphofructokinase; PGM = phosphoglucomutase;
PTA = phosphotransacetylase.

present as a soluble or membrane-bound enzyme or whether it exists
at all in this organism. Were the latter to hold true, it would be
puzzling how the free extracellular glucose and/or the cellobiose-
derived glucose moiety would undergo phosphorylation. However,
very little or no activity has been noted for this organism (88,89;
Lamed, unpublished results). It would thus seem logical to propose
that both glucokinase and cellobiose phosphorylase are plasmalemma-
bound, connected perhaps with the relevant transport system. These
conclusions, of course, are only speculative and additional experimental evidence along these lines would be beneficial.

E. Intracellular Catabolic Pathways and Product Pattern

Several key enzymes of the Embden –Meyerhof –Parnas pathway were
found in extracts of C. thermocellum by Patni and Alexander (90),
and the involvement of this pathway was confirmed by Lamed and
Zeikus (21) both by enzyme activity measurements and by tracer experiments (Fig. 4). The hexose monophosphate pathway was found
not to take part in sugar catabolism. Lamed and Zeikus (21) also
studied the terminal catabolic steps from pyruvate and the regulatory
properties which determine the proportion of the end products. The
typical phosphoroclastic reaction of pyruvate is operative in C. thermocellum.

Intracellular cellobiose-derived glucose-1-phosphate is converted
to glucose-6-phosphate in the conventional manner by the enzyme
phosphoglucomutase. However, since both glucose moieties reportedly undergo equivalent metabolism (85), it is assumed that free glucose is indeed converted directly to glucose-6-phosphate through
the action of glucokinase. It is still debatable whether glucokinase
is constitutive or induced in glucose-adapted cells.

As mentioned above, it is clear that the hexose monophosphate
and Entner –Doudoroff pathways are inoperative in C. thermocellum.
This contention is supported both by radiotracer studies and by the
absence of two initial key enzymes: glucose-6-phosphate dehydrogenase and gluconic acid-6-phosphate dehydrogenase. Production
of NADPH, required for anabolic processes, must therefore be accounted for by other mechanisms as will be discussed later. Thus,
intracellular glucose-6-phosphate is funneled first to fructose-6-
phosphate and then fructose-1,6-diphosphate (FDP) by the enzymes
glucose phosphate isomerase and phosphofructokinase, respectively.
FDP is a key glycolytic intermediate in C. thermocellum. As in
other selected systems (91), the level of FDP is most likely determined by the pool of readily available precursor, and its accumulation plays an important role. Subsequent glycolysis of FDP to pyruvate is carried out in customary fashion accompanied by the generation of 2 mol of NADH and 2 – 3 mol (net) of ATP.

Pyruvate metabolism in *C. thermocellum* proceeds in three major
directions: (1) reduction to L-lactic acid by lactate dehydrogenase,
(2) decarboxylation to acetyl coenzyme A (AcCoA), and (3) participation in a transhydrogenase malate cycle (leading to NADPH generation required for biosynthetic pathways). The NADH-dependent
lactate dehydrogenase in *C. thermocellum* is subject to delicate regulation by FDP, which serves as an essential activator for this enzyme. In this organism, the reaction is unidirectional and the affinity with respect to the activator is pH-dependent. Nevertheless,
the actual in vivo activity levels of lactate dehydrogenase are relatively low, since the levels of intracellular FDP are correspondingly
low, perhaps reflecting a rate-limiting step in an initial stage of
sugar assimilation.

The conversion of pyruvate to AcCoA and CO_2 by pyruvate (de-
carboxylating) dehydrogenase is linked to the reduction of ferredoxin
—a pivotal component of electron flow reactions, the most important
of which is the hydrogenase reaction which channels electrons to
yield molecular hydrogen.

AcCoA is further catabolized in two ways: the contribution of
each is governed by the overall electron balance of the system as
will be discussed presently. The hydrolysis of AcCoA to acetic acid
proceeds by the sequential action of phosphotransacetylase and acetate kinase. Acetate production yields additional ATP, but were
acetate the only product, catabolic pathways would be blocked by
the fact that NAD^+ would not be regenerated from accumulated NADH.
Thus, AcCoA undergoes alternative metabolic conversion to ethanol
via the NADH-linked enzymes, acetaldehyde dehydrogenase (CoA-
deacetylating) (EC1.2.1.10) and alcohol dehydrogenase. Lactate
production from pyruvate serves a similar role, but, as mentioned
above, its contribution is relatively low.

In *C. thermocellum*, the NADH-dependent alcohol dehydrogenase
is a unique enzyme (at least in the two strains studied extensively)
in that it acts unidirectionally. In essence, a more proper designation for this enzyme might be acetaldehyde reductase. In addition,
the enzyme is subject to strong feedback inhibition (by its products,
ethanol and NAD^+), which plays an important regulatory role in limiting the amount of ethanol produced.

The participation of pyruvate in a rather unusual transhydrogenase cycle has been postulated to serve as a biosynthetic shunt
in *C. thermocellum* (92). In this cycle, pyruvate is converted enzymatically to oxaloacetate en route to malate via an NADH-dependent
oxaloacetate reductase. An interesting ammonium-activated malic enzyme (malate/$NADP^+$ oxidoreductase) catalyzes the oxidative decarboxylation of malate back to pyruvate with concomitant production of
NADPH and carbon dioxide at the net expense of NADH. This enzyme has not been detected in other thermophiles and has been used

as a marker for culture purity vis a vis *C. thermocellum* (Lamed, unpublished results) Ammonium activation of the malic enzyme has been proposed to have special physiological significance in this organism in that it may serve as a signal which turns on NADPH generation when sufficient ammonium ions are available for biosynthesis (92). The relatively high catabolic levels (under conditions of full expression) of this enzyme (>0.2 IU/mg protein, depending on the strain employed) are somewhat puzzling in view of the fact that anabolic demands in anaerobes are significantly lower than those of aerobes.

A characteristic feature of *C. thermocellum* is its tendency to produce near-equivalent amounts of ethanol and acetate with relatively low levels of lactic acid. Acetate levels are usually linked to the evolution of molecular hydrogen. Under conditions which yield the above product pattern, the glycolysis-derived NADH (2 mol/mol hexose) is reoxidized by reduction of AcCoA to ethanol. Production of additional ethanol at the expense of acetate would result in a stoichiometric (2 mol/mol) decrease in molecular hydrogen caused by the deflection of electrons from reduced ferredoxin to NADH rather than their participation in the hydrogenase reaction.

Enzyme-mediated electron transfer involving reduced ferredoxin can also give rise to NADPH, which does not have a known catabolic role in *C. thermocellum*. This reaction may be an alternative route (complementary to the malic enzyme cycle) for NADPH formation and subsequent biosynthesis.

Since, from an applied point of view, ethanol is considered to be the desired fermentation product, extensive research has been carried out in order to understand the factors controlling the product pattern in *C. thermocellum*. In addition to the enzymatic regulatory properties that have been discussed above, the more important experimental factors which contribute to the formation rates and final distribution of end products include product and substrate inhibition, product tolerance (cell growth and viability), fermentation conditions, and the formation of stable coculture with other ethanologenic, anaerobic, thermophilic bacteria.

Product inhibitory effects have been studied systematically for gaseous hydrogen (21) and for soluble products, i.e., ethanol and acetate (36) as well as other organic acids (93). The relationship between substrate concentration and product ratios has also been studied (37).

The effect of hydrogen appears to be the most striking regulatory factor and is mediated by the reversible hydrogenase and the ferredoxin NAD^+-reductase reaction, thereby leading to enhanced ethanol/acetate ratios which have been observed in certain strains but not in others (21). Mechanical stirring of fermentation cultures has been found to interfere with the extent of supersaturation levels of hydrogen, the accumulation of which would favor ethanol as final product. The stirring effect appears to be universal in this respect, particularly regarding strains (e.g., YS) which were originally selected (under conditions of cultural growth without stirring) for high ethanol yields. Somewhat confusing is the finding that the ethanol/ acetate ratio may drop in some strains (e.g., LQRI) to 0.3, thus implying that, under stirring conditions, electron flow would ensue from NADH to molecular hydrogen via reduced ferredoxin, a reaction which would presumably be thermodynamically unfavorable (39, 40) unless H_2 were continually removed.

F. Comparative Properties with Other Thermophiles

Although their product ratios differ significantly, *C. thermocellum* is very similar in its catabolic pathway to the noncellulolytic but saccharolytic *Thermanaerobium brockii* (21,94). These two species have been extensively compared with respect to their catabolic enzymes. The differences in product patterns reflect rather delicate differences in regulatory properties. For example, *T. brockii* produces more lactic acid than *C. thermocellum*, whereas the latter produces more H_2 and acetate than the former. Both contain similar FDP-requiring lactate dehydrogenases with similar concentration dependencies. However, for growth on cellobiose, the intracellular FDP levels in *T. brockii* are much higher, possibly due to the rate-limiting initial metabolism of the dissacharide in *C. thermocellum* (21). Similar FDP-requiring lactate dehydrogenases characterize at least two other ethanologenic thermophiles, namely, *Thermoanaerobacter ethanolicus* (95) and *C. thermohydrosulfuricum* (25).

Differences in production of acetate/H_2 levels (higher in *C. thermocellum*) and in the sensitivity to H_2 inhibition (greater in *T. brockii*) are related to the level of expressed activity (reflecting both enzyme levels and regulatory effects) of the key ferredoxin NAD(P)$^+$ reductases, which mediate electron transfer from reduced ferredoxin (obtained from the pyruvate dehydrogenase reaction) to reduced pyridine nucleotides and ultimately to ethanol and lactate, the final reduced end products.

Another important enzymatic difference between the two bacterial species is that the alcohol dehydrogenase in *T. brockii* is a reversible enzyme (94,96) unlike that of *C. thermocellum* (21). Moreover, the enzyme is not subject to the type of potent feedback inhibition which distinguishes the *C. thermocellum* enzyme. The properties of alcohol dehydrogenase and their implications with respect to ethanol tolerance and product patterns have been extensively studied in several anaerobic thermophilic strains by several research groups; this subject was recently reviewed by Slapack et al. (1).

It is interesting to note that *C. thermocellum* is usually not affected by electron acceptors (e.g., ketones, sulfite, and thiosulfate) as much as other thermophilic anaerobes (e.g., *T. brockii* and *C. thermohydrosulfuricum*). Two major factors may account for this observation: On the one hand, disposal of electrons as hydrogen gas occurs with great facility in this organism (in *T. brockii*, for example, this step is usually growth limiting). On the other hand, *C. thermocellum* lacks sufficient levels of enzyme activities which are involved in various common electron transfer reactions which characterize many other microbial systems. Thus, *C. thermocellum* fails to respond to the addition of ketones (25,97) because it lacks secondary alcohol dehydrogenase (unlike *T. brockii*, *T. ethanolicus*, and *C. thermohydrosulfuricum*). Furthermore, *C. thermocellum* is unaffected by sulfite and thiosulfate, owing to the absence of the corresponding reductases.

In conclusion, there is a fine correlation between the net activity levels of the respective catabolic enzymes in a given strain and the observed end-product distribution. Additional strain-specific enzymes of importance are related to interspecies hydrogen transfer, a subject which will be discussed further in the forthcoming section.

IV. COCULTURE FERMENTATIONS

Symbiotic interactions among microorganisms are widespread natural phenomena which constitute a primary ecological axiom. In this context, the establishment of stable, mixed cocultures between a variety of cellulolytic and noncellulolytic (saccharolytic) bacteria is a general phenomenon which occurs readily in nature. Several different factors may in theory dictate the mode of interaction between different bacterial strains which leads to the formation of such cocultures. It is easier, however, to postulate logical alternatives which may account for this phenomenon than to provide concrete evidence in their support.

For the purposes of the present chapter, coculture fermentation refers to the combined interaction of a polymer-degrading microbial species with one or more nondegrading strain(s), such that the former degrades the polymer to a molecular form(s) which can serve as a carbon source for the latter organisms. Alternatively, end products (e.g., H_2, CO_2, ethanol) secreted by the polymer-degrading microorganism may be further utilized in mixed cultures.

For example, noncellulolytic bacteria, such as *C. thermohydrosulfuricum* or *C. thermosaccharolyticum*, are incapable of growing on cellulosic substrates in monoculture. In the presence of a cellulolytic bacterium, however, a competition for the hydrolysis products takes place which leads to the coexistence of the two (or more) strains. With regard to hemicellulose degradation, the situation is even more intriguing. Although *C. thermocellum* is capable of degrading xylan, presumably to xylose and xylobiose (1,25), it is incapable of metabolizing these saccharides. These sugars, however, can be readily utilized by the above mentioned saccharolytic strains.

Under laboratory conditions involving multispecies degradation of cellulose, it is clear that the secondary saccharolytic strain is dependent in absolute terms on the cellulolytic strain. Much less evident is the opposite question, i.e., in what sense does the cellulolytic organism benefit from the presence of the accompanying strain? In fact, starvation and death of the cellulolytic bacterium may result from the more rapid assimilation of soluble sugars by the saccharolytic strain. On the other hand, nutritional requirements of the cellulolytic bacterium may be satisfied by substances (e.g., metabolic intermediates, vitamins, cofactors, and the like) which are secreted by the noncellulolytic strain. The latter may also serve to sequester materials toxic to the polymer-degrading strain, as recently demonstrated for a defined mesophilic coculture (98).

We saw in Sec. III that the cellulosome of *C. thermocellum* hydrolyzes cellulose to cellobiose and glucose. The action of the exocellular protuberance-bound cellulosome may serve to delay or limit diffusional loss of the hydrolyzed sugar to the environment and/or competing bacteria. Together with the adherence and transport phenomena, this may provide *C. thermocellum* with a clear ecological advantage in initial stages of coculture fermentation. In contrast, the extracellular enzyme (in the cell-free, cellulose-bound state) would enrich the environment with diffusable sugars which would then be readily available for competing saccharolytic organisms. In this context, Carreira and Ljungdahl (99) proposed that this latter stage is characterized by an accumulation of fermentation products which inhibit the metabolic activity of the cellulolytic organism. During this phase the character of the fermentation end products would mainly reflect the enzymatic apparatus of the noncellulolytic organism.

Clostridium thermocellum grows in coculture with a variety of other noncellulolytic anaerobic thermophiles. These have been classified as fermentative, methanogenic, or acetogenic bacteria. Of the fermentative bacteria, ethanologenic strains have attracted the most attention owing to their potential industrial application.

In an applicative sense, the purpose of coculture fermentation is essentially to "correct" or to improve deficient or undesirable characteristics in the cellulolytic strain when grown in monoculture. Such defects include poor product pattern (low ethanol yields), insufficient utilization of cellulosic substrates, inability to metabolize hemicellulose-derived pentoses, and unacceptable nutritional

TABLE 1   *Clostridium thermocellum*-based Coculture Fermentation with Various Thermophilic Anaerobic Bacteria

Type of coculture and features	Secondary bacterium	Comments	Selected references
Ethanologenic			
Stable cocultures	*C. thermohydrosulfuricum*	Successful fermentation of pretreated wood	25,26,45, 100,101
Increased cellulose degradation		Final ethanol concentrations <1%	
Increased ethanol yields			
Decreased production of organic acids	*C. thermosaccharolyticum*	"MIT process"	20,41,44, 47,48
Hydrolysis of xylans		Increased ethanol tolerance of both primary and secondary strains	
Utilization of xylose		LDH-negative mutant	

			Ref.
Methanogenic	*Thermoanaerobacter-ethanolicus*	Reduced ethanol yields on cornstover vs. pure cellulose	99,102
Methane production		Higher ethanol yields than monoculture	
Reduced ethanol/acetate ratios		Low ethanol yeilds on natural cellulosics	
	Methanobacterium-thermoautotrophicum	Interspecies hydrogen transfer	103
Acetogenic	*Acetogenium kivui*	Interspecies hydrogen transfer	104
Stable coculture			
Increased cellulose degradation rate			
Homoacetogenic			

requirements. Table 1 provides some representative examples of
the effect of coculture on the propoerties of cellulose fermentation
by *C. thermocellum*. Note that the well-characterized anaerobic
thermophile *T. brockii* has been excluded from the table due to its
inappropriate mixed-product pattern.

One of the most important factors which regulates the product
pattern of mixed cultures is the well-documented phenomenon of interspecies hydrogen transfer. In this regard, a hydrogen-utilizing
bacterium is coupled to a hydrogen-producing bacterial strain such
that the product pattern of the latter is significantly altered. Thus,
coculture of *C. thermocellum* with *Acetogenium kivui* serves to modify
the normal product pattern of the cellulolytic monoculture almost completely to acetate (104). This occurs by virtue of the capacity of
the acetogenic strain to convert H_2 and CO_2 to acetate. In order
to account for the observed stoichiometry, NAD^+ must be regenerated from NADH by formation of molecular hydrogen in the cellulolytic cell rather than via the reduction of pyruvate to lactate or
AcCoA to ethanol. However, such an NADH-linked oxidoreductase
was not detected in the two *C. thermocellum* strains whose catabolic
properties have been extensively studied (21).

The most common natural electron acceptors in anaerobic ecosystems are the hydrogenophilic methagens. Coculture of *C. thermocellum* with a methanogenic strain indeed caused a decrease in the
ethanol/acetate ratio (103), but the effect was less complete than
that observed for the acetogenic strain. This could be due either
to the lack of the required catabolic enzyme apparatus in the specific cellulolytic strain studied or to insufficient coupling between
the two bacterial strains. This phenomenon may be related to the
instability of the coculture, resulting perhaps from insufficient
growth of the methanogen. Complete uncoupling of the hydrogen
transfer reaction occurs when the coculture is grown on cellobiose
as a substrate, whereby the growth rate of *C. thermocellum* is more
than twice that obtained on cellulose and where the pH of the culture medium rapidly drops below the known pH range for growth
of the methanogen.

Thus the coculture approach potentially enables the resolution
of a desired product pattern by appropriate selection of a secondary bacterial strain which exhibits metabolic pathways which complement the primary cellulolytic strain. This approach may in the
long run be advantageous over direct genetic manipulations of the
cellulolytic bacterial strain. This does not, of course, preclude a
mixed approach whereby cocultures consisting of genetically improved primary and/or secondary strains would be employed.

It is clear, however, that more research using additional combinations of potentially attractive cellulolytic and saccharolytic
strains would be desirable. The major problem in conducting such

studies is their laborious and often unproductive nature, which tends to intimidate the majority of researchers.

V. ADDENDUM

The study of the cellulase system of *C. thermocellum* markedly increased as of the end of 1989. Only part of the recent developments can be treated briefly in this short addendum to the present chapter.

A. From Cellulosome to Hydrolysome

New evidence has suggested that the cellulosome concept is a more general phenomenon in many bacterial systems. In this context, cell surface protuberance-like structures have been observed to correlate with cellulase activities in several gram-negative and positive, anaerobic and aerobic, thermophilic and mesophilic, cellulolytic strains (84,105,106). Moreover, the organization of polymer-degrading enzymes into specialized exocellular structures may be an even more common feature of such solid surface-specific bacteria. Thus, surface protuberances have also been observed in noncellulolytic amylolytic anaerobes which form amylosomes (107). As a general descriptive term for such polymer-degrading multienzyme complexes, we propose "hydrolysomes."

B. Structural Aspects of the Cellulosome

The primary structure of the O-glycoside-linked cellulosomal oligosaccharide in *C. therocellum* has been elucidated (108). A novel tetrasaccharide structure was described which contained a unique 3-O-methyl-N-acetylglucopyranoside component and three galactose units:

$$D\text{-}Galp\text{-}\alpha(1\rightarrow3)$$
$$\diagdown$$
$$D\text{-}Galf\text{-}\alpha(1\rightarrow2)\text{-}D\text{-}GalOH$$
$$\diagup$$
$$3\text{-}OMe\text{-}D\text{-}GlcpNAc\text{-}\alpha(1\rightarrow2)$$

Interestingly, a branched galactose in the furanose configuration plays a central part in the tetrasaccharide structure. Also notable is the terminal galactose in α linkage which provides the required specificity for the distinctive interaction with the GSI lectin, described earlier in the chapter. Other cellulolytic strains have been shown to exhibit this specificity, suggesting perhaps a similarity in their surface cellulase-associated saccharides (84,105).

In recent experiments (109), the large glycosylated S1 subunit
of the cellulosome was found to display anomalous electrophoretic behavior in SDS-PAGE, by which the effect of the detergent is reversibly dependent on the conditions prior to sample application. Consequently, the S1 subunit may actually be composed of smaller sub-
components, subject to disassociation by SDS if the sample is maintained below pH 5 or at very low ionic strength. Most interestingly,
the cellulosome also appears to acquire a similar conformational state
upon adsorption to the solid cellulose substrate.

The importance of S1 to "true" cellulase activity of the cellulosome was further supported in reconstitution experiments (110),
wherein recombination of the purified S1 with another purified subunit led to the recovery of a small fraction of the original true cellulase activity. In a more recent study (111), a "subcellulosome"
complex was isolated on the basis of its interaction (presumably via
S1) with a GSI lectin affinity column. The purified complex possessed
a fivefold enhanced specific Avicelase activity compared with that of
the parent cellulosome.

C. Auxiliary Enzyme Systems in *C. thermocellum*

A comparative study of xylanolytic and cellulolytic activities in cellulosomal and noncellulosomal fractions was conducted. At least eight
different polypeptide bands were revealed which displayed xylanase
activity (112). Only a few of these appeared to be exclusive for
xylan degradation; the majority also expressed endoglucanase (carboxymethylcellulase) or cellobiohydrolase-like (methylumbelliferyl-β-
cellobiase) activities. Separate β-glucosidase and β-xylosidase activities were identified in sonicated cell extracts and are believed to
reflect the respective individual cell-bound enzymes.

The β-glucosidase of *C. thermocellum* has been cloned and the
product expressed in *E. coli* (113). The cloned protein was used
to enhance the degradation of Avicel to glucose (114). In a different study (115), the extent of degradation of a concentrated suspension of Avicel (20%) was greatly enhanced by the combined action of the cellulosome and a commercially available β-glucosidase
from *Aspergillus niger*. In an extension of the same study, a similar approach was used to promote the successful degradation of several natural waste cellulosics.

D. Cloning of Cellulases and Xylanases

The explosive development of genetic engineering technique has
been extensively adapted for studies of the cellulase system in
C. thermocellum. Most laboratories that have worked in this area

have concentrated on the cloning and expression of individual cellulase genes. In most cases, the resultant gene products exhibited
endoglucanase and/or xylanase activity. Thus far, efforts to obtain
a cloned cellobiohydrolase in the classical sense (i.e., exclusive hydrolysis of cellulose to yield cellobiose from the nonreducing end of
the polymer) have essentially failed. A survey of the comparative
cellulolytic and xylanolytic activities derived from the cloned genes
was recently reported (116). It is hoped that eventually combined
genetic and biochemical analysis will also provide insight into the
nature of the individual cellulase species regarding their respective
catalytic sites, substrate-binding sites and domains responsible for
cellulosome formation.

REFERENCES

1. G. E. Slapack, I. Russell, and G. G. Stewart, *Thermophilic Microbes in Ethanol Production*, CRC Press, Boca Raton, 1987.
2. A. Margaritis and R. F. J. Merchant, *CRC Crit. Rev. Biotechnol.*, *4*, 327 (1986).
3. J. Wiegel and L. G. Ljungdahl, *CRC Crit. Rev. Biotechnol.*, *3*, 39 (1986).
4. B. Sonnleitner and A. Fiechter, *Trends Biotechnol.*, *1*, 74 (1983).
5. K. Esser and T. Karsch, *Proc. Biochem.*, *19*, 116 (1984).
6. R. H. Madden, *Int. J. Syst. Bacteriol.*, *33*, 837 (1983).
7. M. Taya, Y. Suzuki, and T. Kobayashi, *J. Ferment. Technol.*, *62*, 229 (1984).
8. M. Taya, H. Kinoki, and T. Kobayashi, *Agr. Biol. Chem.*, *49*, 2513 (1985).
9. H. Honda, H. Naito, M. Taya, S. Iijima, and T. Kobayashi, *Appl. Microbiol. Biotechnol.*, *25*, 480 (1987).
10. C. H. Sissons, K. R. Sharrock, R. M. Daniel, and H. W. Morgan, *Appl. Env. Microbiol.*, *53*, 832 (1987).
11. A. MacFayden and F. R. Blaxall, *J. Pathol. Bacteriol.*, *3*, 87 (1896).
12. H. Pringsheim, *Z. Physiol. Chem.*, *78*, 266 (1912).
13. A. H. Lymn and H. Langwell, *J. Soc. Chem. Ind.*, 280T (1923).
14. J. A. Viljoen, E. B. Fred, and W. H. Peterson, *J. Agr. Sci.*, *16*,1 (1926).
15. R. H. McBee, *J. Bacteriol.*, *56*, 653 (1948).
16. L. Enebo, *Nature, 163*, 805 (1949).
17. L. Enebo, Studies on cellulose decomposition by an anaerobic thermophilic bacterium and two associated non-cellulolytic species, Ph.D. dissertation, Royal Institute of Technology, Stockholm, 1954.

18. T.-V. C. Duong, E. A. Johnson, and A. L. Demain, *Topics Enzyme Ferment. Biotechnol.*, *7*, 156 (1983).
19. B. H. Lee and T. H. Blackburn, *Appl. Microbiol.*, *30*, 346 (1975).
20. D. I. C. Wang, I. Biocic, H.-Y. Fang, and S.-D. Wang, Direct microbiological conversion of cellulosic biomass to ethanol, In *Proc. 3rd Annual Biomass Energy Systems Conference*, NTIS, Springfield, VA, 1979, p. 61.
21. R. Lamed and J. G. Zeikus, *J. Bacteriol.*, *144*, 569 (1980).
22. A. Shinmyo, D. V. Garcia-Martinez, and A. L. Demain, *J. Appl. Biochem.*, *1*, 202 (1979).
23. J. Bender, Y. Vatcharapijarn, and T. W. Jeffries, *Appl. Env. Microbiol.*, *49*, 475 (1985).
24. J. Wiegel, C. P. Mothershed, and J. Puls, *Appl. Env. Microbiol.*, *49*, 656 (1985).
25. J. G. Zeikus, A. Ben-Bassat, T. K. Ng, and R. Lamed, In *Trends in the Biology of Fermentations for Fuel and Chemicals* (A. Hollaender and R. Rabson, eds.), Plenum Press, New York, 1981, p. 441.
26. J. G. Zeikus, T. K. Ng, R. Lamed, and A. Ben-Bassat, *Proc. 2nd Int. Symp. Anaerobic Digestion*, New Delhi, 1980, p. 73.
27. T. K. Ng, P. J. Weimer, and J. G. Zeikus, *Arch. Microbiol.*, *114*, 1 (1977).
28. W. S. Park and D. D. Y. Ryu, *J. Ferment. Technol.*, *61*, 563 (1983).
29. J. G. Vidrine and L. Y. Quinn, *Bact. Proc.*, *69*, 135 (1969).
30. N. J. Patni and J. K. Alexander, *J. Bacteriol.*, *105*, 220 (1971).
31. N. J. Patni and J. K. Alexander, *J. Bacteriol.*, *105*, 226 (1971).
32. R. F. Gomez and P. Hernandez, In *Advances in Biotechnology*, Vol. 2 (M. Moo-Young and C. W. Robinson, eds.), Pergamon Press, Toronto, 1981, p. 131.
33. C. Breuil and D. J. Kishner, *Can. J. Microbiol.*, *22*, 1776 (1976).
34. E. A. Johnson, F. Bouchot, and A. L. Demain, *J. Gen. Microbiol.*, *131*, 2303 (1985).
35. P. H. Reynolds, C. H. Sissons, R. M. Daniel, and H. W. Morgan, *Appl. Env. Microbiol.*, *51*, 12 (1986).
36. S. Kundu, T. K. Ghose, and S. N. Mukhopadhyay, *Biotechnol. Bioeng.*, *25*, 1109 (1983).
37. D. Brener and B. F. Johnson, *Appl. Env. Microbiol.*, *47*, 1126 (1984).
38. E. A. Johnson, M. Sakojoh, G. Halliwell, A. Madia, and A. L. Demain, *Appl. Env. Microbiol.*, *43*, 1125 (1982).

39. T. M. Su, R. Lamed, and J. H. Lobos, Effect of stirring and H_2 on ethanol production by thermophilic fermentation, In Technical Information Series Report No. 81CRD090, General Electric CRD, Schenectady, New York, 1981.

40. T. M. Su, R. Lamed, and J. H. Lobos, *Proc. 2nd World Congress on Chemical Engineering and World Chemistry Exposition*, Montreal, 1981, p. 353.

41. G. C. Avgerinos, H. Y. Fang, I. Biocic, and D. I. C. Wang, In *Advances in Biotechnology*, Vol. 2 (M. Moo-Young and C. W. Robinson, eds.), Pergamon, Toronto, 1981, p. 119.

42. J. N. Saddler, M. K.-H. Chan, and G. Louis-Seize, *Biotechnol. Lett.*, *3*, 321 (1981).

43. J. Wiegel, *Experientia*, *36*, 1434 (1980).

44. R. Brooks, T.-M. Su, M. Brennan, and J. Frick, Bioconversion of plant biomass to ethanol, In *Proc. 3rd Annual Biomass Energy System Conference*, Technical Report, SERI/TP, 1979, p. 275.

45. T. K. Ng, A. Ben-Bassat, and J. G. Zeikus, *Appl. Env. Microbiol.*, *41*, 1337 (1981).

46. J. N. Saddler and M. K.-H. Chan, *Eur. J. Appl. Microbiol. Biotechnol.*, *16*, 99 (1982).

47. J. N. Saddler and M. K.-H. Chan, *Can. J. Microbiol.*, *30*, 212 (1984).

48. D. I. C. Wang and H.-Y. Fang, *Am. Chem. Soc. Div. Pet. Chem.*, *25*, 639 (1980).

49. L. R. Lynd, R. H. Wolkin, and H. E. Grethlein, *Biotechnol. Bioeng. Symp. Ser.*, *17*, 265 (1986).

50. E. A. Bayer, E. Setter, and R. Lamed, *J. Bacteriol.*, *163*, 552 (1985).

51. R. Lamed and E. A. Bayer, *Experientia*, *42*, 72 (1986).

52. E. A. Bayer and R. Lamed, *J. Bacteriol.*, *167*, 828 (1986).

53. J. Petre, R. Longin, and J. Millet, *Biochimie*, *63*, 629 (1981).

54. T. K. Ng and J. G. Zeikus, *Biochem. J.*, *199*, 341 (1981).

55. D. Petre, J. Millet R. Longin, P. Beguin, H. Girard, and J.-P. Aubert, *Biochimie*, *68*, 687 (1986).

56. N. Ait, N. Creuzet, and J. Cattaneo, *Biochem. Biophys. Res. Commun.*, *90*, 537 (1979).

57. N. Ait, N. Creuzet, and J. Cattaneo, *J. Gen. Microbiol.*, *128*, 569 (1982).

58. N. Creuzet and C. Frixon, *Biochimie*, *65*, 149 (1983).

59. N. Creuzet, J.-F. Berenger, and C. Frixon, *FEMS Microbiol. Lett.*, *20*, 347 (1983).

60. N. Ait, N. Creuzet, N. and P. Forget, *J. Gen. Microbiol.*, *113*, 399 (1979).

61. R. Lamed, E. Setter, and E. A. Bayer, *J. Bacteriol.*, *156*, 828 (1983).

62. R. Lamed and E. A. Bayer, *Adv. Appl. Microbiol.*, *33*, 1
     (1988).
63. P. Cornet, J. Millet, P. Beguin, and J.-P. Aubert, *Biotech-
     nology*, *1*, 589 (1983).
64. J. Millet, D. Petre, P. Beguin, O. Raynaud, and J.-P. Aubert,
     FEMS Microbiol. Lett., *29*, 145 (1985).
65. G. Joliff, P. Beguin, M. Juy, J. Millet, A. Ryter, R. Poljak,
     and J.-P. Aubert, *Biotechnology*, *4*, 896 (1986).
66. R. Lamed, E. Setter, R. Kenig, and E. A. Bayer, *Biotechnol.
     Bioeng. Symp. Ser.*, *13*, 163 (1983).
67. R. Lamed, R. Kenig, E. Setter, and E. A. Bayer, *Enzyme
     Microb. Technol.*, *7*, 37 (1985).
68. E. A. Johnson, E. T. Reese, and A. L. Demain, *J. Appl.
     Biochem.*, *4*, 64 (1982).
69. E. A. Johnson and A. L. Demain, *Arch. Microbiol.*, *137*, 135
     (1984).
70. G. Halliwell and M. Griffin, *Biochem. J.*, *135*, 587 (1973).
71. E. A. Bayer, R. Kenig, and R. Lamed, *J. Bacteriol.*, *156*, 818
     (1983).
72. J. Wiegel and M. Dykstra, *Appl. Microbiol. Biotechnol.*, *20*, 59
     (1984).
73. E. Morgenstern, E. A. Bayer, J. F. G. Vliegenthart,
     G. Gerwig, H. Kamerling, and R. Lamed, *FEMS Symp. Biochem.
     Gen. Cellulose Degradation*, Abst. P2-06 (1987).
74. J. F. G. Vliegenthart, G. Gerwig, J. P. Kamerling,
     E. Morgenstern, E. A. Bayer, and R. Lamed,
     unpublished results.
75. J. H. D. Wu and A. L. Demain, *Abstr. 85th Ann. Meet. Am.
     Soc. Microbiol.*, K-079, 1985, p. 248.
76. M. P. Coughlan, K. Hon-nami, H. Hon-nami, L. G. Ljungdahl,
     J. J. Paulin, and W. E. Rigsby, *Biochem. Biophys. Res.
     Commun.*, *130*, 904 (1985).
77. K. Hon-nami, M. P. Coughlan, H. Hon-nami, and L. G. Ljung-
     daghl, *Arch. Microbiol.*, *145*, 13 (1986).
78. F. Mayer, Y. Mori, M. P. Coughlan, and L. G. Ljungdahl,
     Abstr. 86th Ann. Meet. Am. Soc. Microbiol., K-112 (1986).
79. P. J. Weimer and W. M. Weston, *Biotechnol. Bioeng.*, *27*,
     1540 (1985).
80. D. Groleau and C. W. Forsberg, *Can. J. Microbiol.*, *27*, 517
     (1981).
81. T. M. Wood, C. A. Wilson, and C. S. Stewart, *Biochem. J.*,
     205, 129 (1982).
82. K. Cavedon, S. B. Leschine, and E. Canale-Parola, *Abstr. 87th
     Meet. Am. Soc. Microbiol.*, K-127 (1987).

83. G. L. Pettipher and M. J. Latham, *J. Gen. Microbiol.*, *110*, 21 (1979).
84. R. Lamed, J. Naimark, E. Morgenstern, and E. A. Bayer, *J. Bacteriol.*, *169*, 3792 (1987).
85. T. K. Ng and J. G. Zeikus, *J. Bacteriol.*, *150*, 1391 (1982).
86. J. K. Alexander, *J. Biol. Chem.*, *243*, 2899 (1968).
87. K. Sheth and J. K. Alexander, *Biochim. Biophys. Acta*, *148*, 808 (1967).
88. N. P. Golovchenko, N. A. Chuvil'skaya, and V. K. Akimenko, *Mikrobiologiya*, *55*, 31−33 (1986).
89. P. E. Hernandez, *J. Gen. Appl. Microbiol.*, *28*, 469 (1982).
90. N. J. Patni and J. K. Alexander, *J. Bacteriol.*, *105*, 220 (1971).
91. T. Yamada and J. Carlson, *J. Bacteriol.*, *124*, 55 (1975).
92. R. Lamed and J. G. Zeikus, *Biochim. Biophys, Acta*, *660*, 251 (1981).
93. A. A. Herrero, R. F. Gomez, B. Snedecor, C. J. Tolman, and M. F. Roberts, *Appl. Microbiol. Biotechnol.*, *22*, 53 (1985).
94. R. Lamed and J. G. Zeikus, *J. Bacteriol.*, *141*, 1251 (1980).
95. L. H. Carreira, L. G. Ljungdahl, F. Bryant, M. Szukzynski, and J. Wiegel, In *Genetics of Industrial Microorganisms* (Y. Ikeda, ed.), American Society for Microbiology, Washington, D.C., 1982, p. 351.
96. R. Lamed and J. G. Zeikus, *Biochem. J.*, *195*, 183 (1981).
97. A. Ben-Bassat, R. Lamed, and J. G. Zeikus, *J. Bacteriol.*, *146*, 192 (1981).
98. W. D. Murray, *Appl. Env. Microbiol.*, *51*, 710 (1986).
99. L. H. Carreira and L. G. Ljungdahl, In *Liquid Fueld Development* (D. L. Wise, ed.), CRC Series in Bioenergy Systems, CRC Press, Boca Raton, 1983, p. 1.
100. J. G. Zeikus, T. K. Ng, A. Ben-Bassat, and R. Lamed, U.S. Patent 4,400,470 (1983).
101. T. K. Kirk and J. G. Zeikus, Developed biological systems for lignocellulose conversion, *4th Progress Report*, Report No. NSF/RA-780891, University of Wisconsin, 1978.
102. L. G. Ljungdahl and J. K. W. Wiegel, U.S. Patent 4,292,406 (1981).
103. P. J. Weimer and J. G. Zeikus, *Appl. Env. Microbiol.*, *33*, 289 (1977).
104. P. LeRuyet, H. C. Dubourguier, and G. Albagnac, *Appl. Env. Microbiol.*, *48*, 893 (1984).
105. R. Lamed and E. A. Bayer, In *Biochemistry and Genetics of Cellulose Degradation* (J. P. Aubert, P. Beguin, and J. Millet, eds.), Academic Press, Orlando, 1988, p. 101.

106. M. A. Bonner and F. J. Stutzenberger, *Lett. Appl. Microbiol.*, *6*, 59 (1988).

107. R. Lamed, E. A. Bayer, B. C. Saha, and J. G. Zeikus, Biotechnological Potential of enzymes from unique thermophiles, In *Proc. 8th Int. Biotechnol. Symp.* (G. Durand and L. Bobichon, eds.), Soc. Franc. de Microb., 1988, pp. 371–383.

108. G. J. Gerwig, P. de Waard, J. P. Kamerling, J. F. G. Vliegenthart, E. Morgenstern, R. Lamed, and E. A. Bayer, *J. Biol. Chem.*, *264*, 1027 (1989).

109. E. Morag (Morgenstern), E. A. Bayer, and R. Lamed, *Appl. Biochem. Biotechnol.*, in press.

110. J. H. D. Wu, W. H. Orme-Johnson, and A. L. Demain, *Biochemistry*, *27*, 1703–1709 (1988).

111. M. Romaniec, T. Kabayashi, U. Fauth, and A. L. Demain, personal communication.

112. E. Morag (Morgenstern), E. A. Bayer, and R. Lamed, *J. Bacteriol.*, in press.

113. S. Kadam, A. L. Demain, J. Millet, P. Beguin, and J.-P. Aubert, *Enzyme Microb. Technol.*, *10*, 9 (1988).

114. S. K. Kadam and A. L. Demain, *Biochem. Biophys. Res. Commun.*, *161*, 706 (1989).

115. R. Lamed, E. Morag (Morgenstern), R. Kenig, F. de Micheo, J. F. Calzada, and E. A. Bayer, *Appl. Biochem. Biotechnol.*, in press.

116. G. P. Hazlewood, M. P. M. Romaniec, K. Davidson, O. Grepinet, P. Beguin, J. Millet, O. Raynaud, and J.-P. Aubert, *FEMS Microbiol. Lett.*, *51*, 231 (1988).

19

Cellulose Degradation by Mesophilic Aerobic Bacteria

KARL-LUDWIG SCHIMZ   *Institut für Biotechnologie, Jülich, Federal Republic of Germany*

I. INTRODUCTION

Studies of cellulose biodegradation are important from the standpoint of basic research [the utilization of an insolbule substrate which is the most abundant of all biopolymers (1)] and commercial application [the potential for cellulosic biomass to partially supplant dwindling fossil fuels as a source of fuels and oxychemicals (2,3)]. Among cellulolytic organisms, one of the most intensively studied groups is the mesophilic aerobic and facultatively anaerobic bacteria.

Mesophilic bacteria are those which grow optimally between 20 and 45°C (4,5) or 25 and 40°C (6). Among them the interval between T_{min} and T_{max} varies widely due to adaptation of the isolates to their habitats. This chapter is focused on these facultative anaerobic and strict aerobic bacteria which are capable of using insoluble cellulose or soluble cellulose derivatives as the sole source of energy and carbon in a mesophilic environment. At the expense of this polymer these bacteria are able to fulfill the following general functions: (1) formation and maintenance of cell structures; (2) formation of energy; (3) performance of metabolism, regulation, and reduplication; and (4) defense against intra- and extracellular stress situations (7). A precondition for all these reactions is the existence of an enzyme or set of enzymes which

transforms the extracellularly localized cellulose polymer to intracellularly metabolizable degradation products such as cellooligomers,
cellobiose, and glucose. Early concepts of enzymatic breakdown of
cellulose by nonhydrolytic (C_1) and hydrolytic (C_x) activities (8,9)
have given way to new models of cellulose decomposition, e.g., that
of Enari and Niku-Paavola (10). The new models take into consideration recent results, including those from molecular biology studies,
and propose the interaction among three different kinds of hydrolytic
enzymes: cellobiohydrolase, endoglucanase, and β-glucosidase.
These models are derived from experimental work with fungal organisms and from the proposal of Wood (11) to differentiate between
"cellulolytic" and "pseudocellulolytic" fungi on the basis of their
ability to degrade native crystalline cellulose.

As some bacteria are also able to degrade native cellulose it is
assumed that their enzyme systems contain principally the same components as the fungal system. Possible reasons for the fact that
cellobiohydrolases (exo-acting enzymes) have not been detected for
many years are discussed in Sec. II.

This chapter, while restricted to mesophilic bacteria, will un-
avoidable mention basic results obtained with other microbial groups
for comparative reasons.

II. AEROBIC, MESOPHILIC, CELLULOLYTIC BACTERIA: A VIEW ON THEIR TAXONOMIC CLASSIFICATION AND ON THEIR METABOLIC DIVERSITY

According to Fahraeus (12), the first discovery of an aerobic,
mesophilic, cellulose-decomposing bacterium was published in 1903
by van Iterson (13). Further early work on decomposition of cellulose by microbes (including bacteria) was summarized by Siu (14)
and by Gascoigne and Gascoigne (15). During recent years bacteria from the following general have been studied: *Bacillus, Cellulomonas, Cellvibrio, Cytophaga, Erwinia, Micromonospora, Pseudomonas, Sporocytophaga*, and *Streptomyces*. The data given in
Table 1 summarize in a rather crude way their association with certain orders and briefly give an overview of their metabolic diversity.

Four of the genera (*Bacillus, Micromonospora, Sporocytophaga,
Streptomyces*) produce a resistant resting stage during their life
cycle, and three of these are well known for the production of secondary metabolites such as toxins (29), antibiotics (21,29,38), and
pigments (37). Moreover, the conversion of carbon sources to intra-
or extracellular materials like trehalose (32), glucogen (18,41), slime

TABLE 1   Taxonomic Classification and Some Properties of Genera Involved in Cellulose Decomposition Under Mesophilic, Aerobic Conditions

Taxa	References regarding taxonomical position		Gram reaction	Production of	
	General	Special		intra-cellular material	extra-cellular material
Gliding bacteria					
Sorangium	16	17, 18	−	+ 18	+ 18
Cytophaga		22 − 25	−		+ 42
Sporocytophaga	19 − 21	23 − 25	−		+ 43
Actinomycetales	19				
Micromonospora	21	26 − 28	+		
Streptomyces		22			+ 29
Eubacteriales					
Bacillus		22, 29	+		
Cellulomonas	19, 21	22, 28, 30, 31, 33, 46	+	+32, 41	
Cellvibrio	19	34	−		+ 34
Erwinia	19, 20	35, 93 − 96	−		
Pseudomonas		22, 36 − 40	−	+ 19	+ 37, 38

material containing glucuronic acid (42), curdlan (34), and unidentified reducing compound (43), and polyhydroxybutyric acid (19) is
expected to influence strongly the yield of any other material produced by cells growing on cellulose or cellulose-containing materials.
    The metabolic diversity of the tabulated genera is further under-
lined by their different metabolic potentials: members of the genus
Sporocytophaga are reportedly unable to use more than four carbon
sources; mannose, glucose, cellobiose, and cellulose (19), while some
species of the genus *Pseudomonas* are known to grow on about 80
different substrates (40).
    It is assumed that all the different properties of these genera
influence in one way or another their cellulolytic capabilities. Cellulose degradation is expected to be regulated with more sophistica-
tion in species with a high potential of carbon utilization than in
those with a low one (e.g., sequential rather than coordinate induction of the enzymes involved). Furthermore, phytopathogenic microorganisms like *Erwinia* may simultaneously induce other polymer-
degrading enzymes (pectinases, proteases) necessary for plant cell
wall destruction.
    In 1983 (44) a novel mesophilic cellulolytic bacterium was iso-
lated from a marine shipworm. This organism is able to fix nitrogen under microaerophilic conditions. Growth of this species under
aerobic and microaerophilic conditions was studied recently (45) but
up to now no taxonomic relation has been established.
    Tables 2 and 3 summarize results on the detection, (minimal)
number, and localization of cellulolytic enzymes from aerobic, mesophilic bacteria grown under different conditions, e.g., on cellulose,
carboxymethylcellulose (CMC), or soluble carbohydrates. The term
"cellulolytic enzymes" is used for all the enzymes which are active
in polymer and oligomer degradation. Bacteria which can grow on
cellobiose but are not able to degrade cellulose, e.g., *Aerobacter
aerogenes* containing an ATP-dependent cellobiose kinase (111) are
not included in this table. The enzymes are named according to
the substrates used by the authors in the assays (cf. Table 4)
thereby avoiding a classification according to the endo/exo nomenclature. Two arguments support this kind of designation. First,
using CMC as a substrate endocellulases are generally characterized
by a rather high quotient of increase in fluidity per unit of time
vs. a simultaneous increase of the concentration of reducing sugars.
This quotient in the case of exocellulases is expected to be rather
low. However, as can be seen by a comparison of four endocellulases from the fungus *Trichoderma koningii* [reviewed by Finch and
Roberts (112); Fig. 3 in Chap. 13] there are large differences between these endo-acting enzymes regarding this quotient. It seems

TABLE 2   Number and Localization of Cellulolytic Enzymes of
Bacteria Detected After Growth Under Different Conditions

Microorganisms	Cellulolytic enzymes localization/substrate hydrolyzed		Ref.
Gliding bacteria			
Sorangium compositum	cbound		17, 18
	cfree	CMC	47
		Whatman	
S. nigrescens	cbound	Filter paper	17, 18
		Cotton	
Cytophaga strain 3	cbound	Cellophan	42
		Cellobiose	
C. globulosa	cbound	Cellophan	42
		Cellobiose	
Cytophaga ATCC 29474	cbound	CMC-liquefying	48
		CMC-saccharifying	
Cytophaga sp.	cfree	CMC	49
Sporocytophaga myxococcoides	cfree	Cellulose	43, 82
		CMC	
		Cellobiose	
Actinomycetales			
Streptomyces antibioticus	cfree	CMC-liquefying	50
S. flavogriseus ATCC 33331	cfree	Avicel	
		Filter paper	
		Acid-swollen cellulose	
		CMC	

TABLE 2   (Continued)

Microorganisms		Cellulolytic enzymes localization/substrate hydrolyzed	Ref.
S. flavogriseus ATCC 33331 (cont.)	cbound	Cellobiose p-nitrophenylglucoside	51–53
S. lividans	cfree	CMC	54
	cbound	Cellobiose	
S. albogriseolus	cfree	Avicel CMC	
S. nitrosporeus	cbound	Cellobiose p-nitrophenylglucoside	55
Micromonospora melanospora	cfree	Avicel CMC	
	cbound	Cellobiose p-nitrophenylglucoside	55
Eubacteriales			
Bacillus strain DLG (*B. subtilis*)	cfree	TNP-CMC	56
B. cereus	cfree	CMC	57
	cbound	CMC	
B. strain N-4 (ATCC 21833)	cfree	⩾2 CMC	58
B. strain N 1139	cfree	1 CMC	59
B. subtilis	cfree	CMC	60
B. subtilis	cfree	CMC	61

TABLE 2   (Continued)

Microorganisms	Cellulolytic enzymes localization/substrate hydrolyzed		Ref.
Eubacteriales (cont.)			
Cellulomonas fimi (ATCC 484)	cfree	~3 CMC	62
	cbound-in	Cellobiose	63
		p-nitrophenylglucoside	
Cellulomonas fimi (ATCC 484)	cfree-S	CMC	64, 65
		Cellopentaose	
		p-nitrophenylcellobioside	
C. fimi	cbound-in	Cellobiose (phosphorolysis)	109
C. fermentans (ATCC 43279)	cfree	3 Avicel	66, 67
		2 CMC	
C. flavigena	cfree	CMC	69
	cfree-S	CMC	
	cbound-in	Cellobiose	
C. flavigena strain IIbc	cfree	4 CMC	77, 78
	cfree-S	2 CMC	
C. sp.	cfree	CMC	68
C. uda CB-4	cfree	Cellulose	70, 71
	cfree	Cellobiose	
C. uda	cfree	CMC	72
	cbound	CMC	
		p-nitrophenylglucoside	

TABLE 2   (Continued)

Microorganisms	Cellulolytic enzymes localization/substrate hydrolyzed		Ref.
Eubacteriales (cont.)			
C. uda (ATCC 21399)	cfree	CMC	73 – 75
	cbound	CMC	
	cbound-in	Cellobiose (phosphorolysis)	
	cfree-S	CMC	76
C. uda	cfree	CMC	79
Cellvibrio fulvus (NCIB 8634)	cfree	3 CMC	80
		1 Avicel	
	cbound	CMC	
Cellvibrio fulvus (NCIB 8634)	cfree-S		81, 82
Cellvibrio fulvus (NCIB 8634)	cbound	CMC	83, 84
C. gilvus	cfree	Cellulose	85
		CMC	
C. gilvus	cbound-in	Cellobiose (phosphorolysis)	86
C. gilvus	cfree	4 Alkali-swollen cellulose	87, 88
		Cellohexaose cellohexaose	
		Reduced cellopentaose	
C. gilvus (ATCC 13127)	cfree	CMC	90

TABLE 2 (Continued)

Microorganisms	Cellulolytic enzymes localization/substrate hydrolyzed		Ref.
Eubacteriales (cont.)			
C. mixtus (UQM 2294)	cfree		91
C. vulgaris	cfree	CMC	92
Erwinia chrysanthemi (INRA 3665)	cfree	TNP-CMC	97–99
	cbound	TNP-CMC	
Pseudomonas strain II	cfree	⩾2 CMC	100
Ps. fluorescens var. *cellulosa*	cfree	2 CMC-liquefying	101–104
	cbound	1 CMC-liquefying	
Ps. fluorescens var. *cellulosa*	cfree	5 CMC-liquefying	105
		CMC-saccharifying	
Ps. fluorescens var. *cellulosa*	cbound	2 *para*-nitrophenylglu-coside	107
	one of them:		
	cound-in	Cellobiose	
		CMC	
		para-nitrophenylcellobioside	
Pseudomonas sp.	cfree	2 CMC-liquefying	108
		2 CMC-saccharifying	
	cbound	CMC-liquefying	
		CMC-saccharifying	
		2 *ortho*-nitrophenylglucoside	
		Cellobiose	

TABLE 2   (Continued)

Microorganisms	Cellulolytic enzymes localization/substrate hydrolyzed		Ref.
Pseudomonas sp. (cont.)	cbound- in	1 *ortho*-nitrophenyl-glucoside	108
		1 Cellobiose	

[a]Available species from a collection are designated by their strain collection numbers; it is not clear if *Pseudomonas fluorescens* var. *cellulosa* obtainable from NCIV 10462 is identical with the strain used by Suzuki's group.

[b]Abbreviations used in this table for the localization of cellulolytic enzymes are given by definition in Table 3.

[c]Instead of using such terms as "endo" and "exo," the names of enzymes in this table are given by the substrate(s) which is(are) cleaved according to the authors. The reasons for this are discussed in the text and the substrates are included in Table 4.

rather arbitrary therefore to differentiate endo- and exocellulases by this criterion alone which is supported by the finding, that catalytic domains of endo- and exoglucanases from different species show a high degree of homology (230). Further classification methods are necessary based on such criteria as the cleavage products obtained with pure cellodextrins (carefully excluding the existence of transfer reactions), or the results and the viscometric behavior of enzymatic degradation of insoluble cellulose in a cellulose solvent (discontinuous assay). Second, the limited number of bacterial exocellulases which have been detected so far (see for example, 53,64,65) reflects either that the endo/exo model is valid only for a certain group of microorganisms or that the enzyme is difficult to detect; it may be cell-bound (surface located) or cell-free but substrate-bound (see for example, 62,65,77,82,107), and in both cases difficult to isolate and to characterize.

Localization of the enzyme activities listed in Table 2 is abbreviated according to the nomenclature given in Table 3. It is only in recent years that attention has been drawn to the polymer-cleaving activities which are cell-bound (surface located) and substrate-bound (cell-free).

With regard to the taxonomic classification (Table 1), two additions to Table 2 merit further discussion. The first concerns *Cellulomonas fimi*, which was reported to be a synonymous with *Cellulomonas flavigena* (19), although significant differences between these

TABLE 3   Localization of Bacterial Enzymes Cleaving β-1,4-Glucosidic Bonds

		Abbreviation used in Table 2	Example (Ref.)
Total enzyme activity — cell-free	free in the culture filtrate	cfree	62. 77, 78
	substrate-bound	cfree-S	62. 64. 65
cell-bound — extracellular	surface located	cbound	83, 84
	periplasmic		
	cell wall-bound		
	membrane-bound		
intracellular	free in the cytoplasm	cbound-in	74
	particle-bound		

TABLE 4   Enzymes Involved in Cellulose Degradation by Bacteria[a]

		Substrates used
(A)	***Cellulose-cleaving enzymes***	
EC 3.2.1.74	exo-1,4-β-glucohydrolase (exoglucanase)	Cotton, filter paper, Avicel, Cellophan,
EC 3.2.1.91	exo-1,4-β-glucan cellobiohydrolase (exoglucanase)	Whatman cellulose, acid- or alkali-swollen cellulose
EC 3.2.1.4	endo-1,4-β-glucan glucanohydrolase (endoglucanase)	Carboxymethylcellulose (CMC), substituted CMC (TNP-CMC), (liquefying and/or saccharifying activity)
(B)	***Cellooligomer-cleaving enzymes***	
EC 2.4.1.49	1,4-β-oligoglucan/orthophosphate β-glucosyltransferase (cellodextrin phosphorylase)	Cellodextrins and reduced cellodextrins (DP $\geqslant$ 3), para-nitrophenylcellobioside
EC	1,4-β-oligoglucan-hydrolase (cellodextrin hydrolase)	
(C)	***Cellobiose-cleaving enzymes***[b]	
EC 2.4.1.20	cellobiose/orthophosphate glucosyltransferase (cellobiose phosphorylase)	Cellobiose, ortho-, para, meta-nitrophenyl-β-D-glucoside
EC 3.2.1.21	cellobiose hydrolase (β-glucosidase or cellobiase)	

[a]Cellobiose dehydrogenase (EC 1.1.99.18, EC 1.1.5.1) and cellobiose kinase (EC 2.7.1.85) are up to now not detected in cellulolytic bacteria and are therefore not included in this table.

[b]Nothing is known on the function of specific aryl-β-glucosidases in the degradation of naturally occurring cellulosic material (110). Therefore, they are not included in this table.

species have been observed (113). The strain of *C. fimi* used by Sato and Takahashi (109) was reportedly isolated from sheep rumen. However, *C. fimi* is generally not considered to be a rumen inhabitant (114), indicating a possible mistake in the classification. This view is strengthened by the observation that this species is reported to utilize succinate under aerobic as well as anaerobic conditions (115,116) whereas Stackebrandt et al. (46) did not detect succinate utilization in 10 different species of the genus *Cellulomonas*, including two strains of *C. fimi* (DSM 20113 and DSM 20114). Nonutilization of succinate was confirmed in another species of *Cellulomonas* (117). Thus the taxonomic status of *C. fimi* remains uncertain.

The second taxonomic point concerns the possible relatedness of *Cellivibrio* and *Pseudomonas* (19). This view is supported by the finding of a gene cluster in *Cellivibrio mixtus* (91) coding for five enzymes which degrade polysaccharides and cellobiose, as well as the ability to attach to surfaces. Clustering of genes is a well-known phenomenon in the genus *Pseudomonas* (36) as well as in some other gram-negative bacteria (cf. 91).

III. FORMATION OF CELLULOLYTIC ENZYMES

An increase in detectable concentrations of cellulolytic enzymes has been reported during growth on insoluble cellulose (51−53,72−75, 80,85,90,101−104,108), on CMC (58−61,69,79), and in some cases during growth on soluble carbohydrates (cellodextrins: 80; cellobiose in presence of CMC: 72; soluble carbohydrates: 97,98, 101−104). Glucose and cellobiose, if simultaneously present with insoluble cellulose, have been reported to prevent the increase of the detectable cellulase concentrations (80,90); however, mutants resistant to this glucose/cellobiose effect have been isolated (118). Furthermore, wild-type forms exist that are not influenced in cellulase formation by soluble carbohydrates either per se (57,77,78) or under appropriate conditions (carbon-limited growth during continuous culture: 97,98,101−104): the increase of the specific activity of cell-free cellulases is reported to be in parallel with the increase in doubling times observed.

Ingestion of insoluble carbon sources like cellulose by microorganisms surrounded by complex (rigid) cell walls has not been demonstrated; therefore, primary events of cellulose degradation which lead to soluble, transportable fragments must occur outside the cell. If so, the following reactions may be involved in this increase or decrease of enzyme concentrations as a response to a particular carbon source: (1) de novo protein biosynthesis; (2) covalent modification leading to activation or inactivation; (3) excretion; and (4) release from insoluble substrate with increasing degree of degradation. Presumably in the case of extracellular enzymes these reactions occur simultaneously and therefore the analysis of the event(s) leading to increased/decreased enzyme concentration is rather complicated.

At our current level of understanding, the use of terms such as induction, repression, derepression, etc., which were originally defined in molecular or mechanistic terms (119–121), are generally not applicable to the description of increases or decreases in detectable concentrations of cellulolytic enzymes.

True induction (according to 119) has been demonstrated in few cellulolytic systems: the bacterium *Cellulomonas uda* (73), and the fungi *Schizophyllum commune* (122) and *Trichoderma viride* (123); however, the mechanism of induction is not yet clear. There is some evidence (124,125) that sophorose, a soluble disaccharide produced from cellobiose by β-glucosidase-catalyzed transglucosylation (124), may be the true inducer of cellulases in *T. viride*. However, up to now no direct investigations [analogous to those for the well-studied *lac* operon (126)] have been reported which proves that the proposed (transglucosylation) "product induction" sufficiently describes the cellulase induction. The same is true for the assumed structures of (product) inducers of polygalacturonic acid-degrading enzymes in *Erwinia* (138–140). Stutzenberger (127, p. 120) discussed several models based on the assumption that "the inducers of cellulase biosynthesis in nature must surely be products (or derivatives of products) from the activity of the same enzymes they are inducing." These models have in common the fact that induction occurs endogenously by a cellulose-derived carbohydrate. However, exogenous induction directly governed by large structures from outside the cells is already well documented in animal cells (128,129) and in some cases it has been shown [phosphoglycerate permease in *Salmonella typhimurium* (130); sugar phosphate transport system of *Escherichia coli* K-1a (208)] or evidence has been presented [extracellular, cell-bound glucose dehydrogenase and gluconate dehydrogenase in *Pseudomonas aeruginosa* (131)] that exogenous induction occurs in microbial systems (132). A precondition for this kind of direct induction by the polymer itself is the existence of signal-transducing receptors at the outermost site of the microbial system investigated. The chemotactic response of a *Pseudomonas* species to cellulose and cellulose degradation products (137) suggests the existence of specific receptors. The well-studied cell envelopes of gram-negative bacteria, as well as the S layers of gram-positive bacteria (133–135), can be assumed to be the site of these receptors. Physical contact between cells and insoluble polymer has been reported as necessary for cellulase formation in *Cellvibrio gilvus* (90). However, more recently, (136) the separation of cells and fibers was reported to be advantageous for polymer degradation.

Results on the formation of polymer-degrading enzyme activities in *Bacillus* species are contradictory in some aspects: an increase of CMCase concentration is reported to be limited to the stationary phase of growth (56) and is reported to occur (like in *Cellvibrio vulgaris*; 92) in parallel to growth (61); furthermore, cellulase formation is reported to occur in the absence of cellulosic material, indicating constitutive synthesis (56,57) and the formation of other polymer-degrading enzymes is reported to be sensitive to cAMP and cGMP, indicating a catabolite repression-like regulatory mechanism (141).

The inhibition of the increase of cellulase concentration by inhibitors of protein biosynthesis has been reported (56,92,116) and their possible effect on both de novo synthesis and excretion has been discussed in principle above. It is the finding of Lampen and Nielsen (142) on the enzyme-catalyzed covalent transformation of the membrane penicillinase from *Bacillus* species as well as the time lag of several hours which is observed between the appearance of cell-bound and cell-free cellulases (75,80) that makes that reasoning plausible. This time lag therefore is not to be taken as argument for the sequential synthesis of cell-bound and cell-free activities.

The detection of CMCase in culture filtrates from *Streptomyces lividans* grown at the expense of xylan (54) can be interpreted to mean that one enzyme with a low degree of specificity is formed; this has been shown to be the case in *Cellulomonas fimi* which produces during growth on cellulose one enzyme containing both cellulase and xylanase activity (65,236). Alternatively, this observation may suggest that CMCase(s) and xylanase(s) are synthesized nonspecifically and simultaneously (by coordinated expression of a gene cluster) during growth on one polymer; this is the case in *Erwinia chrysanthemi* regarding pectate lyases (EC 4.2.2.2) and CMCases (143; Figure 1, pg. 1200).

Experiments with soluble degradation products of cellulose indicate an increase of cell bound CMCase(s) in *Cellvibrio fulvus* (80) and *Pseudomonas fluorescens* var. *cellulosa* (101-104). As pointed out above, and will be discussed further in the following section, this is not an indication for an independent formation of cell-bound and cell-free enzyme species.

Recent work on the regulation of enzyme formation in *Cellulomonas fimi* indicates that two different promoters are involved in the transcription of *cenB*, encoding an extracellular endocellulase (204). One of these promoters is involved during constitutive low-level gene transcription (during growth on carbon sources other than cellulose), and the second promoter is involved in high-level enzyme formation which occurs during growth of *C. fimi* on cellulosic substrate.

IV. LOCALIZATION AND DISTRIBUTION
OF CELLULOLYTIC ENZYMES

Growth on different carbon sources influences not only the concentration of cellulolytic enzymes but also their distribution. Growth
on soluble cellooligomers is reported to lead to a specific increase
in cellulase activity bound to cells in *Cellvibrio fulvus* (80) and in
Pseudomonas fluorescens var. *cellulosa* (101−104) (see above). The
compartmentation of cellulolytic activities of cellulose-grown cells was
investigated in more detail with *Cellvibrio fulvus* (83,84): it was
concluded from experiments with the use of detergents that about
18% of the CMCase is surface-located, 26% is located in the periplasm
(released by polymyxin/lysozyme), and about 55% is located within
the membrane-wall fraction (released by Triton X-100). From experiments performed with different carbon sources (glucose, cellobiose, or cellobiose in the presence of CMC) is appears that the distribution of CMCase in *Cellvibrio fulvus* changes during the growth
cycle (84). The effect of Triton X-100 on the distribution of CMC-
ase was also reported for cellulose-grown cells of *Cellulomonas flavigena* (144). The cell's outer layer, presumably containing surface-
located CMCase(s), is stimulated in its formation by the addition of
cellulose to the culture (145) and ruthenium red staining indicates
that this layer is a glycoprotein calyx (145) of the type described
by Costerton et al. (146). Intact cells of *Cellulomonas uda* contain
cell-bound CMCase(s) even after repeated washings with buffers
(73,75). This enzyme activity can be solubilized by 1.5% (w/v)
sodium dodecyl sulfate (SDS) with an approximate threefold increase
in activity (76). This increase of total cell-bound activity may be
due either to activation by this solubilizer or release of additional
activity which is cryptic in the assay with intact cells. Controls
indicate that no membrane and no cytoplasmic proteins are released
by this technique. Double labeling with [^{3}H]-glycerol and [^{14}C]-
palmitic acid and subsequent electrophoresis of both cell-bound and
cell-free activities clearly show that the CMCases of *Cellulomonas
uda* contain covalently bound glycerol and palmitic acid (76). It is
assumed that these compounds are involved in the anchoring of
CMCase into the surface layer [S-layer according to (133−135), or
"protuberant structures" according to (209)], as was shown to be the
case with penicillinases from *Bacillus licheniformis* and *Staphylococcus
aureus* (142) and with pullulanase from *Klebsiella pneumoniae* (147).

   Thus it is obvious from this discussion that an increase of
CMCase in the supernatant of a culture does not necessarily reflect
immediate de novo enzyme synthesis by the cells.

   Von Hofsten (148) discussed the interaction of cellulolytic enzymes with their substrates and concludes from several models that

cell-bound cellulases are more effective "because they occur in high concentrations and are favorably aligned on the substrate." This conclusion is supported by results in other experimental systems: cellulolytic enzymes of the thermophilic anaerobic bacterium *Clostridium thermocellum* are aggregated into a high molecular weight complex called a "cellulosome" (149; see also Chap. 18). In addition, a yellow substance of unknown chemical structure which mediates binding of cellulases to insoluble substrates was detected in cellulose-grown cultures of this same organism (151). Enzyme complexes containing cellulase, xylanase, and β-glucosidase were reported to exist in the fungus *Trichoderma reesei* (178) and they are probably fixed together by a glycine-rich protein (206). In *Cellvibrio mixtus* the ability to attach to surfaces is genetically clustered with polysaccharide-degrading enzymes (91) and the ability to attach to fibers is stimulated by cellulose in *Cellulomonas* species (145). All these investigations have in common the fact that some factor is produced to concentrate enzyme activities at a defined locus, thereby increasing the efficiency of degradation.

Work with ribosomal preparations of *Cellvibrio gilvus* (150) and of *Pseudomonas fluorescens* var. *cellulosa* (106) has shown that extracellular enzymes are synthesized predominantly by membrane-bound ribosomes whereas intracellular enzymes are synthesized by free ribosomes. As pointed out by Pollock (81), "occurrence of enzymes in the culture supernatant is not sufficient evidence for physiological extracellularity." Cellulases from the supernatant are therefore expected to have been either secreted by the cells without binding to insoluble material or released by autolysis of the cells. If bound to insoluble material they can be released from the substrate either in the course of increasing cellulose degradation or during prolonged cultivation by covalent transformation. Stable binding of cellulases to insoluble substrate was shown by several authors (55,62,69,78) and in some cases the enzymes have been solubilized by 8 M guanidinium chloride and subsequently characterized (62,78). Transformation of cellulolytic enzymes by covalent modification has been detected in vivo (62,105) and in vitro (152), and will be discussed in detail later.

Cellobiose phosphorylase of *Cellulomonas uda* is localized intracellularily in the soluble part of the cytoplasm (74) whereas enzymes with β-glucosidase activity of *Pseudomonas fluorescens* var. *cellulosa* are reported to be membrane-bound or sytosolic (107).

Two questions remain: Are there any supernatant enzymes in cellulose degradation which are not produced by any kind of "artificial" secondary transformation? And, if there are no "true" supernatant enzymes, in what way is the extracellular protein turnover regulated?

V. EXCRETION OF CELLULOLYTIC ENZYMES

Modern techniques of molecular biology have stimulated work and
contributed to an increasing knowledge and understanding of the
mechanism(s) of microbial cellulose degradation. However, little
work has been done up to now on the transport of cellulases through
the membrane and cell wall, the regulation of spatial distribution,
the anchoring of cellulolytic activities, or the secretion of the cellulases into the medium. As discussed above, the inhibition of the
increase of the concentration of cell-free cellulase in Cellvibrio vulgaris (enzyme release parallels growth) (92) as well as in Bacillus,
strain DLG (enzyme release occurs during the stationary phase)
(56) by protein biosynthesis inhibitors clearly demonstrates that in
these cases enzyme release is not caused by cell autolysis.

The possible existence of true supernatant cellulases during
growth on cellulosic material may be decided in the near future
when it is possible to define the true location of an extracellular
enzyme by its (amino terminal) amino acid sequence. This view is
supported by the following facts: (1) research with yeast (153) as
well as with Bacillus (142,154,155) is approaching this level of knowledge; (2) the number of cellulases with known signal sequences
(e.g., 156 − 162,164) has increased in recent years and an excretion
vector has been constructed (pEAP 37, cf. 163) which increases the
extracellular enzyme concentration about 7- to 12-fold in E. coli
(compared with pEA 2); and (3) the limited proteolytic cleavage of
CBH I by papain performed by van Tilbeurgh et al. (152) clearly
demonstrates the importance of a carboxy terminal peptide (10 kD)
for the localization and specificity of CBH I from Trichoderma reesei:
the residual 56-kD protein no longer binds to insoluble cellulose and
thereby has changed its localization as well as its specificity. Specificity for soluble substrate is maintained after limited proteolysis.

VI. HETEROGENEITY OF CELLULOLYTIC ENZYMES:
ISOENZYMES AND MULTIPLE FORMS

As discussed in Chap. 22, many cellulolytic enzymes from several
aerobic mesophilic bacterial species have been purified from culture
filtrates and their properties characterized. However, few enzymes
have been characterized which bind to the extracellular insoluble
substrate (62,77) and no enzyme has been characterized so far which
is attached to the bacterial surface. In contrast to the mesophilic
anaerobic bacterium Ruminococcus albus, which is reported to produce only one endocellulase (165), most other bacterial culture filtrates contain more than one endocellulase (the phenomenon of multiplicity) and the involvement of bacterial exocellulases in the

degradation of cellulose has been established only recently (53,64, 65,158,159,164,166,167). The involvement of other enzymes, non-hydrolytic in their reaction, was described (169) and is under further investigation. Multiplicity of polymer-degrading enzymes is not restricted to cellulose degradation but is also observed in enzymatic pectin degradation (95,143,170). It has been proposed (171) that several isoenzymes of a similar reaction pattern "will allow a microbe to survive and grow in different environments, e.g., will broaden the host range of a pathogen by circumventing specific plant defense mechanisms." Indeed, it has been shown that in some cases multiplicity of enzyme is caused by the multiplicity of genes (cellulases and glycosidases: 65,161,167,172–174; pectate lyases: 175,176), which is a requirement for the designation "isoenzyme" (177). Other mechanisms to increase the number of multiple forms of cellulases occur in vivo (62,105,122) and have been demonstrated in vitro (62, 122,152). Multiplicity produced by the formation of different mRNAs from one gene has been shown in the case of tyrosine hydroxylase (210) and may principally occur during exoenzyme formation by bacteria.

Glycosylation of cellulases has been reported to occur in *Cellulomonas fimi* (62,65,78) and in *Pseudomonas fluorescens* var. *cellulosa* (101). In the latter microorganism, however, covalent binding of carbohydrate(s) to cellulase has not been verified. It is the different degree of glycosylation which may lead to an increased multiplicity of enzymes (for a more detailed discussion, see Chap. 22). Recent investigations on the biological function of covalently bound carbohydrates in cellulases indicate (207) that O-linked but not N-linked glycosylation is necessary for the secretion of endoglucanase I and II in *Trichoderma viride*. Furthermore, it is assumed that glycosylation is necessary for normal hydrolytic activity of the exoglucanase from *Cellulomonas fimi* (164).

The discovery of Sprey and Lambert (178) that a single protein band, obtained by preparative IEF and confirmed by analytical IEF, can be further separated to at least six bands by the titration curve technique in the presence of a detergent clearly demonstrates that protein purification to true homogeneity needs higher priority. The difficulty in applying classical biochemical methodology to the study of cellulases affords a fine opportunity for the use of molecular genetic techniques. The cloning of cellulase genes from different sources and their expression in *E. coli* (99,173,174,179–183) and yeast (156) will probably lead in the near future to homogeneous enzyme preparations useful for both biochemical investigations and biotechnological applications.

However, the possible involvement in cellulose degradation of nonhydrolytic proteins (169,205) and nonprotein factors (e.g., 151) as well as a cellulose-stimulated behavior of intact cells [attachment to insoluble cellulose (91,145)] have not yet been verified by the use of modern methods of genetics. This is also true for the detection of covalent modifications of the cellulases (e.g., glycosylation, lipidation). Some of these difficulties may be overcome by the use of techniques such as selective hybridization (206).

VII. GROWTH PHYSIOLOGY OF AEROBIC, MESOPHILIC, CELLULOLYTIC BACTERIA

The respiration rate of *Sporocytophaga myxococcoides* has been reported to reach $V_{max}/2$ at 1.4 × 10^{-3} M glucose (184,185). This means that metabolic activity cannot be increased with increasing glucose concentrations above $\sim 3 \times 10^{-3}$ M. The actual metabolic rate with insoluble cellulose corresponds to that with a glucose concentration of about 1 × 10^{-3} M. This suggests that the rate of cellulose hydrolysis is metabolic rate limiting, which implies that no transportable soluble carbohydrates accumulate in the culture. From a comparison of growth rates with different carbon sources (crystalline and amorphous cellulose, cellobiose, and glucose) the same conclusion has been drawn. Furthermore, it appears that the degree of cellulose crystallinity strongly influences the rate of glucose equivalent consumption (186,187) in a *Cellulomonas* species [which has been recognized to be a strain of *C. uda* (113)], in *C. fermentans* (67, 188), and in *C. flavigena* (144). During these experiments a preference of cellobiose by *Cellulomonas* species (186,188; exception: 144) as well as by *Cellvibrio gilvus* (85) has been reported, which led to a more detailed investigation of the metabolism of cellobiose and to the detection of the cellobiose phosphorylase in mesophilic, aerobic bacteria (109,189). This enzyme is also detected in *Cellulomonas* species and has been shown to be located within the cytoplasm (73,74).

Radioactively labeled cellobiose has been synthesized from glucose and glucose-1-phosphate by the use of cellobiose phosphorylase to yield two different ^{14}C-labeled disaccharides. With labeled glucose and unlabeled glucose-1-phosphate resulting, cellobiose is labeled at the reducing glucose moiety; with labeled glucose-1-phosphate and unlabeled glucose the resulting cellobiose is labeled at the nonreducing glucose moiety (109,189,190). Resting cells of *Cellvibrio gilvus* produce CO_2 mainly (80%) from the reducing end of cellobiose (189) and resting cells of *Cellulomonas fimi* (its classification is disputed in this chapter; see above) produce higher specific radioactivity in fermentation products as well from the reducing end of cellobiose (109). Cellobiose in both cases is nonequivalently metabolized by resting cells. The biochemical reasons for this preference are not yet clear, but it is assumed that energy metabolism plays a key role. With *Cellulomonas uda* it was recently shown (117) that a shift from aerobiosis to anaerobiosis, which reflects a shift in energy metabolism, increases the specific activity of enzymes from the Embden–Meyerhoff pathway: fructose-6-phosphate kinase is increased about fourfold, and fructose-1,6-diphosphate aldolase is increased about 2.5-fold.

It has been shown that a cellulolytic enrichment culture of *Sporocytophaga* and *Cytophaga* on lignin-free sulfite pulp from spruce contain a nonmotile eubacterium which can use glucose and cellobiose, but is unable to degrade cellulose (191). This finding suggests (in contrast to those described above) that degradation products from cellulose accumulate in the culture in amounts sufficient to permit adequate growth of the noncellulolytic bacteria. Thus, growth may be limited not by the rate of polymer hydrolysis, but by the rate of product uptake or consumption. Uptake measurements with cellodextrins (glucose up to cellohexaose) using resting cells from *Cellvibrio gilvus* (89) led to the conclusion that, independent of the degree of polymerization of substrate, cells are provided with $3.7-4.2 \times 10^7$ molecules glucose equivalents $\times$ cell^{-1} $\times$ min^{-1}.

During growth on different carbon sources two different carbohydrate storage compounds are accumulated in *Cellulomonas uda*: trehalose (32) and glycogen (41). They are degraded at different rates during carbon starvation and it is indicated (192) that it is trehalose which stabilizes the energy charge and the concentration of living cells during starvation. The possible biotechnological relevance of glycogen, which is produced at about 25% of the cells dry weight at high C/N ratios in batch cultivation, is discussed in the last section.

VIII. ATTEMPTS AT APPLICATION OF THE AEROBIC, MESOPHILIC, CELLULOLYTIC BACTERIA AND THEIR ENZYMES FOR COMMERCIAL UTILIZATION OF CELLULOSE

Up to now few investigations have been published on the biotechnological application of cellulolytic, mesophilic, aerobic bacteria. Most of these have concerned the process of single-cell protein production.

Pure culture investigations have included *Cellulomonas cartalyticum* ATCC 21681, grown on cellulose-containing material (193; 15–25% of dry weight is determined as protein) and a *Cellulomonas* species (194) grown on bagasse (195,196). From these experiments it was concluded that the quality of *Cellulomonas* cell protein compares favorably with the qualities proposed by the United Nations Food and Agricultural Organization (195). With glucose as carbon

source, cells cultivated under Zn^{2+} limitation have drastically reduced nucleic acid contents (197). Protein yields from suitably pretreated cellulosic material were superior to those from untreated material (196). Mixed cultures with *Cellulomonas uda* and either *Alcaligenes faecalis* (195,198,199) or *Candida utilis* (200) on pretreated
bagasse (195,199), pretreated Whatman cellulose (198), and pretreated barley straw (200) were reported to contain different relative
amounts of protein — 23% of dry weight (195), 60% of dry weight
(200), and about 80% of dry weight (199) — presumably due to different methods of protein determination used or to different C/N
ratios used during cultivation [which can result in different contents of carbohydrate storage compounds in *Cellulomonas*; cf. (41)].

Mixed cultures without a fixed nitrogen source but with nitrogen-
fixing microorganisms (201,202) were reported to yield 17 – 19 g nitrogen fixed per gram straw consumed (201) and 7.87 mg nitrogen
fixed per g carbon lost (202). A novel nitrogen-fixing cellulolytic
bacterium (29) was reported to produce 7.5 g protein per 100 g
cellulose corresponding to about 12 mg nitrogen fixed per g of cellulose consumed; these values are comparable to the results mentioned above (201).

The expression of *Cellulomonas fimi* cellulases in the photosynthetic nitrogen-fixing *Rhodobacter capsulatus* was reported (203);
however, no excretion into the extracellular compartment was observed up to now.

In vitro transformation of cellulose into glycogen is shown to
occur with an overall yield of about 10% (168). This process may
also be performed in vivo at appropriate C/N ratios with *Cellulomonas uda* (41).

Possibly the overproduction of *Cellulomonas fimi* exoglucanase
in *E. coli* [20% of total cellular protein is expressed as exoglucanase (164)] open up new biotechnological possibilities, especially
for the in vitro hydrolysis of cellulose.

If the biological function of enzyme complexes or attachment of
cells to insoluble cellulose is to concentrate different enzyme activities at a defined locus to increase the efficiency of cellulose hydrolysis (see discussion above), the two different approaches may
lead to more economic in vitro solubilization of cellulose:

1. The selection for enzyme production by microorganisms which
    do not form cellulase-containing enzyme complexes or which do
    not attach to insoluble cellulose, or
2. A more intensive investigation of the mechanistic role of the
    different kinds of nonhydrolytic factors involved in cellulose
    hydrolysis

IX. PROPOSALS AND PERSPECTIVES
FOR FURTHER WORK

As pointed out in the preceding sections of this chapter, a considerable increase in knowledge on the mechanism of cellulose hydrolysis and a great stimulatory effect on further work has resulted
from research during the last few years using molecular biological
and genetic technologies. However, in the following areas many
questions remain open:

1. Purification and characterization of cellulose-bound as well as
   bacterial surface-bound cellulases
2. Investigations on the mechanism of cellulase repression and induction, including the search for the true inducer(s)
3. Work on the mechanism of cellulase export and of cellulase anchoring on microbial surface, presumably including investigations
   on the mechanism of cellulase modification (e.g., glycosylation,
   lipidation) as well as studies on the regulation of the degradation of extracellular cellulases
4. Studies on the involvement in cellulose degradation of proteins
   and factors of hitherto unknown reactions (e.g., hydrogen
   bond-cleaving enzymes) by the application of the method of
   selective hybridization, and finally,
5. Work on the fate of ether linkage in carboxymethylcellulose,
   which in some cases has been used as carbon source for enrichment cultivations (58,59) or which is reported to be consumed after 6 hr of cultivation (105).

X. ADDENDUM

After completion of this chapter (summer 1987) some monographs
(211,212) as well as numerous special contributions have been published improving our knowledge on the mechanism(s) of microbial
cellulose degradation.

To update this chapter a selection of them is given below arranged according to the subtitles (I–IX) of this chapter.

I. Among other regenerative energy sources the use of biomass
   for energy production is discussed by Kleemann and Meliß
   (213).
III. The rate of β-1,4-glucanase mRNA formation in *Bac. subtilis*
   is increased either by (a) so far unknown intracellular me-
   tabolite(s) which is produced during stationary phase of
   growth or by addition of glucose to the medium (214). In
   contrast transcripts of endoglucanase C gene in *Cell. fimi*
   are not detected during growth on glycerol or on glucose

(229). In the case of *Trich. reesei* cellobiohydrolase I induction evidence is presented that a low constitutive cellulase
system catalyzes the formation of a soluble inducer from insoluble cellulose and that this inducer is effective on the
transcriptional level (215).

IV.   Binding of cellulase to insoluble cellulose (216) and subsequent hydrolyses are obtained by different protein domains
(217,238; 218−220) joined by a region consisting of prolyl
and threonyl residues (217,238). Adsorption of cellulase to
its substrate and 'disorganization' of the cellulose fibre by
domains different from the catalytically active site (220) are
of importance with regard to the concept of synergism (239).

V.   Evidence for the existence of an aminoterminal signal sequence necessary for protein export is presented (214,
221−229) and the function of it in different Gram(+) and
Gram (−) species is documented (222−225).

VI.   Multiplicity of cellulases has been shown to be a consequence
of multiple genes in *Ps. fluorescens subsp. cellulosa* (cf.
227, 231) as well as in *Bacillus* (232). In the case of *Cell.
flavigena* proteolytic degradation may lead to multiple forms
(216).

      One enzyme with different specificities towards polymeric
substrates exists in *Clostr. thermocellum* (233) as well as in
Cell. fimi (236).

VII. Conditions of adhesion of cells (234) as well as of enzymes
(218) to insoluble cellulose have been investigated.

VIII. Cellulases are cloned from different species and expressed
in *Zymomonas mobilis* (222,235), in other Gram(−) bacteria
(236) as well as in *Sacch. cerevisiae* (236) thereby extending their possible range of utilizable carbon sources.

      Glycosylation is no necessary condition for activity but
is essential for stability of cellulase (Exg and CenA of *Cell.
fimi*) against heat (236,237) and against protease treatment
(236−238).

      Important for industrial application of purified cellu-
lase(s) is the phenomenon of 'synergism': in the presence
of β-glucosidase the degree of synergism has been shown
to depend on the concentration of the hydrolases but not
on their ratio (239).

      A fusion protein of *Cell. fimi* endo- (CenA) and exo-
glucanase (Exg) does not bind to insoluble cellulose and
shows a different increase in spec. fluidity vs increase of
reducing sugar concentration in comparison to the not fused
proteins (218).

      Truncation of a 92kDa cellulase up to 46kDa is without
influence on activity as well as on pH optimum (221) of the
enzyme.

IX. For the understanding of the mechanism of cell adhesion (234) to insoluble cellulose it is of interest to define and to characterize components of the bacterial surface involved in this process (240,241).

ACKNOWLEDGMENTS

For many stimulating discussions during the last years I thank my colleagues Dr. Benno Sprey and Dr. Peter Fähnrich. Thanks are due also to Prof. Sahm for supporting my work with the cellulolytic *Cellulomonas*.

REFERENCES

1. L. G. Ljungdahl and K.-E. Eriksson, *Adv. Microb. Ecol.*, *8*, 237 (1985).
   D. E. Eveleigh, In *Current Perspectives in Microbial Ecology* (M. J. Klug and C. A. Reddy, eds,), Am. Soc. Microbiol., Washington, D.C., 1984, p. 553.
3. J.-R. Frisch, *Future Stresses for Energy Resources, Energy Abundance: Myth or Reality?* Graham & Trotman, London, 1986.
4. R. Y. Stanier, M. Doudoroff, and E. A. Adelberg, *The Microbial World*, 2nd ed., Prentice-Hall, Englewood Cliffs, N.J., 1963.
5. R. Y. Stanier, E. A. Adelberg, and J. L. Ingraham, *General Microbiology*, 4th ed., Macmillan, London, 1976, p. 305.
6. H. Stolp and M. P. Starr, In *The Procaryotes* (M. P. Starr, H. Stolp, H. G. Trüper, A. Balows, and H. G. Schlegel, eds.), Springer-Verlag, Berlin, 1981, Vol. 1, p. 135.
7. I. M. Campbell, *Adv. Microb. Physiol.*, *25*, 1 (1984).
8. E. T. Reese and H. S. Levinson, *Physiol. Plant.*, *5*, 345 (1952).
9. M. Mandels and E. T. Reese, *Dev. Ind. Microbiol.*, *5*, 5 (1964).
10. T.-M. Enari and M.-L. Niku-Paavola, *CRC Crit. Rev. Biotechnol.*, *5*, 67 (1987).
11. T. M. Wood, *Biochem. Biophys, Acta*, *192*, 531 (1970).
12. G. Fahraeus, Studies on the cellulose decomposition by *Cytophaga*, Inaugural Dissertation, Univeristy of Uppsala, 1947.
13. C. van Iterson, *Verslag. Akad. Wet. Wis. en Natuurkund.*, *11*, 807 (1903).
14. R. G. H. Siu, *Microbial Decomposition of Cellulose*, Reinhold, New York, 1951.
15. J. A. Gascoigne and M. M. Gascoigne, *Biological Degradation of Cellulose*, Butterworths, London, 1960.

16. R. Y. Stanier, E. A. Adelberg, and J. L. Ingraham, *General Microbiology*, 4th ed., Macmillan, London, 1976, p. 630.
17. H. Krzemieniewski and S. Krzemieniewski, *Bull. Int. de l'Acad. Polonaise des Sciences et de Lettres, Classe Sci. Mathem. Natur.*, *Série B: Sciences Naturelles* (I) 11 (1937).
18. H. Krzemieniewski and S. Krzemieniewski, *Bull. Int. de l'Acad. Polonaise des Sciences et de Lettres, Classe Sci. Mathem. Natur.*, *Série B: Sciences Naturelles* (I) 33 (1937).
19. R. E. Buchanan and N. E. Gibbons, *Bergey's Manual of Determinative Bacteriology*, 8th ed., Williams and Wilkins, Baltimore, 1974.
20. N. R. Krieg, *Bergey's Manual of Systematic Bacteriology*, Vol. 1, Williams and Wilkins, Baltimore, 1984.
21. M. P. Starr, H. Stolp, H. G. Trüper, A. Balows, and H. G. Schlegel (eds.), *The Prokaryotes*, Vol. 1 and 2, Springer-Verlag, Berlin, 1981.
22. K. H. Schleifer and O. Kandler, *Bacteriol. Rev.*, *36*, 407 (1972).
23. H. Reichenbach, *Ann. Rev. Microbiol.*, *35*, 339 (1981).
24. H. Reichenbach, H. Behrens, and J. Hirsch, In *The Flavobacterium-Cytophaga Group* (H. Reichembach and O. B. Weeks, eds.), Verlag Chemie, Weinheim, 1981, p. 7.
25. P. J. Christensen, *Can. J. Microbiol.*, *23*, 1599 (1977).
26. M. Goodfellow and S. T. Williams, *Ann. Rev. Microbiol.*, *37*, 189 (1983).
27. L. V. Kalakoutskii and N. S. Agre, *Bacteriol. Rev.*, *40*, 469 (1976).
28. Y. Yamada, G. Inouye, Y. Tahara, and K. Kondo, *J. Gen. Appl. Microbiol.*, *22*, 203 (1976).
29. P. Schaeffer, *Bacteriol. Rev.*, *33*, 48 (1969).
30. E. Stackebrandt, F. Fiedler, and O. Kandler, *Arch. Microbiol.*, *117*, 115 (1978).
31. E. Stackebrandt and O. Kandler, *Zbl. Bakt. Hyg.*, *I. Abt. Orig. A, 228*, 128 (1974).
32. K.-L. Schimz, K. Irrgang, and B. Overhoff, *FEMS Microbiol. Lett.*, *30*, 165 (1985).
33. E. Stackebrandt and O. Kandler, *Zbl. Bakt., I. Abt. Orig. C, 1*, 40 (1980).
34. L. L. Blackall, A. C. Hayward, and L. J. Sly, *J. Appl. Bacteriol.*, *59*, 81 (1985).
35. A. K. Chatterjee and M. P. Starr, *Ann. Rev. Microbiol.*, *34*, 645 (1980).
36. H. Holloway and M. Morgan, *Ann. Rev. Microbiol.*, *40*, 79 (1986).
37. J. M. Turner and A. J. Messenger, *Adv. Microb. Physiol.*, *27*, 211 (1986).
38. T. Leisinger and R. Margraff, *Microb. Rev.*, *43*, 422 (1979).

39. T. Lessie and P. V. Phibbs, *Ann. Rev. Microbiol.*, *38*, 359 (1984).
40. N. J. Palleroni, In *Genetics and Biochemistry of Pseudomonas* (P. H. Clarke and M. H. Richmond, eds.), John Wiley and Sons, London, 1975, p. 1.
41. K.-L. Schimz and B. Overhoff, *FEMS Microb. Lett.*, *40*, 325 (1987).
42. G. Fahraeus, Studies on the cellulose decomposition by *Cytophaga*, Thesis, University of Uppsala, 1947.
43. R.-M. Griffiths and J. H. Parish, In *The Flavobacterium-Cytophaga Group* (H. Reichembach and O. B. Weeks, eds.), Verlag Chemie, Weinheim, 1981, p. 209.
44. J. B. Waterbury, C. B. Calloway, and R. D. Turner, *Science*, *221*, 1401 (1983).
45. R. V. Greene and S. N. Freer, *Appl. Env. Microbiol.*, *52*, 982 (1986).
46. E. Stackebrandt, H. Seiler, and K. H. Schleifer, *Zbl. Bakt. Hyg.*, *I. Abt. Orig. C*, *3*, 401 (1982).
47. P. Coucke and J. P. Voets, *Ann. Inst. Pasteur*, *115*, 549 (1968).
48. W. T. H. Chang and D. W. Thayer, *Can. J. Microbiol.*, *23*, 1285 (1977).
49. J. J. Marshall, *Carboh. Res.*, *26*, 274 (1973).
50. M. D. Enger and B. P. Sleeper, *J. Bacteriol.*, *89*, 23 (1965).
51. C. R. MacKenzie, D. Bilous, and K. G. Johnson, *Biotechnol. Bioeng.*, *26*, 590 (1984).
52. M. Ishaque and D. Kluepfel, *Can. J. Microbiol.*, *26*, 183 (1980).
53. C. R. MacKenzie, D. Bilous, and K. G. Johnson, *Can. J. Microbiol.*, *30*, 1171 (1984).
54. K. Kluepfel, F. Shareck, F. Mondou, and R. Morosoli, *Appl. Microbiol. Biotechnol.*, *24*, 230 (1986).
55. W. H. van Zyl, *Biotechnol. Bioeng.*, *27*, 1367 (1985).
56. L. M. Robson and G. H. Chambliss, *Appl. Env. Microbiol.*, *47*, 1039 (1984).
57. D. W. Thayer, *J. Gen. Microbiol.*, *106*, 13 (1978).
58. K. Horikoshi, M. Nakao, Y. Kurono, and N. Sashihara, *Can. J. Microbiol.*, *30*, 774 (1984).
59. F. Fukumori, T. Kudo, and K. Horikoshi, *J. Gen. Microbiol.*, *131*, 3339 (1985).
60. S. H. Park and M. Y. Pack, *Enzyme Microb. Technol.*, *8*, 725 (1986).
61. Y. Koide, A. Nakamura, T. Uozumi, and T. Beppu, *Agr. Biol. Chem.*, *50*, 233 (1986).
62. M. L. Langsford, N. R. Gilkes, W. W. Wakarchuk, D. G. Kilburn, R. C. Miller, Jr., and R. A. J. Warren, *J. Gen. Microbiol.*, *130*, 1367 (1984).

63. W. W. Wakarchuk, D. G. Kilburn, R. C. Miller, Jr., and
    R. A. J. Warren, *J. Gen. Microbiol.*, *130*, 1385 (1984).
64. S. G. Withers, D. Dombroski, L. A. Berven, D. G. Kilburn,
    R. C. Miller, Jr., R. A. J. Warren, and N. R. Gilkes,
    Biochem. Biophys. Res. Commun., *139*, 487 (1986).
65. N. R. Gilkes, M. L. Langsford, D. G. Kilburn, R. C. Miller,
    and R. A. J. Warren, *J. Biol. Chem.*, *259*, 10455 (1984).
66. C. Bagnara, C. Gaudin, and J.-P. Belaich, *Biochem. Biophys.
    Res. Commun.*, *140*, 219 (1986).
67. C. Bagnara, R. Toci, C. Gaudin, and J. P. Belaich, *Int. J.
    Syst. Bacteriol.*, *35*, 502 (1985).
68. P. Prasertsan and H. W. Doelle, *Appl. Microbiol. Biotechnol.*,
    24, 326 (1986).
69. J. Antheunisse, *Antonie van Leeuwenhoek*, *50*, 7 (1984).
70. N. Nakamura and K. Kitamura, *J. Ferment. Technol.*, *61*, 379
    (1983).
71. K. Nakamura and K. Kitamura, *J. Ferment. Technol.*, *60*, 343
    (1982).
72. W. Stoppok, P. Rapp, and F. Wagner, *Appl. Env. Microbiol.*,
    44, 44 (1982).
73. K.-L. Schimz, B. Broll, and B. John, *Arch. Microbiol.*, *135*,
    241 (1983).
74. K.-L. Schimz and G. Decker, *Can. J. Microbiol.*, *31*, 751
    (1985).
75. P. Fähnrich, K. Irrgang, B. Rütten, and K.-L. Schimz, *Das
    Papier*, *35*, VI (1981).
76. K.-L. Schimz and B. Broll, submitted.
77. P. Béguin, H. Eisen, and A. Roupas, *J. Gen. Microbiol.*, *101*,
    191 (1977).
78. P. Béguin and H. Eisen, *Eur. J. Biochem.*, *87*, 525 (1978).
79. S. P. Peiris, P. A. D. Rickard, and N. W. Dunn, *Eur. J.
    Appl. Microbiol. Biotechnol.*, *14*, 169 (1982).
80. B. Berg, B. von Hofsten, and P. Pettersson, *J. Appl. Bac-
    teriol.*, *35*, 201 (1972).
81. M. R. Pollock, In *The Bacteria*, Vol. 4 (I. C. Gunsalus and
    R. Y. Stanier, eds.), Academic Press, New York, 1962,
    p. 121.
82. B. Berg, B. von Hofsten, and G. Petterson, *J. Appl. Bac-
    teriol.*, *35*, 215 (1972).
83. B. Berg and A. von Hofsten, *Can. J. Microbiol.*, *21*, 386
    (1975).
84. B. Berg, *Can. J. Microbiol.*, *21*, 51 (1975).
85. F. H. Hulcher and K. W. King, *J. Bacteriol.*, *76*, 565 (1958).
86. F. H. Hulcher and K. W. King, *J. Bacteriol.*, *76*, 571 (1958).
87. W. O. Storvick and K. W. King, *J. Biol. Chem.*, *235*, 303
    (1960).

88. F. E. Cole and K. W. King, *Biochim. Biophys. Acta, 81*, 122 (1964).
89. M. L. Schafer and K. W. King, *J. Bacteriol., 89*, 113 (1965).
90. C. Breuil and D. J. Kushner, *Can. J. Microbiol., 22*, 1776 (1976).
91. E. C. Wynne and J. M. Pemberton, *Appl. Env. Microbiol., 52*, 1362 (1986).
92. L. V. Oberkotter and F. A. Rosenberg, *Appl. Env. Microbiol., 36*, 205 (1978).
93. J. N. White and M. P. Starr, *J. Appl. Bacteriol., 34*, 459 (1971).
94. A. F. El-Helaly, M. K. Abo-El-Dahab, M. A. El-Goorani, and M. R. M. Gabr, *Zbl. Bakt. II. Abt., 134*, 187 (1979).
95. J. L. Ried and A. Collmer, *Appl. Env. Microbiol., 52*, 305 (1986).
96. J. Mergaert, L. Verdonck, K. Kersters, J. Swings, J.-M. Boeufgras, and J. de Ley, *J. Gen. Microbiol., 130*, 1893 (1984).
97. M. H. Boyer, J. P. Chambost, M. Magnan, and J. Cattanéo, *J. Biotechnol., 1*, 229 (1984).
98. M. H. Boyer, J. P. Chambost, M. Magnan, and J. Cattanéo, *J. Biotechnol., 1*, 241 (1984).
99. M. H. Boyer, B. Cami, J. P. Chambost, M. Magnan, and J. Cattanéo, *Eur. J. Biochem., 162*, 311 (1987).
100. H. K. Tewari, R. K. Sedha, and D. S. Chahal, *Ind. J. Exp. Biol., 15*, 692 (1977).
101. H. Suzuki, K. Yamane, and K. Nisizawa, *Adv. Chem. Ser., 95*, 60 (1969).
102. K. Yamane, H. Suzuki, M. Hirotani, H. Ozawa, and K. Nisizawa, *J. Biochem., 67*, 9 (1970).
103. K. Yamane, H. Suzuki, and K. Nisizawa, *J. Biochem., 67*, 19 (1970).
104. K. Yamane, T. Yoshikawa, H. Suzuki, and K. Nisizawa, *J. Biochem., 69*, 771 (1971).
105. T. Yoshikawa, H. Suzuki, and K. Nisizawa, *J. Biochem., 75*, 531 (1974).
106. H. Suzuki, Cellulase formation in *Pseudomonas fluorescens* var. *cellulosa*, In *Symposium on Enzymatic Hydrolysis of Cellulose* (M. Bailey, T.-M. Enari, and M. Linko, eds.), SITRA, Helsinki, 1975, p. 155.
107. J.-T. Hwang and H. Suzuki, *Agr. Biol. Chem., 40*, 2169 (1976).
108. K. Ramasamy and H. Verachtert, *J. Gen. Microbiol., 117*, 181 (1980).
109. M. Sato and H. Takahashi, *Agr. Biol. Chem., 31*, 470 (1967).
110. J. Woodward and A. Wiseman, *Enzyme Microb. Technol., 4*, 73 (1982).

111. R. E. Palmer and R. L. Anderson, *J. Biol. Chem.*, *247*, 3415 (1972).

112. P. Finch and J. C. Roberts, In *Cellulose Chemistry and Its Applications* (T. P. Nevell and S. H. Zeronian, eds.), John Wiley and Sons, New York, 1985, p. 312.

113. E. Stackebrandt and O. Kandler, *Int. J. Syst. Bacteriol.*, *29*, 273 (1979).

114. K. Ogimuto and S. Imai, *Atlas of Rumen Microbiology*, Japan Scientific Societies Press, Tokyo, 1981.

115. M. Higuchi and R. Uemura, *J. Gen. Appl. Microbiol.*, *11*, 137 (1965).

116. M. Higuchi and T. Uemura, *J. Gen. Appl. Microbiol.*, *11* 145 (1965).

117. S. Marschoun, P. Rapp, and F. Wagner, *Can. J. Microbiol.*, *33*, 1024 (1987).

118. B. J. Stewart and J. M. Leatherwood, *J. Bacteriol.*, *128*, 609 (1976).

119. M. Cohn, J. Monod, M. R. Pollock, S. Spiegelman, and R. Y. Stanier, *Nature, 172*, 1096 (1953).

120. I. H. Pastan, In *Current Topics in Biochemistry* (C. B. Anfinsen, R. F. Goldberger, and A. N. Schechter, eds.), Academic Press, New York, 1972, p. 65.

121. J. D. Watson, N. H. Hopkins, J. W. Roberts, J. A. Steitz, and A. M. Weiner, *Molecular Biology of the Gene*, 4th ed., Benjamin-Cummings, Menlo Park, CA, 1987.

122. G. W. Willick and V. L. Seligy, *Eur. J. Biochem.*, *151*, 89 (1985).

123. T. Nisizawa, H. Suzuki, and K. Nisizawa, *J. Biochem.*, *70*, 387 (1971).

124. M. Vaheri, M. Leisola, and V. Kauppinen, *Biotechnol. Lett.*, *1*, 41 (1979).

125. T. Nisizawa, H. Suzuki, N. Nakayama, and K. Nisizawa, *J. Biochem.*, *70*, 375 (1971).

126. A. Jobe and S. Bourgeois, *J. Mol. Biol.*, *69*, 397 (1972).

127. F. Stutzenberger, *Ann. Rept. Ferment. Proc.*, *8*, 111 (1985).

128. M. H. Saier and G. R. Jacobson, *The Molecular Basis of Sex and Differentiation*, Springer-Verlag, New York, 1984.

129. P. Abramoff and M. LaVie, *Biology of the Immune Response*, McGraw-Hill, New York, 1970.

130. M. H. Saier, Jr., D. L. Wentzel, B. U. Feucht, and J. J. Judice, *J. Biol. Chem.*, *250*, 5089 (1975).

131. J. C. Hunt and P. V. Phibbs, Jr., *J. Bacteriol.*, *154*, 793 (1983).

132. M. H. Saier, *Mechanisms and Regulation of Carbohydrate Transport in Bacteria*, Academic Press, Orlando, 1985.

133. U. B. Sleytr and P. Messner, *Ann. Rev. Microbiol.*, *37*, 311 (1983).

134. J. Wahlberg, S. Tynkkynen, N. Taylor, J. Uotila, K. Kuusinen, M. Kuusinen, M. Häggblom, J. Viljanen, R. Villstedt, E.-L. Nurmiaho-Lassila, and K. Lounatmaa, *FEMS Microbiol. Lett.*, *40*, 75 (1987).

135. L. O. Lewis, A. A. Yousten, and R. G. E. Murray, *J. Bacteriol.*, *169*, 72 (1987).

136. T. Kauri and D. J. Kushner, *FEMS Microbiol. Ecol.*, *31*, 301 (1985).

137. J. Chet, S. Fogel, and R. Mitchell, *J. Bacteriol.*, *106*, 863 (1971).

138. S. Tsuyumu, *Nature*, *269*, 237 (1977).

139. S. Tsuyumu, *J. Bacteriol.*, *137*, 1305 (1979).

140. A. Collmer, C. H. Whalen, S. V. Beer, and D. F. Bateman, *J. Bacteriol.*, *149*, 626 (1982).

141. R. Esteban, A. R. Nebreda, J. R. Villaneuva, and T. G. Villa, *FEMS Microbiol. Lett.*, *23*, 91 (1984).

142. J. O. Lampen and J. B. K. Nielsen, *Meth. Enzymol.*, *106*, 365 (1984).

143. T. Andro, J.-P. Chambost, A. Kotoujansky, J. Cattanéo, Y. Bertheau, F. Barras, F. van Gijsegem, and A. Coleno, *J. Bacteriol.*, *160*, 1199 (1984).

144. B. H. Kim and J. W. T. Wimpenny, *Can. J. Microbiol.*, *27*, 1260 (1981).

145. M. Vladut-Talor, T. Kauri, and D. J. Kushner, *Arch. Microbiol.*, *144*, 191 (1986).

146. J. W. Costerton, R. T. Irvin, and K.-J. Cheng, *An.. Rev. Microbiol.*, *35*, 299 (1981).

147. A. P. Pugsley, C. Chapon, and M. Schwartz, *J. Bacteriol.*, *166*, 1083 (1986).

148. V. bon Hofsten, Topological effects in enzymatic and microbial degradation of highly ordered polysaccharides, In *Symposium on Enzymatic Hydrolysis of Cellulose* (M. Bailey, T. M. Enari, and M. Linko, eds.), SITRA, Helsinki, 1975, p. 281.

149. R. Lamed, E. Setter, and E. A. Bayer, *J. Bacteriol.*, *156*, 828 (1983 .

150. S. A. Carpenter and L. B. Barnett, *Arch. Biochem. Biophys.*, *122*, 1 (1967).

151. L. G. Ljungdahl, B. Pettersson, K.-E. Eriksson, and J. Wiegel, *Curr. Microbiol.*, *9*, 195 (1983).

152. H. van Tilbeurgh, P. Tomme, N. Claeyssens, R. Bhikhabhai, and G. Pettersson, *FEBS Lett.*, *204*, 223 (1986).

153. R. Schekman, *Ann. Rev. Cell Biol.*, *1*, 115 (1985).

154. A. Pugsley and M. Schwartz, *FEMS Microbiol. Rev.*, *32*, 3 (1985).

155. P. S. F. Mézes and J. O. Lampen, In *The Molecular Biology of Bacilli*, Vol. 2 (D. A. Dubnan, ed), Academic Press, Orlando, 1985, p. 151.

156. N. Skipper, M. Sutherland, R. W. Davies, D. Kilburn,
     R. C. Miller, Jr., A. Warren, and R. Wong, Science, 230,
     958 (1985).
157. W. K. R. Wong, B. Gerhard, Z. M. Guo, D. G. Kilburn, and
     R. C. Miller, Gene, 44, 315 (1986).
158. G. O'Neill, S. H. Goh, R. A. J. Warren, D. G. Kilburn, and
     R. C. Miller, Gene, 44, 331 (1986).
159. G. O'Neill, S. H. Goh, R. a. J. Warren, D. G. Kilburn, and
     R. C. Miller, Gene, 44, 325 (1986).
160. F. Fukumori, T. Kudo, Y. Narahashi, and K. Horikoshi,
     Gen. Microbiol., 132, 2329 (1986).
161. F. Fukumori, N. Sashihara, T. Kudo, and K. Horikoshi,
     J. Bacteriol., 168, 479 (1986).
162. R. M. MacKay, A. Lo, G. Willick, M. Zucker, S. Baird,
     M. Dove, F. Moranelli, and V. Seligy, Nucl. Acids Res., 14,
     9159 (1986).
163. C. Kato, T. Kobayashi, T. Kudo, and K. Horikoshi, FEMS
     Microbiol. Lett., 36, 31 (1986).
164. G. P. O'Neill, D. G. Kilburn, R. A. J. Warren,
     and R. C. Miller, Jr., Appl. Environ. Microbiol., 52, 737
     (1986).
165. T. M. Wood, C. A. Wilson, and C. S. Stewart, Biochem. J.,
     205, 129 (1982).
166. N. Creuzet, J.-F. Berenger, and C. Frixon, FEMS Microbiol.
     Lett., 20, 347 (1983).
167. N. R. Gilkes, D. G. Kilburn, M. L. Langsford, R. C.
     Miller, Jr., W. W. Wakarchuk, R. A. J. Warren, D. J.
     Whittle, and W. K. R. Wong, J. Gen. Microbiol., 130, 1377
     (1984).
168. T. Sasaki and K. Kainuma, Eur. J. Appl. Microbiol. Biotech-
     nol., 18, 64 (1983).
169. H. Griffin, F. R. Dintzis, L. Kruss, and F. L. Baker,
     Biotechnol. Bioeng., 26, 296 (1984).
170. Y. Bertheau, E. Madgidi-Hervan, A. Kotoujansky,
     C. Nguyen-The, T. Andro, and A. Coleno, Anal. Biochem.,
     139, 383 (1984).
171. B. Schink, In Current Perspectives in Microbial Ecology
     (M. J. Klug and C. A. Reddy, eds.), Am. Soc. Microbiol.,
     Washington, D.C., 1984, p. 580.
172. N. Sashihara, T. Kudo, and K. Horikoshi, J. Bacteriol.,
     158, 503 (1984).
173. N. Nakamura, N. Misawa, and K. Kitamura, J. Biotechnol.,
     3, 239 (1986).
174. B. R. Wolff, T. A. Mudry, B. R. Glick, and J. J. Pasternak,
     Appl. Env. Microbiol., 51, 1367 (1986).
175. A. Kotoujansky, A. Diolez, M. Bacara, Y. Bertheau, T. Andro
     and A. Coleno, EMBO J., 4, 781 (1985).

176. F. van Gijsegem, A. Toussaint, and E. Schoonejans, *EMBO J.*, *4*, 787 (1985).
177. IUPAC-IUB Commission on Biochemical Nomenclature (CBN) Recommendations, *Eur. J. Biochem.*, *24*, 1 (1971).
178. B. Sprey and C. Lambert, *FEMS Microbiol. Lett.*, *18*, 217 (1983).
179. D. J. Whittle, D. G. Kilburn, R. A. J. Warren, and R. C. Miller, Jr., *Gene*, *17*, 139 (1982).
180. N. R. Gilkes, D. G. Kilburn, R. C. Miller, Jr., and R. A. J. Warren, *Biotechnology*, *2*, 259 (1984).
181. K. Nakamura, N. Misawa, and K. Kitamura, *J. Biotechnol.*, *3*, 247 (1986).
182. F. Baras, M.-H. Boyer, J.-P. Chambost, and M. Chippaux, *Mol. Gen. Genet.*, *197*, 513 (1984).
183. A. Lejeune, C. Colson, and D. E. Eveleigh, *J. Ind. Microbiol.*, *1*, 79 (1986).
184. A. Hanstweit, Thesis, Undersøkelse av vekst og energistoffskiftet hos *Sporocytophaga myxococcoides*, University of Bergen, Norway, 1971.
185. J. Goksøyr, G. Eidsa, J. Eriksen, and K. Osmundsvag, A comparison of cellulases from different microorganisms, In *Symposium on Enzymatic Hydrolysis of Cellulose* (M. Bailey, T.-M. Enari, and M. Linko, eds.), SITRA, Helsinki, 1975, p. 217.
186. Z. Dermoun, C. Gaudin, and J. P. Belaich, *Appl. Microb. Biotechnol.*, *19*, 281 (1984).
187. Z. Dermoun and J. P. Belaich, *Biotechnol. Bioeng.*, *27*, 1005 (1985).
188. C. Gagnara, C. Gaudin, and J. P. Belaich, *Appl. Microb. Biotechnol.*, *26*, 170 (1987).
189. E. J. Swisher, W. O. Storvick, and K. W. King, *J. Bacteriol.*, *88*, 817 (1964).
190. T. K. Ng and J. G. Zeikus, *Appl. Env. Microbiol.*, *52*, 902 (1986).
191. B. von Hofsten, B. Berg, and S. Beskow, *Arch. Mokrobiol.*, *79*, 69 (1971).
192. K.-L. Schimz and B. Overhoff, *FEMS Microbiol. Lett.*, *40*, 333 (1987).
193. G. R. Carta, U.S. Patent 3,778,349 (1973).
194. Y. W. Han and V. R. Srinivasan, *Appl. Microbiol.*, *16*, 1140 (1968).
195. V. R. Srinivasan and Y. W. Han, *Av. Chem. Ser.*, *95*, 447 (1969).
196. Y. W. Han and C. D. Callihan, *Appl. Microbiol.*, *27*, 159 (1979).

197. R. J. Summers and V. R. Srinivasan, *Appl. Env. Microb.*,
      37, 1079 (1979).
198. V. R. Srinivasan, Production of bio-proteins from cellulose,
      In *Symposium on Enzymatic Hydrolysis of Cellulose* (M. Bailey,
      T.-M. Enari, and M. Linko, eds.), SITRA, Helsinki, 1975,
      p. 393.
199. J. H. Litchfield, *Science*, *219*, 740 (1983).
200. T. P. Kristensen, *Eur. J. Appl. Microbiol. Biotechnol.*, *5*,
      155 (1978).
201. D. M. Hallsall and A. H. Gibson, *Appl. Env. Microbiol.*,
      50, 1021 (1985).
202. D. A. Veal and J. M. Lynch, *Nature*, *310*, 695 (1984).
203. J. A. Johnson, W. K. R. Wong, and J. T. Beatty,
      J. Bacteriol., *167*, 604 (1986).
204. N. M. Greenberg, R. A. J. Warren, D. G. Kilburn, and
      R. C. Miller, Jr., *J. Bacteriol.*, *169*, 4674 (1987).
205. B. Sprey, unpublished results.
206. M. M. Davis, D. J. Cohen, E. A. Nielsen, M. Steinmetz,
      W. E. Paul, and L. Hood, *Proc. Natl. Acad. Sci. USA*, *81*,
      2194 (1984).
207. C. P. Kubicek, T. Panda, G. Schreferl-Kunar, F. Gruber,
      and R. Messner, *Can. J. Microbiol.*, *33*, 698 (1987).
208. L. A. Weston and R. J. Kadner, *J. Bacteriol.*, *169*, 3546
      (1987).
209. R. Lamed, J. Naimark E. Morgenstern, and E. A. Bayer,
      J. Bacteriol., *169*, 3792 (1987).
210. B. Grima, A. Lamouroux, C. Boni, J.-F. Julien, F. Javoy-
      Agid, and J. Mallet, *Nature*, *326*, 707 (1987).

REFERENCES ADDED IN PROOF

211. J.-P. Aubert, et al., Biochemistry and Genetics of Cellulose
      Degradation, Academic Press, New York, 1988.
212. W. A. Wood and S. T. Kellogg, Methods in Enzymology,
      Biomass, Part A, Cellulose and Hemicellulose, Academic
      Press, San Diego, 1988.
213. M. Kleemann and M. Meliß, Regenerative Energiequellen,
      Springer Verlag, Berlin, 1988.
214. L. M. Robson and G. N. Chambliss, *J. Bacteriol.*, *169*, 2017
      (1987).
215. S. El-Gogary, et al., *Proc. Natl. Acad. Sci. USA*, *86*, 6138
      (1989).
216. A. J. Sami, et al., *Enzyme Microb. Technol.*, *10*, 626
      (1988).

217. N. R. Gilkes, et al., *Journ. Biol. Chem.*, *263*, 10401 (1988).
218. R. A. J. Warren, et al., *Gene*, *61*, 421 (1987).
219. P. Tomme, et al., *Eur. Journ. Biochem.*, *170*, 575 (1988).
220. J. Ståhlberg, et al., *Eur. J. Biochem.*, *173*, 179 (1988).
221. F. Fukumori, et al., *FEMS Microbiol. Lett.*, *40*, 311 (1987).
222. N. Brestic-Goachet, et al., *J. Gen. Microbiol.*, *135*, 893 (1989).
223. F. W. Paradis, et al., *Gene*, *61*, 199 (1987).
224. J. B. Owolabi, et al., *Appl. Environm. Microbiol.*, *54*, 518 (1988).
225. N. Din, et al., *Arch. Microbiol.*, *153*, 129 (1990).
226. G. Joliff, et al., *Appl. Environm. Microbiol.*, *55*, 2739 (1989).
227. J. Hall and H. J. Gilbert, *Mol. Gen. Genet.*, *213*, 112 (1988).
228. R. Messner, and C. P. Kubicek, *FEMS Microbiol. Lett.*, *50*, 227 (1988).
229. B. Moser, et al., *Appl. Environm. Microbiol.*, *55*, 2480 (1989).
230. C. A. West, et al., *FEMS Microbiol. Lett.*, *59*, 167 (1989).
231. A. Lejeune, et al., *Biochim. Biophys. Acta*, *950*, 204 (1988).
232. F. Fukumori, et al., *Gene*, *76*, 289 (1989).
233. J. Hall, et al., *Gene*, *69*, 29 (1988).
234. E. J. Morris, *FEMS Microbiol. Lett.*, *51*, 113 (1988).
235. A. Lejeune, et al., *FEMS Microbiol. Lett.*, *49*, 363 (1988).
236. C. Curry, et al., *Appl. Environm. Microbiol.*, *54*, 476 (1988).
237. M. L. Langsford, et al., *FEBS Lett.*, *225*, 163 (1987).
238. N. R. Gilkes, et al., *Journ. Biol. Chem.*, *264*, 17802 (1989).
239. J. Woodward, et al., *Biochem. J.*, *255*, 895 (1988).
240. U. B. Sleytr and P. Messner, *J. Bacteriol.*, *170*, 2891 (1988).
241. S. F. Koval, *Can. Journ. Microbiol.*, *34*, 407 (1988).

20
Cellulose Degradation by Thermophilic Aerobic Bacteria

FRED J. STUTZENBERGER   *Clemson University, Clemson, South Carolina*

I. INTRODUCTION

Cellulose is the world's most abundant organic material (1). Therefore, enzymatic degradation of this polymer is crucial to the cycling of carbon within both the marine and terrestrial habitats. Microorganisms (both aerobic species such as those found in well-aerated soils and composts, and anaerobes such as those active in the herbivore rumen) are responsible for most of the cellulose carbon recycling (2). Although the ecological importance of cellulose decomposition has long been recognized, most of the recent interest in cellulolytic systems has been stimulated by the prospects for commercial utilization of cellulases in bioconversion processes. The majority of these studies on cellulase biosynthesis and activity have dealt with the mesophilic fungi, of which *Trichoderma reesei* has been the center of attention (reviewed in Ref. 3). However, a serious limitation envisioned in the large-scale application of mesophilic fungal cellulases in biomass conversion may be their relatively poor temperature stability under usage and recycling conditions (4,5). The advantages of using thermophiles for cellulase production (reviewed in Ref. 6) can be briefly listed as follows: (1) lower cooling costs in large-scale cultures; (2) increased mass transfer due to decreased fluid viscosity at higher temperatures; (3) reduced risk of contamination; (4) enhanced volitalization of products (e.g.,

ethanol) at higher temperatures; (5) nonpathogenicity of thermophiles, which in most cases cannot grow at 37°C; and (6) a lesser
degree of enzyme inactivation during recovery, manipulation, and
storage of enzymes. Disadvantages include: (1) greater stress on
biotechnology hardware such as control sensors; (2) excessive evap-
oration (may be advantageous for concentration of products); (3)
lower tolerance to products such as alcohol; and (4) decreased oxygen solubility at high temperatures. This last point is of particular
significance for the obligately aerobic thermophiles, since oxygen
solubility at 50°C is only about three-fourths that at room temperature (7).

Contrary to what might be expected, the reaction rates of thermostable enzymes from thermophiles are generally no greater than
those of similar enzymes obtained from mesophiles when both are
measured under optimal conditions (8). However, since most industrial enzymes are employed at temperatures over 50°C (9), the advantages of using thermostable enzymes from thermophilic organisms
outweigh the disadvantages. The biotechnological application of
thermophiles and their thermostable enzymes seem promising, but
there is currently no extensive large-scale use mainly because detailed research on thermophilic systems has not yet been carried out
(10). Therefore a thorough understanding of thermophily at the
molecular level is essential to the selection and development of commercially useful strains. In the next section, some bases for thermophily and thermostability are discussed with emphasis on the aerobic thermophilic cellulolytic bacteria.

II. THERMOPHILY AND THERMOSTABILITY

A thermophile is an organism capable of growth at high temperatures. However, since the maximal temperature for growth depends
on the group of organisms under consideration, the more precise
definition would be an organism capable of growing at temperatures
near the maximum for its taxonomic group (11). These temperatures
maxima are about 50°C for vertebrates, insects, crustaceans, and
plants, about 55°C for the protozoa and eukaryotic algae, and about
60°C for the fungi (12). The thermophilic microbes have classically
been defined as those with an optimal growth temperature at 55°C
or above and a minimal growth temperature above 37°C. However,
the optimal temperatures for thermophilic growth vary widely. Table
1 categorizes thermophiles by cardinal growth temperatures. Since
there are a few fungal and algal species which grow at temperatures
above 55°C, this ecological niche is dominated by the prokaryotes

TABLE 1   Classes of Thermophiles Defined by Cardinal
Temperatures

	Cardinal temperatures (°C)		
Class	Minimal	Optimal	Maximal
Thermotolerant	<30		
Facultative thermophilic		<45	>45
Moderate thermophilic			<70
Thermophilic	⩾40	⩾55	⩾60
Extremely thermophilic	⩾40	⩾65	>70, <110

Source: Data from Refs. 12 – 14.

(15). True thermophily appears most common among the endospore-
forming rods (15 recognized *Bacillus* species and 11 recognized
Clostridium species), the actinomycetes (such as *Thermomonospora*
and *Thermoactinomyces*), the archaebacteria (at least 20 species including the methanogens, thermoacidophiles, and various species
which metabolize inorganic sulfur compounds), and the cyanobac-
teria (such as *Synechococcus* species), formerly classified as "blue –
green algae." However, even bacteria morphologically similar to the
mesophilic pathogens (such as streptococci, staphylococci, and spi-
rochetes) have their thermophilic representatives (16).
      The molecular and structural bases for thermophily have not
been well defined, although the ability of a species to grow at elevated temperatures is obviously a reflection of multiple factors. One
of the major differences between mesophiles and thermophiles lies in
the compositions of their cytoplasmic membranes (recently reviewed
in Ref. 17). Membrane functions, such as nutrient import and exoenzyme export, require this barrier to be in a semifluid state (18,
19). The fluidity of membrane lipids is a function of their chemical
structure and ambient temperature. The lipids largely determine
the molecular architecture of the membrane. They contribute the
hydrophobic characteristics of the matrix and provide the appropriate fluidity for the function of membrane proteins such as enzymes and carriers. The most common apolar chains in thermophilic bacterial lipids are fatty acids. In thermophiles they character-
istically contain longer hydrocarbon chains and more methyl-branched

chains, while unsaturated hydrocarbon chains, which have lower melting points, are rare (20,21). Membrane fatty acid composition can be markedly altered by the growth substrate (22,23) and temperature (24,25). In several thermophiles, growth on acetate or n-heptadecane as sole carbon/energy source caused a doubling in the straight-chain/branched-chain fatty acid ratio compared to growth on glucose or glycerol (26), while in others the percentage of membrane branched-chain fatty acids decreased and chain length increased at higher growth temperatures (27). This ability to synthesize lipids for the alteration of membrane fatty acid composition, thereby maintaining the appropriate membrane fluidity at a particular growth temperature, has been termed "homeoviscous adaptation" (28) and apparently is a characteristic of a variety of microbes (29).

The thermophilic aerobes, notably the genera *Bacillus*, *Thermus*, and *Thermomicrobium*, have provided a number of interesting subjects for membrane studies. *Bacillus stearothermophilus* membrane fatty acids contain 34–64% iso-C_{15} and iso-C_{17} monomethyl-branched chains (30). The more extremely thermophilic *B. caldolyticus*, *B. caldovelox*, and *B. caldotenax* contain iso-C_{15}, -C_{16}, and -C_{17} for about 80% of their total fatty acids. Increasing the growth temperatures causes a shift from iso-C_{15} to iso-C_{17}, and from iso-C_{16} to straight-chain C_{16} (31). In the extremely thermophilic *Thermus* species about 85% of the fatty acids are present as iso-C_{15} and iso-C_{17}; raising the growth temperature also brings about chain elongation (30). However, in *Thermomicrobium roseum* there are few iso-branched chains. About two-thirds of these are a 12-methyl C_{18} fatty acid. Furthermore, this thermophilic aerobe is the only bacterium reported to date which contains long-chain ($C_{18}-C_{23}$) diols rather than glycerolipids (32). Other exceptions occur in the thermoacidophilic bacilli in which a large percentage of the fatty acids possess cyclohexyl or cycloheptyl rings at the terminal ends of the hydrocarbon chains (33,34). The diversity of membrane lipids in the aerobic thermophiles and the relatively small number of species studied make it difficult to generalize about the relationships of composition to temperature stability. A better knowledge of membrane structure and composition is essential not only to our understanding of thermophily in general, but also to clarification of mechanisms by which enzymes such as cellulases are exported to the environment. The action of membrane-altering detergents such as Tween 80 stimulates the liberation of cellulase components from at least one thermophilic bacterium (35). Whether detergent-induced membrane effects stimulate cellulase secretion and whether temperature-induced changes in membrane composition affect the liberation of cellulases from other thermophiles are intriguing questions yet to be answered.

Another explanation of thermophily stems from the thermostability of enzyme systems. While enzymes and other proteins of thermophiles are similar to those of mesophiles in many characteristics (molecular weight, aggregation state, allosteric effectors, and coenzyme requirements), small differences in amino acid composition and sequence apparently result in significant increases in thermostability (reviewed in Ref. 36). Furthermore, the amino acid compositions of thermophilic depolymerizing exoenzymes may be markedly similar within a species. For example, in the cellulolytic actinomycete *Thermomonospora fusca*, two 1,4-β-endoglucanases and a polygalacturonate lyase have compositions within 1% of each other for 10 amino acids; glycine and aspartic acid were the most abundant amino acids in all three of those extracellular enzymes (37,38). Recently, Suzuki et al. (39) reported an interesting correlation between proline content and extracellular enzyme stability. A comparison of oligo-1,6-glucosidases secreted by a variety of *Bacillus* species revealed that increased proline content conferred higher enzyme tolerance not only to heat but to several denaturing agents as well. However, it is difficult to generalize in regard to the stability of the thermophiles' intracellular enzymes vs. those of the mesophiles. Early studies comparing a variety of enzymes in crude extracts from mesophilic and thermophilic species of *Bacillus* (40,41) and *Clostridium* (42−45) established (with some notable exceptions; 46−48) a relationship between thermophily and enzymatic thermostability. This thermostability could not be attributed to the presence of stabilizing factors in crude cell extracts based on a number of observations (49). Characterization of intracellular enzymes in extracted preparations for some thermophiles may be misleading as to their abilities and limitations in their native states. In *Thermonospora curvata*, the β-glucosidase (a largely intracellular enzyme apparently essential for growth on cellulose since the actinomycete does not produce a cellobiose phosphorylase) has a half-life at 65°C of 13 min in intact cells, 8 min in crude cell-free extracts, and 1.5 min in a threefold purified preparation (Rene Bernier, pers. commun.). Clearly, the instability of the essential enzyme even in intact cells offers one explanation why this actinomycete cannot grow above 60°C. Likewise, the cyclic AMP phosphodiesterase isozymes from this actinomycete were thermostable in crude extracts but were rendered very labile when subjected to purification by size exclusion or ion exchange chromatography (50). The lability of these enzymes in the partially purified state indicates that their thermostability depends in large part on their cellular environment rather than on any intrinsic property of their protein structure. Thermostabilization of the intracellular glucose-6-phosphate dehydrogenase from *Bacillus coagulans* by high ionic strength (provided by 8% ammonium

sulfate or 10% sodium chloride) may provide another reflection of
this environment dependence (49). It had earlier been suggested
(46) that rapid resynthesis of essential enzymes could account for
the ability of thermophiles to grow at temperatures which denature
those enzymes. However, subsequent comparisons of protein synthesis and growth rates in mesophiles and thermophiles did not support this suggestion (51–54).

In the endospore-forming aerobic bacilli, thermoadaptation also
apparently plays a role in the synthesis of thermostable enzymes.
Early studies by Campbell (55,56) showed that an "obligate" thermophile could be grown at 36°C in some media and that the thermostability of α-amylase in *Bacillus* was dependent on the temperature
at which the cells were cultured. A later study on the lactate dehydrogenase of *B. caldotenax* showed that thermoadaptation involved
the shift between two forms of the enzyme having the same substrate
specificity but different specific activities and thermostabilities (57).
As pointed out by Amelunxen and Murdock (49), these differences
may be the reflection of induction/repression of separate genes coding for the two (or more) enzymatic forms, or some posttranslational
modification triggered by a temperature shift. The cloning of such
genes into mesophilic hosts should be a useful tool in evaluating the
effects of the two mechanisms and in determining the contributions
of specific amino acid sequences to thermostability. For example,
Gray et al. (58) recently reported the DNA sequences for the thermostable α-amylases of *Bacillus stearothermophilus* and *B. licheniformis* cloned into *Escherichia coli*. The coding and deduced polypeptide sequences were about 60% homologous, with the major difference being a 32-residue COOH terminal tail in the *B. stearothermophilus* sequence. Both genes were expressed in *E. coli* with significant extracellular activity characteristic of those expressed in
the parent organism. A hybrid plasmid constructed in which the
B. stearothermophilus promoter and NH_2 terminal two-thirds of the
coding sequence were fused out of frame to the entire mature coding sequence of the *B. licheniformis* gene. Transformation of *E. coli*
with this plasmid resulted in recipients which expressed hybrid α-
amylases, several of which were characterized as to thermostability.
It was found that increases in the length of the *B. stearothermophilus* NH_2 terminal region decreased hybrid enzyme stability relative to the *B. licheniformis* amylase but increased the relative specific activity. The authors suggested that while site-directed mutagenesis is currently the most widely employed method for changing the catalytic properties, random generation of hybrid genes followed by characterization of their hybrid products may be more
easily controlled for alteration of enzymes within a closely related
family. Such gene hybridization techniques may be particularly in-
formative in terms of the differential properties of closely related

cellulase components which have large differences in temperature optima (such as the two endoglucanases purified from *Thermomonospora fusca* by Calza et al., 37).

In addition to the strategy just described (in which genes coding for thermostable enzymes are hybridized and cloned into mesophilic recipients), an alternate strategy for the investigation of thermostability-conferring characteristics was recently described by Liao et al. (59). Their strategy consists of cloning a gene encoding an enzyme from a mesophile into a thermophile, then selecting for the enzymatic activity at the higher growth temperatures of the host organism. Their model system was based on the enzyme kanamycin nucleotidyltransferase, which is carried on the *Staphylococcus aureus* plasmid pUB110 (60). An initial difficulty was encountered in that the *S. aureus* plasmid was unstable in *B. stearothermophilus* growing above 47°C. This problem was circumvented by constructing a chimeric plasmid containing the heat-stable origin or replication from a large (80-kb) cryptic plasmid of a kanamycin-sensitive *B. stearothermophilus* with the Kan[r] marker of pUB110. Transformants carrying the chimeric plasmid pBST2 retained their kanamycin resistance well at temperatures up to 60°C. Plating of these transformants on kanamycin-containing medium at 63°C yielded a number of mutants without restoring to mutagenic agents. The nucleotide sequences coding for thermostable kanamycin nucleotidytransferases were compared to that of the parent sequence. It was found that the basis for increased thermostability in most of the mutant enzymes was a substitution of tyrosine for aspartate at position 80. In two of the mutants, further thermostability was conferred by the additional change of threonine to lysine at position 130. Both of these mutations occurred in regions predicted to form αhelices. The authors suggested that the separate changes at positions 80 and 130 acted independently and additively to stabilize the enzyme against thermal denaturation. Such a strategy for selecting thermostable varients of cellulase components would obviously be more complicated since a complex of enzymes is required for the effective utilization of cellulose, but an increase in the thermostability of the most labile component by this method may be successfully attempted.

The following discussion of cellulolytic microbes will be limited, as the title of this chapter indicates, to the aerobic, thermophilic bacteria. Table 2 lists the representative characterized species composing this group. Since the cellulolytic fungi far outnumber the cellulolytic bacteria (lists illustrating this inequity have been compiled; 62,63), the majority of the cellulose degraders have been excluded. The bacteria genus most highly graced by scientific attention is undoubtedly *Clostridium* (see Chap. 18). Its aversion to oxygen excludes it from treatment here (although much of our understanding of the thermophily has been, and undoubtedly will be,

TABLE 2   Aerobic and Facultative Anaerobic Eubacteria with
T_{max} $\geqslant 60°C$ and T_{opt} $\geqslant 50°C$

Organism	T_{max}	T_{opt}	T_{min}	Comments; possible application (isolated from)
Thermotrix thiopara	85	72	55	Hot sulfur spring with neutral pH facultative chemolithoautotroph, pH_{opt} 6.8, facultative anaerobe, sulfur metabolism
Bacillus stearo-thermophilus	70–77	50–65	35	Ubiquitous; species includes a wide variety of strains yielding a markedly heterogeneous taxon
Thermus aquaticus[a]	79(85)	70	45	Ubiquitous in hot water lines (hot spring, Yellowstone National Park)
B. schlegelii	80	70	38	H_2 oxidizer, CO oxidizer (sediments of eutrophic lake, Switzerland)
Hydrogenobacter thermophilus	78	70–75	40–43	Soil from hot springs (Japan); H_2 oxidizer
B. acidocaldarius	75	65	35	pH range 2–5 (Yellowstone National Park, Hot Springs UDSST)
Thermomicrobium fosteri	75	60	—	Not accepted as a valid species in recent studies; alkane utilizer
Flavobacterium thermophilum	75	55–60	40	Hydrogen oxidizer (hot spring Almo-Ata UDSSR)
Chloroflexus aurantiacus	70–73	55	35	Phototrophic hot springs (Yellowstone National Park)

TABLE 2   (Cont.)

Organism	T_{max}	T_{opt}	T_{min}	Comments; possible application (isolated from)
Thermoleophilum album	70	60	45	pH_{opt} 7.0, alkane (13−20°C) utilizer (thermal and non-thermal mud, U.S.)
Thermus ruber	70	60	35	Hot springs (Yellowstone National Park)
B. thermoglucosidasium	69	61−63	42	Exo-oligo-1,6-glucosidase producer, pH 6.5−8.5 (soils; Japan)
B. tusciae	65	55	40	H_2 oxidizer, acidophilic, pH_{opt} 4.5 (geothermal manifestation; Tuscani, Italy)
Pseudomonas thermocarboxydovorans	65	50	37	(Aerobic sewage sludge, England)
Methylococcus thermophilus	62	56	37	Obligate methanotroph (UDSSR)
Calderobacterium hydrogenophilum	—	—	—	H_2 oxidizer (UDSSR)
		Actinomycetes[b]		
Thermoactinomyces vulgaris	70	60	37	
Streptosporangium album (var. *thermophilum*)	70	50−55	—	
Thermomonospora curvata[a]	65	50	37	Cellulolytic
T. fusca	65	50−55	35	Cellulolytic

TABLE 2   (Cont.)

Organism	T_{max}	T_{opt}	T_{min}	Comments; possible application (isolated from)
Thermoactino- myces sacchari	65	55 – 60	37	Causes bagassosis (respiratory disease)
T. dichotomicus	65	50 – 58	—	
Micropolyspora rectivigula[c]	65	45 – 55	—	
Streptomyces thermo diastaticus	65	50 – 60	37	
S. eurythemus	65	45 – 55	—	
S. thermo- nitrificans	60	50	37	
Thermoactino- myces candidus	60	—	37	Not starch-hydrolyzing
Microbiospora thermorosea	60	55	—	
M. bispora	60	50 – 60	—	Cellulolytic
M. aerata	60	55	30	
M. thermo- diastatica	60	55	—	
Saccharomono- spora viridis	60	50	—	
Streptomyces thermovulgaris	60	50	—	
S. thermovio- laceus subsp. thermoviolaceus subsp. apingens	60	50	35	
Thermoactino- myces peptono- philus	60	50	—	
T. intermedius	60	—	—	Amylase-negative

TABLE 2 (Cont.)

Organism	T_{max}	T_{opt}	T_{min}	Comments; possible application (isolated from)
Micropolyspora faeni	60	50	70	
Pseudonocardia thermophila	60	40	25	
Streptomyces albus	60	40	25	

[a]Other "thermophilic," validly published *Thermomonospora*, *T. alba*, *T. mesophila*, and *T. mesouviformis* have a T_{max} and T_{opt} below 60 or 50°C, respectively.

[b]The taxonomy of the thermophilic actinomycetes is presently under investigation. The Approved List of validly published organisms at the present time do not represent the diversity and systematic standing of this group.

[c]*Bergey's Manual* (8th ed.) recognized also as thermophilic species: "*M. rubrobrunea*," T_{max} 65°C; "*M viridinigra*," T_{max} 65°C; "*M thermovirida*," T_{max} 57°C; "*M. caesia*," T_{max} 55°C.
Source: Taken in part from Ref. 61.

drawn from studies on the clostridia and other cellulolytic, anaerobic thermophiles). The survivors remaining to fill this chapter are relatively few yet notable in their ecological and metabolic diversity. The single most productive ecosystem for these cellulolytic aerobic thermophiles has been the humble compost pile. In the next section, this microbial melting pot is described to the extent that its complexity and variability allows.

III. COMPOSTING AS AN AEROBIC THERMOPHILIC ENVIRONMENT

The composting environment is the habitat *par excellence* of the aerobic, thermophilic cellulose degraders. During the microbial degradation of large masses of well-aerated organic materials,

temperatures pass through a mesophilic to a thermophilic stage and
then decline to yield a stabilized compost. Compost usually cannot
compete economically with inorganic fertilizer as a source of plant
macronutrients (64). However, the ecological advantages of the
composting process as a method for the bioconversion and stabilization of municipal solid wastes (which contain about 50% cellulose)
and other types of organic refuse have prompted a variety of
studies (reviewed in Ref. 65). The thermophilic phase (which often
exceeds 70°C; 2) kills pathogenic microbes, larvae, and weed seeds
in composting material, while disagreeable odors, insect breeding,
and vermin harborage are eliminated. This pathogen elimination factor is particularly important when concentrated sewage sludge is
mixed with the refuse to be composted (66). Since the bulk density
of compost is higher than that of the starting material, compost
makes a physically stable landfill material which undergoes little re-
heating after relocation (67). The return of compost to the soil re-
cycles mineral plant nutrients which would be lost by other waste
disposal methods. Composting narrows the carbon/nitrogen ratio,
thereby reducing the possibility of damaging crops through soil nitrogen immobilization. In certain specialized agricultural practices
(such as edible mushroom cultivation) compost is an adequate replacement for the traditional manure preparations (68). Mushroom
spawn is a sensitive indicator of metal toxicity in compost (69); although the levels of toxic metals may be raised by the addition of
sewage sludge, municipal solid waste compost moderates metal toxicity in soils (70). Rapid decomposition of certain pesticides is an
added ecological benefit of composting (71).

The chemistry, physics, and microbiology of composting are obviously complex due to the heterogeneity of the starting materials,
the effects of zonation in the composting mass (thus creating micro-
environments), and the variability of natural conditions under which
composting is generally carried out. In most other ecosystems, the
heat liberated from microbial metabolic activity is dissipated too rapidly to cause significant heating of the area. In contrast, decom-
posing organic materials, when present in sufficient volume with
adequate oxygen and water content, generate intense heat and retain it through the self-insulating properties of the composting mass.
The metabolic activity of plant cells may contribute a tiny fraction
to the self-heating of agricultural biomasses, but the establishment
of microorganisms as the prime movers in the composting process
has been acknowledged for over a half celtury (72). While these
properties of self-heating and insulation are useful in the processing of tobacco, cacao, and fodder (73), they are all too obvious
during the spoilage of improperly stored agricultural products such

as hay, grain, and wool. Under enclosed storage conditions, spontaneous combustion with loss of the storage structure is often the result.

Since most naturally composting materials contain a high percentage of cellulose, conditions which promote the rapid establishment of cellulolytic populations lead to the most rapid bioconversion and stabilization of the mass. The relative humidity of the initial mass has an important influence on the bacterial/fungal population mix. An early review by Scott (74) quantitatively summarized the frequent observations that fungi grow better than bacteria at low moisture content. Most bacteria require relative humidities near 100% for maximal growth rate, while the fungi have their optima at 90% or below. The influence of this differential has been studied most closely in the composting of wool (75) and hay (76). At relative humidities around 90%, the fungi are able to slowly bring the temperature of the composting mass up to maxima of 40−53°C. The upper limit for the growth of thermophilic fungi (about 60°C; 77) would limit the potential for self-heating of dry biomass materials, and the attainment of higher temperatures would require increases in relative humidity to those preferred by the thermophilic bacteria. In large masses, some of this water necessary for the transition from fungal to bacterial predominance is a terminal metabolic product of the fungi themselves (75). Once the water necessary for bacterial activity is supplied, temperatures reach 70°C or greater within 48 hr in field scale composting operations (66). Fungal populations are then reduced to below detectable levels in the bulk of the composting material except under circumstances in which the combination of high humidity and moisture lasts only a few hours (78,79). However, at the surface of composting masses, where the material is drier and cooler, the fungi survive as detectable populations (80,81).

The aerobic nature of microbial populations necessary for rapid composting has been revealed by a variety of studies (79,82,83). Withholding air from the composting mass quickly reduces heat output, whereas ample air supply allows the process to proceed to a state in which biodegradable material is at a minimum and the compost is stable as determined by a variety of criteria (84). This requirement for aerobiosis in rapid composting is usually met by mechanical turning several times a week (in field scale operations) or by forced aeration using heated and humidified air (in automatically controlled adiabatic devices). While the typical composting temperature profile (mesophilic → thermophilic → mesophilic) can be readily observed under field conditions, the requirement for mechanical turning can temporarily interrupt the cycle. The use of adiabatic

devices allows a more consistent heat output profile. The profiles are
quite similar for a variety of materials (straw, hay, municipal solid
waste/sewage sludge mixtures, and raw wools) studied under controlled conditions (summarized in Ref. 65). Heat output appears to
occur in two stages —the heating rate maximum occurring at about
40°C (initial mesophilic stage), followed by a minimum at about 50°C
brought about by the activity of the newly established thermophilic
populations.

The pH compost during the initial mesophilic stage is character-
ictically rather acidic (pH 4-6). The consumption of organic acids
such as acetic and butyric and the release of ammonia through mineralization of organic nitrogen takes the pH of composting material
into the slightly alkaline range (pH 7-9). This trend is apparent
for a variety of compostable materials with some notable exceptions
(reviewed in Ref. 65). Trends toward increased pH as well as temperature favor the emergence of the thermophilic bacteria and decline of the initial fungal populations. The aeration rate strongly
influences the pH profile; Suler and Finstein (85) found that weak
aeration or extremely thermophilic temperatures (over 70°C) in a
small-scale controlled compost production unit resulted in material
with acidic pH throughout the process. In large mass composting
under field conditions, distinct oxygen gradients are established.
A detailed study by Wiley and Spillane (83) of large masses of solid
waste compost under field conditions showed that interstitial gas
30-60 cm from the surface contained 7.0 and 1.0% by volume. After
turning and regrinding the material, the oxygen contents were increased to the range of 10-19% with a corresponding decrease in
temperature from 68 to 40°C. However, within 5 hr the metabolic
activity of the thermophilic bacteria had raised the temperature to
56°C and reduced the oxygen content of the interstitial gas within
the center of the windrow to about 1%. The maintenance and rate
of reestablishment of thermophilic temperatures in composting material is directly proportional to oxygen consumption (86,87). The
availability of oxygen and its subsequent influence on the establishment of thermal zones within the composting mass leads to the emergence of well-defined dominant populations of thermophilic actinomycetes. Figure 1 shows such a zone [referred to as a white limelike
coating (88), chalky white encrustation (89), or fire-fang (80), in
earlier reports] in the cross-section of a municipal solid waste windrow at the composting plant operated jointly by the U.S. Public
Health Service and the Tennessee Valley Authority at Johnson City,
Tennessee. Only one actinomycete, identified as *Thermonospora
curvata*, could be consistently isolated in high numbers from this
mycelial zone (90,91), although other cellulolytic bacteria and fungi

FIGURE 1   Zone of thermophilic actinomycete growth (light area
indicated by arrows) in composting lignocellulosic waste.

were isolated from other parts of the windrow cross-section. The
next section describes this and other actinomycetes, an extremely
diverse group of filamentous bacteria. As pointed out by Gottlieb
(92), this morphological feature is tenuous, often requiring imagi-
nation to accept it as a taxonomic basis. To the bacterial taxono-
mists, the actinomycetes have been traditionally a strange group of
organisms, for the most part neglected by medical bacteriologists,
physiologists, and biochemists alike. Alien though they may be,
their recognized role in the decomposition of organic matter in

nature (reviewed in Ref. 93) qualifies them for considerable attention in any discussion of the aerobic, cellulolytic thermophiles.

IV. ACTINOMYCETES

The actinomycetes comprise a morphologically, physiologically, and ecologically diverse group of prokaryotes in which taxonomic relationships remain tenuous at best (94). They can be isolated from a wide range of soil types under various conditions and often outnumber the populations of other bacteria and fungi in soil samples, particularly during dry seasons (95,96). Their predominance is most marked in grasslands and volcanic soils (97), and in such habitats cellulolytic activity is exhibited by one-third to one-half of the isolates (98). Cattle grazing and crop cultivation practices significantly affect both total actinomycete population and prevalance of cellulase production. An important ecological niche for the aerobic thermophilic actinomycetes is in heating plant biomass such as hay, bagasse, timber wastes, and composts (80,88,90,99 – 103). These environments provide high concentrations of lignocellulosic materials. Many actinomycete species are capable of secreting amylases, cellulases, hemicellulases, pectinases, and lignolytic enzymes (103 – 114). This extracellular enzymatic versatility for biopolymer degradation is responsible in large part for the efficiency and predominance of actinomycetes in the composting environment (115).

Because of commercial interest in antibiotic production, the streptomycetes have received more attention than other actinomycete groups. At present there are about 500 recognized *Streptomyces* species (116). Among the thermophilic streptomycetes present in soil, *S. thermodiastaticus* is a prominent cellulose degrader. In an early study by Crawford and McCoy (117), this streptomycete exhibited endoglucanase activity nearly equal to that of *Thermomonospora fusca*. Maximal endoglucanase release occurred at 42°C and pH 7.3 in mineral salts medium while maximal enzyme activity on carboxymethylcellulose was obtained at 55°C and pH 6.5. Reaction rates were linear for at least 1 hr, indicating that the endoglucanase was relatively heat-stable. Chromatographic analysis of CMC degradation products revealed a mixture of glucose, cellobiose; and a mixture of oligosaccharides, indicating a random action for the endoglucanase (118). *Streptomyces thermovulgaris*, another eurythermal thermophile prevalent in fresh and rotted animal manures (119), is also an active cellulose degrader. Rao and Dhala (120) found that this streptomycete produced endoglucanases, exoglucanases, and cellobiase when grown on cellulose powder or CMC. In contrast to responses of most other cellulase producers to noncellulosic carbon/energy sources, arabinose, pyruvate, malate, and citrate were also effective cellulase inducers. Fractionation of the cellulase complex resolved at least five cellulase components. Other cellulolytic streptomycete species are thermotolerant, but the thermostability of their cellulases is relatively poor. For example, *S. lividans* produces an endoglucanase which has a half-life of only 30 min at 60°C (121). In *S. flavogriseus*, endoglucanase activity against CMC is maximal at 50°C, but the thermostability of the enzyme in the absence of the substrate is poor (half-life of 20 min at 50°C; 122). The β-glucosidase is even more thermolabile (half-life of 5 min at 40°C; 123).

During high-temperature composting of animal wastes (which contain about 40% cellulose; 124), at least four *Streptomyces* species can be isolated, but the only *S. thermovulgaris* appeared to play a minor role in the thermophilic phase since its populations were quickly reduced to the undetectable level as the temperatures rose above 40°C (125). Populations of thermophilic fungi were also greatly reduced as temperatures of the composting mass reached 60°C. During the time of maximal heating, the predominant populations were composed of *Micromonospora chalcae* and *Pseudonocardia thermophile*. In related studies (126,127), cellulase production has been characterized in these actinomycetes. During growth on CMC as sole carbon/energy source, endoglucanase was present largely as a true extracellular enzyme in the culture fluid while the β-glucosidase, although detectable in culture fluid, remained mostly cell-bound. These enzymes from both species had pH optima in the range of 6−7. In both organisms, glucose was a potent repressor of cellulase biosynthesis: cellulase production was reduced about 10-fold during rapid growth on the sugar. Cellobiose induced cellulase biosynthesis to yield activities about twice those produced on CMC. The two thermophilic actinomycetes degraded microcrystalline cellulose and filter paper as rapidly as the reference cultures, *Cytophaga hutchinsonii, Cellulomonas flavigena*, and *S. thermoviolaceus*, but only *M. chalcae* was able to degrade lignin and hemicellulose. Evaluating the relative significance of roles played by this or other individual species in a complex heterogeneous system such as a composting mass is difficult. Once an organism is removed from its habitat to the laboratory for closer scrutiny uncluttered by interferences from its former neighbors, it may provide the investigator with a wealth of interesting and reproducible results. However, what those results mean in terms of its effect on its old neighborhood might well tempt even conservative observers into the bog of conjecture.

Comparisons of some characteristics of the actinomycete identified as *T. curvata* with those of minicipal solid waste compost present circumstantial evidence (90,91,128,129) of its role in the

conversion of cellulose in municipal solid waste: (1) it was abun-
dantly apparent in hundreds of samples taken during the thermophilic composting phase; (2) the time of maximal cellulose degradation began when the composting material reached the thermophilic
(53 – 63°C) phase, and this increase in the rate of cellulose disappearance coincided with the emergence of the actinomycete as a
major population; (3) the actinomycete had simple nutritional requirements easily met by the composting environment; (4) the highest temperature at which *T. curvata* can grow (61°C) was also the
average maximum temperature reached by the compost during the
49-day process; and (5) it produced cellulases which in their unfractionated form, had pH and temperature optima for activity and
stability identical to those of the cellulase(s) extracted from the
compost itself, not only in samples taken from the zone of apparent
growth of the thermophile, but also from other sections of the windrow. The prevalence and consistency of cellulases throughout the
composting material suggested that the enzymes remain active even
after turning and displacement from their zone of origin.

The prominence of *Thermomonospora* species in municipal refuse as well as in composts of other origins (80,102) attests to its
catholicity in this conversion process and its potential as an important agent in cellulose degradation. Studies using monoclonal antibody against specific cellulase components of *Thermomonospora* as
probes for their presence in composting materials, together with
comparisons of molecular weights, amino acid compositions, susceptibilities to enzyme inhibitors and inactivators, and proteolytic degradation patterns, would permit a more meaningful evaluation of
the qualitative and quantitative contribution of these organisms to
the process.

Soon after these initial studies on the presence of *Thermomonospora* species in compost and the establishment of these actinomycetes as active cellulose degraders, Bellamy (130) reported the isolation of a thermophilic cellulolytic actinomycete which was termed
MJθr. This actinomycete grew most rapidly at a temperature of
55 – 60°C and pH 7.5 – 8.0. Biotin and thiamine were required for
rapid growth. NaCl (3%) was stimulatory in minimal medium. Addition of salt to the medium also served to control contamination by
thermophilic, noncellulolytic *Sporocytophaga* species which developed
during nonsterile aerated fermentor runs used in the evaluation of
the actinomycete for production of high-protein animal feed from
agricultural wastes. While MJθr and similar actinomycetes produced
no detectable food-borne toxins or other aspects generating human
pathogenesis, a large fraction of total culture protein was extracellular (mainly cellulases and proteases) and therefore protein production by this system could not compete commercially with other
protein sources such as soybeans (131).

This tendency of thermophilic actinomycetes to excrete large amounts of hydrolytic enzymes made them attractive sources of cellulases for the saccharification of waste cellulose to fermentable sugars. Su and Paulavicius (132) described two strains of thermophilic actinomycetes (originally classified as *Thermoactinomyces* species and later reclassified as *Thermomonospora fusca*; 133), which degraded cellulose at a rate faster than that observed for *Trichoderma reesei*. These actinomycetes grew well on buffered mineral salts-cellulose medium at pH 7.5, 55°C. The excreted cellulases bound firmly to the microcrystalline cellulose initially present as sole carbon/energy source and were not released until most of the cellulose was digested. By stationary phase, over 90% of the total culture cellulase was soluble in culture fluid. These soluble enzymes were most active against crystalline cellulose at pH 5.5−6.5 and 60−65°C. Amorphous cellulose was rapidly hydrolyzed (complete hydrolysis within 2 hr), but the crystalline cellulose was much more recalcitrant to attack: at 1% (w/v) concentration, only about 20% was degraded even after 48 hr. This reduced degradation of crystalline cellulose during prolonged incubation was apparently not due to thermal inactivation since the enzymes retained about 75% activity even after 24 hr, nor was it due to end-product inhibition because the total accumulation of reducing sugar (glucose and cellobiose) caused less than a 10% inhibition. Repeated treatment of the residual cellulose with fresh enzyme renewed the initial rapid rate of hydrolysis. These results indicated that the adsorption/desportion process, in which enzyme binds to the substrate and is then released as the binding site on the substrate is degraded, may cause inactivation of the enzymes. If irreversible binding of enzyme to substrate were the cause of decreased rates on prolonged incubation, it would seem probable that this binding would be reflected as a permanent decrease in soluble protein of the reaction mixture; such a permanent decrease has been observed in the binding of *Trichoderma reesei* cellulases to ground newspaper (134) or steam-exploded wheat straw (135), substrates which have high residual lignin contents.

Growth characteristics of the *Thermomonospora* species have been studied in both continuous and batch cultures. The filamentous nature of these actinomycetes present some problems in continuous culture in that clumping and wall growth makes representative sampling difficult. Lee and Humphrey (136) obtained a growth yield of 0.42 g cells/g of glucose and a maximum specific growth rate of 0.36 hr^{-1} (calculated from the dilution rate in which washout occurred) in continuous culture at 55°C. This calculation correlates well with a maximum value of 0.41 hr^{-1} obtained during exponential growth of *T. curvata* at the optimal temperature of 59°C in batch culture with cellobiose as carbon/energy source (Rene

Bernier, pers. commun.). Little cellulase is produced under either
batch or continuous culture conditions in which the cells are not severely limited by carbon/energy supply. In continuous culture, cellulase production, measured as endoglucanase units per mg cells,
was maximal at a dilution rate which limited growth rate to the range
of $0.05-0.08$ hr^{-1} (137). This is the growth rate range to which
the actinomycete is limited during growth on cotton fibers, the best
carbon/energy source for cellulase production (138).

The minimal nutritional requirements for growth of these actinomycetes have been determined (128,131,139). For the few strains
that have been studied, a simple mineral salts medium, buffered to
pH $7.5-8.5$, containing biotin and thiamine with a single carbon/
energy source is sufficient for near-maximal growth rates. The
study of T. fusca strain 190th by Crawford (139) revealed a rather
distinctive physiology. The actinomycete was quite versatile in its
ability to degrade polysaccharides and to utilize sugars as carbon/
energy source. Some organic acids, such as acetate and pyruvate,
were as suitable as the sugars. Ammonium compounds were by far
the best nitrogen source. Amino acids were not utilized nor could
they serve as a nitrogen source in the absence of ammonia. However, it should be pointed out that the early work by Desai and
Dhala (140,141) clearly established that these thermophiles are capable of excreting highly active, thermostable proteases which can
digest a wide variety of cells of other microbial general as well as
soluble proteins such as casein. Even if the proteolytic degradation
products are not usable by the actinomycetes, their abilities to kill
other organisms may be of selective advantage. Alternatively, these
proteases may have specific functions as agents of posttranslational
modification in the cellulases. In this regard, Calza et al. (37)
showed tha a serine protease from T. fusca modified the extracel-
lulase endoglucanases from that species. Evidence for proteolytic
activation and degradation of extracellular enzymes from other microbes is being reported with increased frequency (142−145). Perhaps these extracellular proteases serve a regulatory function akin
to the regulation via proteolysis observed in E. coli during rapid
shifts in nutrient availability (146).

Whether manipulations of culture conditions will result in significant increases in cellulase production by the thermophilic actinomycetes has yet to be thoroughly evaluated. The cellulolytic fungi
have been subjected to novel inducers (147,148), pH and temperature profiling (149), closely controlled two-stage continuous culture (150), various surfactants (151), and a variety of other manipulations (reviewed in Ref. 152) to wring out the maximal cellulase
of which each strain is genetically and physiologically capable. In
contrast, beyond the determinations of nutritional requirements for
growth, relatively little attention has been given to the use of

stimulants of cellulase production in *Thermomonospora*. The effects
of several metals have been reported to alter production by *T. curvata* in minimal medium; surprisingly, calcium, which stabilizes the
cellulases against thermal denaturation, also depresses cellulase biosynthesis (153). A recently completed study (35) showed that
Tween 80 causes a component-specific stimulation of cellulase secretion in *T. curvata*, but the overall effect has been hardly as dramatic as the 20-fold stimulations achieved when fungi were subjected
to that surfactant (reviewed in Ref. 151). Since these actinomycetes
are alkalophilic [growth range of pH 7 – 12 (154)], control of pH
above neutrality may be of benefit in raising cellulase production,
particularly in light of a recent study (155) that showed that at
least one of the endoglucanases undergoes a pH-dependent thermal
activation; this activation provides a significant increase in activity
if carried out at pH 8, while rapid inactivation occurs at the same
temperature (70°C) at pH 6.

The mechanisms regulating rates of cellulase biosynthesis in
these thermophilic actinomycetes remain relatively undefined (as
they are for the cellulolytic aerobic bacteria in general). Extracellular enzymes which degrade large insoluble polymers compose a
rather unique class in which products of the enzyme activity are
necessarily the inducers of biosynthesis. The recalcitrance of cellulose to degradation in nature would require a relatively high level
of constitutive cellulase biosynthesis in order to provide sufficient
product, such as cellobiose, to trigger induction. *Thermomonospora*
is very sensitive to cellobiose; even when the dissaccharide is present as only a small fraction of total soluble sugar in the medium,
it is preferentially taken up (156). Cellobiose availability also alters
the surface structure of the actinomycete. When grown on glucose
as sole carbon/energy source, the mycelial surface is quite smooth
(Fig. 2a). When cellobiose (a cellulase inducer) is provided, surface topology is altered by the formation of multiple protuberances,
which give the mycelia a ragged appearance (Fig. 2b). These protuberances, which appear to be similar to surface cellulase complexes
(cellulosomes) recently described by Lamed et al. (157) in other bacteria, may provide the mechanism by which the actinomycets adhere
firmly to their insoluble lignocellulosic substrates (example shown
in Fig. 3).

Cellobiose is clearly a cellulase inducer in *Thermomonospora* and
other bacteria (157,158), but catabolite repression during rapid
growth on cellobiose severely limits the rate of biosynthesis. Catabolite repression of catabolic enzyme systems is at least in part dependent on intracellular levels of cyclic adenosine-3',5'-monophos-
phate (cAMP). The mechanism for cAMP control of catabolic enzyme
systems appears to be effected through a binding of the nucleotide
to a cAMP receptor protein; the resulting complex binds to DNA at

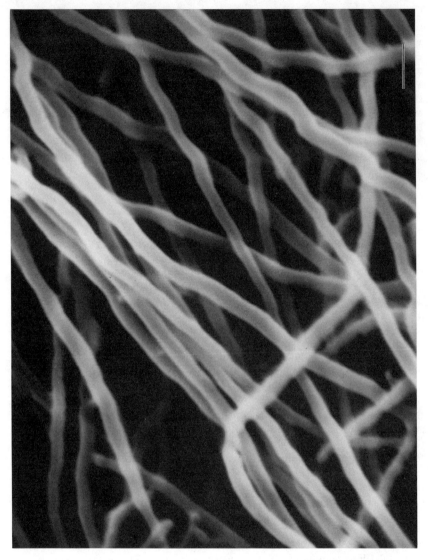

(a)

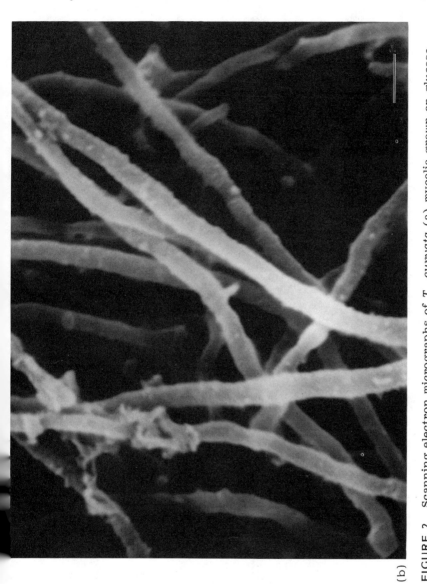

(b)

FIGURE 2   Scanning electron micrographs of *T. curvata* (a) mycelia grown on glucose as sole carbon cource. Note the smooth surface of the glucose-grown mycelia compared to that of the cellobiose-grown mycelia, which is covered by multiple small protuberances.

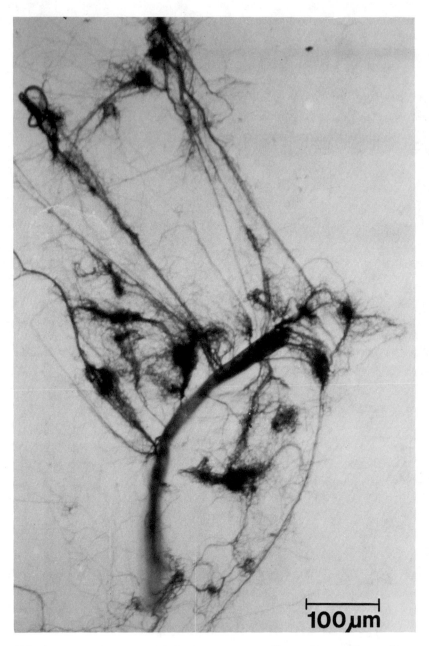

FIGURE 3   Adherence of *T. curvata* mycelia to the surface of a
cotton fiber.   The surface protuberances formed during cellulase induction (as shown in Fig. 2b) may act as cellulosomes, containing
a multicomponent complex of cellulolytic enzymes and adherence factors which enable the mycelia to maintain themselves close to the
substrate surface for maximal cellulose breakdown and product
uptake.

the promoter regions of cAMP receptor protein; the resulting complex binds to DNA at the promoter regions of cAMP-dependent operons and stimulates transcription (159). In *Thermomonospora*, rates of cellulase biosynthesis correlate with intracellular cAMP levels. During growth on relatively recalcitrant substrates such as cotton fibers, intracellular cAMP reaches a maximum of about 20−30 pmol/mg cell dry weight; these levels decline as the culture reaches late exponential phase, and the rate of cellulase biosynthesis slows during that time. During rapid growth on glucose or other soluble sugars, cAMP levels are reduced 10-fold and cellulase biosynthesis is reduced 200-fold (138). However, in catabolite-resistant mutants, which have elevated levels of cAMP compared to wild-type cells, these sugars cannot repress cellulase biosynthesis (160,161). The addition of exogenous cAMP to wild-type cells stimulates cellulase biosynthesis in toluene-treated cultures which are permeable to the nucleotide, and this stimulation is increased when theophylline (a potent cAMP phosphodiesterase inhibitor) is present (138).

Intracellular cAMP levels are clearly not the only factor regulating the rate of biosynthesis of cellulase in *Thermomonospora* or of other catabolic enzymes in other organisms (reviewed in Ref. 162). Furthermore, the importance of cAMP in other cellulolytic flora is questionable. Prabakaran and Dube (163) reported a marked effect of exogenously added cAMP on catabolite repression of the cellulase complex in the fungus *Penicillium islandicum*. Addition of the nucleotide to cultures growing on glucose, cellobiose, and a variety of other sugars almost completely reversed catabolite repression of the cellulolytic system. However, Montenecourt and associates could find no evidence for participation of cAMP in the regulation of cellulase biosynthesis in *Trichoderma* (164,165). In *Bacillus*, its lack of detectable cAMP rules out the possibility that the nucleotide is a controlling factor in the synthesis of any of its catabolic enzymes (166). The only safe general statement at this stage of our understanding of cellulase regulation at the gene level is that the degrees to which induction and catabolite repression (and the interplay of agents thereof) control cellulase biosynthesis vary greatly with the organism studied.

The cellulolytic enzymes of *Thermomonospora* species, unlike those of *Trichoderma*, have not been subjected to extensive testing as to their activity on a wide range of potentially important commercial lignocellulosics. The cellulases of the actinomycetes can surely degrade recalcitrant substrates such as cotton fibers; indeed, that is the best sole carbon/energy source for cellulase induction in *T. curvata* (128). However, substrates which contain large amounts of lignin will be very resistant to attack by the actinomycetes if such substrates have not been pretreated in some chemical or physical manner (see Refs. 167−169 for reviews of pretreatments).

Lignin is a complex aromatic polymer composed largely of phenylpro-
pane units (170). The number of bonds susceptible to enzymatic
attack are few and consequently it is slowly degraded by even the
most lignolytic of microbes (reviewed in Ref. 171). The lignin content of plant material ranges from about 7 to 32% (172), with the
highest lignin content in the primary cell wall (173). Not only do
lignin and cellulose combine to produce a chemically and physically
resistant layer, but residual lignin after partial cellulose digestion
appears to irreversibly bind the cellulases so that they are in effect inactivated (135). Although some lignolytic activity has been
demonstrated in *Thermomonospora* (174,175) and the interaction of
actinomycetes with other flora in lignin utilization during wood decomposition is important (176), these actinomycetes appear to be less
able to extensively degrade this resistant polymer than some of the
saprophytic fungi such as *Phanerochaete chrysosporium* (177,178).
However, there are obviously factors other than lignin content
which control susceptibility to attack by the thermophilic actinomycete cellulases. For example, barley straw contains less than one-
fourth the lignin found in hardwood sawdust (7% vs. 32%), yet it
is equally resistant to attack by the cellulases of *T. curvata* (179).

Organisms of the genus *Thermomonospora* are generous sup-
pliers of exoenzymes. During growth in minimal medium with a polysaccharide as sole carbon/energy source, about one-half of the total
culture protein is secreted into the culture fluid (104,128). With
such a copious supply available, it is rather remarkable that the
majority of those enzymes directed toward the degradation of cellulose and other natural polymers have escaped extensive purification
and detailed characterization until quite recently. In the initial
study (129) on partially purified cellulases from *T. curvata* culture
fluid, a multiplicity of components was apparent. At least two cellulases having activity against crystalline cellulose were separable
on size exclusion chromatography. Maximal activity against both
crystalline cellulose and CMC was obtained in unfractionated preparations at pH 6.0–6.2 and 65°C. The sole detectable product of
their action on cotton fibers was cellobiose. The enzymes were
stable at 60°C in the absence of substrate, but rapidly inactivated
(half-life of 3–5 min) at 75°C. Enzyme replacement studies, in
which fibers were treated repeatedly with fresh cellulase, showed
that the enzymes could rapidly cleave away a small portion of the
cellulose (obstensibly the more amorphous regions), leaving the remainder of the fiber as a relatively recalcitrant substrate.

A variety of other superficial studies provided some additional
information about *Thermomonospora* cellulases which indicated that
they had significant potential in industrial application. The crude
enzymes were active against a wide range of natural and man-made
cellulosic substrates (179). Although end-product inhibition of their

activity of cellobiose was quite severe (2% concentrations of the dissaccharide caused a 50% reduction in activity), they were much less
sensitive to glucose accumulation. They were marginally activated
and stabilized by calcium (153,180) as are a variety of other extracellular enzymes from this genus (38) and others (181-184). The
use of the *Thermomonospora* cellulases would probably be effective
in a simultaneous saccharification/fermentation system since they
were only slightly inhibited by ethanol accumulations up to 6% w/v;
however, the cellobiase associated with cell solids was markedly more
ethanol-sensitive and therefore the actinomycete preparation would
have to be supplemented with a more resistant cellobiase for maximal rates of cellulose conversion to glucose (180).

A major difference (first reported by Hagerdal et al., 185,186)
between the organization of the *Thermomonospora* cellulolytic system
and that of the well-characterized fungal systems was that the cellulases of the actinomycete could not be distinguished as having
exo- or endoglucanase activities. In fungi such as *Botrydiplodia
theobromae* (187), *Chrysosporium lignorum* (188), *Talaromyces emersonii* (189), and *Trichoderma reesei* (190) a multicomponent system
was characteristic of cellulolytic organization. This system contains
one or more endoglucanases (which cleave β-1,4 bonds randomly
along the polymer), one or more exoglucanases (which cleave cellobiose units from the nonreducing free ends liberated by the endoglucanases), and a β-glucosidase which removes the cellobiose (a
strong inhibitor of both endo- and exoglucanase action; 191). However, even preparative isoelectric focusing could not separate endo
and exo activities of *Thermomonospora* extracellular proteins; although
three major fractions with cellulolytic activity could be distinguished
(isoelectric points of pH 3.5, 3.7, and 4.6), their rates on CMC and
crystalline cellulose were parallel.

In further contrast to the fungal systems, Hagerdal and colleagues also confirmed an earlier observation (128) that activities
against CMC and crystalline cellulose appeared to be induced coordinately during growth. All of the proteins present in the stationary phase extracellular milieu were already evident in isoelectric
focusing by early exponential phase when cellulolytic activity became
detectable. This continuity in the protein pattern not only appeared
to dismiss the possibility of sequential induction of cellulase components during growth, but also indicated that proteolytic processing of cellulase proenzymes play a minor, if any, role in the fashion-
ing of the cellulolytic system of the actinomycete. While this lack of
evidence appears to dismiss proteolytic involvement in posttranslational modification of *Thermomonospora* cellulases, proteases may still
play a role in the degradation of extracellular enzymes in response
to different carbon/energy sources in this (38) and other microbes
(142-144). Furthermore, the detailed studies described in the next

472   Stutzenberger

paragraph establish the potential for proteolytic modification of
Thermomonospora cellulases.

The only thorough purification and characterization of Thermomonospora endoglucanases reported to date has been the work of
Calza et al. (37). Through a protocol including size exclusion, ion
exchange and/or hydroxylapatite chromatography followed by preparative polyacrylamide gel electrophoresis, they were able to purify
two 1,4-β-glucanases from T. fusca to near-homogeneity. Attempts
at further purification employing immobilized concanavalin A, CMC,
or phosphocellulose were unsuccessful. Although their preparations
contained at least one other endoglucanase with similar properties,
the two major enzymes (designated E1 and E2) accounted for about
90% of the total endoglucanase activity. E1, the most active of the
two forms, was a monomeric protein with a molecular weight of 94
kD. It contained less than 1% carbohydrate and had as isoelectric
point of pH 3.5. Maximal activity on CMC was obtained at 74°C,
pH 6.5. The K_m for the soluble substrate was 360 μg/ml. E1 activity on acid-swollen amorphous cellulose yielded cellobiose with a
trace of glucose; the enzyme's ability to cleave cellobiose was extremely low. E2 was also a monomeric protein with a molecular
weight of about 46 kD. The carbohydrate content of this endoglucanase was quite high (25%) and the isoelectric point was pH 4.5.
Its temperature optimum was only 58°C (20% loss of activity at 56°C
for 18 hr). The pH optimum was similar to E1. E2 produced a mixture of glucose, cellobiose, cellotriose, and high oligomers from cellulose, and its affinity for CMC was somewhat higher (K_m of 120
μg/ml). E1 and E2 were markedly similar in their amino acid compositions: four amino acids (aspartic, glutamic, glycine, and alan-
ine) contributed nearly one-half the total residues in their sequences.
In view of the high protease activity in T. fusca culture fluids, the
similarity in amino acid composition, and the twofold differences in
molecular weight, it would be tempting to surmise that E2 might represent the product of a symmetrically located limited proteolytic
cleavage; however, no antigenic similarity between E1 and E2 could
be detected either in inhibition studies or in immunodiffusion gels.
Other interesting differences between the two forms centered around
heavy-metal effects. For example, $CoCl_2$ was inhibitory to E2 but
not E1, while $CaCl_2$ was stimultory to E1 and not E2.

β-Glucosidase, the terminal enzyme in the cellulolytic sequence,
is present in very low concentration in Thermomonospora culture
fluids (185). In this regard, the actinomycete cellulase system is
similar to that of Trichoderma reesei. Allen and Sternberg (192)
estimated that only about 0.2% of the total fungal extracellular protein is contributed by the presence of that enzyme. This lack of
β-glucosidase in Thermomonospora culture fluids was reflected in the
fact that cellobiose was the only soluble sugar detectable in the hydrolysate after prolonged incubation with crystalline cellulose (129).

β-Glucosidase deficiency in cell-free culture fluids of cellulolytic microbes has generally been taken to mean that the enzyme is truly intracellular and what little extracellular activity is detectable is merely the result of cell lysis. Recent results obtained in our lab indicate that this may not be entirely true in the case of *Thermomonospora*. Although the actinomycete grows at temperatures up to 61°C and produces endoglucanases which are quite stable at that temperature, the β-glucosidase is much more labile. The highest temperature at which the extracellular form(s) is stable on prolonged incubation appears to be 45°C. The half-life under simulated culture conditions (sterile filtered culture fluid shaken at 55°C) was only 2 hr. Cell growth in the presence of the serine protease inhibitor phenylmethylsulfonylfluoride did not increase the extracellular β-glucosidase activity (Rene Bernier, unpublished). Therefore it appears that the lack of β-glucosidase activity in actinomycete culture fluid may be the result, at least in part, of thermal denaturation. The use of a protease-negative strain [a strategy employed by Dunne (193) to determine the effect of proteolysis on *Trichoderma* cellulases] might be used to evaluate the relative contributions of thermal denaturation and proteolysis.

In any case, the inability of fungal or bacterial exoproteins to take cellulose completely to glucose poses some problems in their commercial use as agents in converting biomass to fermentable sugars. Cellobiose, and even glucose at concentrations far below commercial feasibility, inhibits cellulase activity (180,194), so that prolonged incubation to obtain maximal sugar yield per mass of substrate would take hydrolysis rates to unbearably low levels. Even if cellobiose accumulations were not terribly inhibitory, the majority of yeasts suitable for production of industrial alcohol cannot ferment cellobiose (195). This would necessitate an additional step in which the hydrolysate was treated with β-glucosidase or, alternatively, cellulase mixtures could be supplemented with β-glucosidase from other sources. A summary of suggested strategies was included in the recent review by Coughlin (196).

Until quite recently, the attention paid to the cellulolytic abilities of the thermophilic actinomycetes was focused almost entirely on the *Thermomonospora* species. That era ended with the reports of Waldron et al. (197) and Waldron and Eveleigh (198) on a search for other stable thermophilic actinomycete strains. These workers collected soil samples from hot springs, streams, pools, composts, volcanic fumeroles, and geysers in a variety of countries around the world. Cellulolytic thermophiles were isolated after enrichment in moist microcrystalline cellulose incubated at 60°C. Forty different strains of thermophilic, aerobic actinomycetes were recovered by solid substrate enrichment. However, of that number, only 11 would grow in a liquid medium of mineral salts, yeast extract, and cellulose; only eight produced easily detectable amounts of

extracellular cellulase in liquid culture (range of about 2−6 endoglucanase IU/ml); and only one was stable on prolonged cultivation.
These ratios point up the instability of the thermophilic actinomycetes as reflected by the frequently encountered inability of apparently promising isolates to grow in sequential transfer or to continue
biosynthesis of the desired product. It was shown in other microbes
that mutation rates are accelerated by continued culture at subop-
timally high temperatures (199,200). In the thermophile, *Bacillus
stearothermophilus* (which has an optimal growth temperature of
about 65°C), gene expression and plasmid stability decreases dramatically when cultured above 55°C (201). Whether mutation rates
in thermophiles growing at their temperature optima exceed those of
mesophiles at their optima has not, to the knowledge of the author,
been quantitatively determined.

The stable actinomycete strain, isolated from a compost sample,
was identified as *Microbispora bispora*. The actinomycete, originally identified by Henssen (202) as *Thermopolyspora bispora* and later
redesignated by Lechevalier (203), produces a distinctive chalk-
white aerial mycelium brandishing many lateral paired spores, which
may contribute to the fire-fang layer in compost described by Fergus
(80). It has a rather narrow growth temperature range of 50−60°C
(204). The optimal conditions for the production of cellulolytic enzymes
by *M. bispora* were determined by Waldron and colleagues as follows:
carbon and nitrogen source, microcrystalline cellulose (2% Avicel), and
ammonium sulfate (1.32% g N/liter), respectively, pH 7.2, incubated
for 2 days at 60°C in liquid mineral salts−yeast extract medium. A
decrease in incubation temperature from 60°C to 55°C resulted in a
seven-fold lowering of extracellular endoglucanase activity; however,
most of this decrease could be accounted for by cell-associated endoglucanase activity. Therefore, the small change in incubation temperature may have its major effect on cellulase excretion rather than biosynthesis. β-Glucosidase activity was largely cell-associated (the low
β-glucosidase activity in culture fluid may have been the result of
thermal denaturation since that enzyme was unstable at temperatures
above 55°C). Cellulase production was completely repressed during
growth on 1% glucose but was stimulated as much as 70% by the addition of 0.5% lactose or galactose to cultures growing on cellulose.

The genus *Thermomonoactinomyces* may or may not have an important role in cellulose degradation. This genus encompasses at
least six species of actinomycetes which form single spores on both
the aerial and substrate mycelia. These spores are true heat-
resistant bacterial endospores, the majority of which can withstand
heating at 100°C for 20 min (205). The properties of this spores
(206), their menaquinone composition (207), their DNA base ratios
(208), and their 16S ribosomal RNA sequences (94) place them taxonomically closer to the family Bacillaceae than to the true

actinomycetes. However, the thermoactinomycetes and the white thermomonosporas cannot be easily distinguished morphologically (154). Figure 4 shows a diagrammatic comparison of the major morphological differences between them, although there appears to be an almost continuous range of morphological variation found between the extremes represented by these two genera (209). This close similarity in appearance has led to the initial identification of cellulolytic thermomonosporas as thermoactinomycetes (90,136,210). Because of this confusion, it is difficult to attribute a significant role in cellulose degradation to the thermoactinomycetes. While *Thermoactinomyces* species contribute to many soil populations to only a negligible extent (less than 0.2% of total isolates; 211), they form very high spore populations in lignocellulosic materials such as heating hay, composts, stored grain, and heaps of decaying vegetation in nature (212). In soils subjected to high temperatures, *Thermoactinomyces* species may play a significant role in cellulose degradation. Mahmoud et al. (213) recovered 229 isolates from various types of Egyptian soils. Sixty-five percent of these isolates hydrolyzed cellulose. *Thermoactinomyces vulgaris* demonstrated higher cellulolytic activity than did the *Bacillus* and *Thermomonospora* species. The most active (*T. vulgaris* strain 145) produced maximal cellulase (measured as hydrolysis of CMC) during growth at 60°C, pH

FIGURE 4 Morphological comparison of spore arrangements in *T. curvata* (left) and *T. vulgaris* (right). (Drawings courtesy of Dr. Tom Cross, University of Bradford, England.)

6.5 – 7.5, with cellodextrins and yeast extract as carbon and nitrogen sources, respectively. Cellulase production was inducible; low
or undetectable activity was produced during growth on glucose,
maltose, acetate, glycerol, fructose, sucrose, and starch. This report of a cellulolytic Thermoactinomyces species is unusual in light
of the finding by McCarty and Cross (154) that all strains classified
under that genus could neither hydrolyze CMC nor solubilize cellulose powder.

V. BACILLUS

The genus Bacillus encompasses many species. The broad range of
DNA (guanosine + cytosine) content (32 – 52%) is an indication of
their genetic heterogeneity (16). Most species are mesophilic and
aerobic. Some grow well on a single organic carbon/energy source
in inorganic salts while others require amino acids and vitamins.
Their major habitat is soil and their primary ecological role appears
to be the degradation of biomass polymers. Their proteolytic and
amylolytic activities were recognized very early (reviewed in Ref.
46), and later studies have shown that all of the recognized Bacillus
species secrete extracellular enzymes (reviewed in Ref. 214). Currently, production of Bacillus proteases and amylases accounts for
over one-half the total commercial enzyme volume (215). Despite
the recognized capability of these bacilli to degrade biopolymers,
the possibility of a significant role for them in cellulose decomposition has been largely dismissed. Until recently, these organisms
were assigned a secondary role (if any) compared to the prominent
cellulolytic, obligately anaerobic thermophiles. However, Zemek
et al. (216) showed that 11 of 25 Bacillus species secrete 1,4-β-
glucanases. All of the B. stearothermophilus strains, 85% of the
B. coagulans strains, and 50% of the B. licheniformis strains tested
(three species which grow well at 50 – 60°C; 217) produce cellulase
components. Thermostable cellulases from Bacillus species are currently receiving increased attention. Robson and Chambliss (218)
reported the following characteristics for an endoglucanase of a
group I Bacillus strain from heat-shocked soil samples: (1) maximal production during growth on cellobiose; (2) release in stationary phase of growth; (3) optimal pH and temperature of 4.8 and
58°C; (4) molecular weight between 10 and 100 kD; (5) no metal
cofactor requirement; and (6) no detectable activity against filter
paper or crystalline cellulose. Endoglucanase production appeared
to be constitutive.
    Dhillon et al. (219) recently reported that B licheniformis pro-
duced an endoglucanase during growth on crystalline cellulose, cellulase acetate, and carboxymethylcellulose, but that maximal

secretion occurred when the cells were provided soluble sugars such as sucrose, maltose, glucose, mannose, or mannitol. The pH and temperature optima were 6.1 and 55°C, respectively. Cellulase production in *Bacillus* provides an interesting contrast to the highly cellulolytic microbes since production appears constitutive and is obviously not under the control of cAMP-mediated catabolite repression (220,221).

The most detailed study of cellulase production by a *Bacillus* species was the recent report of Chan and Au (222). *Bacillus subtilis* T-14 was isolated from cotton waste compost, mutagenized with N-nitrosoguanidine, and screened for maximal cellulase production. A mutant, designated AU-1, was isolated and characterized. This mutant had additional growth requirements which necessitated the addition of amino acids, peptone, and yeast extract to the minimal medium; even this plethora of additives resulted in rather low cell yield (less than 0.4 mg cell dry wt per ml). Despite the metabolic handicaps, AU-1 produced four times the endoglucanase observed for the wild-type *B. subtilis* in a previous study (223). Maximal endoglucanase production was obtained from cells grown on the trisaccharide raffinose; a variety of other sugars such as rhamnose, arabinose, and cellobiose induced endoglucanase less effectively. It is interesting that while raffinose supported the best endoglucanase production by this mutant, microcrystalline cellulose or filter paper was necessary as a carbon/energy source for the highest exoglucanase production. However, growth on crystalline cellulose was severely limited (cell yields about one-third that observed on soluble sugars) due mainly to the low activity of the *Bacillus* enzymes on that substrate (less than 2% of that observed with carboxymethylcellulose as substrate).

With raffinose as carbon/energy source, endoglucanase secretion was maximal when cultures were shaken at 50°C, pH 6.0. Secretion ceased by stationary phase. This pattern is converse to that observed earlier (218) in which little enzyme was released until cessation of growth. Strain AU-1 was also unusual in its sensitivity to endoglucanase induction by low concentrations of cellobiose or raffinose and repression by high concentrations of glucose or cellobiose. Preliminary characterization of the endoglucanase in crude culture fluid yielded the following results: (1) the enzyme had a broad pH optimum (5.0-8.5) within which it was stable at 55°C in the absence of substrate; (2) the temperature optimum was 65°C, with half-maxima at about 40 and 76°C; and (3) incubation at 80°C for 30 min resulted in total loss of activity.

Although the aerobic, thermophilic *Bacillus* species have not distinguished themselves as highly cellulolytic, they currently appear to be excellent sources of exoenzymes which degrade other materials associated with cellulose in nature. Okazaki et al. (224) reported

production of thermostable xylanases by four alkalophilic thermophilic *Bacillus* species. The optimal temperature and pH for activity of these enzymes was 65−70°C and 6.0−7.0, respectively.
Gruninger and Fiechter (225) recently described some characteristics of a highly thermostable D-xylanase produced by *B. stearothermophilus*-like isolates from heat-treated (50−67°C) sludge of a
pilot scale, aerobic, thermophilic sewage digestor. In a complex
medium, maximal xylanase activity was produced within 2−4 days
at 65°C in shaken flasks. Biosynthesis was apparently inducible
(in contrast to the constitutive nature of endoglucanase synthesis
in the *Bacillus* species) since no xylanase was detected during
growth on starch. The optimal temperature for activity was 78°C
but the limit for stability over a 5-day period was 64°C. The enzyme was stabilized by the presence of substrate (half-life of 30
min at 80°C in its absence). Its pH optimum was 6.5−7.0 and was
somewhat dependent on the buffer used. With larchwood xylan as
substrate, the enzyme appeared to be an endoxylanase, yielding
xylose as the major product, with traces of xylobiose, xylotriose,
and larger fragments on prolonged incubation.

What role such xylanases play in the destruction of plant material in nature would be a matter of some dispute. Cellulose, hemicellulose, pectin, and lignin are the main components of the plant
cell wall. Combinations of enzymes degrading polysaccharides and
lignin act synergistically in the degradation of the cell wall matrix
(226−228). Therefore xylanases, in their clearance of hemicellulose
from the cell wall matrix, should conceivably make other components
such as cellulose more accessible to attack. Furthermore, the xylanases from *Bacillus* species may have a rather broad substrate
specificity which allows some activity on cellulose. Uchino and
Nakane (229) purified a thermostable xylanase to electrophoretic
purity from a thermophilic acidophilic *Bacillus* isolated from a hot
spring. The temperature optimum for activity against larchwood
xylan was 80°C at pH 4.0. The maximal specific velocity with xylan and CMC were 8.9 μmol and 19.3 μmol/min/mg purified enzyme,
respectively. As the enzyme exhibited endo activity against xylan,
CMC, and insoluble cellulose, the authors suggested that the enzyme be considered as a cellulase with broad substrate specificity
for 1,4-β-glucosyl linkages. Such enzymes may be rather common
in nature; other enzymes showing this cross-specificity have been
found in other bacteria (230,231) and in fungi (232,233), although
most microbial cellulases lack this broad substrate specificity.

VI. AQUATIC THERMOPHILES

The early work of Brock and others (234−236), in which was reported the isolated of *Thermus aquaticus* from hot-water taps and

thermally polluted streams, stimulated studies on thermophilic bacterial ecology and physiology. Aerobic incubation in a liquid mineral salts medium of low organic content at 70°C yielded *T. aquaticus* cultures from over one-half of the commercial water heater samples tested. In the decade following the initial description of *T. aquaticus*, a variety of newly discovered aquatic thermophiles were isolated and characterized from sources around the world (reviewed in Ref. 11).

One of the major sources of aquatic thermophiles has been the hot springs area of Yellowstone National Park in the United States. The temperatures of these springs range from 30°C to boiling (237), although consistency in temperature and chemical composition are subject to a variety of influences such as earthquakes or silica decomposition. The pH distribution among the Yellowstone springs is biomodal: The majority are rather highly acidic, but others are neutral-to-alkaline. The acidic (pH 2−5) hot springs have yielded both *Bacillus acidocaldarius* and *Sulfolobus* species but neither have been shown to possess any cellulolytic activity. Recently, Mohagheghi and coworkers (181,238) reported the isolation of 12 obligately aerobic, acidophilic, cellulolytic thermophiles from the upper Norris Geyser basin area of Yellowstone. The acidic springs which yielded these isolates had temperature and pH ranges of 45−65°C and 4−5.5, respectively. All of the isolates had identical morphology (gram-variable, nonpigmented, nonflagellated, non-spore-forming slender rods tending to filament). The isolates shared some characteristics in common with *Thermus* and *Thermomicrobium* species (morphology, lysozyme sensitivity, and catalase production) but were distinct in their patterns of carbon/energy sources, pH, and temperature optima for growth (5.5 and 58°C, respectively), antibiotic sensitivities, and DNA G + C content. A variety of mono-, di-, and polysaccharides including cellulose, xylan, and starch (but not pectin) could be utilized as sole carbon/energy source. The amino acid composition of cell wall hydrolysates showed composition typical of peptidoglycan (which sets them apart from the thermoacidophilic archaebacteria such as *Sulfolobus acidocaldarius* and *Thermoplasma acidophilus*; 239). Mohagheghi et al. (238) pointed out that these isolates, which have been classified as *Acidothermus cellulolyticus*, do not represent major microbial populations in geothermal hot springs; however, it was noted that no thermophilic fungi or actinomycetes were detected by their enrichment and isolation procedures.

The cellulase activity of *A. cellulolyticus* has been partially characterized to date. Growth on crystalline cellulose yielded culture fluids with activity equivalent to 0.4 filter paper units (as defined by Mandels et al., 240). The optimal temperatures for filter paper and endoglucanase activities are 75 and 83°C, respectively. Both the major cellulase component (molecular weight of 188 kD) and a minor component (molecular weight of 30 kD) degrade CMC and

filter paper. These cellulases appear to be the most thermostable
yet reported for the aerobic thermophiles.

VII. CONCLUSION

Our knowledge of cellulose degradation by the aerobic thermophilic
bacteria lags by an appalling margin behind that known of the mesophilic fungi. Surely the persistent and productive efforts of Elwyn
Reese, Mary Mandels, and colleagues (chronicled in Ref. 3) have
contributed much more significantly to this disparity than any lack
of activity by the few bacteriologists who study the cellulolytic thermophiles. The Natick group showed us that we have a long way to
go. While it was to be expected that the cloning of cellulase genes
from the thermophilic bacteria would wait to ride the crest of the
genetic engineering wave of this past decade, it is unfortunate that
we still know very little about the actions and interactions of bacterial
cellulases. It is more unfortunate that we have allowed the black
boxes to virtually remain shut around the molecular mechanisms which
govern the rates of cellulase biosynthesis in the aerobic thermophiles,
while mechanisms governing other catabolic enzymes in other microbes
have been threshed and gleaned (examples reviewed, 152, 241).
When such detailed studies include the cellulolytic microflora, their
exploitation in a variety of beneficial processes may be economically
feasible.

ACKNOWLEDGMENTS

I thank Alex Bonner, Rene Bernier, and Michael Himmel for unpublished data; June Huff and Bobbie Griffith for typing of the manuscript; and Ben Paynter for critical review. Work in our lab was
funded by the U.S. Army Research Office and the South Carolina
Energy Research and Development Center.

REFERENCES

1. T. D. Brock, *Principles of Microbial Ecology*, Prentice-Hall,
   Englewood Cliffs, NJ, 1966, p. 123.
2. R. M. Atlas and R. Bartha, *Microbial Ecology: Fundamentals
   and Applications*, Addison-Wesley, Reading, MA, 1981, p. 355.
3. E. T. Reese and M. Mandels, In *Annual Reports on Fermentation Processes*, Vol. 7 (G. T. Tsao, ed.), Academic Press,
   Orlando, 1984, p. 1.

4. E. T. Reese and M. Mandels, *Biotechnol. Bioeng.*, *22*, 323 (1980).
5. H. Durand, P. Sourcaille, and G. Tiraby, *Enzyme Microb. Technol.*, *6*, 175 (1984).
6. A. Margaritis and R. F. J. Merchant, *CRC Crit. Rev. Biotechnol.*, *4*, 327 (1986).
7. R. C. Weast (ed.), *Handbook Chem. Phys.*, 51st ed., Chemical Rubber Co., Cleveland, 1971, p. B-116.
8. T. K. Ng and W. R. Kenealy, In *Thermophiles: General, Molecular and Applied Microbiology* (T. D. Brock, ed.), John Wiley and Sons, New York, 1986, p. 197.
9. T. Godfrey and J. Reichelt, *Industrial Enzymology*, Nature Press, London, 1983.
10. T. D. Brock, *Science*, *230*, 132 (1985).
11. T. D. Brock, In *Thermophiles: General, Molecular and Applied Microbiology* (T. D. Brock, ed.), John Wiley and Sons, New York, 1986, p. 1.
12. K. O. Stetter, H. Koenig, and E. Stackebrandt, *Syst. Appl. Microbiol.*, *4*, 535 (1983).
13. L. G. Ljungdahl, *Adv. Microb. Physiol.*, *19*, 149 (1979).
14. B. Sonnleitner, *Adv. Biochem. Eng. Biotechnol.*, *28*, 69 (1983).
15. M. R. Tansey and T. D. Brock, *Proc. Natl. Acad. Sci. USA*, *69*, 2426 (1972).
16. P. J. Vandemark and B. L. Batzing, *The Microbes*, Benjamin-Cummings, Menlo Park, CA, 1987, p. 153.
17. T. A. Langworthy and J. L. Pond, In *Thermophiles: General, Molecular and Applied Microbiology* (T. D. Brock, ed.), John Wiley and Sons, New York, 1986, p. 107.
18. R. N. McElhaney and K. A. Souze, *Biochim. Biophys. Acta*, *443*, 348 (1976).
19. C. A. Abbas and G. L. Card, *Biochim. Biophys. Acta*, *602*, 469 (1980).
20. M. Oshima and A. Miyagawa, *Lipids*, *9*, 476 (1974).
21. R. N. McElhaney, In *Extreme Environments* (M. R. Heinrich, ed.), Acaemic Press, New York, 1976, p. 255.
22. C. E. Cerniglia and J. J. Perry, *J. Bacteriol.*, *118*, 844 (1974).
23. D. H. King and J. J. Perry, *Can. J. Microbiol.*, *21*, 85 (1975).
24. P. H. Ray, D. C. White, and T. D. Brock, *J. Bacteriol.*, *106*, 25 (1971).
25. J. M. F. G. Aerts, A. M. Lauwers, and W. Heinen, *Antonie van Leeuwenhoek*, *51*, 155 (1985).
26. G. J. Merkel and J. J. Perry, *Appl. Env. Microbiol.*, *34*, 626 (1977).

27. J. L. Pond and T. A. Langworthy, *J. Bacteriol.*, *169*, 1328 (1987).
28. M. Sinesky, *Proc. Natl. Acad. Sci. USA*, *71*, 522 (1974).
29. A. R. Cossins and M. Sinesky, In *Physiology of Membrane Fluidity* (M. Shinitzky, ed.), CRC Press, Boca Raton, 1984, p. 1.
30. M. Oshima and A. Miyagawa, *Lipids*, *9*, 476 (1974).
31. Y. Hasewaga, N. Kawada, and Y. Nosoh, *Arch. Microbiol.*, *126*, 103 (1980).
32. J. L. Pond, T. A. Langworthy, and G. Holzer, *Science*, *231*, 1134 (1986).
33. M. Oshima, Y. Sakaki, and T. Oshima, In *Biochemistry of Thermophily* (S. M. Friedman, ed.), Academic Press, New York, 1978, p. 31.
34. K. Pralla and W. A. Konig, *FEMS Microbiol. Lett.*, *16*, 303 (1983).
35. F. J. Stutzenberger, *J. Appl. Bacteriol.*, *63*, 239 (1987).
36. A. M. Klibanov, *Adv. Appl. Microbiol.*, *29*, 1 (1983).
37. R. E. Calza, D. C. Irwin, and D. B. Wilson, *Biochemistry*, *24*, 7797 (1985).
38. F. J. Stutzenberger, *J. Bacteriol.*, *169*, 2774 (1987).
39. Y. Suzuki, K. Oishi, H. Nakano, and T. Nagayama, *Appl. Microbiol. Biotechnol.*, *26*, 546 (1987).
40. W. Militzer, T. B. Sonderegger, L. C. Tuttle, and C. E. Georgi, *Arch. Biochem.*, *24*, 75 (1949).
41. R. E. Amelunxen and M. Lins, *Arch. Biochem. Biophys.*, *125*, 765 (1968).
42. N. Howell, J. M. Akagi, and R. H. Himes, *Can. J. Microbiol.*, *15*, 461 (1969).
43. L. Ljungdahl, E. Irion, and H. G. Wood, *Biochemistry*, *4*, 2771 (1965).
44. J. M. Poston, K. Kuratomi, and E. R. Stadtam, *J. Biol. Chem.*, *241*, 4209 (1966).
45. M. R. Moore, W. E. O'Brien, and L. G. Ljungdahl, *J. Biol. Chem.*, *249*, 5250 (1974).
46. M. B. Allen, *Bacteriol. Rev.*, *17*, 125 (1953).
47. A. Weerkamp and R. D. MacElroy, *Arch. Microbiol.*, *85*, 113 (1972).
48. J. W. Crabb, A. L. Murdock, and R. E. Amelunxen, *Biochemistry*, *16*, 4830 (1977).
49. R. E. Amelunxen and A. L. Murdock, *CRC Crit. Rev. Microbiol.*, *6*, 343 (1978).
50. L. Gerber, D. G. Neubauer, and F. J. Stutzenberger, *J. Bacteriol.*, *169*, 2267 (1987).

51. S. Hashizume, T. Sekiguchi, and Y. Nosoh, *Arch. Microbiol.*, *107*, 75 (1976).
52. T. D. Brock, *Science*, *158*, 1012 (1967).
53. B. Bubela and E. S. Holdsworth, *Biochim. Biophys. Acta*, *123*, 364 (1966).
54. B. Bubela and E. S. Holdsworth, *Biochim. Biophys. Acta*, *123*, 376 (1966).
55. L. L. Campbell, *J. Bacteriol.*, *68*, 505 (1954).
56. L. L. Campbell, *Arch. Biochem. Biophys.*, *54*, 154 (1955).
57. H. V. Haberstich and H. Zuber, *Arch. Microbiol.*, *98*, 275 (1974).
58. G. L. Gray, S. E. Mainzer, M. W. Rey, M. H. Lamsa, K. L. Kindle, C. Carmona, and C. Requadt, *J. Bacteriol.*, *166*, 635 (1986).
59. H. Liao, T. McKenzie, and R. Hageman, *Proc. Natl. Acad. Sci. USA*, *83*, 576 (1986).
60. Y. Sadaie, K. C. Burtis, and R. H. Doi, *J. Bacteriol.*, *141*, 1178 (1980).
61. J. Wiegel and L. G. Ljundahl, *CRC Crit. Rev. Biotechnol.*, *3(1)*, 39 (1986).
62. M. Mandels and R. E. Andreotti, *Proc. Biochem.*, *13*, 6 (1978).
63. M. P. Coughlan, *Biotechnol. Gen. Eng. Rev.*, *3*, 39 (1985).
64. D. A. Mays, G. L. Terman, and J. C. Duggan, *J. Env. Qual.*, *2*, 89 (1973).
65. M. S. Finstein and M. L. Morris, In *Adv. Appl. Microbiol.* (D. Perlman, ed.), *19*, 113 (1975).
66. R. P. Poincelot, *Conn. Agr. Exp. Sta., New Haven, Bull.*, 727 (1972).
67. K. L. Schulze, *Compost. Sci.*, *5*, 5 (1965).
68. S. S. Block, *Appl. Microbiol.*, *13*, 5 (1965).
69. C. W. P. Francis, *Water Pollut. Control*, *66*, 19 (1967).
70. G. L. Terman, J. M. Soileau, and S. E. Allen, *J. Env. Qual.*, *2*, 84 (1973).
71. W. W. Rose and W. A. Mercer, *Fate of Pesticides in Composted Agricultural Waste*, Progress Report Pt. 1, West. Res. Lab. Natl., Canners Assoc., Berkeley, 1968. •
72. C. Browne, *Science*, *77*, 223 (1933).
73. D. G. Cooney and R. Emerson, *Thermophilic Fungi*, W. H. Freeman, San Francisco, 1964, p. 140.
74. W. J. Scott, *Adv. Food Res.*, *7*, 83 (1957).
75. M. H. Dye and H. P. Rothbaum, *N. Zeal. J. Sci.*, *7*, 97 (1964).
76. H. P. Rothbaum, *J. Appl. Chem.*, *13*, 291 (1963).

77. M. R. Tansey and T. D. Brock, In *Microbial Life in Extreme Environments* (D. Kushner, ed.), Academic Press, London, 1978, p. 159.

78. D. J. Eastwood, *Trans. Br. Mycol. Soc.*, *35*, 215 (1952).

79. G. N. Festenstein, J. Lacy, F. A. Skinner, P. A. Jenkins, and J. Pepys, *J. Gen. Microbiol.*, *41*, 389 (1965).

80. C. L. Fergus, *Mycologia*, *56*, 267 (1964).

81. U. Chang and H. J. Hudson, *Trans. Br. Mycol. Soc.*, *50*, 649 (1967).

82. L. K. Walker and W. J. Harrison, *N. Zeal. J. Agr. Res.*, *3*, 861 (1960).

83. J. S. Wiley and J. T. Spillane, *J. Sanit. Eng. Div. Am. Soc. Civil Eng.*, *87* (SA5), 33 (1961).

84. G. J. Jann, D. H. Howard, and A. G. Sallee, *Appl. Microbiol.*, *7*, 271 (1959).

85. D. J. Suler and M. S. Finstein, *Appl. Env. Microbiol.*, *33*, 345 (1977).

86. K. L. Schulze, *Proc. Ind. Waste Conf.*, *13* (96), 541 (1958).

87. C. L. Cooney, D. I. C. Wang, and R. I. Mateles, *Biotechnol. Bioeng.*, *11*, 269 (1968).

88. W. G. Forsyth and D. M. Webley, *Proc. Soc. Appl. Bacteriol.*, 34 (1948).

89. D. Erikson, *J. Gen. Microbiol.*, *6*, 286 (1952).

90. F. J. Stutzenberger, A. J. Kaufman, and R. D. Lossin, *Can. J. Microbiol.*, *16*, 553 (1970).

91. F. J. Stutzenberger, *Appl. Microbiol.*, *22*, 147–172 (1971).

92. D. Gottlieb, In *Actinomycetales: Characteristics and Practical Importance* (G. Sykes and F. A. Skinner, eds.), Academic Press, London, 1973, p. 1.

93. S. T. Williams, *Int. Biodeter. Bull.*, *2*, 125 (1966).

94. E. Stackebrandt and C. R. Woese, *Curr. Microbiol.*, *5*, 197 (1981).

95. S. Ishizawa, T. Suzuki, T. Koda and O. Sato, *Bull. Natl. Inst. Agr. Sci.*, *B8*, 73 (1958).

96. H. Nonomura and Y. Ohara, *J. Ferment. Technol.*, *49*, 895 (1971).

97. E. Kuster, In *Actinomycetes: The Boundary Microorganisms* (T. Arai, ed.), Toppan, Tokyo, 1976, p. 109.

98. S. Isizawa and M. Araragi, In *Actinomycetes: The Boundary Microorganisms* (T. Arai, ed.), Toppan, Tokyo, 1976, p. 97.

99. J. Pepys, P. A. Jenkins, G. N. Festenstein, M. E. Lacy, P. H. Gregory, and F. A. Skinner, *The Lancet*, *7308*, 607 (1963).

100. J. Lacy, *J. Gen. Microbiol.*, *66*, 327 (1971).

101. D. L. Crawford and J. B. Sutherland, *Dev. Ind. Microbiol.*, *20*, 143 (1978).
102. G. Niese, *Arch. Microbiol.*, *34*, 285 (1959).
103. P. H. Gregory and M. E. Lacy, *J. Gen. Microbiol.*, *30*, 75 (1963).
104. F. J. Stutzenberger and R. Carnell, *App. Env. Microbiol.*, *34*, 234 (1977).
105. D. Kluepfel and M. Ishaque, *Dev. Ind. Microbiol.*, *23*, 389 (1982).
106. D. L. Crawford, *Biotechnol. Bioeng. Symp. Ser.*, *11*, 275 (1986).
107. A. J. McCarthy, E. Peace, and P. Broda, *Appl. Microbiol. Biotechnol.*, *21*, 238 (1985).
108. S. K. C. Obi and F. J. C. Odibo, *Can. J. Microbiol.*, *30*, 780 (1984).
109. M. J. Kuo and P. A. Hartman, *J. Bacteriol.*, *92*, 723 (1966).
110. L. J. Glymph and F. J. Stutzenberger, *Appl. Env. Microbiol.*, *34*, 391 (1977).
111. M. Shimizu, M. Kanno, M. Tamura, and M. Suekane, *Agr. Biol. Chem.*, *42*, 1681 (1978).
112. H. Hidaka, Y. Koaze, K. Yoshida, and T. Niwa, *J. Jap. Soc. Starch Sci.*, *25*, 148 (1974).
113. M. B. Phelan, D. L. Crawford, and A. L. Pometto, III, *Can. J. Microbiol.*, *25*, 1270 (1979).
114. D. L. Crawford, *Appl. Env. Microbiol.*, *35*, 1041 (1978).
115. C. Catton, *New Scientist*, *100*, 38 (1983).
116. T. G. Pridham and H. D. Tresner, In *Bergey's Manual of Determinative Bacteriology*, 8th ed. (R. E. Buchanan and N. E. Gibbons, eds.), Williams and Wilkins, Baltimore, 1974, p. 747.
117. D. L. Crawford and E. McCoy, *Appl. Microbiol.*, *24*, 150 (1972).
118. B. Norkans, *Physiol. Plant*, *9*, 198 (1956).
119. E. B. Shirling and D. Gottlieb, *Int. J. Syst. Bacteriol.*, *19*, 391 (1969).
120. M. R. Rao and S. A. Dhala, *Ind. J. Microbiol.*, *21*, 216 (1981).
121. D. Kluepfel, F. Shareck, F. Mondou, and R. Morosoli, *Appl. Microbiol. Biotechnol.*, *24*, 230 (1986).
122. C. R. MacKenzie, D. Bilous, and K. G. Johnson, *Biotechnol. Bioeng.*, *26*, 590 (1984).
123. N. Moldoveanu and D. Kluepfel, *Appl. Env. Microbiol.*, *46*, 17 (1983).
124. K. Gray, K. Sherman, and A. J. Biddlestone, *Proc. Biochem.*, *6*, 1 (1971).

125. B. Godden and M. J. Penninchx, *Ann. Microbiol.* (Inst. Pasteur), *135B*, 69 (1984).

126. M. Malfait, B. Godden, and M. J. Penninckx, *Ann. Microbiol.* (Inst. Pasteur), *135B*, 79 (1984).

127. B. Godden, M. Malfait, and M. Penninchx, *Physiol. Biochim.*, *90*, B192 (1982).

128. F. J. Stutzenberger, *Appl. Microbiol.*, *24*, 77 (1972).

129. F. J. Stutzenberger, *Appl. Microbiol.*, *24*, 83 (1972).

130. W. D. Bellamy, *Biotechnol. Bioeng.*, *16*, 869 (1974).

131. W. D. Bellamy, *ASM News*, *45*, 325 (1979).

132. T. M. Su and I. Paulavicius, In *Proc. 8th Cellulose Conference. I. Wood Chemicals: A Future Challenge* (T. E. Timell, ed.), Applied Polymer Symp., *28*, John Wiley and Sons, New York, 1975, p. 230.

133. J. D. Ferchak, B. Hagerdal, and E. K. Pye, *Biotechnol. Bioeng.*, *22*, 1527 (1980).

134. M. Castanon and C. R. Wilke, *Biotechnol. Bioeng.*, *22*, 1037 (1980).

135. M. V. Deshpande and K. E. Eriksson, *Enzyme Microb. Technol.*, *6*, 338 (1984).

136. S. E. Lee and A. E. Humphrey, *Biotechnol. Bioeng.*, *21*, 1277 (1979).

137. F. J. Stutzenberger, In *Ann. Reports Ferment. Proc.* G. T. Tsao, ed.), *8*, 111 (1985).

138. W. E. Wood, D. G. Neubauer, and F. J. Stutzenberger, *J. Bacteriol.*, *160*, 1047 (1984).

139. D. L. Crawford, *Can. J. Microbiol.*, *21*, 1842 (1975).

140. A. J. Desai and S. A. Dhala, *Antonie van Leeuwenhoek*, *37*, 56 (1967).

141. A. J. Desai and S. A. Dhala, *J. Bacteriol.*, *100*, 149 (1969).

142. B. Friebe and A. W. Holldorf, *J. Bacteriol.*, *122*, 818 (1975).

143. K. E. Eriksson and B. Pettersson, *Eur. J. Biochem.*, *124*, 635 (1982).

144. A. Paszczynski, J. Fiedurek, Z. Ilczuk, and G. Ginalska, *Appl. Microbiol. Biotechnol.*, *22*, 434 (1985).

145. S. P. Howard and J. T. Buckley, *J. Bacteriol.*, *163*, 336 (1985).

146. S. Gottesman, P. Trisler, A. Torres-Cabassa, and M. R. Maurizi, In *Microbiology – 1985* (L. Leive, ed.), Am. Soc. Microbiol., Washington, D.C., 1985, p. 350.

147. D. Sternberg and G. R. Mandels, *J. Bacteriol.*, *139*, 761 (1979).

148. D. Rho, M. Desrochers, L. Jurasek, H. Driguez, and J. Defaye, *J. Bacteriol.*, *149*, 47 (1982).

149. S. N. Mukhopadhyay and R. Malik, *Biotechnol. Bioeng.*, *22*, 2237 (1980).
150. D. Ryu, R. Andreotti, M. Mandels, B. Gallo, and E. T. Reese, *Biotechnol. Bioeng.*, *21*, 1887 (1979).
151. E. T. Reese, *Biotechnol. Bioeng. Symp. Ser.*, *3*, 43 (1972).
152. H. P. Hohn and H. Saham, In *Enzyme Technology* (R. M. Lafferty, ed.), Springer-Verlag, Berlin, 1982, p. 55.
153. F. J. Stutzenberger and I. Sterpu, *Appl. Env. Microbiol.*, *36*, 201 (1978).
154. A. J. McCarthy and T. Cross, *J. Gen. Microbiol.*, *130*, 5 (1984).
155. F. Stutzenberger and D. Lupo, *Enzyme Microb. Technol.*, *8*, 205 (1986).
156. R. F. Bernier and F. J. Stutzenberger, *Appl. Env. Microbiol.*, *53*, 1743 (1987).
157. R. Lamed, J. Naimark, E. Morgenstern, and E. A. Bayer, *J. Bacteriol.*, *169*, 3792 (1987).
158. W. Stoppok, P. Rapp, and F. Wagner, *Appl. Env. Microbiol.*, *44*, 44 (1982).
159. A. Peterkofsky, In *Advances in Cyclic Nucleotide Research*, Vol. 7 (P. Greengard and G. A. Robinson, eds.), Raven Press, New York, 1976, p. 1.
160. G. Fennington, D. Neubauer, and F. Stutzenberger, *Biotechnol. Bioeng.*, *25*, 2271 (1983).
161. G. Fennington, D. Neubauer, and F. Stutzenberger, *Appl. Env. Microbiol.*, *47*, 201 (1984).
162. A. Ullmann and A. Danchin, In *Advances in Cyclic Nucleotide Research*, Vol. 15 (P. Greengard and G. A. Robinson, eds.), Raven Press, New York, 1983, p. 1.
163. K. Prabakaran and H. C. Dube, *Curr. Sci.*, *54*, 1183 (1985).
164. B. S. Montenecourt, *Trends Biotechnol.*, *1*, 156 (1983).
165. B. S. Montenecourt, S. D. Nhlaop, H. Trimino-Vazques, S. Cuskey, D. H. J. Schamhart, and D. E. Eveleigh, In *Trends in the Biology of Fermentations for Fuels and Chemicals* (A. Hollaender, ed.), Plenum Press, New York, 1981, p. 33.
166. P. Setlow, *Biochem. Biophys. Res. Commun.*, *52*, 365 (1973).
167. M. R. Ladisch, K. W. Lin, M. Voloch, and G. T. Taso, *Enzyme Microb. Technol.*, *5*, 82 (1983).
168. L. T. Fan, Y. H. Lee, and M. M. Charpuray, *Adv. Biochem. Eng.*, *23*, 157 (1982).
169. M. M. Change, T. Y. C. Chou, and G. T. Taso, *Adv. Biochem. Eng.*, *20*, 15 (1981).
170. T. K. Kirk, *Biotechnol. Bioeng. Symp. Ser.*, *5*, 139 (1975).

171. R. L. Crawford and D. L. Crawford, *Enzyme Microb. Technol.*, *6*, 434 (1984).
172. M. G. Jackson, *Anim. Feed Sci. Technol.*, *2*, 105 (1977).
173. A. J. Panshin, C. deZeeuw, and H. P. Brown, *Textbook of Wood Technology*, Vol. 1, 2nd ed., McGraw-Hill, New York, 1964, p. 17.
174. A. J. McCarthy and P. Broda, *J. Gen. Microbiol.*, *130*, 2905 (1984).
175. A. J. McCarthy, A. Paterson, and P. Broda, *Appl. Microbiol. Biotechnol.*, *24*, 347 (1986).
176. H. Greaves, *Mater Organismem*, *5*, 265 (1970).
177. M. Tien and T. K. Kirk, *Science*, *221*, 661 (1983).
178. P. J. Kersten, M. Tien, B. Kalyaharaman, and T. K. Kirk, *J. Biol. Chem.*, *260*, 2609 (1985).
179. F. J. Stutzenberger, *Biotechnol. Bioeng.*, *21*, 909 (1979).
180. J. D. Ferchak and E. K. Pye, *Biotechnol. Bioeng.*, *25*, 2865 (1983).
181. R. Seltzer, *Chem. Eng. News*, May 24, p. 23 (1987).
182. S. K. C. Obi and F. J. C. Odibo, *Can. J. Microbiol.*, *30*, 780 (1984).
183. J. Hsiu, E. H. Fischer, and E. A. Stein, *Biochemistry*, *3*, 61 (1964).
184. M. Arai, Y. Minoda, and K. Yamada, *Agr. Biol. Chem.*, *33*, 922 (1969).
185. B. G. R. Hagerdal, J. D. Ferchak, and E. K. Pye, *Appl. Env. Microbiol.*, *36*, 606 (1978).
186. B. Hagerdal, J. Ferchak, and E. K. Pye, *Adv. Chem. Ser.*, *181*, 331 (1979).
187. G. M. Umezurike, *Biochem. J.*, *177*, 9 (1979).
188. K. E. Eriksson and B. Pettersson, In *Biodeterioration of Materials*, Vol. 2 (A. Harry Walters and E. H. Hueck-van der Plas, eds.), Elsevier, Amsterdam, 1968, p. 116.
189. A. McHale and M. P. Coughlan, *Biochim. Biophys. Acta*, *662*, 152 (1981).
190. G. Beldman, M. F. Searle-van Leeuwen, F. M. Rombouts, and F. G. H. Voragen, *Eur. J. Biochem.*, *146*, 301 (1985).
191. D. Sternberg, P. Vijayakumar, and E. T. Reese, *Can. J. Microbiol.*, *23*, 139 (1977).
192. A. L. Allen and D. Sternberg, *Biotechnol. Bioeng. Symp. Ser.*, *10*, 180 (1980).
193. C. P. Dunne, In *Enzyme Engineering*, Vol. 6 (I. Chibata, S. Fukui, and L. B. Wingard, Jr., eds.), Plenum Press, New York, 1982, p. 355.
194. T. K. Ng and J. G. Zeikus, *Appl. Env. Microbiol.*, *42*, 231 (1981).

195. J. Woodward and A. Wiseman, *Enzyme Microb. Technol.*, *4*, 73 (1982).

196. M. P. Coughlin, *Biotechnol. Gen. Eng. Rev.*, *3*, 39 (1985).

197. C. R. Waldron, C. A. Becker-Vallone, and D. E. Eveleigh, *Appl. Microbiol. Biotechnol.*, *24*, 477 (1986).

198. C. R. Waldron and D. E. Eveleigh, *Appl. Microbiol. Biotechnol.*, *24*, 487 (1986).

199. S. Zamenhof and S. Greer, *Nature*, *182*, 884 (1958).

200. L. L. Larcom and N. H. Thaker, *Biophys. J.*, *19*, 299 (1977).

201. T. Imanaka and S. Aiba, In *Thermophiles: General, Molecular and Applied Microbiology* (T. D. Brock, ed.), John Wiley and Sons, New York, 1986, p. 159.

202. A. Henssen, *Arch. Mikrobiol.*, *26*, 373 (1957).

203. H. A. Lechevalier, *Int. Bull. Bacteriol. Nomenclat. Taxon.*, *15*, 139 (1965).

204. T. Cross, In *Bergey's Manual of Determinative Bacteriology*, 8th ed. (R. E. Buchanan and N. E. Gibbons, eds.), Williams and Wilkins, Baltimore, 1974, p. 860.

205. J. C. Ensign, *Ann. Rev. Microbiol.*, *32*, 185 (1978).

206. T. Cross, P. D. Walker, and G. W. Gould, *Nature*, *220*, 352 (1968).

207. M. D. Collins, G. C. Mackillop, and T. Cross, *FEMS Microbiol. Lett.*, *13*, 151 (1982).

208. T. Craveri, P. L. Manachini, F. Aragozzini, and C. Merendi, *J. Gen. Microbiol.*, *74*, 201 (1973).

209. T. Cross and J. Lacey, In *The Actinomycetales* (H. Prauser, ed.), Gustav Fischer, Hena, 1970, p. 211.

210. B. Haerdal, J. D. Ferchak, and E. K. Pye, *Biotechnol. Bioeng.*, *22*, 345 (1979).

211. J. Lacy, In *Actinomycetales: Characteristics and Practial Importance* (G. Sykes and F. A. Skinner, eds.), Academic Press, London, 1973, p. 231.

212. M. P. Lechevalier and H. Lechevalier, In *Biology of Industrial Microorganisms* (A. L. Demain and N. A. Solomon, eds.), Benjamin-Cummings, London, 1985, p. 315.

213. S. A. Z. Mahmoud, F. M. Thabet, A. Hazem, and M. A. El-Borollosy, *J. Monoufeia Agr. Res.*, *8*, 13 (1984).

214. F. G. Priest, *Microbiol. Rev.*, *41*, 711 (1977).

215. K. Aunstrup, In *Applied Biochemistry and Bioengineering*, Vol. 2 (L. B. Wingard, E. Katchalski-Katzir, and L. Goldstein, eds.), Academic Press, New York, 1979, p. 27.

216. J. Zemek, J. Augustin, R. Borriss, L. Kuniak, M. Svabova, and Z. Pacova, *Folia Microbiol.*, *26*, 403 (1981).

217. T. Gibson and R. E. Gordon, In *Bergey's Manual of Determinative Bacteriology*, 8th ed. (R. E. Buchanan and N. E. Gibbons, eds.), Williams and Wilkins, Baltimore, 1974, p. 540.

218. L. M. Robson and G. H. Chambliss, *Appl. Env. Microbiol.*, 47, 1039 (1984).
219. N. Dhillon, S, Chhibber, M. Saxena, S. Pajni, and D. V. Vadehra, *Biotechnol. Lett.*, 7, 695 (1985).
220. B. Setlow and P. Setlow, *J. Bacteriol.*, 136, 433 (1978).
221. K. H. Yeung, G. Chaloner-Larsson, and K. Yamazaki, *Can. J. Biochem.*, 54, 854 (1976).
222. K. Y. Chan and S. Au, *Antonie van Leeuwenhoek*, 53, 125 (1987).
223. K. S. Au and K. Y Chan, *Microbes*, 48, 93 (1986).
224. W. Okazaki, T. Akiba, H. Horikoshi, and R. Akahoski, *Appl. Microbiol. Biotechnol.*, 19, 335 (1984).
225. H. Gruninger and A. Fiechter, *Enzyme Microb. Technol.*, 8, 309 (1986).
226. K. Eriksson, *Pure Appl. Chem.*, 53, 33 (1981).
227. A. G. J. Voragen, R. Heutink, and W. J. Piknik, *Appl. Biochem.*, 2, 452 (1980).
228. G. Beldman, F. M. Rombouts, A. G. J. Voragen, and W. Pilnik, *Enzyme Microb. Technol.*, 6, 503 (1984).
229. F. Uchino and T. Nakane, *Agr. Biol. Chem.*, 45, 1121 (1987).
230. S. P. Peiris, P. A. D. Rickard, and N. W. Dunn, *Eur. J. Appl. Microbiol. Biotechnol.*, 14, 169 (1982).
231. G. L. Pettipher and M. J. Latham, *J. Gen. Microbiol.*, 110, 21 (1979).
232. M. John, B. Schmidt, and J. Schmidt, *Can. J. Biochem.*, 57, 125 (1979).
233. S. Toda, H. Suzuki, and K. Nisizawa, *J. Ferment. Technol.*, 49, 499 (1971).
234. T. D. Brock and H. Freeze, *J. Bacteriol.*, 98, 289 (1969).
235. T. D. Brock and I. Yoder, *Proc. Ind. Acad. Sci.*, 80, 183 (1971).
236. R. F. Ramaley and J. Hixson, *J. Bacteriol.*, 103, 527 (1970).
237. E. T. Allen and A. L. Day, *Hot Springs of the Yellowstone National Park*, Carnegie Inst., Washington, D.C., 1935, Publ No. 466.
238. A. Mohagheghi, K. Grohmann, M. Himmel, L. Leighton, and D. M. Updegraff, *Int. J. Syst. Bacteriol.*, 36, 435 (1986).
239. H. G. Schlegel, *General Microbiology*, Cambridge University Press, New York, 1986, p. 104.
240. M. Mandels, R. Adreotti, and C. Roche, *Biotechnol. Bioeng. Symp. Ser.*, 6, 21 (1976).
241. P. W. Postma, In *Regulation of Gene Expression* (I. R. Booth and C. F. Higgins, eds.), Cambridge University Press, Cambridge, 1986, p. 21.

21

Fungal Cellulases

THOMAS M. WOOD   *Rowett Research Institute, Aberdeen, Scotland*

I. INTRODUCTION

In catalyzing the decay of lignocellulosic residues, cellulolytic fungi are of major importance in our ecosystem. The cellulase enzymes synthesized by the fungi have a crucial role to play in recycling nutrients, in maintaining soil fertility, and in preserving the carbon cycle in nature. Not surprisingly, fungal cellulases have been the subject of many investigations over the years. However, what may once have been regarded as an esoteric subject investigated by a few academics is now of major commercial interest (1). The growing need for energy, food, and chemicals and the ever present problem of disposing of industrial wastes has focused attention on the importance of cellulases, and fungal cellulases in particular, for the generation of glucose feedstocks for further biological conversion.

The intense interest in fungal cellulase stems from the fact that several fungi provide extracellular cellulases in good yield (2); bacterial cellulases, with one or two notable exceptions (3–5), are cell wall-bound. Because it is generally felt that realization of the full industrial potential of fungal cellulases is dependent on the development of a good understanding of the basic biochemistry involved in the bioconversion, a large number of investigations have been carried out in this area. Significant advances have undoubtedly been made in the last 10–15 years and the most notable of these up until 1986 are reviewed briefly in this chapter. The reader is referred to reviews by Reese (6), Reese and Mandels (7), Mandels (8), Wood and McCrae (2), Bisaria and Ghose (9), Goksoyr and Eriksen (10), Eriksson (11), Ryu and Mandels (12), Eriksson and Wood (13), Wood (14), Coughlan (15), Enari (16), Enari and Niku-Paavola (17), Ljungdahl and Eriksson (18), Henrissat and Chanzy (19), and Wood (20) for further consideration of these and other aspects of the biochemistry of fungal cellulolysis.

In this review, β-glucosidases are also discussed. Strictly speaking, these enzymes are not cellulases. However, since the β-glucosidases have an extremely important role to play in the rate and extent of degradation of cellulose by fungal cellulases, their properties are also reviewed. Consideration is also given to oxidative enzymes which are important in some cellulases (11).

II. CELLULASE COMPLEX

The cell-free cellulase that can degrade crystalline cellulose is often described as a "cellulase complex," a "true cellulase," or a "cellulase system." Relatively few such cell-free cellulases are known, but notable in this respect are those from *Trichoderma viride* (21,22), *T. reesei* (formerly *T. viride* QM6a) (23), *T. koningii* (24,25), *Fusarium solani* (26), *Penicillium funiculosum* (27), *Penicillium pinophilum/funiculosum* (28), *Sporotrichum pulverulentum* (11), *Talaromyces emersonii* (29), and *Neocallimastix frontalis* (30).

As currently understood, the cellulase complex consists of a number of different enzymes, which by acting sequentially and in concert effectively render cellulose soluble (2). There are many reported characterizations of the component enzymes, but it is still not known with any degree of certainty how the components arise, how they interact, and what factors control the interactions that result in the solubilization of native cellulose. These uncertainties have resulted in continuing speculation and debate regarding the mechanism of cellulase action, and have raised questions regarding the purity and substrate specificity of the enzymes (31).

Notwithstanding these uncertainties, there is some consensus that the enzymes of the complex comprise three major types: endo-1,4-β-glucanases (endo-1,4-β-D-glucan-4-glucanohydrolases, EC 3.2.1.4), exo-1,4-β-glucanases (normally 1,4-β-D-glucan cellobiohydrolase, EC 3.2.1.91), and β-D-glucoside glucohydrolase (EC 3.2.1.21). Some cellulase systems also contain a glucohydrolase (1,4-β-D-glucan glucohydrolases, EC 3.2.1.74), but these are minor constituents (2). It seems that the systematic nomenclature has now completely displaced that used previously, which described the enzyme system in terms of a nonhydrolytic chain-disaggregating enzyme (so-called C_1) and hydrolytic enzymes (so-called C_x) (32). However, the use of the systematic names implies that the various enzymes fall neatly into one or another of the categories, and the recent literature reveals that this is very far from being the case. Indeed, if some of the most recent views (13,33,34) propounded on the mode of action of the enzymes described as cellobiohydrolase and endoglucanase are proved to be tenable, a complete reclassification of the enzymes may be necessary.

III. ENZYME SUBSTRATES

Unequivocal and reproducible evidence for the substrate specificity of the individual enzymes appears to be very difficult to obtain. Apart from the problem of obtaining homogeneous protein from an extremely complex mixture, the cellulase investigator is faced with the problem of using substrates that are very poorly defined. Cellulose in which the chains are highly hydrogen bond-ordered (crystalline), cellulose in which the chains are partially degraded and partially hydrated (amorphous), and cellulose in which the chains are completely hydrated and water-soluble are all used to measure cellulase activity. All these substrates are very variable in composition, and enzyme activity can only be compared if exactly the same substrate is used. Recommendations for measuring the various types of activity have been drawn up recently by a IUPAC Commission on Biotechnology (35) in an attempt to permit meaningful comparison of data obtained in various laboratories, However, it seems that these recommendations, which have been circulated for some time to the various workers in the area, have largely been ignored: only the assay to measure filter paper-hydrolyzing activity is widely used (36).

Cotton fiber is favored by some investigators (23,25,26) as the archetypal crystalline cellulose because in its native state it is virtually pure cellulose and can therefore be used with a minimum of pretreatment. By most physical measurements, cotton is highly crystalline, but *Valonia* microcrystals are reported to be the most perfect native cellulose known (37). Indeed, in *Valonia* each microfibril is reported to be an individual crystal of cellulose (38). In addition to cotton fiber (23,25,26), filter paper (that prepared from cotton) (36), *Valonia* cellulose crystals (34,39), and the microcrystalline cellulose Avicel (22,40) are all used as examples of highly hydrogen bond-ordered cellulose to measure true cellulase activity. Avicel has a relatively short cellulose chain length of approximately

200, which compared with values of 10,000 – 15,000 quoted for native cotton; Avicel also has a much lower crystallinity index than either cotton or *Valonia* cellulose (34). Because Avicel has more nonreducing end groups per unit weight than other crystalline substrates, it is useful for measuring endwise acting enzymes (2).

Amorphous cellulose is prepared by treatment of native cellulose with phosphoric acid (Walseth cellulose) (41) or by ball-milling (42, 43), and soluble cellulose is prepared by introducing either carboxymethyl or hydroxyethyl substituents into the chain. The extent to which the crystallinity is disrupted by the physical treatment and the extent of the substitution all have a marked effect on the rate and degree of hydrolysis effected by the various enzymes. The difficulty in preparing these substrates in a reproducible fashion creates problems for workers trying to compare data.

[1-^{3}H]Cellooligosaccharides (44) and 4-methylumbelliferylglucosides (45) have been used recently to determine the mode of action of the cellulase and β-glucosidases of *T. reesei*.

IV. PURIFICATION AND PROPERTIES

A. Isolation and Purification of Individual Components

Publications on the separation and purification of the individual components of the cellulase complex continue to appear. Each of these demonstrates the difficulty of obtaining a homogeneous protein from the cellulase system, which contains several proteins of very similar physicochemical properties. Earlier methods of separation were dependent on differential adsorption and on ion exchange, but all the normal methods of protein purification including gel filtration, electrophoresis, isoelectric focusing, chromatofocusing, and hydrophobic interaction chromatography are now used. Normally repeated application of one or another of these methods is required to obtain a single cellulase protein species. Invariably, these methods involve repeated concentration, precipitation, or desalting steps as preliminary to chromatography or electrophoresis, and clearly there is a risk that this may result in partial denaturation of the enzyme. These denaturations are not always apparent unless recovery of activity is measured using a wide variety of cellulose substrates. It seems that this factor must be considered when attempting to evaluate the significance of claims of the isolation of cellulase components of unusual substrate specificity.

More gentle methods of separation involving affinity chromatography have been used with success. Differential affinity of the various components for cellulose, first used by Gilligan and Reese (46) and later by Li et al. (47), Nummi et al. (48), and Halliwell and Griffin (49), was used to effect some partial separation of the cellulase components. However, the complete separation of β-glucosidase of *T. viride* (50) or *T. koningii* (51) from the rest of the cellulase system was obtained using the lectin concanavalin A immobilized on agarose or Sepharose. The β-glucosidase was not adsorbed, but the other components, which were glycoproteins, were bound and later eluted with α-methyl-D-mannoside (50,51).

More elegant procedures involving affinity chromatography have been used to provide highly purified cellobiohydrolase from *T. reesei*. These include (1) the sequential use of biospecific adsorption and immobilized antibodies to cellobiohydrolase (43) and (2) a column of the ligand p-aminobenzyl-1-thio-β-D-cellobioside coupled to Affigel 10 (Biorad) (52). Method (1) was claimed to yield a highly purified cellobiohydrolase with unique substrate specificities; method (2) resulted in a complete separation of cellobiohydrolase activity from endoglucanase and made possible the separation of two immunologically distinct cellobiohydrolases which are now known to be a feature of *T. reesei* cellulase (Fig. 1). In method (2) the cellobiohydrolases which were adsorbed were displaced selectively —cellobiohydrolase I with lactose and cellobiohydrolase II with cellobiose.

Using these various methods of purification, a large number of components have been judged to be pure by criteria which included isoelectric focusing, electrophoresis in denaturing conditions, ultracentrifugation, and immunoelectrophoresis. However, the demonstration by Sprey and Lambert (53) that cellulase components of *T. reesei* apparently homogeneous by isoelectric focusing could be separated into at least six proteins after treatment with

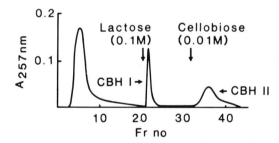

FIGURE 1   Affinity chromatography of crude cellulase of *Trichoderma reesei*. The biospecific adsorbent was p-aminobenzyl-1-thio-β-D-cellobioside coupled to Affigel-10. The crude cellulase was dissolved in acetate buffer pH 5.0 containing 1 mM gluconolactone to inhibit β-glucosidase activity which would damage the gel. [Reproduced with permission from *FEBS Lett. 169*, 215 (1084).]

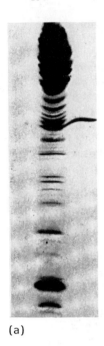

(a)

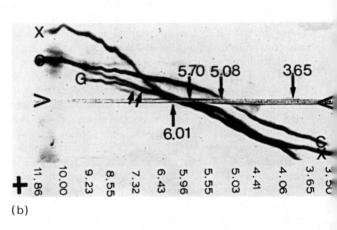

(b)

FIGURE 2   Multicomponent enzyme complexes in *Trichoderma reesei*.   (a)   Isoelectric focusing in ultrathin polyacrylamide gels of the total extracellular cellulase system (left lane) and a cellulase component purified by preparative flat bed isoelectric focusing (right lane).   The gel was stained for protein using Coomassie blue. (b)   Titration curve of the apparently homogeneous cellulase component (right lane, Fig. 2a) after treatment with 6 M urea-octylglucoside.   The gel was stained for protein using Coomassie blue. X = xylanase, C = cellulase (endoglucanase), G = β-glucosidase. Numbers indicate isoelectric points.   Arrowheads show the application trench.   [Reproduced with permission from *FEMS Microbiol. Lett.*, *18*, 217 (1983).]

urea-octylglucoside shows clearly that the criteria normally used for purity may not be sufficient when working with cellulases (Fig. 2). The implication of this study in terms of the published substrate specificities are clearly alarming, if such enzyme—enzyme complexes are indeed a feature of all cellulase preparations. The existence of enzyme—enzyme complexes may explain many of the anomalous results and unusual substrate specificities recorded in the literature.

It was suggested that the more gentle affinity methods of purification may result in the isolation of enzymes with substrate specificities more closely akin to those found in culture filtrates (17). However, there is clearly also a greater danger of isolating complexes which may be artifacts. Perhaps the more drastic the purification procedures used, the more likely that the complexes will be disrupted.

B. Properties of Individual Components

1. Cellobiohydrolase

It has been held for some time that fungal cellobiohydrolases act by removing cellobiose from the nonreducing end of the cellulose chain (2). This was demonstrated by the fact that the degree of polymerization of acid-treated cellulose decreased very slowly during hydrolysis by *T. viride* cellobiohydrolase (54). Similar results were obtained with the cellobiohydrolase of *F. solani* (28), *T. koningii* (28), and *P. pinophilum/funiculosum* (28,55), using cotton which had been swollen in phosphoric acid. Recently, however, it was suggested that some cellobiohydrolases may attack some linkages some distance from the nonreducing end of the chain (17): some workers indeed interpret their data to indicate that an enzyme, often described as a cellobiohydrolase, is in fact an endoglucanase (17,34,56). Thus, interpretation of the mode of action of the cellobiohydrolases is currently in some state of confusion.

Inversion of configuration of the hydroxyl on the anomeric carbon during hydrolysis is a characteristic property of exoglucanases (57), but so far only an exoglucanase of *T. viride* (58) and the exoglucanase of *S. pulverulentum* (59) have been demonstrated to act in this way.

Most "pure" cellobiohydrolases are reported to also release small quantities of glucose along with the cellobiose. In some cases, the amount of glucose released is a significant proportion of the total soluble sugars released, but the enzyme is still termed a cellobiohydrolase. It is not apparent in all cases if the lack of specificity for the penultimate glycosidic link is a property of the enzyme or the result of some associated hydrolase which is difficult to separate; yet a solution to this problem is of crucial importance before unequivocal conclusions on the mechanism of cellulase action and the

498   Wood

substrate specificity of the enzyme can be drawn. An enzyme re-
leasing cellotetraose, cellotriose, cellobiose, and glucose in the ratio
of 1:1:2:1 was isolated from the cellulase of S. pulverulentum (59),
and the enzyme was described as an exoglucanase.

When present in culture filtrates, cellobiohydrolase constitutes
the major protein. In T. reesei cellulase it amounts to approximately
80% of the extracellular protein (60): in P. pinophilum/funiculosum
cellulase the figure is almost as high (65%) (61). Cellobiohydrolase
has as yet been found in the cellulase of relatively few fungi (2).

Opinions as to the substrate specificity of the cellobiohydrolase
enzyme differ markedly. Thus degradation of cotton fiber is reported to be extensive (33,43) by the cellobiohydrolase of T. reesei
acting alone but negligible in T. koningii cellobiohydrolase (2,62).
Whatman no. 1 filter paper which is prepared from cotton fiber, and
cotton fiber itself, were also broken down into short fragments by
T. reesei cellobiohydrolase (17), but no breakdown was observed
with either T. koningii (62) or P. pinophilum/funiculosum (63,64).
T. koningii (24), P. pinophilum/funiculosum (63,64), and F. solani
(65) cellohydrolases were unable to effect changes in the cotton
fiber that resulted in an increase in the amount of alkali absorbed;
T. reesei cellobiohydrolase again differed in this respect (17,43).
These differences in opinion may point to the existence of a variety
of enzyme–enzyme complexes, as indicated by Sprey (53).

Avicel is reported to be a substrate for most cellobiohydrolases
isolated, although the rate and extent of hydrolysis varies widely.
Avicelase is now regarded to be synonymous with cellobiohydrolase,
and even the smallest release of reducing sugar is considered to indicate the presence of cellobiohydrolase. This is unfortunate, however, for some endoglucanases can also hydrolyze Avicel. Indeed,
the specific activity for Avicel of some endoglucanases of T. reesei
and T. viride was reported to be higher than the cellobiohydrolase.
Equally regrettable is the tendency to equate Avicelase activity with
cellobiohydrolase when a crude cellulase preparation is being assayed
In this circumstance, synergistic activity between cellobiohydrolase
and endoglucanase (see later) would give completely erroneous
results.

Carboxymethylcellulose is reported not to be a substrate for
cellobiohydrolase (2), and this characteristic is widely held as being
the property that distinguishes these enzymes from endoglucanases.
However, a scan of the literature reveals that this distinction has
become rather blurred lately. Thus, cellobiohydrolases isolated
from T. reesei (43,56) and T. viride (66) culture filtrates are reported to be able to hydrolyze CMC to an extensive degree.

Cellooligosaccharides are hydrolyzed by cellobiohydrolases with
increasing rate as the degree of polymerization increases (13).

Cellobiose, which is not hydrolyzed by cellobiohydrolase, is a potent inhibitor (13), but gluconolactone is not. The reported 83% inhibition of the exoglucanase of *S. pulverulentum* by 1 μM gluconolactone is exceptional (67). Glucose is less inhibitory to cellobiohydrolase than is cellobiose (68).

The physiochemical properties of a number of selected purified cellobiohydrolases are listed in Table 1. Molecular weights range from 41–65 kD, and isoelectric points are acidic. All except the exoglucanase of *S. pulverulentum* are glycoproteins with proportions of carbohydrate ranging from nil to 33%: mannose is the principal sugar present (63).

A characteristic of cellobiohydrolase is that it is found in some cellulase preparations in a multitude of forms: two of these forms appear to be immunologically unrelated (so-called cellobiohydrolase I and II) (22,64,69); the nature and origin of these various forms will be discussed later.

Extensive structural analysis has been carried out on cellobiohydrolase I of *T. reesei*, which consists of a single polypeptide chain. The complete amino acid sequence of the enzyme, which comprises 497 residues, was determined by Edman degradation (70). The amino terminal was found to be blocked by a pyroglutamyl residue (71), and the cellobiohydrolase II enzyme was found to be the same in this respect (71). In cellobiohydrolase I, glucosamine was present (22) and was attached to three of the five asparagine residues (70) found in the sequences that serve as recognition sites for the glycosylating enzymes (70). In addition, it was estimated that 22 serine/threonine residues contained O-glycosidically linked carbohydrate, and 10 of these were localized near the C-terminal end of the enzyme (70,71). Recently, the nucleotide sequence of cellobiohydrolase I was reported (73): the structure of the protein derived from this sequence was in excellent agreement with the amino acid sequence.

All of these structural studies are aimed at developing a good understanding of the structure–function relationships of the enzymes, with site-specific mutagenesis being the long-term aim. Currently the analysis is directed at the determination of the three-dimensional structure. Twelve disulfide bridges were discovered in cellobiohydrolase I and six have been located (74). The secondary structure was predicted using three different algorithms (75).

2. Endoglucanases

Endoglucanases hydrolyze cellulose chains at random to effect a rapid change in degree of polymerization with a slow increase in reducing power (2). Substrates include carboxymethylcellulose and cellulose swollen in phosphoric acid or alkali, but neither cotton fiber (2) nor Avicel is hydrolyzed extensively, in the main

TABLE 1 Properties of Some Fungal Cellobiohydrolases

Fungus	Enzyme	$M_r \times 10^{-3}$	pI	Carbohydrate content (%)	Ref.
Trichoderma koningii	CBH I	62	3.80	9.0	62
	CBH II	62	3.95	33.0	
Trichoderma viride	CBH I	46	3.79	3.3	54,146
		42		9.2	
	CBH A	53	—	1.4	121
	CBH B	53	—	5.8	
	CBH C	53	—	10.4	
	CBH D	53	—	6.7	
Trichoderma reesei	CBH I	—	3.95	—	69
	CBH II	—	5.0−5.6	—	

Trichoderma reesei	CBH I	65	3.6 – 4.2	10.0	43
	CBH II	—	5.9	—	56
Fusarium solani	CBH A	45	4.75	21.0	120
	CBH B	45	4.90	10.0	
	CBH C	45	4.82	12.0	
	CBH D	45	4.95	1.0	
Penicillium funiculosum	CBH I	46.3	4.36	9.0	63
(syn. pinophilum)	CBH I	51.0	4.7	9.0	64
	CBH II	51.0	5.0	19.0	
Sporotrichum pulverulentum	Exoglucanase	48.6	4.3	NIL	129
Sclerotium rolfsii	CBH	41.5	4.32	7.0	184

TABLE 2   Properties of Some Fungal Endoglucanases

Fungus	Enzyme	$M_r \times 10^{-3}$	pI	Carbohydrate content	Temperature optimum (°C)	Optimum pH	Ref.
Trichoderma koningii	E_1	13.0	4.73	—	—	—	51
	E_{3a}	48.0	4.32	—	—	—	
	E_{3b}	48.0	4.34	—	—	—	
	E_4	31.0	5.09	—	—	—	
Trichoderma viride	II	37.2	—	4.5	—	4.0–4.5	76
	III	52.0	—	15.0	—	—	
	IV	49.5	—	15.2	—	—	
Trichoderma viride	I	12.5	4.60	21.0	—	—	54
	II	50.0	3.39	12.0	—	—	
Talaromyces emersonii	I	35.0	3.19	27.7	75–80	5.5–5.8	119
	II	35.0	3.08	29.0	—	—	
	III	35.0	3.0	44.7	—	—	
	IV	35.0	2.86	50.8	—	—	

Organism							
Sporotrichum pulverulentum	T$_1$	32.3	5.32	10.5	—	—	128
	T$_{2a}$	36.7	4.72	0.0	—	—	
	T$_{2b}$	28.3	4.40	7.8	—	—	
	T$_{3a}$	37.5	4.65	4.7	—	—	
	T$_{3b}$	37.0	4.20	2.2	—	—	
Fusarium solani	1	37.0	4.75	—	—	—	124
	2	37.0	4.8	—	—	—	
	3	37.0	5.15	—	—	—	
Aspergillus fumigatus		12.5	7.1	0.0	60	4.8	185
Sclerotium rolfsii	A	50.0	4.55	—	74	4.0	186
	B	27.5	4.2	—	50	2.8 – 3.0	
	C	78.0	4.51	—	50	4.0	
Trichoderma reesei	I	20.0	7.5	—	—	—	71
	II	40.0	4.6	—	—	—	

(2,22,76). However, a purified endoglucanase from T. *reesei* was reported to have no capacity for degrading an amorphous cellulose prepared by ball-milling (77), while T. *viride* endoglucanase could solubilize Avicel extensively (78).

Cellobiose is not attacked by endoglucanases. Cellotriose is attacked by some endoglucanases (79), but not by all. The rate of hydrolysis of longer chain cellooligosaccharides is high, and the rate increases with the degree of polymerization (2). Optimal activity for the endoglucanase of *Aspergillus niger* (80) and *Thermoascus aurantiacus* (81) is observed on substrates that have at least five or six glucose residues. Glucose, cellobiose, and longer chain oligosaccharides are the normal products of the reaction of endoglucanases. However, as transglycosylation has been reported to be a property of T. *viride* endoglucanase (76), in some conditions other oligosaccharides will be found in the solution. Transglycosylation would not appear to be a property common to all endoglucanases.

Cellobiose is normally an inhibitor of the action of endoglucanase on carboxymethylcellulose (79). Remarkably, however, a low molecular weight endoglucanase from T. *koningii* cellulase was reported to be stimulated by cellobiose (79); similar observations were recorded for a T. *viride* endoglucanase (46).

A feature of most fungal cellulases is the multiplicity of forms in which the endoglucanase activity exists. These various forms differ in their mode of action (79,76), in the extent to which they are adsorbed on cellulose (82–85), and in their capacity for acting synergistically with cellobiohydrolase in solubilizing cellulose (79). These aspects are discussed in detail later in the chapter.

Some physicochemical properties of selected endoglucanases are listed in Table 2.

The endoglucanases are normally glycoproteins of acidic pI and range in size from 12 to approximately 60 kD.

3. β-Glucosidases

Fungal β-glucosidases can catalyze the hydrolysis of both alkyl- and aryl-β-D-glucosides as well as glucosides containing only carbohydrate (86,87). There are, however, notable exceptions to this generalization. Thus, β-glucosidases were isolated from *Stachbotrys atra* (86), *Stereum sanguinolentum* (86), and T. *viride* (78) with no activity to cellobiose. It was shown that aryl-β-glucosidase and cellobiase activity may originate in different enzymes in T. *reesei* (88). β-Glucosidases hydrolyze long-chain cellooligosaccharides at rates which decrease with an increasing degree of polymerization (13).

β-Glucosidases are not specific for the β-1,4 linkage, and β-1, 2, β-1,3, and β-1,6 linkages are attacked with ease (86). The β configuration is retained on hydrolysis and this property distinguishes them from exoglucanases, which act by inversion (89). β-Glucosidases can also be transferases acting on glucose units and forming alcohols, or other sugar molecules such as dimers, trimers, as well as higher oligosaccharides (51,22): this property may be important in the induction of cellulases (90,91) or in the biosynthesis of specialized sugar derivatives.

Most β-glucosidases are strongly inhibited by glucose and gluconolactone, and by excess substrate (13,86,87). End-product inhibition was reported to be either competitive (13,22,86,92−96) or noncompetitive (97−99) depending on the source, but both types were reported in a culture filtrate of *Monilia* species (100,101). Nojirimycin is a powerful inhibitor of β-glucosidase of *T. reesei* (102).

Because β-glucosidase enzymes are the only enzymes of the cellulase system that can hydrolyze cellobiose, they have a vital role to play in cellulose degradation. In cellulase systems deficient in β-glucosidase activity, the concentration of cellobiose increases during hydrolysis, and this results in inhibition of the cellobiohydrolase and endoglucanase enzymes. *Trichoderma reesei*, although a good producer of cellulase, produces β-glucosidase in only small quantities (8). The black aspergilli are the best sources of β-glucosidase (87), and enzymes from these sources, particularly from *Aspergillus phoenicis*, have been used to supplement the *T. reesei* cellulase system in attempts to produce an enzyme system that could be used commercially for the production of glucose (102−105).

The physicochemical properties of selected β-glucosidases are listed in Table 3. Several of the purified enzymes have been found to be glycoproteins with molecular weights ranging from 40 to 400 kD. Where tested, the β-glucosidases appear to exist as single polypeptides; however, that from *Aspergillus fumigatus* (106) is a dimer, and the 332 kD enzyme from *Botryodiplodia theobromae* (107) is reported to exist as an octamer. The octamer could be further dissociated into electrophoretically identical noncatalytic polypeptides of molecular weight 10−12 kD.

4. Oxidases and Dehydrogenases

In most fungal cellulase systems only β-glucosidase is synthesized to catalyze the hydrolysis of cellobiose and relieve end-product inhibition of cellulases. However, in other enzyme systems cellobiose oxidase and/or cellobiose dehydrogenase are also produced. As yet only the white rot fungus *S. pulverulentum* (11,109,110) was reported to synthesize cellobiose oxidase, but there is some indirect evidence that *T. koningii* (108) and *T. reesei* (109,111) may also produce this enzyme.

TABLE 3   Properties of Some Fungal β-Glucosidases

Fungus	Enzyme	$M_r \times 10^{-3}$	Carbohydrate content (%)	pI	K_m (mM) p-nitrophenyl-β-D-glucoside	K_m (mM) cellobiose	K^i (mM) glucose	Ref.
Botryodiplodia theobromae	—	332.0			0.33	1.0	—	107
Chaetomium thermophile	1[a]	—	—	—	1.1	1.1	91.0	95
	2[a]	—	—	—	0.09	—	0.5	
	3	—	—	—	0.07	—	0.3	
Monilia sp.	1[a]	480.0	—	4.4	1.67	20.0	1.6	100
	2[a]	37.5	—	8.3	0.08	5.7	0.67	
	3	37.5	—	8.3	0.08	5.7	0.67	
Penicillium funiculosum					0.4	2.1	1.7	93
Sclerotium rolfsii	1	95.5	+	4.10	1.07	3.65	—	141
	2	95.5	+	4.55	1.38	3.07	—	
	3	106.0	+	5.10	0.89	5.84	—	
	4	95.5	+	5.55	0.79	4.15	—	

Sporotrichum pulverulentum	A_1	165.0	—	4.80	0.15	4.5	—	130
	A_2	172.0	—	4.52	0.15	4.5	—
	B_1	165.0	—	5.15	—	—	—
	B_2	172.0	—	4.87	—	—	—
	B_3	182.0	—	4.56	—	—	—
Talaromyces emersonii	I	138.0	51.0	3.4 – 4.2	0.14	0.58	0.17	131
	III	40.0	26.0	3.60	1.03	23.70	—
	IV[a]	52.5	12.0	4.4 – 4.5	0.81	1.47	52.0
Trichoderma koningii	1	39.8	0.0	5.53	0.37[b]	1.18	1.05	51
	2	39.8	—	5.82	0.85[b]	0.86	0.66
Trichoderma viride		51.2	1.03	—	0.17	1.8	—	125
Trichoderma viride	1	76.0	—	—	—	2.65	—	50
	2	76.0	—	—	—	2.56	1.22
	3	76.0	—	—	—	2.70	4.26

[a] Intracellular or cell-bound.
[b] o-Nitrophenyl-β-D-glucoside.

The cellobiose oxidase of S. pulverulentum is a hemoprotein and contains a flavin adenine dinucleotide (FAD) group (113). It oxidizes cellobiose and higher cellodextrins to their corresponding onic acids using molecular oxygen. It was suggested that cellobiose oxidase may oxidize the reducing groups formed on the cellulose chain by endoglucanase (109).

Cellobiose dehydrogenase is found in culture filtrates of several types of fungi, namely, S. pulverulentum (112,113), the thermophilic fungus S. thermophile (114), Sclerotium rolfsii (115), and a species of the imperfect fungus Monilia (116).

Cellobiose dehydrogenase of S. pulverulentum is highly specific for cellobiose, which it oxidizes to cellobionolactone in the presence of electron acceptors such as quinones or phenoxy radicals, produced by the oxidation of lignin by phenol oxidase. The natural electron acceptors for the cellobiose dehydrogenase of Monilia sp. and S. thermophile have not been identified, but some artificial electron acceptors have been found (114,116) for the in vitro assay. In all three fungi the cellobiose dehydrogenase is induced simultaneously with the cellulase activity and may therefore be important in cellulose utilization.

V. MULTIPLICITY IN CELLULASE ACTIVITY
AND ITS SIGNIFICANCE

Detailed fractionation studies showed that multiple forms of all types of enzyme exist in all fungal cellulases. The nature and origin of, and the need for, the various forms has been the subject of much discussion in the literature. Some hold the view that much of the multiplicity is the result of partial proteolysis during culture or storage (97,117,118), differential glycosylation of a common polypeptide chain (62,119 − 121), interaction and aggregation of the enzymes with each other and with impurities in culture (53), or manipulation of the enzyme during purification (77). Others believe that the multiplicity may in some cases be genetically determined (121 − 123). At present it would seem that there is no consensus emerging; indeed, all possibilities seem to operate.

Heterogeneity in the cellulases of Trichoderma species has been particularly well studied. These cellulases have been shown to be heterogeneous with respect to β-glucosidase, endoglucanase, and cellobiohydrolase. Thus the cellulase of T. koningii (2,31,51) was found to contain two β-glucosidases, four major endoglucanases, and two cellobiohydrolases. The cellulases of P. funiculosum/ P. pinophilum (2,63,108), F. solani (120,124), T. viride (125 − 127), T reesei (7,17,71), S. pulverulentum (13,128 − 130), and Talaromyces emersonii (29,92,119,131) were shown to be equally heterogeneous.

A. Cellobiohydrolase Multiplicity

Heterogeneity with respect to cellobiohydrolase is of particular current interest. On the basis of several reports it would seem that immunologically related and unrelated cellobiohydrolases exist in cellulases of *T. viride* and *T. reesei* (17,22,69). The two cellobiohydrolases isolated so far from *P. funiculosum/pinophilum* were of the immunologically distinct type (55,64). The cellobiohydrolase with the more acidic pI were in each case designated cellobiohydrolase I; the more basic was called cellobiohydrolase II.

Studies on the heterogeneity of cellobiohydrolase I have been the most extensive. Three forms have been isolated (121). They were similar in molecular weight, heat stability, amino acid composition, and C-terminal residues, indicating approximately identical polypeptide portions. The enzymes also showed immunological cross-reactivity. The principal differences between the two enzymes were in the isoelectric pH and in the composition and content of covalently bound carbohydrate, which was mostly mannose. Studies on two cellobiohydrolases from *T. koningii* (62) and four from *F. solani* (120) (Table 1) were less exhaustive, but in both cases it was tentatively concluded that the heterogeneity could in all probability be ascribed to differential glycosylation of a common polypeptide chain. It was suggested that differences in the structures of the carbohydrate moieties may affect dissociable groups in neighboring amino acid residues in a different manner and therefore give rise to isoelectric multiplicity (132).

The recent discovery that cellobiohydrolase of *T. reesei* (17, 22,69) and *P. pinophilum/funiculosum* (30,64) can exist in two immunologically distinct forms (cellobiohydrolase I and II) is very important. That cellobiohydrolase I and II were markedly different in each system was supported by amino acid analysis (17,64,69), which showed clearly that there was no apparent relationship between the two enzymes (Table 4). Unexpected, however, was the finding that a clear homology existed between the main endoglucanase (EGI) and cellobiohydrolase I (71). This finding was of interest in that it immediately raised the possibility that if the homology extended to the active sites then both endoglucanase and cellobiohydrolase might use the same mechanism of action. The recent report revealing a similarity between the active site region of an endoglucanase of the fungus *Schizophyllum commune* and a *sequence* in *T. reesei* cellobiohydrolase (133) gives further credence to this possibility.

An extension of these sequence analyses has given further insight into the active sites of the cellulase components. Thus some evidence was obtained (133,134) demonstrating a similarity between the respective catalytic sites of lysozyme, a cellobiohydrolase of *T. reesei* and an endoglucanase (EG-1) of *S. commune*. By

TABLE 4   Amino Acid Composition of the
Cellobiohydrolases of *Trichoderma reesei*

	CBH II		CBH I	
	N[a]	mol%	N	mol%
CM-Cys	10	2.3	24	4.8
Asp	52	11.6	56	11.3
Thr	34	7.6	57	11.5
Ser	45	10.2	56	11.3
Glu	30	6.9	41	8.2
Pro	31	7.0	28	5.6
Gly	39	8.8	62	12.5
Ala	57	12.6	29	5.8
Val	24	5.4	23	4.6
Met	4	0.8	6	1.2
Iso	13	3.0	12	2.4
Leu	32	7.2	28	5.6
Tyr	21	4.6	24	4.8
Phe	12	2.8	15	3.0
His	4	0.9	5	1.0
Lys	10	2.2	13	2.6
Trp[b]	13	2.8	9	1.8
Arg	15	3.4	9	1.8
Total	446[c]	100.0	497[d]	100.0

[a]N = number of moles of residues per mole of
protein.

[b]Determined spectrophotometrically.

[c]Estimated value.

[d]From sequence analysis.

Source: Reprinted with permission from *CRC
Critical Reviews in Biotechnology*, 5(1), 67
(1987). Copyright CRC Press, Inc., Boca
Raton.

determining the N-terminal sequence of endoglucanase EG-1 of
S. commune, it was found that the sequence from Glu-33 to Tyr-51
was homologous with the active site of hen egg-white lysozyme, including the lysozyme catalytic residues (Glu-35, Asp-52) and the
substrate-binding residue Asn-44. The same homology was established in the active site of another endoglucanase (EG-2) of the same
fungus (134). It was also demonstrated that in the active site region, residues Glu-65 to Asp-74 in a cellobiohydrolase of *T. reesei*
were homologous with the active site sequence Glu-11 to Asp-22 in
the lysozyme produced to phage T4. In establishing these homologies, the hypothesis that cellulases and lysozyme use a common catalytic mechanism seems to be justified. The involvement of carboxyl
groups in the active site of *S. commune* endoglucanase was confirmed by chemical modification and kinetic experiments (135).

Further interesting homologies have now been established between cellobiohydrolase I and II and the main endoglucanase (EG-I)
of *T. reesei* cellulase (17,136) (Table 4). This was derived from
sequence analysis of the cloned genes. All three genes contained
two or three introns, but these were situated in the genes at different locations (Fig. 3). A common amino acid sequence of 34
residues was found in all the three enzymes at the end of the molecule. This sequence appeared to be joined to the other part of
the enzyme by a short region in which the amino acids were highly
O-glycosylated. Surprisingly, however, whereas in cellobiohydrolase II the 34-long amino acid sequence was found at the N-terminal
end of the enzyme, in both cellobiohydrolase I and endoglucanase
II it was localized at the C-terminal end.

FIGURE 3   Comparison of cellulase genes from *Trichoderma reesei*.
The basic structure of the genes of the three principal cellulase
components is shown. The hatched areas show the approximate positions and sizes of the introns associated with each gene. Cellobiohydrolase I and endoglucanase I show approximately 40% amino
acid homology but there is no apparent similarity with the location
of their introns. There is little homology in the amino acid sequence
between cellobiohydrolase I and II. [Reprinted with permission from
CRC Crit. Rev. Biotechnol., 5(1), 67 (1987). Copyright CRC Press,
Inc., Boca Raton.]

These structural features suggest clearly that the 34-amino-acid region and possibly the carbohydrate region may be important determinants of enzyme action on crystalline cellulose (17). Support for this conclusion was provided by the discovery that cellobiohydrolase I when treated with papain released a 10-kD, highly glycosylated fragment from the C-terminal end (137). As the loss of the peptide with its associated carbohydrate resulted in a concomitant loss in the capacity of the high molecular weight peptide (56-kD) to adsorb to crystalline cellulose but no loss in its activity to soluble chromophoric glycosides, it was suggested that cellobiohydrolase I consisted of two domains. The domain corresponding to the C terminal was considered to be involved in the binding of enzyme to insoluble cellulose, while the other contained the active site. Similar results have been obtained with cellobiohydrolase II (138).

B. Endoglucanase Multiplicity

As already stated, multiplicity in terms of endoglucanase is also the rule in most cellulase preparations investigated so far (2). However, it seems that opinion as to the nature and origin of the heterogeneity is particularly diverse and controversial. The various endoglucanases found in any one cellulase are reported to differ in their isoelectric points, their molecular weights, and their association with carbohydrate (2,31). In their catalytic properties, each endoglucanase is reported to give a characteristic slope when the fluidity of a solution of carboxymethylcellulose which has been incubated with enzyme is plotted against reducing power (2). By this criterion, several endoglucanases which differ in the randomness of their attack have been found in *T. viride* (127) and *T. koningii* (79) cellulases; this appears to be the norm. However, a notable exception in this regard was found in the cellulase of *Talaromyces emersonii*, when Moloney et al. (119) established that the four endoglucanases purified from the culture filtrate, although differing in the pI and carbohydrate content, appeared to effect changes in the carboxymethylcellulose solution that resulted in an identical fluidity/reducing sugar relationship. For these and other reasons it was concluded that differential glycosylation of a single enzyme form rather than genetically determined differences in primary structure accounted for the heterogeneity.

Evidence that heterogeneity may result either from proteolytic modification or from the formation of artifacts during purification is provided by (1) Nakayama et al. (117), who observed that partial proteolysis of endoglucanases yielded enzymes with altered substrate specificities; (2) Eriksson and Pettersson (118), who reported increased specific activity of one of the endoglucanases of *S. pulverulentum* after treatment with a protease isolated from the same

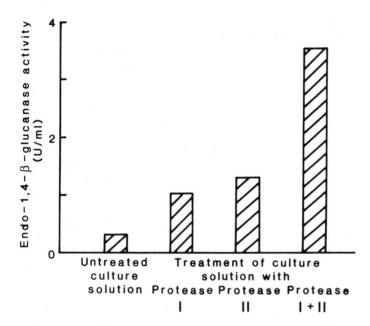

FIGURE 4   Activation of endo-1,4-β-glucanase from *Sporotrichum
pulverulentum* after treatment with protease I and II.   Two acidic
proteases were isolated from cultures of *S. pulverulentum* grown
using cellulose as the carbon source.   [Reprinted by permission
from *Eur. J. Biochem., 124*, 635 (1982).]

culture filtrate (Fig. 4); (3) Deshpande et al. (139), who observed
that the endoglucanase activity found in 2-day-old culture filtrates
of *Penicillium janthinellum* was enhanced fourfold on addition of 10-
day filtrates of a *Penicillium funiculosum* culture.   In the same context, Gong et al. (97) found that the heterogeneity of each type of
enzyme in cultures of *T. reesei* increased with the duration of culture and with increasing protease activity; the heterogeneity was
markedly less when the culture conditions were carefully controlled,
and Petterson et al. (71) support the latter conclusion.   Using the
same fungus, Niku-Paavola and her colleagues (77) conclude that
the various endoglucanases in a culture filtrate all originate from the
same enzyme and are artifacts caused by faulty manipulation of the
enzyme during purification.
    All of these data would suggest that the fungi produce an endo-
glucanase enzyme system much simpler than that found in most culture filtrates.   In contrast, Labudova and Farkas (122) and Dunne
(123) interpret their data to indicate that the ability to produce

multiple endoglucanase is a characteristic property of *T. reesei*.
Kolar et al. (140) reached the same conclusion after studying the
release of endoglucanase from protoplasts of *T. reesei*.

Thus, the nature and origin of the heterogeneity continues in
both endoglucanase and cellobiohydrolase activity to be uncertain.
On balance it would seem that there is much evidence for postsecre-
tional modification of the enzymes synthesized by the fungus, and
it seems certain that some of the heterogeneity at least is artifactual.
However, it is not known as yet with any degree of certainty just
how many enzymes are genetically determined, and it is likely that
this problem will remain unresolved until more detailed analysis has
been carried out on a variety of cloned genes and on the enzymes
obtained by expression of these genes. Wood (55,68) argued from
theoretical consideration of the stereochemistry of cellulose chains
in the cellulose crystallite that two stereospecific endoglucanase and
two stereospecific cellobiohydrolases may be required for maximum
efficiency in terms of solubilizing crystalline cellulose.

C. β-Glucosidase Multiplicity

Multiple forms of β-glucosidase exist in many fungi. For example,
both *Sclerotium rolfii* (141) and *Sporotrichum pulverulentum* (130)
produce up to five extracellular forms, but *T. koningii* (51) and
A. fumigatus (142) produce only two. Both β-glucosidases obtained
from *Sporotrichum thermophile* (143) were intracellular, but *Monilia*
species (100) produced at least one intracellular and two extracellular forms. One of the intracellular β-glucosidases of *Monilia* sp. and
the extracellular form were inducible: the inducible intracellular enzyme was shown to be the nascent form of the extracellular enzyme
prior to its secretion into the extracellular medium (144).

VI. MECHANISM OF CELLULASE ACTION

A. Initial Attack

Over the years, one of the most controversial aspects of the mechanism of action has been which enzyme actually initiated the attack
on the crystalline cellulose. The hypothesis put forward by Reese
et al. (32) in 1950 (so-called C_1C_x hypothesis) envisaged that attack was initiated by a nonhydrolytic chain-separating enzyme (C_1)
which produced a reactive cellulose which could then be attacked by
the hydrolytic enzymes (C_x) (Fig. 5). This concept, which dominated the thinking and guided the experimental approach for many
years, was abandoned by most workers with the isolation of a cellobiohydrolase (62,145−147) which had some of the properties of the
hypothetical C_1-type enzyme. Reese (148), however, was not

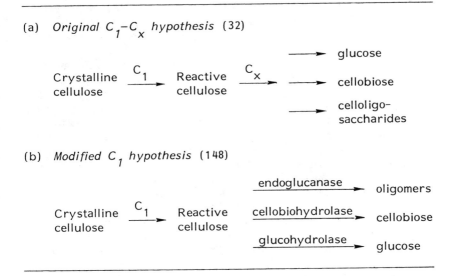

FIGURE 5   Original and amended hypotheses for the mechanism of cellulase action involving a C_1 component.

convinced that the cellobiohydrolase was the C_1 factor, and he later modified his hypothesis to suggest that C_1 might be a randomly acting endoglucanase capable of cleaving a few covalent links but having no detectable hydrolytic activity. In this new hypothesis, cellobiohydrolase was relegated to the role of hydrolyzing the partially degraded cellulose along with the endoglucanase.

There is no good evidence as yet for the existence of a C_1-type enzyme, but the recent isolation from a commercial cellulase (*T. reesei*) by Griffin et al. (149) of a nonenzymic factor capable of generating microfibrils and short fibers from filter paper may renew the controversy. Koenigs (150) suggested that brown rot fungi, which do not appear to use a cellobiohydrolase-endoglucanase system for hydrolysis, initiate attack on crystalline cellulose via an H_2O_2/Fe^{2+} system; and other workers (13,151,152) reported evidence suggesting the existence of a low molecular weight factor that can diffuse into cellulose and depolymerize it. Griffin et al., however, found that the microfibril-generating factor present in *T. reesei* cellulase does not involve H_2O_2/Fe^{2+} (149).

An interesting feature of the involvement of cellobiose oxidase (109,110) in cellulose degradation by *S. pulverulentum* cellulase is the generation of superoxide anion as a product of the reaction

(153). Superoxide anion may be involved in initiating an attack on cellulose (153). However, it is also possible that the H_2O_2 which is produced from the anion by superoxide anion dismutase may also be involved in this way (153).

Vaheri (91) also considers that oxidation of some kind may be associated with early attack on the cotton fiber, as manifested by the generation of short fibers. Moreover, the oxidation would appear to be enzymic in origin and to result in the formation of onic acids. Vaheri suggests that in the initial stage of breakdown of cotton, oxidative reactions increase the accessibility of the fibers to the endoglucanases, which then generate the short fibers. The involvement of oxygen in short-fiber formation was earlier shown by Marsh (154); Wood (155) is of the opinion that it is the endoglucanases that effect the changes leading to the generation of these fragments. There does, however, seem to be some evidence in the literature that the generation of short fibers is also effected by some purified cellobiohydrolase enzymes (43). Thus, the formation of short fibers is reported to be a property of both cellobiohydrolase I (43) and II (17) of *T. reesei*; but neither cellobiohydrolase I (63) nor II (64) of *P. pinophilum/funiculosum* nor cellobiohydrolase I of *T. koningii* (62) was found to produce short fibers.

All of these results are of interest in developing an understanding of all the degradation processes involved in the breakdown of some cellulosic materials. However, it is doubtful if short-fiber formation has any real significance in the context of attack on the cellulose crystallite. It is known that the cotton fiber is characterized by the presence of areas of weakness that occur at intervals along the length of the cotton fiber (156) and that these traverse the whole of the fiber. It is likely that these areas, which contain cellulose chains that are less hydrogen bond-ordered than in the crystallite, will be preferentially attacked. Attack on some areas of weakness would result also in a dramatic fall in the tensile strength of the fiber; Wood (155) showed that such changes in tensile strength are more a property of the endoglucanase than the cellobiohydrolase.

Undoubtedly of more significance as far as degrading crystalline cellulose is concerned is the fragmentation of the ribbon of cellulose produced by the bacterium *Acetobacter xylinum*. The ribbon, which is seen to be a composite of microfibrils in the electron microscope, was reported by White and Brown (157) to split along its long axis into bundles of microfibrils when incubated with a purified endoglucanase from *T. reesei* cellulase: purified cellobiohydrolase I from the same fungus produced no visible change in the cellulose structure when acting alone, but when acting in conjunction with endoglucanase it brought about the rapid dissolution of the cellulose (158). These results, however, are in complete contrast to those of Chanzy

et al. (33), who claimed that their studies in the electron microscope showed that cellobiohydrolase I from *T. reesei* could effect
some disaggregation of the cellulose into microfibrils. They also interpreted their electron micrographs to show that cellobiohydrolase
I, when acting alone, could cause the complete dissolution of crystals of cellulose from the alga *Valonia macrophyta* (33) and that endoglucanase action was confined to the amorphous areas of the cellulose (159). As a result of these and other studies (33,39), it was
suggested that cellobiohydrolase I indeed had an endoglucanase mode
of action, and only cellobiohydrolase II (159) acted as a typical exoglucanase. However, Enari and Niku-Paavola (17) are of the opinion
that cellobiohydrolase II from the same fungus has some of the properties of an endoglucanase.

Thus, the questions regarding which enzyme initiates the attack
on the cellulose crystallite and the nature of that attack are very
far from being resolved.

B. Synergism Between Enzyme Components

Synergism between the enzymes found in the cellulases of fungi was
first demonstrated by Gilligan and Reese in 1954 (46). Despite the
passage of time, the understanding of the various interactions involved in the synergistic process is still poor, although there has
been no shortage of published data on the subject.

The degree of synergistic activity shown by various "purified"
components is obviously largely dependent on the extent to which
the individual components are capable of solubilizing crystalline cellulose when acting in isolation. As stated above, opinions on this
particular aspect differ widely. Some maintain that only one enzyme
is required for extensive hydrolysis (47,54). Wood and McCrae (2),
however, showed that when the enzymes are highly purified and the
substrate is dewaxed cotton fiber which has been subjected to the
minimum of pretreatment, degradation of crystalline cellulose in this
form is virtually completely dependent on the cooperative action of
the various enzymes (Table 5). Synergistic activity in this circumstance is apparent between the enzymes classified as endoglucanase
and cellobiohydrolase, and, to a limited extent, between β-glucosidase and cellobiohydrolase, but the complete recovery of activity of
the original unfractionated enzyme is dependent on the cooperation
of all three types of enzyme. Synergistic effects between endoglucanase, cellobiohydrolase, and β-glucosidase were also demonstrated
with enzymes isolated from extracellular cellulases of *P. funiculosum*
(27), *T. viride* (27), *P. funiculosum/pinophilum* (63), *S. pulverulentum* (59), and *T. emersonii* (29), but the degree of synergistic
activity observed between the various enzymes varied widely.

TABLE 5   Cellulase Activity of the Components of
Trichoderma koningii and *Fusarium solani* Cellulase When
Acting Alone and in Combination[a]

	Cellulase activity (solubilization of cotton) (%)	
Enzyme	*T. koningii*	*F. solani*
Cellobiohydrolase (CBH)	1	2
Endo-1,4-β-glucanase (Endo)	1	1
β-Glucosidase	Nil	Nil
Endo + CBH	55	58
Endo + β-glucosidase	6	4
CBH + β-glucosidase	20	18
Endo + CBH + β-glucosidase	72	71
Original unfractionated enzyme	71	71

[a]All enzymes were present in the same proportions in which
they were present in the original unfractionated enzyme.
Source: From Ref. 187.   Reproduced with permission from
Kodansha Ltd., Tokyo.

From these and other results with these fungal cellulases it is
apparent that synergism is most marked when crystalline cellulose
(particularly cotton fiber) is the substrate, but it is low or nonex-
istent with amorphous, extensively hydrated cellulose, and it is absent with soluble cellulose derivatives (2).   Further, it would appear that the synergistic effect is dependent on the ratio of cellobiohydrolase to endoglucanase (155) and on the type of endoglucanase acting in concert with the cellobiohydrolase (79).   Thus, in
this latter respect it was observed (79) that two of the four major
endoglucanases in *T. koningii* cellulase had no apparent capacity
for acting synergistically with the cellobiohydrolase.   In contrast,
all four endoglucanases of *T. emersonii* (119) acted equally efficiently with the cellobiohydrolases in solubilizing crystalline cellulose.
     Other interesting observations on the synergistic cooperation
between the enzymes have been made upon mixing cellobiohydrolase
from one cellulase with endoglucanase from another (2,27,160,161).

It was found by Wood and his colleagues (2,160,161) that some —
but not all — endoglucanases and cellobiohydrolases acted synergistically to solubilize crystalline cellulose and that synergism was only
significant when the enzymes were obtained from a fungal source
that was capable of estensive hydrolysis of crystalline cellulose
(160).

The question of why all endoglucanases are not able to act synergistically with the cellobiohydrolase to the same extent was addressed by Wood and his colleagues (2,62,155) in a number of publications and resulted in the presentation of several hypotheses. All
of these hypotheses are based on the model put forward several
years ago in which it was suggested that synergism is likely to be
explained by a sequence of events in which endoglucanases cleave
cellulose chains thereby creating new chain ends for the cellobiohydrolase to attack (2,62,155).

The original model has now been modified to account for the
"anomalous" results, and two postulates have been made by Wood
and his colleagues. Thus, they have suggested that (1) rapid sequential action between endoglucanase and cellobiohydrolase is necessary to prevent the reformation of glycosidic linkages between glucose residues rigidly held in position by hydrogen bonds, and that this
can best be effected by those endoglucanases and cellobiohydrolases
that can form a "loose" endoglucanase–cellobiohydrolase complex on the
face of the cellulose crystallite (108,79), and (2) the end group generated by the endoglucanase in the first phase of the hydrolysis must
have the correct configuration for attack by the stereospecific cellobiohydrolase involved in the second stage of hydrolysis (Fig. 6) (160,161).
As yet there is no evidence to support any of these possibilities, but
theoretical considerations of the stereochemistry of the cellulose chains
in the cellulose crystallite give some credence to the latter (68).

Another factor relating to the stereospecificity of the enzymes,
and hence synergistic action, is adsorption of the enzymes to the
cellulose crystallite. Adsorption of endoglucanase is very variable.
Ryu et al. (162) examined the differences in adsorption of the separated components of the cellulase complex of *T. reesei* MCG 77 on
Avicel and interpreted their observations in terms of synergistic
action. They suggested that the endoglucanases and cellobiohydrolase adsorb at different sites and that these sites corresponded to
the sites of hydrolysis. Endoglucanase consisted of adsorbable and
nonadsorbable components, but cellobiohydrolase had the strongest
adsorption affinity. Significant was the suggestion that adsorption
of the cellulase components was competitive and that this could explain the synergistic interaction of the components in solubilizing
crystalline cellulose. Thus, the addition of cellobiohydrolase to substrate to which endoglucanase was already bound affected the endoglucanase in some indeterminate way so that endoglucanase action

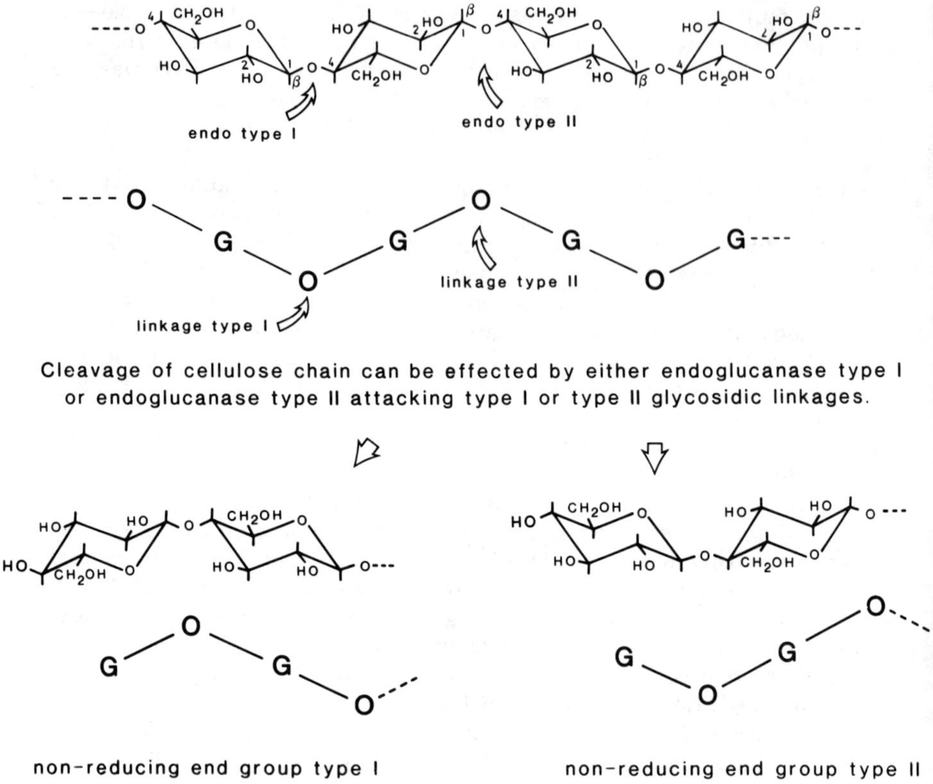

Cleavage of cellulose chain can be effected by either endoglucanase type I or endoglucanase type II attacking type I or type II glycosidic linkages.

non-reducing end group type I   non-reducing end group type II

FIGURE 6   Proposed mechanism of synergistic action of endoglucanase and cellobiohydrolase on cellulose. Because cellobiose is the
repeating unit in the cellulose chain which is rigidly held in position by intra- and intermolecular hydrogen bonds, it is possible to
predict that two classes of stereospecific endoglucanase and cellobiohydrolase would be required for effective hydrolysis. If it is
assumed that endoglucanase initiates the attack, it is clear that for
stereochemical reasons only one of the two contiguous glycosidic
linkages in the chain would be cleaved and that only one of two
possible configurations (type I or II) of nonreducing end group
would be generated. Synergism between endoglucanase and cellobiohydrolase will only be apparent when the end group generated
by a stereospecific endoglucanase has the correct configuration for
attack by a stereospecific cellobiohydrolase. This hypothesis can
explain the synergistic action between endoglucanase and cellobiohydrolase and also the instances where no synergism is apparent
between these two enzymes. [After Wood (160).]

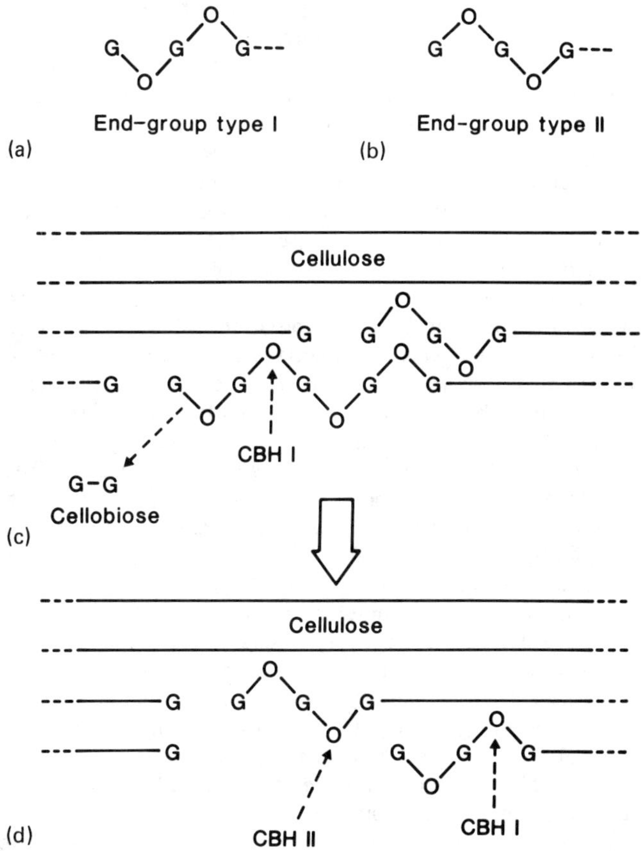

FIGURE 7 Proposed explanation for synergism between cellobiohy-drolase I and II. Because cellobiose is the repeating unit in the cellulose chain, theoretically two types of nonreducing end groups (I and II; see (a) and (b)] will exist in the cellulose crystallite. These end groups will be held rigidly in position by noncovalent bonds and will require two different stereospecific cellobiohydrolases for hydrolysis. (c) Cellobiohydrolase I attacking the nonreducing chain end of configuration I; (d) Cellobiohydrolase II attacking the nonreducing end of configuration II exposed by cellobiohydrolase I action. [From Wood (14). Reprinted by permission from *Biochem. Soc. Trans.*, *13*, 407 (1985). Copyright (C) 1985. The Biochemical Society, London.]

was enhanced and desorption facilitated. Addition of endoglucanase
had a similar effect on cellobiohydrolase bound to cellulose. The
rate and extent of desorption and hence the synergistic activity
varied according to the substrate used. Henrissat et al. (34) also
showed synergism to vary according to substrate. Interestingly,
the degree of competitive adsorption observed by Ryu et al. (162)
was highest when cellobiohydrolase and endoglucanase components
were present in the ratio in which they were present in the crude
culture filtrate. These data could be interpreted to support the
hypothesis that endoglucanase −cellobiohydrolase complexes are required for hydrolysis of crystalline cellulose.

In all of these hypotheses, it is the synergism between endoglucanase and cellobiohydrolase that is discussed. However, the
recent isolation of two immunologically unrelated cellobiohydrolases
from *T. reesei* and *P. pinophilum*, and the demonstration of synergistic activity between the enzymes (55,69), opened up a new and
important aspect of the mechanism of cellulase action. Some workers
interpret their data to indicate that most of the synergistic activity
in *T. reesei* cellulase may indeed be attributed to the cooperation
of the two immunologically unrelated cellobiohydrolase (71,163), but
this requires further confirmation.

Synergism between two endwise-acting enzymes is difficult to
explain, but Wood (31,55) suggested that a reasonable speculation
might be that cellobiohydrolase I and II could exhibit different substrate stereospecificities, each attacking one of the two types of
naturally-occuring nonreducing group that might be found in the
cellulose crystallite. The synergistic interaction could then be understood if one envisages the successive removal of cellobiose from
one type of nonreducing chain end could expose on a neighboring
chain an end group with the correct configuration for attack by the
other stereospecific cellobiohydrolase (Fig. 7).

Thus the actual mechanism of cellulase action continues to be
uncertain although progress has clearly been made.

VII. REGULATION OF ENZYME ACTIVITY

Cellulase is an inducible enzyme system and induction is effected
by a number of carbohydrate molecules. Cellulose is recognized to
be the best inducer, but cellobiose (164−167) and, in some instances,
sophorose (168) or lactose (97) are very effective. It is not known
how an insoluble polymer such as cellulose can initiate cellulase synthesis, but it has been suggested that close physical contact between the fungus and the cellulose could trigger induction (169−
171), the implication being that there are suitable recognition sites
on the cellulose. An alternative explanation is that the fungus

constitutively synthesizes basal levels of cellulase which release cellobiose which in turn effects the induction (166). However, the natural inducer is not really known, and it is felt that this may be a product of transglycosylation, which is reported to be a property of extracellular and cell-bound β-glucosidases (22,51,89,91): β-glucosidases are constitutively produced in some fungi (95,172).

Sophorose, which could be produced by transglycosylation, has been shown to be a very potent inducer at very low concentrations in *T. reesei*, *T. viride*, and *T. koningii* (8,22,168,173,174), but it does not induce cellulase in other fungi (8). Sophorose is found in trace amounts in some commercial glucose preparations (175) and its presence is believed to explain the induction of cellulases reported by glucose (90). Sophorose has also been detected in cellobiose solution treated with fragmented cell wall suspensions of *T. reesei* grown on glycerol as carbon source and containing no cellulase activity (91). It has been shown that cellulase appears only after the sophorose has been taken up by the cell, but the presence of the inducer is required to maintain synthesis (90,173). Cellulase synthesis ceases after the separation of the mycelium from the induction medium. Interestingly, while sophorose induces synthesis of cellulase, it represses the synthesis of β-glucosidase which would hydrolyze the disaccharide to glucose (168).

Sophorose induces both exo- and endoglucanases, but the number of enzyme components induced is less than that synthesized in the presence of cellulose (90). Thus, it is possible that sophorose may not be the true inducer in *T. reesei*. However, the levels of enzyme produced are such that cellulose can be effectively solubilized (22).

Cellobiose is an inducer of cellulase in *T. reesei* (164−167), *S. pulverulentum* (167), *Sporotrichum thermophile* (176), and *Neurospora crassa* (177). However, there appears to be considerable differences in the levels of induction effected by different concentrations of cellobiose. Thus, endoglucanase in *S. pulverulentum* is induced by cellobiose at concentrations as low as 1 mg/ml, but the same concentration of cellobiose does not induce the cellulase in *T. reesei* (167).

Cellulase biosynthesis is repressed by the catabolite glucose in the presence of cellulose and other inducers (12,178,179). The addition of glucose to a culture of *T. reesei* was reported to effectively stop synthesis until residual glucose falls below a critical concentration of 100 mg/liter (12). Catabolite repression of endoglucanase synthesis is obtained in *S. pulverulentum* (167) cultures at glucose concentrations of only 50 mg/liter. Glucose inhibition of cellulase synthesis was also reported in many other fungi (180,181). The precise mechanism by which glucose inhibits is not known.

It would appear that both cellobiohydrolase and endoglucanase are coordinately regulated in *T. reesei* (164) and *P. pinophilum/ funiculosum* (182). This has been established using both wild-type and mutant strains. However, β-glucosidase and xylanase in *P. pinophilum/funiculosum* seem to be controlled separately (182).

Regulation of enzyme activity can also be controlled at the secretion stage. By studying the effect of glycosylation, Merivouri et al. (169) concluded that O-glycosylation of cellulase enzymes but not N-glycosylation is required for secretion.

A model for the regulation of cellulase biosynthesis has been proposed (164).

The activity of cellulase enzymes in culture filtrates of the fungi *S. pulverulentum* and probably *T. reesei* is dependent not only on the mechanisms regulating their biosynthesis but also on the presence of specific inhibitors of the enzymes. Gluconolactone is such an inhibitor, and was shown to be produced by hydrolytic cleavage of cellobionolactone. The gluconolactone has been shown to be important for the regulation of β-glucosidase which is found in both extracellular and cell-bound forms in *S. pulverulentum* (130). The extracellular enzyme comprises two components which are separable by chromatography. The K_i values for gluconolactone inhibition for the two extracellular β-glucosidases were 3.5×10^{-7} and 15×10^{-7} M, respectively. The K_i value for the same inhibitor acting on *T. reesei* QM 9414 β-glucosidase was much higher (3.2×10^{-5} M).

Specific inhibitors therefore may play a major role in some cellulase systems. In most cellulases, however, the effectiveness of the enzyme is carefully regulated by inhibition of the end products produced. Glucose has been shown to be a potent inhibitor of β-glucosidase (51,183), the primary function of which is to hydrolyze cellobiose, which is reported to be a powerful inhibitor of cellobiohydrolase in some cellulases.

Endoglucanase is normally also inhibited by increasing concentrations of cellobiose, but it is less affected than is cellobiohydrolase (79).

REFERENCES (reviewed until 1986)

1. D. E. Eveleigh, In *Biomass Utilization* (W. Cote, ed.), Plenum Press, New York, 1982, p. 365.
2. T. M. Wood and S. I. McCrae, *Adv. Chem. Ser.*, *181*, 181 (1979).
3. R. Lamed, R. Kenig, E. Setter, and E. A. Bayer, *Enzyme Microb. Technol.*, 7, 37 (1985).

4. N. Creuzet, J.-F. Berenger, and C. Frixon, *FEMS Microbiol. Lett.*, *20*, 347 (1983).
5. B. G. R. Hagerdal, J. D. Ferchak, and E. K. Pye, *Appl. Env. Microbiol.*, *36*, 606 (1978).
6. E. T. Reese, In *Recent Advances in Phytochemistry* (F. Loewus and V. C. Runeckles, eds.), *11*, 311 (1977).
7. E. T. Reese and M. Mandels, *Ann. Rep. Ferment. Proc.*, 7, 1 (1984).
8. M. Mandels, *Ann. Rep. Ferment. Proc.*, *5*, 35 (1982).
9. V. S. Bisaria and T. K. Ghosh, *Enzyme Microb. Technol.*, *3*, 90 (1981).
10. J. J. Goksoyr and J. Eriksen, In *Microbial Enzymes and Bioconversions*, Vol. 5, *Economic Microbiology* (A. H. Rose, ed.), Academic Press, London, 1980, p. 283.
11. K.-E. Eriksson, In *Trends in the Biology of Fermentation for Fuels and Chemicals* (E. A. Hollaender, ed.), Plenum Press, New York, 1981, p. 19.
12. D. D. U. Ryu and M. Mandels, *Enzyme Microb. Technol.*, *2*, 91 (1980).
13. K.-E. Eriksson and T. M. Wood, In *Biosynthesis and Biodegradation of Wood Components* (T. Higuchi, ed.), Academic Press, London, 1985, p. 469.
14. T. M. Wood, *Biochem. Soc. Trans.*, *13*, 407 (1985).
15. M. P. Coghlan, *Biotechnol. Genet. Eng. Rev.*, *3*, 39 (1985).
16. T.-M. Enari, In *Microbial Enzymes and Biotechnology* (W. M. Fogarty, ed.), Applied Science, London, 1983, p. 183.
17. T.-M. Enari and M.-J. Niku-Paavola, *CRC Crit. Rev. Biotechnol.*, *5*, 67 (1987).
18. L. G. Ljungdahl and K.-E. Eriksson, In *Adv. Microb. Ecol.* (K. C. Marshall, ed.), *8*, 237 (1985).
19. B. Henrissat and H. Chanzy, In *Cellulose Chemistry Technology* (R. Young, ed.), Wiley, New York, 1986, p. 281.
20. T. M. Wood, In *Cellulose and Its Derivatives* (J. F. Kennedy, ed.), Ellis-Horwood, Chichester, 1985, p. 173.
21. K. Ogawa and N. Toyama, *J. Ferm. Technol.*, *50*, 236 (1972).
22. M. Gritzali and R. D. Brown, *Adv. Chem. Ser.*, *181*, 237 (1979).
23. M. Mandels and E. T. Reese, *Dev. Ind. Microbiol.*, *5*, 5 (1964).
24. T. M. Wood, *Biochem. J.*, *109*, 217 (1968).
25. G. Halliwell, *Biochem. J.*, *95*, 270 (1965).
26. T. M. Wood, *Biochem. J.*, *115*, 457 (1969).
27. K. Selby, *Adv. Chem. Ser.*, *95*, 34 (1969).
28. T. M. Wood and S. I. McCrae, The mechanism of cellulase action with special reference to the C_1 component. In *Proc.*

Bioconversion Symposium, Indian Institute of Technology, Delhi, 1977.
29. A. McHale and M. P. Coughlan, *FEBS Lett.*, *117*, 319 (1980).
30. T. M. Wood, C. A. Wilson, and K. Joblin, *FEMS Microbiol. Lett.*, *34*, 37 (1986).
31. T. M. Wood, *Biochem. Soc. Trans.*, *13*, 407 (1985).
32. E. T. Reese, R. G. H. Siu, and H. S. Levinson, *J. Bacteriol.*, *59*, 485 (1950).
33. H. Chanzy, B. Henrissat, R. Vuong, and M. Schulein, *FEBS Lett.*, *153*, 113 (1983).
34. B. Henrissat, H. Driguez, C. Viet, and M. Schulein, *Biotechnology*, *3*, 722 (1985).
35. T. K. Ghose, H. J. Bailey, V. S. Bisaria, and T.-M. Enari, *Pure Appl. Chem.*, *59*, 247 (1987).
36. M. Mandels, R. E. Andreotti, and C. Roche, *Biotechnol. Bioeng. Symp. Ser.*, *6*, 21 (1976).
37. A. K. Kulshreshtha and N. E. Dureltz, *J. Polym. Sci. Phys. Ed.*, *11*, 487 (1973).
38. A. Bourret, H. Chanzy, and R. Lazaro, *Biopolymers*, *11*, 893 (1972).
39. H. Chanzy, B. Henrissat and R. Vuong, *FEBS Lett.*, *172*, 193 (1984).
40. Y. Tomita, H. Suzuki, and K. Nisizawa, *J. Ferm. Technol.*, *52*, 233 (1974).
41. C. S. Walseth, *Tappi*, *35*, 233 (1952).
42. M. Nummi, P. C. Fox, M.-L. Niku-Paavola, and T.-M. Enari, *Anal. Biochem.*, *116*, 133 (1981).
43. M. Nummi, M.-L. Niku-Paavola, A. Lappalainen, T.-M. Enari, and V. Raunio, *Biochem. J.*, *215*, 677 (1983).
44. W. Chirica and R. D. Brown, *Eur. J. Biochem.*, *165*, 348 (1987).
45. H. Van Tilbeurgh, M. Claeyssens, and K. de Bruyne, *FEBS Lett.*, *149*, 152 (1982).
46. W. Gilligan and E. T. Reese, *Can. J. Microbiol.*, *1*, 90 (1954).
47. L. H. Li, R. M. Flora, and K. W. King, *Arch. Biochem. Biophys.*, *111*, 439 (1965).
48. M. Nummi, M.-L. Niku-Paavola, T.-M. Enari, and V. Raunio, *Anal. Biochem.*, *116*, 137 (1981).
49. G. Halliwell and M. Griffin, *Biochem. J.*, *169*, 713 (1978).
50. C. S. Gong, M. R. Ladisch, and G. T. Tsao, *Biotechnol. Bioeng.*, *19*, 959 (1977).
51. T. M. Wood and S. I. McCrae, *J. Gen. Microbiol.*, *128*, 2973 (1982).
52. H. van Tilbeurgh, R. Bhikhabhai, L. G. Pettersson, and M. Claeyssens, *FEBS Lett.*, *169*, 215 (1984).

53. B. Sprey and L. Lambert, *FEMS Micribiol. Lett.*, *18*, 217 (1983).
54. L. E. R. Berghem, L. G. Pettersson, and U.-B. Axio-Fredricksson, *Eur. J. Biochem.*, *61*, 621 (1976).
55. T. M. Wood and S. I. McCrae, *Biochem. J.*, *234*, 93 (1986).
56. A. Kyriacou, C. R. MacKenzie, and R. J. Neufeld, *Enzyme Microb. Technol.*, *9*, 25 (1987).
57. E. T. Reese, H. Maguire, and F. W. Parrish, *Can. J. Biochem.*, *46*, 25 (1968).
58. L. H. Li, R. M. Flora, and K. W. King, *Arch. Biochem. Biophys.*, *121*, 416 (1965).
59. M. Streamer, K.-E. Eriksson, and B. Pettersson, *Eur. J. Biochem.*, *59*, 607 (1975).
60. B. S. Montenecourt, T. J. Kelleher, and D. E. Eveleigh, *Biotechnol. Bioeng. Symp. Ser.*, *10*, 15 (1980).
61. T. M. Wood and S. I. McCrae, unpublished data.
62. T. M. Wood and S. I. McCrae, *Biochem. J.*, *128*, 1183 (1972).
63. T. M. Wood, S. I. McCrae, and C. M. Macfarlane, *Biochem. J.*, *189*, 51 (1980).
64. T. M. Wood and S. I. McCrae, *Carboh. Res.*, *148*, 331 (1986).
65. T. M. Wood, *Biochem. J.*, *115*, 457 (1969).
66. G. Beldman, M. F. Searle-Van Leeuiren, F. M. Rombouts, and F. G. J. Voragen, *Eur. J. Biochem.*, *146*, 305 (1985).
67. K.-E. Eriksson and B. Pettersson, *Eur. J. Biochem.*, *51*, 213 (1975).
68. T. M. Wood, Enzyme interactions involved in fungal degradation of cellulosic materials, In *Proc. Ekman—Days Int. Symposium on Wood and Pulping Chemistry*, Vol. 3, SPCI, Stockholm, 1981, p. 31
69. L. G. Fagerstam and L. G. Pettersson, *FEBS Lett.*, *119*, 97 (1980).
70. L. G. Fagerstam, L. G. Pettersson, and J. A. Engström, *FEBS Lett.*, *167*, 309 (1984).
71. G. Pettersson, L. Fagerstam, R. Bhikhabhai, and K. Leandoer, The cellulase complex of *Trichoderma reesei* QM 9414, In *Proc. Int. Symposium on Wood and Pulping Chemistry*, Vol. 3., Ekman—Days, SPCI, Stockholm, 1981.
72. E. K. Gum and R. D. Brown, *Biochim. Biophys. Acta*, *446*, 371 (1976).
73. S. Shoemaker, V. Schweickart, M. Ladner, D. Gelfand, S. Kirok, M. Myambo, and M. Innis, *Biotechnology*, *1*, 691 (1983).
74. R. Khikhabhai and G. Pettersson, *Biochem. J.*, *222*, 729 (1984).

75. R. Bhikhabhai, G. Johannsson, and G. Pettersson, *Int. J. Peptide Protein Res.*, *25*, 368 (1985).
76. S. P. Shoemaker and R. D. Brown, *Biochim. Biophys. Acta*, *148*, 808 (1977).
77. M.-L. Niku-Paavola, A. Lappalainen, T.-M. Enari, and M. Nummi, *Biochem. J.*, *231*, 75 (1985).
78. G. Beldman, M. F. Searle-Van Leeuwen, F. M. Rombouts, and F. G. J. Voragen, *Eur. J. Biochem.*, *146*, 308 (1985).
79. T. M. Wood and S. I. McCrae, *Biochem. J.*, *171*, 61 (1978).
80. P. L. Hurst, P. A. Sullivan, and M. G. Shepherd, *Biochem. J.*, *169*, 389 (1978).
81. M. G. Shepherd, A. L. Cole, and C. C. Tong, *Biochem. J.*, *193*, 67 (1981).
82. M. L Rabinovich, N. Van Viet, and A. A. Klyosov, *Biokhimiya*, *47*, 465 (1982).
83. A. A. Klyosov, V. M. Chernoglazou, M. L. Rabinovich, and A. P. Sinitsym, *Bioorganisheskaya Khimiya*, *8*, 643 (1982).
84. M. L. Rabinovitch, V. M. Chernoglazov, and A. A. Klyosov, *Biokhimiya*, *48*, 369 (1983).
85. A. A. Klyosov, O. V. Mitkevich, and A. P. Sinitsyn, *Biochemistry*, *25*, 540 (1986).
86. J. G. Shewale, *Int. J. Biochem.*, *14*, 435 (1982).
87. J. Woodward and A. Wiseman, *Enzyme Microb. Technol.*, *4*, 73 (1982).
88. B. S. Montencourt, S. M. Cuskey, S. D. Nhlapo, H. Trimino-Vazquez, and D. E. Eveleigh, Strain development for the production of microbial cellulases, In *Proc. Ekman—Days Int. Symposium on Wood and Pulping Chemistry*, Vol. 3, SCPI, Stockholm, 1981.
89. E. T. Reese, A. H. Maguire, and F. W. Parrish, *Can. J. Biochem.*, *46*, 25 (1968).
90. D. Sternberg and G. R. Mandels, *J. Bacteriol.*, *139*, 761 (1979).
91. M. Vaheri, M. Leisola, and V. Kauppinen, *Biotechnol. Lett.*, *1*, 41 (1939).
92. A. McHale and M. P. Coughlan, *J. Gen. Microbiol.*, *128*, 2327 (1982).
93. S. R. Parr, *Enzyme Microb. Technol.*, *5*, 457 (1982).
94. C. S. Evans, *Appl. Microbiol. Biotechnol.*, *21*, 452 (1985).
95. A. J. Lusis and R. R. Becker, *Biochim. Biophys. Acta*, *329*, 5 (1973).
96. M. H. Smith and M. H. Gold, *Appl. Env. Microbiol.*, *37*, 938 (1979).
97. C. S. Gong, M. R. Ladisch, and G. T. Tsao, *Adv. Chem. Ser.*, *181*, 261 (1979).

98. C. S. Gong, M. R. Ladisch, and G. T. Tsao, *Biotechnol. Bioeng.*, *19*, 959 (1977).
99. D. Herr, F. Baumer, and H. Dellweg, *Eur. J. Appl. Microbiol. Biotechnol.*, *5*, 29 (1978).
100. R. F. H. Dekker, *J. Gen. Microbiol.*, *127*, 177 (1981).
101. R. K. Berry and R. F. H. Dekker, *Carboh. Res.*, *157*, 1 (1986).
102. E. T. Reese and F. W. Parrish, *Carboh. Res.*, *18*, 381 (1971).
103. D. Sternberg, P. Vijayakumar, and E. T. Reese, *Can. J. Microbiol.*, *23*, 139 (1977).
104. F. Bissett and D. Sternberg, *Appl. Env. Microbiol.*, *35*, 750 (1978).
105. A. L. Allen and D. Sternberg, *Biotechnol. Bioeng. Symp. Ser.*, *10*, 189 (1980).
106. M. J. Rudrick and A. D. Elbein, *J. Bacteriol.*, *124*, 534 (1975).
107. G. M. Umezurike, *Biochem. J.*, *145*, 361 (1975).
108. T. M. Wood and S. I. McCrae, The mechanism of cellulase action with particular reference to the C_1 component, In *Proc. Bioconversion Symposium*, Indian Institute of Technology, Delhi, 1977.
109. K.-E. Eriksson, Enzyme mechanisms involved in the degradation of wood components, In *Symposium on Enzymatic Hydrolysis of Cellulose*, SITRA, Helsinki, 1975.
110. A. R. Ayers, S. B. Ayers, and K.-E. Eriksson, *Eur. J. Biochem.*, *90*, 1 (1978).
111. M. P. Vaheri, *J. Appl. Biochem.*, *4*, 153 (1982).
112. U. Westermark and K.-E. Eriksson, *Acta Chem. Scand.*, *B29*, 419 (1975).
113. U. Westermark and K.-E. Eriksson, *Acta Chem. Scand.*, *B28*, 204 (1974).
114. M. R. Coudray, G. Canavascini, and H. Meier, *Biochem. J.*, *203*, 277 (1982).
115. J. C. Sadana and R. V. Patil, *J. Gen. Microbiol.*, *131*, 1917 (1985).
116. R. F. H. Dekker, *J. Gen. Microbiol.*, *120*, 309 (1980).
117. M. Nakayama, Y. Tomita, H. Suzuki, and K. Nisizawa, *J. Biochem.* (Tokyo), *79*, 955 (1976).
118. K.-E. Eriksson and B. Pettersson, *Eur. J. Biochem.*, *124*, 635 (1982).
119. A. C. Moloney, S. I. McCrae, T. M. Wood, and M. P. Coughlan, *Biochem. J.*, *225*, 365 (1985).
120. T. M. Wood and S. I. McCrae, *Carboh. Res.*, *57*, 117 (1977).

121. E. K. Gum and R. D. Brown, *Biochim. Biophys. Acta, 492*, 225 (1977).

122. I. Labudova and V. Farkas, *Biochim. Biophys. Acta, 744*, 135 (1983).

123. C. P. Dunne, *Enzyme Eng., 6*, 355 (1971).

124. T. M. Wood, *Biochem. J., 121*, 353 (1971).

125. G. M. Emert, E. K. Gum, J. A. Lang, T. H. Liu, and R. D. Brown, *Adv. Chem. Ser., 136*, 79 (1974).

126. K. Nisizawa, *J. Ferment. Technol., 51*, 267 (1973).

127. S. P. Shoemaker and R. D. Brown, *Biochim. Biophys. Acta, 523*, 133 (1978).

128. K.-E. Eriksson and B. Pettersson, *Eur. J. Biochem., 51*, 193 (1975).

129. K.-E. Eriksson and B. Pettersson, *Eur. J. Biochem., 51*, 213 (1975).

130. V. Deshpande, K.-E. Eriksson, and B. Pettersson, *Eur. J. Biochem., 90*, 191 (1978).

131. A. McHale and M. P. Coughlan, *Biochim. Biophys. Acta, 662*, 152 (1981).

132. S. Hayashi and H. Yoshioka, *Agr. Biol. Chem., 44*, 481 (1980).

133. M. G. Paice, M. Desrochers, D. Rho, L. Jurasek, C. Roy, C. F. Rolin, E. De Miguel, and M. Yaguchi, *Biotechnology, 2*, 535 (1984).

134. M. Yaguchi, C. Roy, C. F. Rollin, M. G. Paice, and L. Jurasek, *Biochem. Biophys. Res. Commun., 116*, 408 (1983).

135. A. J. Clarke and M. Yaguchi, *Eur. J. Biochem., 149*, 233 (1985).

136. M. Pentilla, P. Lehtovaara, H. Nevelaineau, R. Bhikhabhai, and J. Knowles, *Gene, 45*, 253 (1986).

137. H. Van Tilbeurgh, P. Tomme, M. Claeyssens, R. Bhikhabhai, and G. Pettersson, *FEBS Lett., 204*, 223 (1986).

138. P. Tomme, H. van Tilbeurgh, G. Pettersson, J. van Damm, J. Vandekerckhove, J. Knowles, T. Teeri, and M. Claeyssens, *J. Biochem.* (Tokyo), *63*, 591 (1968).

139. V. Deshpande, M. Rao, S. Keskar, and C. Mishra, *Enzyme Microb. Technol., 6*, 371 (1984).

140. H. Kolar, H. Mischak, W. P. Kammel, and C. P. Kubicek, *J. Gen. Microbiol., 131*, 1339 (1985).

141. J. G. Shewale and J. C. Sadana, *Arch. Biochem. Biophys., 207*, 185 (1981).

142. M. J. Rudrick and A. D. Elbein, *J. Biol. Chem., 248*, 6506 (1973).

143. M. P. Meyer and G. Canavascini, *Appl. Env. Microbiol.*, *41*, 924 (1981).

144. R. K. Berry and R. F. H. Dekker, *FEMS Microbiol. Lett.*, *21*, 309 (1984).

145. G. Halliwell, M. Griffin, and R. Vincent, *Biochem. J.*, *127*, 43 (1972).

146. L. E. R. Berghem and L. G. Pettersson, *Eur. J. Biochem.*, *37*, 21 (1973).

147. K.-E. Eriksson and B. Pettersson, *Eur. J. Biochem.*, *51*, 213 (1975).

148. E. T. Reese, In *Biological Transformation of Wood by Micro-organisms* (W. Leise, ed.), Springer-Verlag, New York, 1975, p. 165.

149. H. Griffin, F. R. Dintzis, L. Krull, and F. C. Bakers, *Biotechnol. Bioeng.*, *26*, 296 (1984).

150. J. W. Koenigs, *Wood Fiber*, *6*, 66 (1974).

151. T. L. Highley, *For. Prod. J.*, *25*, 38 (1975).

152. T. Nilson, *Mater. Org.*, *9*, 173 (1974).

153. K.-E. Eriksson, Microbial degradation of cellulose and lignin, In *Proc. Ekman–Days Int. Symposium on Wood and Pulping Chemistry*, SCPI, Stockholm, 1981.

154. C. Marsh, *Biochim. Biophys. Acta*, *122*, 367 (1966).

155. T. M. Wood, *Biotechnol. Bioeng. Symp. Ser.*, *5*, 111 (1975).

156. J. O. Warwicker, R. Jeffries, R. L. Colbran, and R. N. Robinson, A Review of the Literature on the Effect of Caustic Soda and Other Swelling Agents on the Fine Structure of Cotton, Shirley Institute, Didsbury, Manchester, 1966, Pamphlet Nol. 93.

157. A. R. White and R. M. Brown, *Proc. Natl. Acad. Sci. USA*, *78*, 1047 (1981).

158. A. R. White, In *Cellulose and Other Natural Polymer Systems* (R. M. Brown, ed.), Plenum Press, New York, 1982, p. 489.

159. H. Chanzy and B. Henrissat, *FEBS Lett.*, *184*, 285 (1985).

160. T. M. Wood, Co-operative action between enzymes involved in the degradation of crystalline cellulose, In *Colloque Cellulolyse Microbienne*, CNRS, Marseille, 1980.

161. T. M. Wood, Enzymic and other factors affecting the microbial generation of soluble sugars from cellulosic materials, In *Proc. Tappi Research and Development Division Conference*, Asheville, North Carolina, 1982.

162. D. D. Y. Ryu and M. Mandels, *Enzyme Microb. Technol.*, *26*, 488 (1984).

163. T.-M. Enari, Biotechnology a challenge and an opportunity for the forest product industry, In *Proc. Marcus Wallenberg*

Foundation Symposium, New Horizons for Biotechnological
Utilization of the Forest Resource, Falum, Sweden, 1985.

164. C.-S. Gong and G. T. Tsao, *Ann. Rep. Ferment. Proc.*, *3*,
     111 (1979).

165. F. Stutzenberger, *Ann. Reports Ferment. Proc.*, *8*, 112
     (1985).

166. M. Mandels and E. T. Reese, *J. Bacteriol.*, *79*, 816 (1960).

167. K.-E. Eriksson and S. G. Hamp, *Eur. J. Biochem.*, *90*, 183
     (1978).

168. D. Sternberg and G. R. Mandels, *J. Bacteriol.*, *144*, 1197
     (1980).

169. H. Merivuori, K. M. Siegler, J. A. Sands, and B. S. Mon-
     tenecourt, *Biochem. Soc. Trans.*, *13*, 411 (1985).

170. B. Berg and L. G. Pettersson, *J. Appl. Bacteriol.*, *42*, 65
     (1977).

171. A. Binder and T. K. Ghose, *Biotechnol. Bioeng.*, *20*, 1187
     (1978).

172. R. L. D. Gussen, G. M. Aerts, M. Claeyssens, and
     C. K. Debruyne, *Biochim. Biophys. Acta*, *525*, 142 (1978).

173. J. R. Lowenberg and C. M. Chapman, *Arch. Microbiol.*, *113*,
     61 (1977).

174. T. Nisizawa, H. Suzuki, and K. Nisizawa, *J. Biochem.*
     (Tokyo), *71*, 999 (1972).

175. M. Mandels, F. W. Parrish, and E. T. Reese, *J. Bacteriol.*,
     83, 400 (1960).

176. G. Canavascini, M. R. Coudray, J. P. Rey, R. J. G. South-
     gate, and H. Meier, *J. Gen. Microbiol.*, *110*, 291 (1979).

177. B. M. Eberhart, R. S. Beck, and K. M. Goolsby, *J. Bac-
     teriol.*, *130*, 181 (1977).

178. B. J. Gallo, R. Andreotti, C. Roche, D. D. Y. Ryu, and
     M. Mandels, In *Utilization of Agricultural Crop Residues*
     (Y. W. Han and S. K. Smith, eds.), U.S. Dept, Agriculture,
     ARS W-53, 1979, p. 89.

179. N. Peiterson, In *Bioconversion of Cellulosic Substances into
     Energy, Chemicals and Microbial Proteins* (T. K. Ghose, ed.),
     Indian Institute of Technology, Delhi, 1977, p. 281.

180. P. Rapp, U. Knobloch, and R. Wagner, *J. Bacteriol.*, *149*,
     783 (1982).

181. D. Sternberg, *Biotechnol. Bioeng. Symp. Ser.*, *6*, 35 (1976).

182. R. M. Hoffman and T. M. Wood, *Biotechnol. Bioeng.*, *22*,
     81 (1985).

183. T. M. Wood and S. I. McCrae, The cellulase complex of
     Trichoderma koningii, In *Proc. Symposium on Enzymatic
     Hydrolysis*, SITRA, Aulanko, Finland, 1975.

184. R. V. Patil and J. C. Sadana, *Can. J. Biochem. Cell. Biol.*, *62*, 920 (1984).

185. J. B. Parry, J. C. Stewart, and J. Hepinstall, *Biochem. J.*, *213*, 437 (1983).

186. J. C. Sadana, A. H. Lachke, and R. B. Patil, *Carboh. Res.*, *133*, 297 (1984).

187. T. M. Wood and S. I. McCrae, In *Development in Food Science* (H. Chiba and M. Fujmaki, eds.), Elsevier, Amsterdam, 1979, p. 257.

22
Bacterial Cellulases

PETER RAPP* and ASTRID BEERMANN   *Institut für Biochemie und
Biotechnologie, Technische Universität Braunschweig, Braunschweig,
Federal Republic of Germany*

I. INTRODUCTION

Three classes of enzymes are found both in fungal and bacterial cellulase systems: (1) 1,4-β-D-glucan-4-glucanohydrolases or endo-1,
4-β-glucanases (EC 3.2.1.4), which randomly cleave the internal cellulosic bonds; (2) 1,4-β-D-glucan cellobiohydrolases (EC 3.2.1.91)
and 1,4-β-D-glucan glucohydrolases (EC 3.2.1.74); these two exo-
1,4-β-D-glucanases both act from the nonreducing end of cellulose
chains or cellooligosaccharides, removing cellobiose or glucose; and
(3) 1,4-β-D-glucosidases or cellobiases (EC 3.2.1.21), which hydrolyze cellobiose and specifically cleave glucosyl units from the nonreducing ends of cellooligosaccharides. Some bacteria do not have cellobiases and cleave cellobiose and cellodextrins by cellobiose phosphorylases (EC 2.4.1.20) and cellodextrin phosphorylases (EC
2.4.1.49), respectively.

However, cellulases can be classified not only by their exo and
endo action or phosphorolytic character but also by the configuration
of their products. Endo splitting of a β-glucan results in retention
of the configuration of the products. The configuration of the products
is inversed when exoglucanases act on their corresponding substrates.
β-Glucosidases, on the other hand, act in such a way that the

---

*Current affiliation: Gesellschaft für Biotechnologische Forschung,
Braunschweig, Federal Republic of Germany.

configuration is retained. Finally the configuration of the products
of cellobiose and cellodextrin phosphorylases is inverted (1).

Apart from aryl glucosides and disaccharides, trimers and higher
oligosaccharides are hydrolyzed by both glucosidases and exoglucanases, but to different extents. Dimers and trimers are readily hydrolyzed by glucosidases, but the hydrolytic rate decreases with increasing chain length. Exoglucanases show the opposite effect:
long chains are readily hydrolyzed, and dimers are quite resistant
(1).

Although much is known of the components of multienzyme systems of fungal cellulases, our understanding of the mechanism of
cellulose degradation has remained rather superficial. The present
concept (discussed in more detail in Chap. 21) is that hydrolysis
of crystalline cellulose occurs in a sequential attack on the substrate
by the different fungal hydrolytic enzymes. The creation of nonreducing ends by endoglucanases which are sites of attack for exoglucanases leads to a synergism in the rate of cellulose degradation.
Maximum rate and extent of hydrolysis requires the simultaneous
presence of not only endo- and exoglucanases but also β-glucosidases. Furthermore, considerable doubt still exists on the mechanism of initiation of cellulose degradation.

Although a detailed discussion of cellulase assay methods is
beyond the scope of this chapter, a few points regarding assay
methods are worth mentioning, because they impact the interpretation of data provided below. Endoglucanase activity is often measured by the amount of reducing end groups formed by hydrolysis
of carboxymethylcellulose (CMC); however, it must be born in mind
that the other enzymes of the cellulase system also form reducing
end groups. A more sensitive assay method, free from confusing
interferences, is the viscometric method using water-soluble CMC or
hydroxyethylcellulose as substrate (2). A good means for characterization of individual endoglucanases seems to be the determination of the relationship between fluidity and reducing power during
hydrolysis of CMC by endoglucanases (3). This ratio is an indication of the degree of randomness in cellulose hydrolysis. However,
it has not been widely employed in the characterization of bacterial
cellulases.

In contrast to cultures of fungi, exoglucanase activity has been
detected in only a few bacterial cultures (Table 3). Even some of
these results are questionable, since at present no reliable method
exists for the determination of exoglucanase activity in the presence
of other cellulase components. Direct measurements of exoglucanase
activity can only be done with purified enzymes, since no specific
substrates for exoglucanases are available. The method of Rabinowitch et al. (4) using carboxymethyl-substituted cellodextrins is

limited to the determination of exoglucanases splitting off only
glucose units, and the method of Desphande et al. (5) requires the
purification of an endoglucanase from the studied mixture of cellulases for standardization of the assay procedure.

The apparent deficiency of exoglucanase activity and the high
ratio of endo- to exoglucanase activity in cultures of bacteria as well
as the occurrence of cellobiose and cellodextrin phosphorylases in
some cellulolytic bacteria suggest that bacterial cellulolytic systems
differ considerably from those of fungi. Furthermore a large difference appears to exist in the location of the cellulase systems of bacteria and fungi. The bacterial β-glucosidases are always cell-bound
in both aerobic or anaerobic bacteria. Many of the anaerobically
formed 1,4-β-glucanases are found to be organized in multienzyme
complexes, which again are often associated with the cells. Only a
few 1,4-β-glucanases formed by aerobic and facultatively anaerobic
bacteria have up to now been found to be organized in multienzyme
complexes.

II. PROPERTIES OF CELLULASES FROM ANAEROBIC BACTERIA

A. Cellulases from *Acetivibrio*

Acetivibrio cellulolyticus is a cellulolytic, gram-negative, non-spore-
forming, anaerobic bacterium which has been placed in the family
Bacteroidaceae (6). Two extracellular endoglucanases and another
extracellular 1,4-β-glucanase, which was considered to be a 1,4-β-
glucan cellobiohydrolase, have been partially characterized (7,8).
Some properties of these three enzymes are shown in Tables 1 and
3. The endoglucanase of a molecular weight of 33,000 (33 kD) can
be reversibly dissociated into polypeptide subunits similar to the low
molecular weight endoglucanase (8). *Acetivibrio cellulolyticus* also
contains a cell-associated β-glucosidase (8,9), the characteristics of
which are represented in Table 5. The enzyme exhibits a strong
preference for p-nitrophenyl-β-D-glucopyranoside (pNPGlc) over
cellobiose and 70% of its activity is readily released by a simple
freeze-thaw procedure (9). The extracellular cellulase of *A. cellulolyticus* requires Ca^{2+} and possibly sulfhydryl groups for efficient
hydrolysis of microcrystalline cellulose. The effect of these agents
on p-nitrophenyllactopyranoside hydrolysis suggests that they are
required by an exoglucanase component (140). The protein profile
of cellulase complex adsorbed on cellulose is very similar to that of
Clostridium thermocellum cellulosome (10−12), including a major protein species with a molecular weight in the 200-kD range. Ninety
percent of Avicel-hydrolyzing activity and 15% of endoglucanase activity of the total cellulase adsorb to 2% (w/v) Avicel at 40°C (13).

TABLE 1   Properties of Endo-1,4-β-Glucanases from Anaerobic Bacteria

Organism	Molecular weight (kD)	Activity optima	
		pH	T (°C)
Acetivibrio cellulolyticus CD2 (NRCC 2248)	33.0 10.4	5.0 (crude preparation)	50
Fibrobacter succinogenes S85	9–13% of extrac. endogluc. act. assoc. with non-sedimentable material of >4000.0; 28–38% of extrac. endogluc. act. assoc. with molecules of 45.0	5.6–6.6 (Partially purified endoglucanase)	50
F. succinogenes S85	65.0 (E$_1$)	6.4	39
	118.0 (E$_2$)	5.8	39
	94.0 (proteolytic degradation product of the 118-kD enzyme)		
Clostridium acetobutylicum (NRRL B527) (ATCC 824) (P270)		5.2 5.2 4.6	37
C. cellulolyticum H10 (ATCC 35319)	40.0 (expression of endogluc. gene in E. coli)		

pH and temperature stability	Apparent K_m (mg/ml) (on indicated substrate)	P_I	Ref.
			8, 135
50% of max. act. at pH 5.1 and 7.7; 15.1% of max. act. at 60°C after 30 min (Partially purified endoglucanase)			16
90% of max. act. at pH 5.9 and 7.1; 25% of max. act. at 45°C after 30 min	3.6 (CMC)	4.8	173
90% of max. act. at pH 5.4 and 6.2; 93% of max. act. at 45°C after 30 min	12.2 (CMC)	9.4	
Endogluc. of both strains unstable < 5.2			33, 35
			189

TABLE 1   (Continued)

Organism	Molecular weight (kD)	Activity optima	
		pH	T (°C)
C. josui FERMP-9684	45.0	6.8	60
C. stercorarium	100.0		
C. stercorarium	91.0−99.0	6.4	80
C. thermocellum LQRI	83.0−94.0	5.2	62
C. thermocellum (NCIB 10682)	56.0	6.0	80
C. thermocellum (NCIB 10682, ATCC 27405)	49.0 52.0 (expression of *celA* gene in *E. coli*)	5.5−6.5 5.5−6.5	75 75
C. thermocellum (NCIB 10682)	125.0 ± 10.0 (complex of at least 5 proteins)		
C. thermocellum (ATCC 27405)		6.1 (CMC) 5.0 (TNP-CMC)	70 (CMC)
C. thermocellum YS	170.0 98.0 75.0 } constituents of cellu- 60.0 some of 54.0 MW 2100 48.0		
C. thermocellum JW 20 (ATCC 31449)	50.0− 60.0 60.0− 75.0 85.0−100.0 130.0−150.0 (of OBL, OBS, and FB complexes[a])	6.0 (Originally bound large complex)	65

pH and temperature stability	Apparent K_m (mg/ml) (on indicated substrate)	P_I	Ref.
80% of max. act. below 50°C after 1 hr	15.4 mM [(Glc)$_4$] 6.7 mM [(Glc)$_5$]		175
		4.6	191
50% of max. act. at 85°C after 1 hr	7.14 (CMC, DS 0.75)	3.85	36
	2.3 mM [(Glc)$_5$] 0.56 mM [(Glc)$_6$]	6.72	38
No act. at 85°C after 30 min		6.2	39
Stable for several hours at 60°C			40
			136
	3.6 (CMC) 3.3 (TNP-CMC)		55, 137
			11
10% of max. act. at 80°C after 1 hr (Originally bound large complex)			63, 182

TABLE 1   (Continued)

Organism	Molecular weight (kD)	Activity optima	
		pH	T (°C)
C. thermocellum (NCIB 10682)	65.0 (expression of *celD* gene in *E. coli*)	6.0	
C. thermocellum (NCIB 10682)	78.0 57.0−73.0 55.0−90.0 80.0−90.0 (endogluc. genes expressed in *E. coli*)		65
C. thermocellum	38.0 (expression of *celC* gene in *E. coli*)	6.0	60
C. thermocellum (ATCC 27405)	48.0 52.0 (expression of *celA* gene in *B. subtilis*)		75 75
Clostridium strain M7		6.5	67
C. thermocopriae JF3-3	46.0	6.5	
Ruminococcus albus F-40	50.0	6.7	44
R. albus SY-3	>1500.0 30.0		
R. flavefaciens strain 67	>3000.0 89.0	6.4 6.4	45 45

pH and temperature stability	Apparent K_m (mg/ml) (on indicated substrate)	P_I	Ref.
The enzyme was denatured above 65°C		5.4	146, 214
			215
		6.2	145, 192
			216
		4.5	37
		6.3	
16% of max. act. at 75°C after 7 hr			200
Almost no activity at 70°C after 10 min	7.2 (CMC, DS 0.6)		142
	0.7 (CMC, DS 0.95)		
	0.4 (CMC, DS 1.4)		
		—	3
		6.0 – 6.1	143, 217
			21

TABLE 1   (Continued)

| | | | Activity optima |
| | | | T |
Organism	Molecular weight (kD)	pH	(°C)
Thermoanaerobacter cellulolyticus (IFO 14436)	(expression of β-glucanase gene in *E. coli*)	7.0	80

[a]OBL, originally bound large; OBS, originally bound small; FB, free bindable.

B. Cellulases from *Bacteroides*

Fibrobacter succinogenes is a cellulolytic, gram-negative, strictly anaerobic bacterium present in the rumen of many ruminants (14). Although *F. succinogenes* grows well in media with filter paper or microcrystalline cellulose as carbon source, up to now no exoglucanase activity has been detected. During growth on cellobiose and glucose, endoglucanase activity is found to be cell-bound (15). In cellulose-grown cultures 70–80% of endoglucanase activity is present in the culture supernatant. Of this extracellular endoglucanase activity, 50–62% is associated with sedimentable membrane fragments, 9–13% with nonsedimentable material of a molecular weight higher than 4×10^6 D (4 MD), and 28–38% with molecules having a molecular weight of approximately 45 kD (16). Furthermore, the membrane-associated endoglucanase activity is readily solubilized with Triton X-100 (16) or with bovine pancreatic trypsin (17). Some properties of a partially purified endoglucanase are shown in Table 1. Forsberg et al. (18) suggest that membranous endoglucanases may be released by bleb formation from the outer membrane of intact cells, primarily in pockets between cells and cellulose.
    Fibrobacter succinogenes also possesses a largely cell-bound cellobiase. The only product detected after cellobiose hydrolysis is glucose. No transglycosylation products are observed. Using a crude enzyme preparation an apparent K_m value of 4.3 mM for cellobiose hydrolysis was determined. The fact that the bulk of β-glucosidase activity is associated with the membrane fraction of disrupted cells while the endoglucanase activity is largely released associated with outer membrane fractions indicates that cellobiase is more firmly bound, perhaps to the inner cytoplasmic membrane (15). The enzyme is sensitive to air and requires sulfhydryl groups for activity (19). Cellobiase seems to be constitutive whereas the

pH and temperature stability	Apparent K_m (mg/ml) (on indicated substrate)	P_I	Ref.
75% of max. act. at 70°C after 5 hr		221	

endoglucanase activity is eight times higher in cells grown on cellulose than on cellobiose or glucose (15).

Bacteroides polypragmatus is capable to ferment simultaneously glucose and cellobiose to ethanol. The formation of the intracellular β-glucosidase is induced by cellobiose and not repressed by glucose (141). Other properties are shown in Table 5.

C. Cellulases from *Ruminococcus*

Ruminococcus albus and *Ruminococcus flavefaciens* occur along with *F. succinogenes* in the rumen in proportions that vary according to the diet (20). Properties of endoglucanases, exoglucanases, and β-glucosidases of both ruminococci, shown in Tables 1, 3, and 5, are similar. Most of the glucanase activity of *R. albus* and *R. flavefaciens* is cell-associated and of very high molecular weight (3,21, 22). The cell wall-bound enzymes of *R. albus* of a molecular weight higher than 1.5 MD can readily be removed from the cell wall by washing with phosphate buffer or water. Under some cultural conditions it appears that the high molecular weight complex dissociates into low molecular weight fragments of 25 to 30 kD. Rumen fluid in the medium seems to enhance the stability of the high molecular weight material. Both high molecular weight and low molecular weight cellulases show the same relationship between fluidity and reducing power, suggesting that they are composed of the same enzymes (3). In contrast to Leatherwood (23,24), who postulated that an affinity factor of high molecular weight and a hydrolytic factor of low molecular weight are essential features of the cellulase system of *R. albus*, Wood et al. (3) suggested that *R. albus* cellulase exists as an aggregate of low molecular weight cellulase compounds on the bacterial cell wall and in solution under certain conditions. A typical endoglucanase from *R. albus* having a thiol functional

TABLE 2 Properties of Endo-1,4-β-Glucanases from Aerobic and Facultatively Anaerobic Bacteria

Organism	Molecular weight (kD)	Activity optima	
		pH	T (°C)
Bacillus sp. strain N-4	50.0 50.0	10.0 10.0	
Bacillus sp. strain N-4	50.0 58.0	5−10.9 5−10.9	
Bacillus sp. KSM-635		9.5	40
Bacillus sp. No. 1139	92.0 94.0 (expression of cellulase gene in *E. coli*)	9.0 9.0	
Bacillus strain KSM-522	35.0	7−10	50−60
B. subtilis DLG	35.2 (extracellular) 51.5 (expression of 1,4-β-glucanase gene in *E. coli*)	4.8	58
B. subtilis AU-1	23.0	5.5	
B. subtilis	33.0 (expression of cellulase gene in *B. megaterium*)	5.5	60
B. subtilis IFO 3034	(expression of cellulase gene in *E. coli*)	6.0−6.5	55

pH and temperature stability	Apparent K_m (mg/ml) (on indicated substrate)	P_I	Ref.
Stable up to 60°C. Stable up to 80°C; both enzymes are stable from pH 6−10			65
Both enzymes are stable from pH 5−11 and up to 75°C			66
80−90% of max. act. at pH 11 after 10 min; 50% of max. act. at 65°C after 20 min			204
Stable from pH 6−11 and up to 40°C; inactive at 60°C	0.48 (CMC, DS 0.65) 0.48 (CMC, DS 0.65)	3.1 3.1	67, 154
70% of max. act. at 60°C after 30 min			203
Stable up to 45−50°C			64, 150, 218
Stable at 65°C for 1 hr	4.0 (CMC)		151
80% of max. act. at 60°C after 24 hr	1.6 (CMC)	7.23	205
Stable at pH 5.0−8.5 at 55°C for 1 hr; 35% of max. act. at 70°C after 1 hr			152

TABLE 2   (Continued)

Organism	Molecular weight (kD)	Activity optima pH	T (°C)
Cellulomonas sp.	62.9	–	–
UQM 2903	44.3	7.0	40
	62.9	–	–
	76.9	7.0	50
	142.9	7.0	50
	120.6	7.0	40–50
Cellulomonas sp.	118.0		
strain IIbc	49.0–52.0		
	53.5		
Cellulomonas sp.	78.0		
ATCC 21399	63.0		
	56.0		
C. fermentans	40.0		
strain M	57.0		
(ATCC 43279)			
C. fimi	110.0 (expression of *cenB* gene in *E. coli*)		
C. fimi	53.0 48.7 (expression of *cenA* gene in *E. coli*)		
C. flavigena (ATCC 482)	40.0	7.0	
C. uda		6.8	
C. uda CB4	39.0	6.0–8.0	55–60
C. uda CB4	40.5 (expression of endoglucanase gene in *Zymomonas mobilis*)		

pH and temperature stability	Apparent K_m (mg/ml) (on indicated substrate)	P_I	Ref.
(Half-life at 60°C)	−		138
0.9 hr	6.67 (CMC)		
−	−		
<1.0 hr	5.55 (CMC)		
0.4 hr	24.39 (CMC)		
0.35 hr	3.13 (CMC)		
			69
			155
	1.5 (CMC)		156
	59.0 (CMC)		
			174
			206
	7.5 (CMC, DS 0.65−0.85)		76
		4.7	70, 71
		4.8	
		4.9	
			126, 139
			208

TABLE 2   (Continued)

Organism	Molecular weight (kD)	Activity optima pH	Activity optima T (°C)
Cellvibrio fulvus (NCIB 8634)		7.0	
Erwinia chrysanthemi strain 3665	45.0 (endoglucan- ase Z)	6.2 – 7.5	52
E. chrysanthemi strain 3665	35.0 (expression of *celY* gene in *E. coli*)	5.5	
Pseudomonas sp. -1		6.5 – 7.0	40
Pseudomonas sp. (NCIB 8634)	23.0		
Pseudomonas fluorescens var. *cellulosa*	A B C	8.0 8.0 7.0	
	I-(1) >200.0 (con- sists only of component B)		
	II 65.0 ⎫ consists of III 60.0 ⎬ components IV 50.0 ⎭ A and B		
	V 40.0 (contains only component A)		
P. fluorescens var. *cellulosa*	38.0 (expression of endoglucanase gene in *E. coli*)	7.0	

pH and temperature stability	Apparent K_m (mg/ml) (on indicated substrate)	P_I	Ref.
50% of max. act. at 50°C after 10 min			92
50% of max. act. at 32°C after 1 hr (without shaking)		4.3	103
		8.2	163
83.3% of max. act. at 50°C after 10 min	1.8 (CMC)		112
			164
A, B, and C most stable at pH 7−8; completely inactivated at 60°C after 10 min			105
All nearly completely inactivated at 60°C after 10 min			108, 109
			219, 220

TABLE 2 (Continued)

Organism	Molecular weight (kD)	Activity optima	
		pH	T (°C)
P. fluorescens var. *cellulosa*	106.0 (expression of an endoglucanase gene in *E. coli*)		
Sporocytophaga myxococcoides Q.M.B. 482	52.0 46.0	5.5 – 7.5 6.5 – 7.5	
Streptomyces strain KSM-9	32.0 32.5 92.0	8.5 6.0 6.0	45 – 55 45 – 55 45 – 55
Thermomonospora curvata	46.0 106.0		
T. curvata		6.0	65
T. curvata	22.0		
T. fusca YX	108.0 42.0 71.0 106.0 45.0	6.5 6.5 – – –	
Thermomonospora sp. strain YX		5.9	70
T. fusca strain YX	94.0 46.0	6.0 6.0	74 58
T. fusca strain YX	43.0 (expression of E$_2$ gene in *S. lividans*)		

pH and temperature stability	Apparent K_m (mg/ml) (on indicated substrate)	P_I	Ref.
			209
		4.75	100
		7.5	
			168
			222
	3.5 (CMC, DS 0.7)		118
	2.3 (CMC)	3.7	223
	0.36 (CMC)	3.2	224
	0.12 (CMC)	4.7	
	—	3.1	
	—	3.6	
	—	4.5	
85% of max. act at pH 6.0−7.3 after 24 hr at 55−60°C; stable at 60°C for 24 hr; 50% of max. act. at 65°C after 24 hr			117
	0.36 (CMC)	3.5	170
	0.12 (CMC)	4.5	
			225

TABLE 3   Properties of Exo-1,4-β-Glucanases for Anaerobic and
Aerobic Bacteria

Organism	Molecular weight (kD)	Activity optima pH	Activity optima T (°C)
Acetivibrio cellulolyticus CD2 (NRCC 2248)	38.0		
Fibrobacter succinogenes S85	75.0	6.2	39 (without Cl⁻) 45 (with 0.2 M Cl⁻)
	50.0 (periplasmic)	6.1	45−50
	50.0 (extra-cellular)	6.3−6.7	45−50
Clostridium stercorarium (NCIB 11745)	87.0	−	−
Ruminococcus albus	100.0 (SDS-PAGE) 200.0 (gel filtr.)	6.8 6.8	37 37
R. flavefaciens strain 67	>3000.0 89.0	6.4−6.6 6.4−6.6	39 39
R. flavefaciens FD-1	118.0 (SDS-PAGE) 230.0 (gel filtr.)	5.0 5.0	39−45 39−45
Cellulomonas fimi	56.0 (expression of *cex* gene in *E. coli*)		
C. fimi	(native enzyme)		
	(expression of *cex* gene in *S. cerevisiae*)		

pH and temperature stability	Apparent K_m (mg/ml) (on indicated substrate)	P_I	Ref.
			8
30% of max. act. at 39°C after 24 hr	0.1 mM [pNP(Glc)$_2$]	6.7	176
20% of max. act. at 39°C after 4 hr	0.27 mM [pNP(Glc)$_2$]	4.9	177, 183
–	0.21 mM [pNP(Glc)$_2$]	4.9	
–	–	–	191
Stable between pH 5.5 and 8.0; 50% of max. act. at 40°C after 15 hr		5.3 5.3	30
			21
	3.08 mM (pNP(Glc)$_2$] 3.08 mM [pNP(Glc)$_2$]		144
			158, 160, 226
50% of max. act. [pNP(Glc)$_2$] at 75°C after 30 min	3.18 ± 0.214 (CMC) 0.641 ± 0.027 mM [pNP(Glc)$_2$]		207
58% of max. act. [pNP(Glc)$_2$] at 75°C after 30 min	3.156 ± 0.256 (CMC) 0.706 ± 0.012 mM [pNP(Glc)$_2$]		

TABLE 3   (Continued)

Organism	Molecular weight (kD)	Activity optima	
		pH	T (°C)
C. fimi	49.3 47.3 (expression of cex gene in E. coli)		
C. uda CB 4	81.0	5.5 − 6.5	45 − 50
Steptomyces flavogriseus IAF-45-CD (ATCC 33331)	45.0		
Thermomonospora curvata		6.0 − 6.5	65
Thermomonospora[a] sp. strain YX		7.0	65

[a]Exoglucanase character questionable.

group was purified by Ohmiya et al. (142). The enzyme degrades water-insoluble cellulose to water-insoluble fragments. It seems to be somewhat different from the endoglucanase reported by Wood and Wilson (143), which releases large quantities of cellotriose and smaller amounts of cellotetraose from phosphoric acid-swollen cellulose. Some other properties of both endoglucanases are summarized in Table 1.

Stack and Hungate (25) reported that R. *albus* grown in the presence of 3-phenylpropanoic acid (PPA) produced considerable amounts of cell-bound as well as extracellular high molecular weight cellulase and small amounts of two low molecular weight enzymes. In the absence of PPA R. *albus* produces only large amounts of the two low molecular weight enzymes. PPA does not affect the kind of

pH and temperature stability	Apparent K_m (mg/ml) (on indicated substrate)	P_I	Ref.
	—		206
	0.7 mM [pNP(Glc)$_2$]		
Stable between pH 5.5 and 8.0; act. completely lost at 60°C after 10 min	2.9 (Avicel)	4.4	77, 227
		4.15	167
50% of max. act. at pH 5.0 or 8.0; 50% of max. act. at 75°C after 10 min			118
50% of max. act. at 60°C after 24 hr			117

of proteins produced but influences the distribution of the two major proteins of 102 and 85 kD. These two proteins are primarily cell-associated when *R. albus* is cultured in the presence of PPA but are mainly in the culture supernatants in the absence of PPA. Since rumen fluids normally contain relatively high concentrations of PPA (26,27), the results cited above correspond to some extent to those of Wood et al. (3). PPA seems to be involved in the formation of lobed ruthenium red-stainable capsule surrounding the cell wall and in the formation of small vesicular structures, which appear to aggregate into large spherical units. Stack and Hungate (25) suggest that the vesicular structures or their aggregates may represent the high molecular weight enzyme while the lobes represent stages in the process of vesicle formation.

TABLE 4   Properties of Cellobiose Phosphorylase from Anaerobic and Aerobic Bacteria

Organism	Molecular weight (kD)	Activity optima		pH and temperature stability	Apparent K_m (mM) (on indicated substrate)	Ref.
		pH	T (°C)			
Clostridium thermocellum strain 651		6.5		Active from pH 4.5 to 9.0	9.2 (6-deoxy-Glc); 73.0 (2-deoxy-Glc); 35.0 (D-Xyl); 9.5 (Glcn); 85.0 (D-Man); 240.0 (D-Ara); 7.3 [(Glc)2]; 160.0 (L-Fuc); 2.9 (P_i); 2.1 (Glc-1-P)	47
Cellvibrio gilvus (ATCC 13127)	Four subunits of 72.0	7.6		Completely inactivated at 60°C after 10 min	1.25 [(Glc)2]; 0.77 (P_i)	90

Cellulase and xylanase activity in *R. flavefaciens* both occur in a large enzyme complex of a molecular weight higher than 3 MD as well as in a smaller protein of molecular weight 89 kD. Although present in the same enzyme complex, both enzyme activities probably have different active sites (21). Latham et al. (28) showed that *R. flavefaciens* possesses a glycoprotein coat with rhamnose, glucose, and galactose as its principal carbohydrates. The bacteria adhere strongly by means of this coat to cellulosic materials. Ljungdahl and Eriksson (29) suggest that the large enzyme complex with its cellulase and xylanase activity is possibly associated with the glycoprotein coat. The cellulases of *R. flavefaciens* require Ca^{2+} or Mg^{2+} and thiol groups for activity (21). They appear to be constitutive enzymes and their synthesis seems to be under catabolite repression (22). Cellobiose, cellotriose, and glucose in a ratio of 6.2:5.8:1.0 are the major products of CMC hydrolysis by cellulases from *R. flavefaciens*. This and the rapid release of reducing sugars relative to the slight increase of specific fluidity indicates weak endoglucanase and very active exoglucanase activity in the cellulase complex of *R. flavefaciens* (21).

An exoglucanase from *R. flavefaciens* was purified by Gardner et al. (144). Its N-terminal amino acids and extracellular location in the later stages of growth suggest that the enzyme is secreted. The catalytic properties of the exoglucanase from *R. flavefaciens* FD-1 are similar to those reported for the cellulase complex of *R. flavefaciens* 67 (21). However, the exoglucanase is stimulated only by Ca^{2+} and not by Mg^{2+}. Other properties are shown in Table 3.

The exoglucanase from *R. albus*, which is released from the cells during the stationary growth phase, was studied extensively by Ohmiya et al. (30). They purified and characterized an enzyme active against *p*-nitrophenyl-β-D-cellobioside [pNP(Glc)$_2$]. Its properties are summarized in Table 3. The molecular weight of the enzyme was estimated by SDS-gel electrophoresis to be 100 kD and by gel filtration to be 200 kD. Thus the native enzyme seems to consist of two 100-kD subunits. The rather high activity against pNP(Glc)$_2$ and the low activity toward cellulose makes its function appear unclear as a cellulolytic enzyme (30).

Ohmiya et al. (31) purified another enzyme from *R. albus* which is active toward pNPGlc. Its properties are shown in Table 5. The molecular weight is estimated to be 82 kD by polyacrylamide gel electrophoresis (PAGE) and gel filtration. The molecular weight of 116 kD determined by SDS-gel electrophoresis is probably overestimated due to the high carbohydrate content of 12% in the enzyme. The highest enzymatic rates are measured with pNP(Glc)$_2$ and pNPGlc. The rate of enzyme reaction against cellooligomers (DP 2–5) are only

TABLE 5   Properties of β-Glucosidases from Anaerobic and Aerobic
Bacteria

Organism	Molecular weight (kD)	Activity optima	
		pH	T (°C)
Acetivibrio cellulolyticus CD2 (NRCC 2248, ATCC 33288)	81.0	6.5 (pNPGlc) 6.0 [(Glc)$_2$] (crude cell-free extract)	
Bacteroides polypragmatus strain GP4 (NRC 2288)	100.0	7.0	
Caldocellum saccharolyticum	52.0 (expression of β-glucosidase gene in *E. coli*)	6.25	85
Clostridium acetobutylicum (NRRL B527) (ATCC 824)		5.6 [(Glc)$_2$] 4.8−6.0 [(Glc)$_2$]	
C. stercorarium	85.0	5.5	65
C. thermocellum (NCIB 10682)	43.0	6.0−6.5 [(Glc)$_2$ and pNPGlc]	65
C. thermocellum	84.1		
C. thermocopriae JT3-3	46.0	6.5	80
Ruminococcus albus F-40	82.0−116.0	6.5 (pNPGlc)	30−35
	85.0 (expression of β-glucosidase gene in *E. coli*)	6.0−6.5	37

pH and temperature stability	Apparent K_m (mM) (on indicated substrate)	P_I	Ref.
	4.0 [pNPGlc) 40.0 [(Glc)$_2$] (crude cell-free extract)		8, 9
50% of max. act. at 55°C after 4 hr	0.73 (pNPGlc) 100.0 [(Glc)$_2$]	4.2	141
50% of max. act. at 80°C after 14 hr	0.66 (pNPGlc) 21.0 [(Glc)$_2$]		185
			33
Most stable at pH 5.2; Stable from pH 4.8 – 6.5			
20% of max. act. at 60°C after 1 hr	0.8 (pNPGlc) 33.0 [(Glc)$_2$]	4.8	179
60% of max. act. at 60°C after 7 hr and 50% of max. act. at 68.5°C after 62.5 min with pNPGlc as substrate	2.6 (pNPGlc) 83.0 [(Glc)$_2$]		44
			198
21% of max. act. at 75°C after 7 hr	2.3 (pNPGlc)		200
80% of max. act. at 37°C after 10 min	2.2 (pNPGlc) 26.0 [(Glc)$_2$]		31
			188

TABLE 5 (Continued)

		Activity optima	
Organism	Molecular weight (kD)	pH	T (°C)
R. flavefaciens		6.4	40
		(oNPGlc)	
Extremely thermo-phific anaerobic bacterium Wai21W2	43.0	6.2	
Agrobacterium sp. (ATCC 21400)	50.0−52.0 120.0−160.0	− 6.0−7.0 (pNPGlc)	
Cellulomonas cartalyticum (DSM 20106)			
C. uda		6.0−7.0 (pNPGlc)	
Streptomyces flavogriseus IAF-45CD (ATCC 33331)		6.5−7.5 [(Glc)$_2$]	
Streptomyces sp. strain CB-12		6.0−6.5 [(Glc)$_2$]	
Pseudomonas fluorescens var. *cellulosa*	A:	7.0 (pNPGlc)	
	B:	7.0 (pNPGlc)	

pH and temperature stability	Apparent K_m (mM) (on indicated substrate)	P_I	Ref.
	0.267 mg/ml (oNPGlc)		21
50% of max. act. at 75°C after 47 min	0.15 (pNPGlc) 0.73 [(Glc)$_2$]	4.55	180
	–		178, 228
Stable between pH 6.5 and 7.8 at 30°C for 2 hr; 10% of max. act. at 60°C after 6.5 min	0.125 (pNPGlc)		134
	45.5 [(Glc)$_2$] 2.67 (oNPGlc)		80
50% of max. act. at 50°C after 19 min	103.5 (pNPGlc)		70
10% of max. act. at 40°C after 10 min	8.05 [(Glc)$_2$]		166
70% of max. act. at 40°C after 30 min	2.65 [(Glc)$_2$]		166
50% of max. act. at 40°C after 5 min	1.9 (pNPGlc)		110
50% of max. act. at 60°C after 5 min	6.1 (pNPGlc) 5.8 [(Glc)$_2$] 0.162 [(Glc)$_3$] 0.145 [(Glc)$_4$] 0.140 [(Glc)$_5$] 0.143 [(Glc)$_6$]		

TABLE 5   (Continued)

Organism	Molecular weight (kD)	Activity optima	
		pH	T (°C)
Thermomonospora *curvata*	66.0 (intracellular)		
	66.0 (extracellular)		
Thermomonospora sp. strain YX		6.5 (pNPGlc)	55
Thermus strain Z-1		4.5–6.5	85

about 3% of that against pNPGlc and pNP(Glc)$_2$.   Glucose is the product released from all susceptible substrates.

Pettipher and Latham (21,22) found an aryl-β-glucosidase in *R. flavefaciens* which for the most part is located intracellularly in the soluble fraction of the cell lysate.   The few available data of the properties of this enzyme are represented in Table 5.   However, cellobiose is more likely cleaved in vivo by a cellobiose phosphorylase, detected in cell-free extracts of *R. flavefaciens* by Ayers (32).   This enzyme catalyzes the reversible phosphorolysis of cellobiose to α-D-glucose-1-phosphate and glucose.

D. Cellulases from *Clostridium*

Clostridia are strictly anaerobic, rod-shaped, spore-forming bacteria.   Among cellulolytic clostridia the best described cellulase systems are those from the mesophilic *C. acetobutylicum* and the thermophilic *C. stercorarium* and *C. thermocellum*.

pH and temperature stability	Apparent K_m (mM) (on indicated substrate)	P_i	Ref.
20% of max. act. at 65°C after 5 min	5.6 (pNPGlc) 30.3 [(Glc)$_2$]		212
90% of max. act. at 65°C after 5 min	1.0 (pNPGlc) 0.7 [(Glc)$_2$]		
50% of max. act. at 55°C after 8 hr (cell-assoc. β-glucosidase). 50% of max. act. at 60°C after less than 1 hr (cell-free β-glu-cosidase obtained by sonication of the cells)			117
Stable between pH 4.5 and 7.0 at 70°C for 2 hr; 50% of max act. at 75°C after 5 days	0.28 (pNPGlc) 2.00 [(Glc)$_2$]		202

Lee et al. (33) report that *C. acetobutylicum* possesses mainly extracellular endoglucanase activity and a chiefly cell-bound cellobiase activity. Some data of both enzyme activities of the three strains NRRL B527, ATCC 824, and P270 are shown in Tables 1 and 5. Strain B527 exhibits some activity toward Avicel. The strains B527 and P270 are both able to degrade acid-swollen cellulose but not to grow on it (33). Cellobiase from strain B527 is inducible and, like that of strain ATCC 824, is mainly cell-bound, while cellobiase from strain P270 is extracellular and constitutive. The enzyme activities produced by the former two strains are subject to substrate inhibition (33–35). Cellobiase from strain P270 is not specific for cellobiose but also releases glucose from salicin (35).

More interesting are the two thermophilic bacteria *C. thermocellum* and *C. stercorarium*. Creuzet and Frixon (36) purified an extracellular endoglucanase from *C. stercorarium*. They determined molecular weights of 91 and 99 kD by SDS-PAGE and gel filtration, respectively. The enzyme contains 2% of carbohydrate. Inhibition of the enzyme by SH reagents indicates that thiol groups are essential for activity. Cellobiose significantly inhibits the endoglucanase activity with 1.25 mM causing an inhibition of 50%. Substrate specificity is highest for CMC followed by amorphous (Walseth) cellulose, microcrystalline cellulose (MN 300 or Avicel), and crystalline cellulose (cotton or filter paper). The end products of hydrolysis of cellulose MN 300 are glucose and cellobiose. Other properties are shown in Table 1.

The cellulase system of *C. thermocellum* has been under intense study over the past 9 years. Several strains from different sources are currently used. The *Clostridium* species strain M7, an isolate from Lee and Blackburn (37), is considered to belong to *C. thermocellum* (Table 1). Despite great efforts of various groups to purify and characterize the cellulase components of *C. thermocellum*, the purification of only two endoglucanases and one β-glucosidase has been reported.

Ng and Zeikus (38) purified an extracellular endoglucanase which accounts for over 25% of the total extracellular endoglucanase activity. The molecular weight determined by amino acid composition analysis is 88 kD. The higher apparent molecular weight of 94 kD obtained by polyacrylamide gel electrophoresis may be a result of the common lower electrophoretic mobility of glycoproteins compared with the protein standards (38), since the enzyme contains 11.2% carbohydrate. The enzyme also contains 2% methionine and lacks cysteine. The endoglucanase shows high activity toward CMC, celloheptaose, cellohexaose, and cellopentaose; low activity toward Avicel and cellotetraose; and no activity toward cellotriose and cellobiose. Some other characteristics are shown in Table 1.

Petre et al. (39) purified a second extracellular endoglucanase from *C. thermocellum*. It is most active with CMC as substrate, whereas amorphous depolymerized cellulose powder (DP 30) is hydrolyzed at a slower rate, resulting in cellotriose and cellobiose as major end products and smaller amounts of glucose and cellotetraose. Cellobiose (and, to a lesser extent, glucose) at high concentration partially inhibits the endoglucanase activity. Other properties are summarized in Table 1.

A cellulase gene (*celA*) of *C. thermocellum* was cloned and expressed in *Escherichia coli* (40). The gene product was purified and characterized as an endoglucanase, the properties of which are shown in Table 1. They resemble those of the endoglucanase purified by Petre et al. (39). The endoglucanase expressed in *E. coli* degrades, in addition to cellulosic substrates, glucans with alternating β-1,4 and β-1,3 linkages such as barley β-glucan and lichenan (40). Béguin, Cornet, and coworkers (41,42) cloned two structural cellulase genes, *celA* and *celB*, coding for endoglucanases A and B, respectively, of *C. thermocellum* into *E. coli*. The expressed

products are the known endoglucanase of molecular weight 56 kD
and a previously undiscovered endoglucanase of molecular weight
66 kD. Two other endoglucanases of 38- and 65-kD molecular weight
were characterized by cloning and expressing the corresponding
genes *celC* and *celD* in *E. coli* (145,146). Endoglucanase D has been
crystallized and studied by X-ray diffraction (146). Endoglucanase
A and D hydrolyze CMC more randomly than the endoglucanases B
and C. Cellulose MN300, Avicel, xylan, pNPGlc, cellobiose, and
cellotriose are not hydrolyzed by endoglucanase D (146). Comparison of the amino acid sequences deduced from the nucleotide sequences of *celA*, *celB*, and *celD* revealed the presence of N-terminal
signal sequences required for protein transport across the plasma
membrane (125,147,148). Furthermore, the C-terminal regions of
the three endoglucanases share a highly conserved reiterated domain
(146,148).

A cell-bound β-glucosidase was purified from *C. thermocellum*
by Aït et al. (43,44). It appears to be located in the periplasmic
space and formed constitutively. Some of its characteristics are
shown in Table 5. The enzyme is active on pNPGlc and on cellobiose. These two activities appear to be properties of the same enzyme, but the affinity of the β-glucosidase for cellobiose is very
much lower than for pNPGlc. The enzyme is inhibited competitively
by glucono-δ-lactone, laminaribiose, and sophorose. The effect of
thiol reagents suggests the existence of essential thiol groups in the
protein. The β-glucosidase is specific for substrates with β configuration, particularly β-1,3 and β-1,2 linkages such as laminaribiose,
laminaritriose, and sophorose. It does not hydrolyze CMC or cellulose, but cleaves cellooligosaccharides. Lactose and p-nitrophenyl-
β-D-galactoside carrying a C-4 hydroxyl group epimerized relative
to glucosides are also hydrolyzed, thus indicating that the stereospecificity for this C-4 hydroxyl group is not strict. Based on these
results and on the finding that *C. thermocellum* contains both a cellobiose phosphorylase and a cellodextrin phosphorylase, it is unlikely
that the β-glucosidase functions in vivo as a cellobiase. Rather it
may aid in hydrolyzing saccharides having β-1,3 and β-1,2 linkages
(29). Cellobiose and cellodextrins may instead enter the cell, where
they are cleaved phosphorolytically to yield glucose-1-phosphate and
glucose (45−48).

Some of the properties of cellobiose phosphorylase of *C. thermocellum* are shown in Table 4. In addition to cellobiose, the enzyme
catalyzes the phosphorolysis of 4-O-β-D glucopyranosyl-D-altrose.
It also cleaves cellobiose arsenolytically. Furthermore, cellobiose
phosphorylase is active on the following glucosyl acceptors: D-glucose, 2-deoxyglucose, 6-deoxyglucose, D-glucosamine, D-mannose,
D-altrose, L-galactose, L-fucose, D-arabinose, and D-xylose (47).

The cellodextrin phosphorylase of *C. thermocellum* is active over a pH range of 5.5 – 9.0 with an optimum at about 7.5. It shows an absolute requirement for reducing compounds such as cysteine, dithiothreitol, 2-mercaptoethanol, reduced glutathione, or sodium sulfite (48). The phosphorolysis of cellodextrins proceeds with an inversion in the formation of α-D-glucose-1-phosphate. The enzyme cleaves phosphorolytically cellohexaose, cellopentaose, cellotetraose, and cellotriose, but not cellobiose or cellulose. Celloheptaose apparently can be phosphorolyzed since cellohexaose is active as a glucosyl acceptor. Cellobiose, cellotriose, cellotetraose, cellopentaose, cellohexaose, and some other di- and trisaccharides can serve as glucosyl acceptors. The glucosyl units added in these reactions presumably are linked to the acceptor through a 1,4-β-glucosidic bond (48). The apparent K_m values for the glucosyl acceptors and donors were reported by Alexander (48).

The formation of glucose-1-phosphate is an advantage for a microorganism, since the energy of the 1,4-β-glycosidic bond is conserved. Phosphoglucomutase converts glucose-1-phosphate to glucose-6-phosphate. The metabolic fate of the second glucose in cellobiose grown cells of *C. thermocellum* is apparently not very clear (49), since no hexokinase could be measured by conventional methods in extracts of cellobiose-grown cells. Evidence of hexokinase-type activity was provided by the detection of [^{14}C]glucose-6-phosphate from [^{14}C]cellobiose. This activity may be a key regulatory enzyme that couples the phosphorolytic cleavage of cellobiose with glucose phosphorylation in the cell membrane and is required for cellobiose phosphorylase activity because the reaction equilibrium of cellobiose phosphorylase strongly favors cellobiose formation (47,50).

Although the ability of *C. thermocellum* cellulase to successfully cleave crystalline cellulose suggests that a distinct and perhaps unique exoglucanase may be present in this enzyme complex (51), no exoglucanase has yet been purified and characterized. Wu and Demain (52) demonstrated the presence of both Avicel- and CMC-hydrolyzing activity in a partially purified complex of a molecular weight of about 6.5 MD. The Avicelase fraction was resolved into more than six proteins with molecular weights ranging from 60 to 220 kD. This indicates that the Avicelase activity resides in a rather large enzyme complex. Several fractions with CMC-hydrolyzing activity were detected on treatment of the complex with SDS (52).

Ten distinct EcoRI fragments of *C. thermocellum* DNA have been cloned in *E. coli* (53). Apart from seven cloned fragments coding for endoglucanases, three separate clones hydrolyze methylumbelliferyl-β-cellobioside but not CMC. Millet and coworkers (53) suggest that they may express three different cellobiohydrolase genes.

The crude extracellular cellulase from *C. thermocellum* has a high specific activity toward crystalline cellulose in the presence of dithiothreitol and Ca^{2+} comparable with the enzymes from *T. reesei* (54,55). Johnson and Demain (54) assume that the component of *C. thermocellum* which contains essential sulfhydryl groups is probably an exoglucanase, since endoglucanase activity is found to be unaffected by oxidation or thiol reagents. The cellulase system from *C. thermocellum* is strongly inhibited by cellobiose and, to a much lesser extent, by glucose when acting on microcrystalline cellulose. The enzyme complex is less or not inhibited at all using amorphous cellulose and trinitrophenylcarboxymethylcellulose (TNP-CMC), respectively (56). Other properties of the endoglucanase activity of the extracellular cellulase system from *C. thermocellum* are shown in Table 1.

The observation that the cellulolytic enzymes of *C. thermocellum* exist in large aggregates, which in turn are found either associated with cells or with cellulosic substrates or both, has led to the characterization of these large complexes and compounds responsible for these adherence phenomena.

The group of Bayer, Lamed et al. showed that virtually all of the major enzymes involved in the cellulose degradation by *C. thermocellum* are arranged in a cellulose-binding multicellulase complex, which they designated the cellulosome (10–12,49,57). This complex is responsible for the adherence of the cells to cellulose but can also be present extracellularly in the late stationary growth phase of *C. thermocellum* (10).

A spontaneous adherence-defective mutant has been isolated from the wild-type strain YS (10). Cellobiose-grown mutant cells lack the cellulosome on the surface and produce only minor quantities of extracellular cellulosome accompanied by other low molecular weight cellulases (58). Since both the cell-associated and extracellular forms of the cellulosome appear to be immunochemically similar (10), the cellulosome was isolated from the cell-free culture supernatant. It was found to be a multiprotein complex with a molecular weight of about 2.1 MD. Electron microscopic studies of the purified complex reveal particulate structures of relatively uniform size of about 18 nm. Urea fails to break the complex in its components, but SDS-PAGE yields 14 distinct polypeptide bands with molecular weights ranging from 48 to 210 kD. Using a gel overlay assay, most of these cellulosome-associated polypeptides are shown to exhibit activity against CMC. Subunits with molecular weights of 170 and 75 kD show the highest cellulase activity. Four subunits with molecular weights of 98, 60, 54, and 48 kD also appear to be cellulases (Table 1) and some cellulase activity may be associated with four other subunits (11).

Only one subunit with the highest molecular weight of 210 kD is found to be antigenically active toward the cellulosome-specific antibody preparation. In this subunit S1, which accounts for a quarter of the total cellulosomal protein, no cellulolytic activity could be detected. Its highly antigenic activity may be related to the rather external position on the cell surface and/or perhaps to its carbohydrate content (49).

The above mentioned dependence of crude cellulase activity on Ca^{2+} and thiols as well as its inhibition by cellobiose (55,56) is observed too for the purified cellulosome by Lamed and coworkers (12). They also suggest that Ca^{2+} and thiols act on a cellobiohydrolase component of the cellulosome rather than on an endoglucanase component. Moreover, Lamed and coworkers claim that the cellulosome essentially accounts for the total activity toward Avicel. On the contrary only about 70−80% of the endoglucanase activity appears to be associated with the cellulosome. Furthermore, the cellulose-adsorbed cellulosome can be detached at low ionic strength (12). This behavior is similar to the desorption of endoglucanase activity, bound to yellow cellulose, with distilled water (59).

The cell-associated form of the cellulosome was partly purified from cellulose-grown cells of *C. thermocellum*. The biochemical properties of both the cell-associated and the extracellular form of cellulosome are strikingly similar. Both forms react with cellulosome-specific antibodies. They have similar molecular weights. Size, shape, and ultrastructure of both complexes appear to be identical. Moreover, the polypeptide compositions of the two forms are very similar (57). It is noteworthy that the molecular weight of 54 kD of one of the subunits resembles the molecular weight of the endoglucanase purified by Petre et al. (39,57) (Table 1).

The specific function of the antigenic S1 subunit is still unknown. Bayer et al. (58) suggest that it may be responsible for the organization of the components into the complex or it may serve to anchor the complex onto the cell surface. They also considered a cellulose-binding role for the S1 subunit.

It is necessary to add that the cellulosome purified from another *C. thermocellum*, the strain NCIB 10682, essentially contained the same components as strain YS, except that the subunit S1 in strain NCIB 10682 has a molecular weight of about 250 kD. Like the S1 subunit of 210 kD, the 250-kD S1 subunit interacts with the cellulosome antibody and contains large amounts of carbohydrate (49).

When the wild-type cells of *C. thermocellum* are grown on cellobiose and labeled with an anion-specific marker, cationized ferritin, a great number of novel protuberant surface structures become visible. After additionally labeling with antibodies specific for the S1 subunit the immunochemical stain is mainly restricted to the outer surface of the protuberances, while the cationized ferritin completely penetrates these structures. This indicates that the interior of the protuberances contains anionic compounds and the cellulosomes are arranged on the surfaces (60). Lamed and Bayer (60) suggest that several hundred cellulosomes are located in one of these protuberances.

When the cells come into contact with cellulose some of the protuberances seem to protract yielding a fibrous network. A distance of up to 400−500 nm separates the cells from the cellulose surface and the fibrous network appears to connect both (60). It is assumed by Bayer and Lamed (60) that in the absence of cellulose the fibers are closely packed together in the lumen of the cell surface protuberances. However, in the presence of cellulose the material in the lumen forms the extended fibers. The resulting contact zones may serve to direct products of cellulose degradation toward the cell surface.

To sum up, Lamed and Bayer (49) suggest that the hydrolysis is mediated by cellulosome clusters at the surface of the insoluble substrate. The transfer of the products toward the cell may be a highly ordered process achieved by the fibrous components in the contact zones. Moreover, the complex may be built up in such a way that the various product intermediates are protected and their transfer to other cellulase components for further hydrolysis is facilitated. Simultaneous multiple-point cleavage of cellulose would prevent extensive restoration of newly cut glycosidic bonds and would take place within individual cellulosome, within polycellulosome clusters, and along the entire length of the bacterium.

It is very interesting that Ljungdahl and his group using a different experimental approach came to a similar concept of the cellulolytic enzyme system of *C. thermocellum*. They showed that *C. thermocellum* JW20 (ATCC 31449) produces a yellow substance that is attached to cellulose early in the fermentation. This yellow substance effectively binds endoglucanase to cellulose fibers (59). Whether this yellow substance is involved in the adhesion of the cells themselves to the cellulose fibers is as yet not known (61). Coughlan et al. (62) detected by electron microscopy three major globular complexes with diameters of 21 nm [similar perhaps to the 18 nm of the cellulosome of the YS strain (11,57)] and larger forms of 35−45 and 61 nm. Hon-nami et al. (63) showed that *C. thermocellum* JW20 produces at the early stages of growth on cellulose a cellulase complex which is largely bound to the substrate. This bound enzyme fraction, extracted with distilled water from cellulose, contains two major components, a very large complex with a molecular weight of 100 MD and a small complex with a molecular weight of 4.5 MD. As cellulose is consumed the bound enzyme is released as free enzyme, which is resolved by affinity chromatography into a complex that binds to the column and into a nonbindable mixture of proteins. All

four fractions exhibit endoglucanase activity but only the two bound
complexes and the free bindable complex hydrolyze crystalline cellulose to cellobiose as the main product. Each of these three complexes contains about 20 different proteins with molecular weights of
45–210 kD. Of these proteins 10–12 are major components while
others exist in smaller amounts. The protein composition of the
originally bound large complex is qualitatively, but not quantitatively,
almost identical to that of the originally bound small complex. While
the protein with the molecular weight of 210 kD is a major component
of the originally bound large complex, it is a rather minor component
of the originally bound small complex. At least four polypeptides of
the complexes have endoglucanase character (Table 1).

Wiegel and Dykstra (61) reported that attachment of *C. thermocellum* cells to cellulose occurred via a fibrous ruthenium red staining material. Ruthenium red staining suggests that the material is
composed of acid mucopolysaccharides. This tight attachment allows
C. thermocellum to sporulate on the surface of the cellulose fibers.

Hon-nami et al. (63) developed the following concept of the cellulolytic enzyme system of *C. thermocellum*: the originally bound
large complex is initially bound to the cell and the cellulose. As
the fermentation proceeds and the bacterium sporulates and the
spores leave the cellulose surface, the originally bound large complex disaggregates. It loses some proteins, such as the above mentioned polypeptide with the molecular weight of 210 kD, and the
originally bound small complex is formed. This complex may desorb
from cellulose to become the free bindable complex, which may read-
sorb to cellulose and again participate in cellulose degradation. Alternatively, it may further disaggregate into its components of which
some maintain endoglucanase activity as found in the culture supernatant.

After detailed transmission electron microscopic examination of
the cellulase complexes of *C. thermocellum* JW20 and YM4, Mayer
et al. (149) expanded the above described concept. In the early stages
of growth of strain JW20 clusters of tightly packed cellulosomes
(polycellulosomes) are located on the cell surface and are bound to
cellulose. The polycellulosomes have a particle mass of 5×10^4 to
8×10^4 kD. The cellulosome of strain JW20 was found to have a
molecular weight of 2×10^3 to 2.5×10^3 kD and to contain about
35 polypeptides, ranging from 20 to 200 kD. The cellulosome of
strain YM4 has a particle mass of 3.5×10^3 kD with 45–50 polypeptides of 20- to 200-kD molecular weight. In the early stages of
cultivation, the cellulosomes from both strains exist as tightly packed
complexes (tight cellulosomes). They subsequently decompose to
loosely packed complexes (loose cellulosomes) and ultimately to free
polypeptides. The loose cellulosomal particles contain rows of equi-
distantly spaced polypeptide subunits with apparently identical

orientation, arranged parallel to the major axis of the loose cellulosome. Mayer et al. (149) postulated that a simultaneous multicutting
event takes place along a cellulose fiber aligned beside such a row
of subunits. This leads to the release of cellooligosaccharides of
four cellobiose units in length. Cellotetraose and cellobiose might
then be formed from cellooctaose or from cellulose, either by a similar multicutting event, mediated by chains of smaller subunits with
smaller center-to-center distances, or by single cuts.

III. PROPERTIES OF CELLULASES FROM AEROBIC AND FACULTATIVELY ANAEROBIC BACTERIA

A. Cellulases from *Bacillus*

The genus *Bacillus* contains spore-forming aerobic or facultatively
anaerobic bacteria. Cellulases from *Bacillus* species have not been
characterized in detail. Properties of some endoglucanases from
B. subtilis and some alkalophilic *Bacillus* strains are summarized in
Table 2.

The *Bacillus subtilis* DLG isolated by Robson and Chambliss (64)
resembles other cellulolytic *Bacillus* species in that it apparently
lacks a complete cellulase system. No Avicelase activity and no cellobiase activity could be detected. It is proposed that cellobiose is
hydrolyzed by a cellobiose phosphorylase. Another sharp contrast
to many other cellulase systems is that the endoglucanase activity
is not inhibited and its formation is not catabolite-repressed by glucose and cellobiose. The endoglucanase activity does not appear in
the culture supernatant until the cells have reached the stationary
phase of growth. Kinetic experiments indicate that more than one
endoglucanase is present in the culture supernatant (64). From the
culture supernatant of *B. subtilis* DLG an endoglucanase of 32.5-kD
molecular weight was purified (150). Au and Chan (151) purified
another extracellular endoglucanase of 23 kD from *B. subtilis* AU-1.
Thiol groups are involved in its active site.

An endoglucanase gene from *B. subtilis* was cloned and expressed
in *E. coli* (152). The nucleotide sequence of an endoglucanase gene
from *B. subtilis* was determined and compared with the amino acid
sequence of the purified enzyme. It was shown that the mature
protein is extended by a signal sequence of 36 amino acids (153).

A cellulolytic alkalophilic *Bacillus* sp. N-4 was isolated from soil
and found to be similar to *B. pasteurii*, except for its ability to
grow at high pH values. Two extracellular endoglucanases of molecular weight 50 kD and having an optimal pH of 10.0 were partially
purified. Both enzymes are associated with a multienzyme system
and hydrolyze cellotetraose to cellobiose (95%) and glucose and cellotriose (5%) (65). The cellulase genes of the alkalophilic *Bacillus*

sp. strain N-4 were cloned in *E. coli* with the plasmic vector
pBR322 and two endoglucanases with molecular weights of 50 and
58 kD were expressed (66). Furthermore, an endoglucanase with a
molecular weight of 92 kD was purified from the culture supernatant
of an alkalophilic *Bacillus* sp. strain 1139. Its formation is repressed
by glucose. Transglycosylation is observed during hydrolysis of
CMC and cellooligosaccharides by the endoglucanase. A high aspartic acid content was found by amino acid composition analysis
(67).

A cellulase gene from *Bacillus* sp. strain 1139 was cloned in
E. coli. The putative signal peptide consists of 29–30 amino acid
residues (154).

B. Cellulases from *Cellulomonas* Species

Much less is known about the location, organization, and multiplicity of *Cellulomonas* cellulases than about these properties of the cellulase systems of *Clostridium* species.

Béguin et al. (68) found that about 80% of the endoglucanase
activity produced during the lag and exponential growth phase of
Cellulomonas flavigena are tightly associated with cellulose. Cellulose-bound activity reaches a maximum at the beginning of the stationary growth phase and then declines, whereas the free endoglucanase activity increases and becomes highest in the late stationary
phase of growth. The affinity of the cellulose-bound enzymes for
their substrates is high, since the activity remains bound even after
washing with salt solutions of concentrations up to 3.6 M NaCl at pH
values between 3.5 and 11.5 (69).

Although endoglucanases from *C. uda* are found to be extracellular, a very small amount of cell-bound endoglucanase activity is
always measured throughout the cultivation with cellulose or mono-
and disaccharides as carbon source (70). At the beginning of cellulose degradation 60–70% of the total endoglucanase activity is associated with the cellulosic substrate (71). This portion of cellulose-
adsorbed endoglucanase activity declines simultaneously with cellulose
degradation reaching a value of 25% at the beginning of the stationary growth phase. It was also observed that endoglucanases once
desorbed from the residual cellulose could not be readsorbed on
newly added cellulosic substrate (71). Antheunisse (72) made the
opposite observation in that free 1,4-β-glucanase activity found in
the supernatant of *C. flavigena* cultures was readsorbed on filter
paper.

As shown in Table 2 Béguin and Eisen (69) purified three extracellular endoglucanases from *Cellulomonas* sp. strain IIbc. One
was found to be free in the culture supernatant and two others were
found to be bound to residual cellulose. The latter two are glycosylated.

A cellulose-binding endoglucanase of *Cellulomonas* sp. ATCC 21399 was purified to immunological homogeneity by affinity chromatography on phosphoric acid-swollen cellulose (155). This enzyme is similar to the two endoglucanases from *Cellulomonas* sp. strain IIbc of 49−52 kD and 53.5 kD, respectively (69). It shows low specific activity against CMC, an ability to hydrolyze Avicel when acting alone, affinity to Con A, and a strong binding to microcrystalline cellulose (155).

Bagnara et al. (156) purified two extracellular endoglucanases from *C. fermentans*. These enzymes rapidly hydrolyze CMC and amorphous cellulose, but show very low activity toward Avicel.

Despite the detection of up to 10 components with endoglucanase activity in supernatants of *C. fimi* cultures grown on Avicel, Langsford et al. (73) assumed that the *C. fimi* cellulase system contains no more than three or four components, at least two of which are glycosylated. They found their results consistent with the proposal and suggest further that these enzymes adsorb on Avicel and are stabilized by this substrate. Once the concentration of the residual cellulose falls below a certain level or the substrate has been sufficiently altered by enzymatic attack, newly synthesized enzymes do not bind to the residual substrate and some bound enzyme is released by degradation of the substrate or by proteolysis. The free enzyme in the supernatant can be altered by proteolysis or by deglycosylation. This results in a diversity of products, some of which are still enzymically active but reduced in their substrate affinity.

Two *C. fimi* genes were cloned in *E. coli*, the *cenA* gene encoding an endoglucanase (157−159) and the *cex* gene encoding an exoglucanase (158,160). The mature exoglucanase of *C. fimi* is a glycoprotein with a molecular weight of 56 kD and the glycosylated endoglucanase purified from *C. fimi* cultures has a molecular weight of 58 kD (158). From the sequence of the *cex* gene a molecular weight of 49 kD for the mature and 53 kD for the precursor form was calculated (160). A 51.8-kD protein was predicted from the sequence of the *cenA* gene (159). The discrepancy between the molecular weights of the native enzymes and those derived from the nucleotide sequences may be due to glycosylation of the native enzymes. A leader sequence of 31 and 41 amino acids has been proposed for the endo- and exoglucanase, respectively (159,160). Both the endo and the exoglucanases contain three distinct regions: a short sequence of about 20 amino acids containing only proline and threonine (the Pro-Thr box); an irregular region, rich in hydroxyl-amino acids, of low charge density, which is predicted to have little secondary structure; and, finally, an ordered region of higher charge density, which contains a potential active site and which is predicted to have secondary structure (161). Furthermore, it was proposed that both enzymes bind to cellulose through discrete cellulose-binding domains (162). Glycosylated cellulases from *C. fimi* were compared with their nonglycosylated counterparts synthesized in *E. coli* and found that the glycosylated enzymes are protected from attack by a *C. fimi* protease when bound to cellulose (162).

An endoglucanase gene from *C. uda* CB4 was cloned and expressed in *E. coli*. The signal sequence of the endoglucanase consists of 23 amino acids (126).

Vladut-Talor et al. (74) have shown by electron microscopy that a *Cellulomonas* sp. grown in the presence of cellulose adheres to the cellulose fibers. These cellulose-grown cells develop a thicker outer layer of material that stains with ruthenium red and may help in the adhesion to the substrate.

Properties of endoglucanases from *Cellulomonas* sp. strain IIbc, *C. fermentans* (156), *C. flavigena*, *Cellulomonas* sp. UQM 2903, *Cellulomonas* sp. ATCC 21399 (155), *C. uda*, and *C. uda* CB4 are summarized in Table 2. Common to almost all these species is the optimum pH value of about 7.0 for endoglucanase activity. Furthermore, endoglucanase activities from *Cellulomonas* sp. (75), *C. uda* (70), and *C. flavigena* (76) are found to be inhibited by cellobiose, the latter competitively. Endoglucanase formation in *C. uda* (70), *Cellulomonas* sp. strain IIbc (68), and *Cellulomonas* sp. (75) is found to be induced by cellobiose and to be catabolite-repressed.

An extracellular 1,4-β-glucan cellobiohydrolase was isolated from Avicel-grown cultures of *C. uda* CB4 and purified to apparent homogeneity (77). The enzyme rapidly hydrolyzes microcrystalline cellulose with cellobiose as end product. It exhibits little activity toward CMC and filter paper. The enzyme hydrolyzes cellotetraose to cellobiose but does not cleave cellotriose. Other properties of the enzyme are shown in Table 3.

Within the genus *Cellulomonas* two mechanisms of cellobiose cleavage are known: hydrolytic cleavage of cellobiose yields glucose and phosphorolytic cleavage of cellobiose results in α-glucose-1-phosphate and glucose. Both types of enzymes are present in *C. fimi*. Sato et al. (78) identified cellobiose phosphorylase activity in crude extracts and Wakarchuk et al. (79) found two intracellular β-glucosidases. One of these glucosidases is constitutive and hydrolyzes only pNPGlc but not cellobiose, while the other is induced by cellobiose and hydrolyzes both substrates. An intracellular β-glucosidase was also found in *C. cartalyticum* (80). Although it hydrolyzes both cellobiose and oNPGlc it has a greater affinity for the latter. The enzyme activity is inhibited competitively by N-methyl-1-deoxynojirimycin with a K_i value of 5.6 and 7.5 mM when cellobiose and oNPGlc, respectively, are used as substrates. Both β-glucosidase and cellobiose phosphorylase are found in *C. uda*. Stoppok et al. (70) characterized a constitutively formed, cell-bound β-glucosidase activity, the properties of which are shown in Table 5. The β-glucosidase activity, when pNPGlc is

used as substrate, is inhibited noncompetitively by glucose with a
K_i value of 0.667 mM. Stoppok et al. (70) also suggested the existence of a cellobiose phosphorylase. This enzyme activity was clearly
demonstrated in *C. uda* and *C. flavigena* by Schimz et al. (81).
Moreover, Schimz and Decker (82) found that cellobiose phosphorylase is located in the soluble fraction of *C. uda*.

C. Cellulases from *Cellvibrio* species

The classification of *Cellvibrio* species seems to be not clear (83;
see also Chap. 00 of this volume). Therefore Blackall et al. (84)
propose a redescription of the genus *Cellvibrio*. The salient properties of *Cellvibrio* strains are gram-negative, aerobic, slightly
curved rods which have an oxidative metabolism of glucose and the
ability to degrade cellulose.

Four distinct cellulase components have been isolated from culture filtrates of *C. gilvus*. There appears to be no major difference
in the functions of the four components. The constitutive cellulase
system degrades cellulose to chiefly cellobiose and to some cellotriose
(85-87). The mode of action of one of the four electrophoretically
distinguishable exo-1,4-β-glucanases was studied in greater detail
by Storvick et al. (88). They clearly demonstrated that cellooligosaccharides are preferentially attacked at the second and third glucosyl bond from the nonreducing end of the oligosaccharide chain.
Both the release of α-cellobiose and the absence of transferase activity suggest that the enzyme acts by single displacement mechanism. Hulcher and King (85,89) demonstrated the presence of a
cellobiose phosphorylase in *C. gilvus*. Sasaki et al. (90) purified
and characterized this enzyme. It has a specificity for cellobiose
but not for cellodextrins and requires P_i and Mg^{2+} for phosphorylation. Its activity is competitively inhibited by nojirimycin with a
K_i value of 45 μM when cellobiose is used as substrate. Furthermore, it is inhibited by thiol reagents. Other properties are shown
in Table 4. Schafer and King (91) observed that cellooligosaccharides as large as cellohexaose are taken up and utilized by *C. gilvus*
prior to hydrolysis. They found that the efficiency of the cells
metabolizing cellodextrins increases from cellobiose to cellohexaose,
thus indicating that a phosphorolytic cleavage of these bonds is involved. Since only a cellobiose phosphorylase seems to be present
in *C. gilvus*, King and Vessal (87) suggested that cellotriose to cellohexaose are metabolized by transglycosylation to yield cellobiose
which in turn is cleaved phosphorolytically.

The cellulase system of *C. fulvus* seems to consist of several
components with different substrate specificities (92). The location
of cellulase in this bacterium depends on the carbon source for
growth and the age of the culture. When the cells are grown on

glucose or cellobiose, all CMC-hydrolyzing activity is cell-bound,
but only part of the activity is bound to the cell surface. Cellulase
is also found to be located in the periplasmic space and bound to a
membrane fraction (93). Growth on cellulose results in cell-free
CMC-hydrolyzing activity. The formation of this enzyme activity is
repressed by glucose but remains unaffected by cellobiose (93).

Whereas Berg (93) suggested that the phenomenon of extracellular cellulase is largely accounted for by cell lysis during the long
cultivation, Oberkotter and Rosenberg (94) found an actively secreted, true extracellular endoglucanase from *C. vulgaris.*

D. Cellulases from the *Cytophaga* group

The genera *Cytophaga* and *Sporocytophaga* belong to the family *Cy-
tophagaceae* (95). They are gram-negative, motile by gliding, nonfruiting, and known to degrade agar and cellulose.

From a *Cytophaga* species WTHC 2421 (ATCC 29474) two cellulases were isolated and purified with remarkably low molecular
weights of 8.65 and 6.25 kD (96). The latter appears to be an
exoglucanase located in the periplasmic space and having an optimum
pH value for activity of 7.0. The enzyme is active against microcrystalline cellulose and produces glucose from CMC or dewaxed cotton. This probable 1,4-β-glucan glucohydrolase is slightly inhibited
by glucose, lactose, and cellobiose. The second 8.65-kD cellulase
appears to be an endoglucanase with an optimum pH value for activity of 8.0. A β-glucanase with a molecular weight of 18 kD was isolated and partially purified from another *Cytophaga* species (NCIB
9497). It hydrolyzes CMC and cellodextrins to oligosaccharides and
lichenan to tri- and tetrasaccharides as well as higher oligosaccharides (97).

Electron microscopic studies revealed that *Sporocytophaga myxococcoides* grows in close contact with cellulose creeping both on
and within the fibers. During growth on cellulose large amounts of
a slimy polysaccharide are produced (98).

Charpentier and Robic (99) extracted from *S. myxococcoides*
with Triton X-100 an exoglucanase originally located in the periplasmic space. They clearly demonstrated the release of α-glucose
from soluble and insoluble cellodextrins and cellobiose. This 1,4-
β-glucan glucohydrolase has a molecular weight of 67 kD. Two extracellular endoglucanases have been isolated from culture supernatant of *S. myxococcoides* Q.M.B. 482 (100). Some properties of
the enzymes are shown in Table 2. They have rather similar amino
acid compositions. Furthermore, a cell-associated cellulase was
partly purified with similar properties as the extracellular endoglucanase with a molecular weight of 52 kD. Osmundsvåg and Goksøyr
(100) found that 10−20% of the total cellulase activity is loosely

bound to the cells. Furthermore, they suggest that an additional
10% of the activity is rather firmly bound to the cell wall or to the
cytoplasmic membrane.

E. Cellulases from *Erwinia*

Erwinia chrysanthemi, a gram-negative, facultatively anaerobic rod
is a soft-rot pathogen which has a marked capability to macerate
plant tissues. Cellulases, pectinases, and proteases are produced
by this bacterium (101,102).

An extracellular endoglucanase designated as endoglucanase Z
was isolated from *E. chrysanthemi*, the properties of which are
shown in Table 2. Its formation appears to be constitutive and
under catabolite repression. The enzyme hydrolyzes CMC and phosphoric acid-swollen cellulose. The latter is degraded to mainly cellobiose and cellotriose. The endoglucanase Z exhibits only a very
low activity toward microcrystalline cellulose (102,103). A *celY*
gene encoding the endoglucanase Y of *E. chrysanthemi* strain 3665
was cloned and expressed in *E. coli*. The purified protein of 35-
kD molecular weight is significantly different from the major endoglucanase Z secreted by *E. chrysanthemi* strain 3665 (163).

F. Cellulases from *Pseudomonas* species

The genus *Pseudomonas* is the catabolically most complex group of
gram-negative bacteria. It consists of straight or curved rods,
motile by polar flagella, non-spore-forming, and strictly aerobic.
Pseudomonas fluorescens var. *cellulosa* strain 107 was isolated by
Ueda et al. (104) from soil, but the relationship of this strain to
the genus *Pseudomonas* is questionable (28).

The development of the present picture of the cellulase system
of *P. fluorescens* var. *cellulosa* is somewhat puzzling. Yamane et al.
(105) showed that this bacterium produces at least three cellulase
components (Table 2) designated A, B, and C. They appear to be
constitutive enzymes whose formation is regulated by catabolite repression (106). A and B are extracellular and C is a periplasmic
enzyme (107). All three components contain a considerable amount
of carbohydrate. They hydrolyze CMC, cellooligosaccharides, and
amorphous and microcrystalline cellulose. For the action of C three
consecutive glucosyl units and for the action of A and B four consecutive glucosyl residues are necessary (105). The two extracellular components differ electrophoretically, with A moving faster
toward the cathode than B (108). But Yoshikawa et al. (108,109)
found that the two cellulase components A and B are not homogeneous. They separated five extracellular cellulases according to their
molecular weight (Table 2). A cellulase component of a molecular

weight higher than 200 kD [peak I(1)] was isolated from the supernatant of an Avicel-grown culture. It was shown to be a single protein by gel filtration and electrophoresis and corresponds with B. Enzymes of peak II-IV are almost homogeneous with regard to molecular size. However, they are still composed of A and B in a characteristic manner. On the contrary, cellulase of Peak V consists only of A. The specific endoglucanase activity of these cellulases increases in reverse proportion to molecular weight. Only the fractions of peak II show (besides endoglucanase activity) a very high activity toward Avicel. Molecular weights and some other properties of endoglucanases from *P. fluorescens* var. *cellulosa* are summarized in Table 2. Yoshikawa et al. (109) showed that A and B differ not only in electrophoretic mobility but also in carbohydrate content. Component B in the cellulase fractions is converted to A by a fungal β-glucosidase without remarkable change of molecular size. Moreover, treating the fractions with pronase results in conversion of larger into smaller components without noticeable changes in electrophoretic mobility. During cultivation of *P. fluorescens* var. *cellulosa*, the ratio of A to B increases in the extracellular cellulases but the ratio of low to high molecular weight components increases as well (108). Yoshikawa et al. (109) conclude from these results that the occurrence of multiple cellulases during cultivation of *P. fluorescens* var. *cellulosa* may be due to the action of proteases and glycosidases (particularly the former) during bacterial growth. This conclusion is supported by the finding that only high molecular weight cellulase corresponding to that of cellulase in peak I is found in the intrawall fraction during early stages of cultivation, when cellulases are actively secreted (108). The isolation of a single endoglucanase gene from *P. fluorescens* var. *cellulosa* by Wolff et al. (164) supports the suggestion of Yoshikawa et al. (109) that the multiplicity is a result of proteolysis and glycosylation. On the other hand, Gilbert et al. (165) described the isolation of four distinct endoglucanase genes, which show no homology with each other. Two of the genes are closely linked on the chromosome and could be controlled by the same operon.

In addition to the above mentioned 1,4-β-glucanases, two forms of β-glucosidase occur in *P. fluorescens* var. *cellulosa*. One is membrane-bound while the other is cytoplasmic. Some properties of the two enzymes are shown in Table 5. The cytoplasmic β-glucosidase B was partially purified. The enzyme not only attacks pNPGlc and pNP(Glc)$_2$ but also splits the terminal β-glucosyl units from cello-oligosaccharides (DP 2−6), CMC, sophorose, and laminaribiose. These results indicate that the enzyme may be an atypical β-glucosidase which possesses certain exoglucanase character (110).

A *Pseudomonas* sp. isolated from activated sludge produces two extracellular endoglucanases during growth on cellulose (111).

Moreover, during growth on cellulose as well as on cellobiose a periplasmic and a cytoplasmic endoglucanase are present. Three β-glucosidases are also formed during cultivation on both substrates. Two of them are located in the periplasmic space and one in the cytoplasm. All three β-glucosidases but neither of the endoglucanases are able to hydrolyze cellobiose. Ramasamy and Verachtert (111) suggest the following scheme for cellulose degradation of *Pseudomonas* sp.: cellulose is degraded to long cellooligosaccharides by the extracellular endoglucanases. These in turn are hydrolyzed by the periplasmic endoglucanase to short cellooligosaccharides and by the first periplasmic β-glucosidase to very short oligosaccharides, mainly cellobiose. The latter compounds are finally hydrolyzed to glucose by the second periplasmic β-glucosidase.

From another *Pseudomonas* strain isolated from soils of sugar cane fields an extracellular cellulase was partially purified (112). Some of its properties are shown in Table 2.

G. Cellulases from the Actinomycete Group

Among mycelium- and spore-forming bacteria which are able to degrade cellulose are the genera *Streptomyces* and *Thermomonospora*.

Enger and Sleeper (113) isolated from crude extracellular enzyme preparations of *S. antibioticus* five electrophoretically distinct endoglucanases. Endoglucanases I–III are immunologically identical whereas endoglucanases IV and V have an antigenic moiety not common to each other or to endoglucanases I–III. *Streptomyces flavogriseus* strain IAF 45-CD, isolated by Ishaque and Kluepfel (114), produces during growth on Avicel considerable amounts of extracellular endoglucanases. Aryl-β-glucosidase activity and a very small cellobiase activity are found in the mycelial fraction. In Table 5 some properties of the β-glucosidase of *S. flavogriseus* strain IAF 45-CD are compared with those of the β-glucosidase of *Streptomyces* strain CB-12 (166). From *S. flavogriseus* strain IAF 45-CD an extracellular exoglucanase of 45-kD molecular weight was purified (167). The enzyme extensively hydrolyzes acid-swollen cellulose but shows insignificant activity toward cotton fibers and filter paper. Cellobiose and cellotriose in a ratio of about 6:1 are the only products of acid-swollen cellulose, indicating the cellobiohydrolase character of the enzyme. Three endoglucanases were purified from the alkalophilic *Streptomyces* strain KSM-9 (168). Some of their properties are shown in Table 2.

Hägerdal et al. (115–117) reported that a thermophilic, filamentous bacterium of the genus *Thermomonospora* (originally classified in the genus *Thermoactinomyces*) produces extracellular CM-cellulase and Avicelase activity as well as cell-bound β-glucosidase activity. During exponential growth phase up to 50% of both extracellular

1,4-β-glucanase activities is adsorbed to residual cellulose and released during stationary growth phase when the major part of the
cellulosic substrate is degraded. The two 1,4-β-glucanase activities
adsorbed to cellulose could be desorbed with water. Optima and stabilities of the three enzyme activities with respect to pH value and
temperature are shown in Tables 2, 3, and 5. It should be added
that the highest pH value studied (7.3) is the most destabilizing for
all three enzyme activities. The cell-bound β-glucosidase activity is
most probably intracellular during the entire growth of the bacterium.
The same β-glucosidase probably accounts for the hydrolysis of both
pNPGlc and cellobiose.

   Thermomonospora curvata produces cellulases active against both
cotton fibers and CMC (118). Cotton-hydrolyzing activity was separated from CMC-hydrolyzing activity by partial purification of the
cellulase complex of *T. curvata*. Cellobiose appears to be the only
product of ground cotton fibers degraded by the purified cotton-
hydrolyzing cellulase fractions, indicating that the enzyme preparation exhibits exoglucanase activity. Other properties of endo and
exoglucanase activities of *T. curvata* are shown in Tables 2 and 3.
Bernier and Stutzenberger (169) showed that *T. curvata* utilizes
cellobiose in preference to glucose, but no evidence of a cellobiose
phosphorylase was observed. The cellulase formation is under catabolite repression (119). However, Fennington et al. (119) obtained
a mutant which is resistant to catabolite repression.

   Two endoglucanases, E_1 and E_2, were isolated from the culture
supernatant of *T. fusca* YX (170). E_1 contains less than 1% carbohydrate and E_2 25%. The products of hydrolysis of CMC and phosphoric acid-swollen cellulose catalyzed by E_1 are mainly cellobiose
and small amounts of glucose. E_2 produces glucose, cellobiose, cellotriose, and larger oligosaccharides from the same substrates. Neither enzyme is inhibited appreciably by glucose but both by cellobiose. Other properties are shown in Table 2. The endoglucanase
synthesis in *T. fusca* YX is induced by cellobiose and catabolite repressed (171).

IV. REGULATION OF CELLULASE SYNTHESIS

The regulation of protein synthesis in bacteria takes place on the
level of transcriptional and translational control. Furthermore, activities of cell-associated or extracellular cellulolytic enzymes are
regulated on the secretory and catalytic level. Induction, repression,
and catabolite repression are mechanisms controlling transcription
rather than translation (120). Although there have been some reports on the constitutive formation of cellulases (29,102,106), most
bacterial 1,4-β-glucanases are found to be inducible. Since cellulose

cannot enter the cells and not act itself as inducer, a low molecular weight product of cellulose degradation or a derivative thereof has to take over this function. In order to generate this inducing compound, operons for cellulose degradation must be expressed constitutively at a low basal level. As shown in the preceding sections (II and III) the formation of many endoglucanases is found to be induced by cellobiose and repressed by glucose (7,67,68,70,75,112).

The intracellular concentration of cyclic nucleotides such as cAMP has been shown to play a major role in mediating catabolite repression of enzyme biosynthesis (121). The currently accepted model for the mechanism of action of cAMP in bacteria is described by Botsford (122). Fennington et al. (119,121), in an attempt to apply this concept of catabolite repression to cellulase synthesis by *Thermomonospora curvata*, found that there is at least a temporal relationship between cAMP levels and the rates of cellulase synthesis. Furthermore the repression of endoglucanase synthesis in *T. fusca* is partially relieved when cAMP is added with the repressing carbon source (171).

In a recent review (123), Stutzenberger developed a model for the regulation of cellulase synthesis. It has many features in common with the system of activator-controlled induction of the arabinose catabolism in *E. coli*. In the *ara* operon, the regulatory protein acts as activator in the presence of the inducer and as repressor in its absence. In the model proposed by Stutzenberger the repressing function is omitted, since in the cellulase operon a relatively high constitutive level of enzyme formation is required.

Intracellularly formed cellulases have to be translocated through the cytoplasmic membrane to a place where they can bind to their polymeric substrate. The translocation of proteins was recently treated in an excellent review (124). The precursors of the protein contain a signal sequence, a peptide extension at the NH_2 terminus. This is removed during the translational export process by the signal peptidase, which is bound to the outside of the cytoplasmic membrane. The amino acid sequence of the signal peptides of the pre-endoglucanases from *C. thermocellum* (125) and *C. uda* CB4 (126) have been determined. The latter was deduced from the DNA sequence. It consists of 23 amino acids, of which one is Arg in the region near the NH_2 terminus and 14 are hydrophobic amino acids (126).

In contrast to the synergistic action of fungal hydrolytic enzymes on crystalline cellulose (127–129), synergism among bacterial cellulases has not been well studied. Hydrolysis of crystalline cellulose by fungal cellulases is suggested to occur in a sequential manner (1,28,127). Synergism having become synonymous with cooperative cellulase action on insoluble cellulosic substrates (130) is extended by Fujii and Shimizu (131) on soluble cellulosic substrates.

The organization of cellulases into large complexes, which has been shown for *Clostridium* (49) and *Ruminococcus* (3) and has been suggested for a variety of gram-negative and gram-positive bacteria (172), appears to be a prerequisite for synergistic action of bacterial cellulase components.

V. PURIFICATION OF CELLULASES

Many cellulase components hydrolyze cellulose by acting in a complementary fashion toward each other. Therefore, even a trace of another cellulolytic enzyme activity could influence significantly the specific action of the supposedly purified enzyme.

The purification techniques reported in the literature (see references in Tables 1–5) include the use of ultrafiltration, fractional precipitation, gel filtration, ion exchange chromatography, gel electrophoresis, and isoelectric focusing (chromatofocusing). Surprisingly, affinity chromatography has rarely been used in the purification of bacterial cellulases. The same is true for the use of immunosorbent chromatography using monoclonal antibodies. All these purification methods are rather involved and only small amounts of enzymes can be employed at one time (132). A procedure employing three-phase partitioning with *t*-butanol as organic cosolvent, DEAE-Sephadex chromatography, and hydrophobic chromatography using salicin and spacer arm – Sepharose enabled rapid processing of several grams of an extracellular crude enzyme mixture from *T. reesei* (133). A β-glucosidase, a cellobiohydrolase, and two endoglucanases were separated and isolated to 96–98% purity in 50–60% overall yield. Methods employing the partition of enzymes between two or three liquid phases can possibly also apply to the purification of bacterial cellulases.

In the following some examples of purification of exo- and endoglucanases, β-glucosidases, and cellobiose phosphorylases to apparent homogeneity are cited. Endoglucanases were purified from *Bacillus* sp. strain 1139 (67), *B. subtilis* DLG (150), *B. subtilis* AU-1 (151), *Fibrobacter succinogenes* S85 (173), *Cellulomonas* sp. (69), *C. fermentans* (156), *C. fimi* (174), *C. uda* CB4 (126), *Clostridium josui* (175), *C. stercorarium* (36), *C. thermocellum* (38,39), *Cytophaga* sp. WTHC 2421 (97), *Erwinia chrysanthemi* strain 3665 (103), *Ruminococcus albus* (142), *Streptomyces* strain KSM-9 (168), and *Thermomonospora fusca* YX (170). Apparently pure exoglucanases were isolated from *F. succinogenes* S85 (176,177), *C. uda* CB4 (77), *R. albus* (30), *R. flavefaciens* FD-1 (144), and *S. flavogriseus* (167). Purified β-glucosidases were obtained from *Agrobacterium* sp. (134,178), *C. stercorarium* (179), *C. thermocellum* (43,44), *R. albus* (31), and the extremely thermophilic anaerobic bacterium

Wai 21W2 (180). From *Cellvibrio gilvus* a cellobiose phosphorylase
was isolated and purified to apparent homogeneity (90). The isolation of cell-free and cell-associated forms of the cellulosome from
C. thermocellum (49,181,182) is an example of the isolation of a discrete multienzyme cellulase complex.

VI. ADDENDUM

This additional material is intended to bring to the reader's attention the major developments from the beginning of 1988. Since these
developments have to be treated very concisely, the reader will be
referred more frequently than in the previous sections to Tables
1-5. For easy identification, the following comments are marked
by the same section designations to which they pertain in the text
of the chapter.

(II.B) Two endoglucanases of 65- and 118 kD molecular weight
and different cellulose-binding properties were purified from supernatants of *Fibrobacter succinogenes* S85 cultures (173) (Table 1).
From the same organism an extracellular cellobiosidase of 75-kD molecular weight was isolated and characterized (176) (Table 3). It is
activated by chloride and other anions and releases cellobiose from
cellotriose, and cellobiose and cellotriose from longer chain cellooligosaccharides and acid-swollen cellulose. A 50-kD cellodextrinase
was purified from both the culture supernatant and the periplasmic
space of *F. succinogenes* S85 (177,183) (Table 3). The latter was
cloned and expressed in *E. coli* (184).

A β-glucosidase gene from *Caldocellum saccharolyticum*, a thermophilic, anaerobic bacterium, was cloned and expressed in *E. coli*
(185,186) (Table 5).

(II.C) Cloned and expressed in *E. coli* was also the gene for
the *Ruminococcus albus* β-glucanase (187) and β-glucosidase (188)
(Table 5).

(II.D) A gene coding for an endoglucanase from *Clostridium
cellulolyticum* was cloned in *E. coli*. The synthesized enzyme has
a molecular weight of 40 kD (189).

An endoglucanase of 45-kD molecular weight was purified from
the culture supernatant of *C. josui*. The enzyme hydrolyzes cellotetraose, cellopentaose, and cellohexaose to cellobiose and cellotriose, but does not cleave cellobiose or cellotriose. The CMCase
activity was about 30 times higher than the Avicel hydrolyzing activity. The N-terminal amino acid sequence of the enzyme was determined (175) (Table 1). A *C. josui* endoglucanase gene was
cloned and expressed in *E. coli* (190).

Cultures of *C. stercorarium* grown on cellobiose show both extracellular and cell-bound β-glucosidase activity. The β-glucosidase

present in the culture supernatant was purified and found to be identical with the cell-bound enzyme. Thiol groups are essential for its activity and it is considered as a β-glucosidase with a broad substrate specificity (179) (Table 5). From the same organism an extracellular endoglucanase and an exoglucanase were partially characterized (191) (Tables 1 and 3).

Endoglucanase C encoded by the *celC* gene of *C. thermocellum* was purified from a recombinant *E. coli* strain. It efficiently degrades glucans with alternating β-1,4 and β-1,3 linkages but does not hydrolyze acid-swollen cellulose (192) (Table 1). The nucleotide sequence of the *celC* gene was determined and the N-terminal signal sequence consists most probably of 21 amino acid residues. Most of the protein bears no resemblance to the endoglucanases A, B, and D of the same organism (193). The nucleotide sequence of the *celE* gene coding for an endoglucanase with xylan-hydrolyzing activity was also determined (194).

Two β-glucosidase genes *bglA* and *bglB* from *C. thermocellum* were cloned and expressed in *E. coli* (195-197). The properties of the β-glucosidase A, expressed in *E. coli*, correspond to those of the major β-glucosidase activity of *C. thermocellum* cultures (44), whereas the β-glucosidase B seems to be different. The nucleotide sequence of the *bglB* gene was determined and the derived amino acid sequence corresponds to a protein of 84.1 kD (198).

From the recently found thermophilic *C. thermocopriae* (199) an extracellular endoglucanase and β-glucosidase were isolated (200). The endoglucanase hydrolyzes CMC faster as the molecular weight increases (Tables 1 and 5).

Acidothermus cellulolyticus (ATCC 43068), a recently isolated, moderately thermophilic, aerobic bacterium, produces extracellular CMC degrading enzymes with half-lives of 60 and 12 min at 85 and 90°C, respectively (201).

From an extremely thermophilic aerobic bacterium of the genus *Thermus* a β-glucosidase was isolated and partially characterized (202) (Table 5).

(III.A) An extracellular endoglucanase from a *Bacillus* strain, taxonomically related to *B. pumilus*, was partially purified. Its activity is stable in the presence of surfactants, chelating agents, and proteolytic enzymes used as components of laundry detergents (203) (Table 2). From another *Bacillus* species an endoglucanase was isolated with an optimum pH value for activity of 9.5. Its activity is also not affected by various detergent components (204) (Table 2).

An endoglucanase encoded by a *B. subtilis* gene was cloned and expressed in *B. megaterium* (205) (Table 2).

(III.B) Nonglycosylated endoglucanase and exoglucanase from *Cellulomonas fimi* were purified from cultures of *E. coli* expressing recombinant *cex* and *cenA* genes. Both enzymes could be cleaved in vivo in a highly specific manner by an extracellular *C. fimi* protease. The affinity of the parent enzyme for cellulose is contained in a 20 amino acid residues comprising N-terminal fragment of the endoglucanase (*cenA*) and in a 8 amino acid residues containing carboxy terminal fragment of the exoglucanase (*cex*). These fragments contain homologous amino acid sequences, which were proposed to comprise the cellulose-binding domains. Corresponding fragments of 30 and 35 amino acid residues from endo and exoglucanase, respectively, which are unable to bind to cellulose contain the catalytic domains. In both enzymes the two functional domains are joined by a hinge region consisting solely of prolyl and threonyl residues (206) (Tables 2 and 3).

The gene *cenB* from *C. fimi* for a second endoglucanase of 110-kD molecular weight was also cloned and expressed in *E. coli*. It was purified in one step by affinity chromatography on Avicel. The enzyme is preceded by a putative signal sequence of 33 amino acids (174).

Curry et al. (207) compared the properties of the native exoglucanase (*cex*) of *C. fimi* with that expressed in *S. cerevisiae* (Table 3).

An endoglucanase gene of *C. uda* CB4 was cloned and expressed in *Zymomonas mobilis* (208) (Table 2).

(III.F) Cloned and expressed in *E. coli* was an endoglucanase from *Pseudomonas fluorescens* var. *cellulosa* (209) (Table 2). The complete nucleotide sequence of the gene encoding this endoglucanase was determined (209).

(III.G) From the alkalophilic *Streptomyces* strain KSM-9 a gene encoding an endoglucanase was cloned and expressed in *S. lividans*. The enzyme has an unusually long signal sequence of 70 amino acid residues. It is suggested that the endoglucanase is processed in two steps during maturation (210).

Five distinct endoglucanases appear in the culture supernatant of *Thermomonospora curvata*, the appearance of which is dependent on culture age (211).

An intra- and extracellular β-glucosidase of the same molecular weight was partially purified from *T. curvata* (212) (Table 5). The intracellular enzyme is stabilized by a small molecular weight factor (213).

REFERENCES

1. E. T. Reese, *Rec. Adv. Phytochem.*, *11*, 311 (1977).

2. K. Buchholz, P. Rapp, and F. Zadrazil, In *Methods of Enzymatic Analysis*, 3rd ed., Vol. 4 (H. Bergmeyer, ed.), Verlag Chemie, Weinheim, 1984, p. 178.

3. T. M. Wood, C. A. Wilson, and C. S. Stewart, *Biochem. J.*, *205*, 129 (1982).

4. M. L. Rabinowitch, V. A. Mart' yanov, G. A. Chumak, and A. A. Klesov, *Bioorg. Khim.*, *8*, 204 (1982).

5. M. V. Deshpande, K.-E. Eriksson, and L. G. Pettersson, *Anal. Biochem.*, *138*, 481 (1984).

6. G. B. Patel, A. W. Khan, B. J. Agnew, and J. R. Colvin, *Int. J. Syst. Bacteriol.*, *30*, 179 (1980).

7. J. N. Saddler, A. W. Khan, and S. M. Martin, *Microbios*, *28*, 97 (1980).

8. J. N. Saddler and A. W. Khan, *Can. J. Microbiol.*, *27*, 288 (1981).

9. C. R. MacKenzie and D. Bilous, *Can. J. Microbiol.*, *28*, 1158 (1982).

10. E. A. Bayer, R. Kenig, and R. Lamed, *J. Bacteriol.*, *156*, 818 (1983).

11. R. Lamed, E. Setter, and E. A. Bayer, *J. Bacteriol.*, *156*, 828 (1983).

12. R. Lamed, R. Kenig, E. Setter, and E. A. Bayer, *Enzyme Microb. Technol.*, *7*, 37 (1985).

13. C. R. MacKenzie, D. Bilous, and G. B. Patel, *Appl. Env. Microbiol.*, *50*, 243 (1985).

14. R. E. Hungate, *The Rumen and Its Microbes*, Academic Press, New York, 1966.

15. D. Groleau and C. W. Forsberg, *Can. J. Microbiol.*, *27*, 517 (1981).

16. D. Groleau and C. W. Forsberg, *Can. J. Microbiol.*, *29*, 504 (1983).

17. D. Groleau and C. W. Forsberg, *Can. J. Microbiol.*, *29*, 710 (1983).

18. C. W. Forsberg, T. J. Beveridge, and A. Hellstrom, *Appl. Env. Microbiol.*, *42*, 886 (1981).

19. C. W. Forsberg and D. Groleau, *Can. J. Microbiol.*, *28*, 144 (1982).

20. D. E. Akin, *Appl. Env. Microbiol.*, *39*, 242 (1980).

21. G. L. Pettipher and M. J. Latham, *J. Gen. Microbiol.*, *110*, 21 (1979).

22. G. L. Pettipher and M. J. Latham, *J. Gen. Microbiol.*, *110*, 29 (1979).

23. J. M. Leatherwood, *Adv. Chem. Ser.*, *95*, 53 (1969).

24. J. M. Leatherwood, *Fed. Proc.*, *32*, 1814 (1973).

25. R. J. Stack and R. E. Hungate, *Appl. Env. Microbiol.*, *48*, 218 (1984).

26. R. E. Hungate and R. J. Stack, *Appl. Env. Microbiol.*, *44*, 79 (1982).

27. T. W. Scott, P. F. V. Ward, and R. M. C. Dawson, *Biochem. J.*, *90*, 12 (1964).
28. M. J. Latham, B. E. Brooker, G. L. Pettipher, and P. J. Harris, *Appl. Env. Microbiol.*, *35*, 156 (1978).
29. L. G. Ljungdahl and K.-E. Eriksson, *Advances in Microbial Ecology*, Vol. 8, Plenum Press, New York, 1985, p. 237.
30. K. Ohmiya, M. Shimizu, M. Taya, and S. Shimizu, *J. Bacteriol.*, *150*, 407 (1982).
31. K. Ohmiya, M. Shirai, Y. Kurachi, S. Shimizu, *J. Bacteriol.*, *161*, 432 (1985).
32. W. A. Ayers, *J. Biol. Chem.*, *234*, 2819 (1959).
33. S. F. Lee, C. W. Forsberg, and L. N. Gibbins, *Appl. Env. Microbiol.*, *50*, 220 (1985).
34. S. F Lee, C. W. Forsberg, and L. N. Gibbins, In *5th Canadian Bioenergy Research and Development Seminar*, Elsevier, Amsterdam, 1984, p. 569.
35. E. R. Allcock and D. R. Woods, *Appl. Env. Microbiol.*, *41*, 539 (1981).
36. N. Creuzet and C. Frixon, *Biochimie*, *65*, 149 (1983).
37. B. H. Lee and T. H. Blackburn, *Appl. Microbiol.*, *30*, 346 (1975).
38. T. K. Ng and J. G. Zeikus, *Biochem. J.*, *199*, 341 (1981).
39. J. Petre, R. Longin, and J. Millet, *Biochimie*, *63*, 629 (1981).
40. W. H. Schwarz, F. Gräbnitz, and W. L. Staudenbauer, *Appl. Env. Microbiol.*, *51*, 1293 (1986).
41. P. Béguin, P. Cornet, and J. Millet, *Biochimie*, *65*, 495 (1983).
42. P. Cornet, J. Millet, P. Béguin, and J.-P. Aubert, *Biotechnology*, *1*, 589 (1983).
43. N. Aït, N. Creuzet, and J. Cattanéo, *Biochem. Biophys. Res. Commun.*, *90*, 537 (1979).
44. N. Aït, N. Creuzet, and J. Cattanéo, *J. Gen. Microbiol.*, *128*, 569 (1982).
45. J. K. Alexander, *J. Biol. Chem.*, *243*, 2899 (1968).
46. K. Sheth and J. K. Alexander, *Biochim. Biophys. Acta*, *148*, 808 (1967).
47. J. K. Alexander, *Meth. Enzymol.*, *28*, 944 (1972).
48. J. K. Alexander, *Meth. Enzymol.*, *28*, 948 (1972).
49. R. Lamed and E. A. Bayer, *Adv. Appl. Microbiol.*, *33*, 1 (1988).
50. T. K. Ng and J. G. Zeikus, *J. Bacteriol.*, *150*, 1391 (1982).
51. T.-V. C. Duong, E. A. Johnson, and A. L. Demain, *Topics in Enzyme and Fermentation Biotechnology*, Vol. 7, John Wiley and Sons, New York, 1983, p. 156.

52. J. H. D. Wu and A. L. Demain, *Abstr. Am. Soc. Microbiol.* *85th Ann. Meet.*, 1985, p. 248.
53. J. Millet, D. Pétré, P. Béguin, O. Raynaud, and J.-P. Aubert, *FEMS Microbiol. Lett.*, *29*, 145 (1985).
54. E. A. Johnson and A. L. Demain, *Arch. Microbiol.*, *137*, 135 (1984).
55. E. A. Johnson, M. Sakajoh, G. Halliwell, A. Madia, and A. L. Demain, *Appl. Env. Microbiol.*, *43*, 1125 (1982).
56. E. A. Johnson, E. T. Reese, and A. L. Demain, *J. Appl. Biochem.*, *4*, 64 (1982).
57. R. Lamed, E. Setter, R. Kenig, and E. A. Bayer, *Biotechnol. Bioeng. Symp. Ser.*, *13*, 163 (1983).
58. E. A. Bayer, E. Setter, and R. Lamed, *J. Bacteriol.*, *163*, 552 (1985).
59. L. G. Ljungdahl, B. Pettersson, K.-E. Eriksson, and J. Wiegel, *Curr. Microbiol.*, *9*, 195 (1983).
60. R. Lamed and E. A. Bayer, *Experientia*, *42*, 72 (1986).
61. J. Wiegel and M. Dykstra, *Appl. Microbiol. Biotechnol.*, *20*, 59 (1984).
62. M. P. Coughlan, K. Hon-nami, H. Hon-nami, L. G. Ljungdahl, J. J. Paulin, and W. E. Rigsby, *Biochem. Biophys. Res. Commun.*, *130*, 904 (1985).
63. K. Hon-nami, M. P. Coughlan, H. Hon-nami, and L. G. Ljungdahl, *Arch. Microbiol.*, *145*, 13 (1986).
64. L. M. Robson and G. H. Chambliss, *Appl. Env. Microbiol.*, *47*, 1039 (1984).
65. K. Horikoshi, M. Nakao, Y. Kurono, and N. Sashihara, *Can. J. Microbiol.*, *30*, 774 (1984).
66. N. Sashihara, T. Kudo, and K. Horikoshi, *J. Bacteriol.*, *158*, 503 (1984).
67. F. Fukumori, T. Kudo, and K. Horikoshi, *J. Gen. Microbiol.*, *131*, 3339 (1985).
68. P. Béguin, H. Eisen, and A. Roupas, *J. Gen. Microbiol.*, *101*, 191 (1977).
69. P. Béguin and H. Eisen, *Eur. J. Biochem.*, *87*, 525 (1978).
70. W. Stoppok, P. Rapp, and F. Wagner, *Appl. Env. Microbiol.*, *44*, 44 (1982).
71. W. Stoppok, Bildung, Lokalisation und Regulation der cellulolytischen Enzyme von *Cellulomonas uda*, Ph.D. dissertation, Technische Universität Braunschweig, FRG, 1981.
72. J. Antheunisse, *Antonie van Leeuwenhoek*, *50*, 7 (1984).
73. M. L. Langsford, N. R. Gilkes, W. W. Wakarchuk, D. G. Kilburn, R. C. Miller, Jr., and R. A. J. Warren, *J. Gen. Microbiol.*, *130*, 1367 (1984).

74. M. Vladut- Talor, T. Kauri, and D. J. Kushner, *Arch. Microbiol.*, *144*, 191 (1986).
75. B. J. Stewart and J. M. Leatherwood, *J. Bacteriol.*, *128*, 609 (1976).
76. D. W. Thayer, S. V. Lowther, and J. G. Phillips, *Int. J. Syst. Bacteriol.*, *34*, 432 (1984).
77. K. Nakamura and K. Kitamura, *J. Ferment. Technol.*, *61*, 379 (1983).
78. M. Sato and H. Takahashi, *Agr. Biol. Chem.*, *31*, 470 (1967).
79. W. W. Wakarchuk, D. G. Kilburn, R. C. Miller, Jr., and R. A. J. Warren, *J. Gen. Microbiol.*, *130*, 1385 (1984).
80. H. Sahm and K.-L. Schimz, *Appl. Microbiol. Biotechnol.*, *20*, 54 (1984).
81. K.-L. Schimz, B. Broll, and B. John, *Arch. Microbiol.*, *135*, 241 (1983).
82. K.-L. Schimz and G. Decker, *Can. J. Microbiol.*, *31*, 751 (1985).
83. M. Doudoroff and N. J. Palleroni, In *Bergey's Manual of Determinative Bacteriology*, 8th ed., Williams and Wilkins, Baltimore, 1974, p. 217.
84. L. L. Blackall, A. C. Hayward, and L. I. Sly, *J. Appl. Bacteriol.*, *59*, 81 (1985).
85. F. H. Hulcher and K. W. King, *J. Bacteriol.*, *76*, 565 (1958).
86. W. O. Storvick and K. W. King, *J. Biol. Chem.*, *235*, 303 (1960).
87. K. W. King and M. I. Vessal, *Adv. Chem. Series*, *95*, 7 (1969).
88. W. O. Storvick, F. E. Cole, and K. W. King, *Biochemistry*, *2*, 1106 (1963).
89. F. H. Hulcher and K. W. King, *J. Bacteriol.*, *76*, 571 (1958).
90. T. Sasaki, T. Tanaka, S. Nakagawa, and K. Kainuma, *Biochem. J.*, *209*, 803 (1983).
91. M. L. Schafer and K. W. King, *J. Bacteriol.*, *89*, 113 (1965).
92. B. Berg, B. v. Hofsten, and G. Pettersson, *J. Appl. Bacteriol.*, *35*, 201 (1972).
93. B. Berg, *Can. J. Microbiol.*, *21*, 51 (1975).
94. L. V. Oberkotter and F. A. Rosenberg, *Appl. Env. Microbiol.*, *36*, 205 (1978).
95. E. R. Leadbetter, In *Bergey's Manual of Determinative Bacteriology*, 8th ed., Williams and Wilkins, Baltimore, 1974, p. 99.
96. W. T. H. Chang and D. W. Thayer, *Can. J. Microbiol.*, *23*, 1285 (1977).
97. J. J. Marshall, *Carboh. Res.*, *26*, 274 (1973).

98. B. Berg, B. v. Hofsten, and G. Pettersson, *J. Appl. Bacteriol.*, *35*, 215 (1972).
99. M. Charpentier and D. Robic, *Comptes rendus Acad. Sci.*, Sér. D, *279*, 863 (1974).
100. K. Osmundsvåg and J. Goksøyr, *Eur. J. Biochem.*, *57*, 405 (1975).
101. M. P. Starr and A. K. Chatterjee, *Ann. Rev. Microbiol.*, *26*, 389 (1972).
102. M. H. Boyer, J. P. Chambost, M. Magnan, and J. Cattanéo, *J. Biotechnol.*, *1*, 229 (1984).
103. M. H. Boyer, J. P. Chambost, M. Magnan, and J. Cattanéo, *J. Biotechnol.*, *1*, 241 (1984).
104. K. Ueda, S. Ishikawa, and T. Asai, *J. Agr. Chem. Soc. Jap.*, *26*, 35 (1952).
105. K. Yamane, H. Suzuki, and K. Nisizawa, *J. Biochem.*, *67*, 19 (1970).
106. H. Suzuki, *Symposium on Enzymatic Hydrolysis of Cellulose*, SITRA, Helsinki, 1975, p. 155.
107. K. Yamane, T. Yoshikawa, H. Suzuki, and K. Nisizawa, *J. Biochem.*, *69*, 771 (1971).
108. T. Yoshikawa, H. Suzuki, and K. Nisizawa, *J. Biochem.*, *75*, 531 (1974).
109. T. Yoshikawa, H. Suzuki, and K. Nisizawa, *Sc. Rep. Tokyo Kyoiku Daigaku*, *B16*, 87 (1975).
110. J.-T. Hwang and H. Suzuki, *Agr. Biol. Chem.*, *40*, 2169 (1976).
111. K. Ramasamy and H. Verachtert, *J. Gen. Microbiol.*, *117*, 181 (1980).
112. H. K. Tewari and D. S. Chahal, *Ind. J. Microbiol.*, *11*, 88 (1977).
113. M. D. Enger and B. P. Sleeper, *J. Bacteriol.*, *89*, 23 (1965).
114. M. Ishaque and D. Kluepfel, *Can. J. Microbiol.*, *26*, 183 (1980).
115. B. G. R. Hägerdal, J. D. Ferchak, and E. K. Pye, *Appl. Env. Microbiol.*, *36*, 606 (1978).
116. B. Hägerdal, H. Harris, and E. K. Pye, *Biotechnol. Bioeng.*, *21*, 345 (1979).
117. B. Hägerdal, J. D. Ferchak, and E. K. Pye, *Biotechnol. Bioeng.*, *22*, 1515 (1980).
118. F. J. Stutzenberger, *Appl. Microbiol.*, *24*, 83 (1972).
119. G. Fennington, D. Neubauer, and F. Stutzenberger, *Appl. Env. Microbiol.*, *47*, 201 (1984).
120. A. L. Lehninger, *Biochemistry*, 2nd ed., Worth, New York, 1981, p. 977.

121. G. Fennington, D. Neubauer, and F. Stutzenberger, *Biotechnol. Bioeng.*, *25*, 2271 (1983).
122. J. L. Botsford, *Microbiol. Rev.*, *45*, 620 (1981).
123. F. Stutzenberger, *Ann. Rep. Ferment. Proc.*, *8*, 111 (1985).
124. T. J. Silhavy, S. A. Benson, and S. D. Emr, *Microbiol. Rev.*, *47*, 313 (1983).
125. P. Béguin, P. Cornet, and J.-P. Aubert, *J. Bacteriol.*, *162*, 102 (1985).
126. K. Nakamura, N. Misawa, and K. Kitamura, *J. Biotechnol.*, *4*, 247 (1986).
127. T. M. Wood and S. I. MacCrae, *Adv. Chem. Ser.*, *181*, 181 (1979).
128. L. G. Fägerstam and L. G. Pettersson, *FEBS Lett.*, *119*, 97 (1980).
129. A. R. White and R. M. Brown, Jr., *Proc. Natl. Acad. Sci. USA*, *78*, 1047 (1981).
130. M. R. Ladisch, K. W. Lin, M. Voloch, and G. T. Tsao, *Enzyme Microb. Technol.*, *5*, 82 (1983).
131. M. Fujii and M. Shimizu, *Biotechnol. Bioeng.*, *28*, 878 (1986).
132. C.-S. Gong and G. T. Tsao, *Ann. Rep. Ferment. Proc.*, *3*, 111 (1979).
133. B. H. Odegaard, P. C. Anderson, and R. E. Lovrien, *J. Appl. Biochem.*, *6*, 156 (1984).
134. Y. W. Han and V. R. Srinivasan, *J. Bacteriol.*, *100*, 1355 (1969).
135. J. N. Saddler and A. W. Khan, *Can. J. Microbiol.*, *26*, 760 (1980).
136. N. Aït, N. Creuzet, and P. Forget, *J. Gen. Microbiol.*, *113*, 399 (1979).
137. A. Shinmyo, D. V. Garcia-Martinez, and A. L. Demain, *J. Appl. Biochem.*, *1*, 202 (1979).
138. P. Prasertsan and H. W. Doelle, *Appl. Microbiol. Biotechnol.*, *24*, 326 (1986).
139. K. Nakamura, N. Misawa, and K. Kitamura, *J. Biotechnol.*, *3*, 247 (1986).
140. C. R. MacKenzie, G. B. Patel, and D. Bilous, *Appl. Env. Microbiol.*, *53*, 304 (1987).
141. C. R. MacKenzie and G. B. Patel, *Arch. Microbiol.*, *145*, 91 (1986).
142. K. Ohmiya, K. Maeda, and S. Shimizu, *Carboh. Res.*, *166*, 145 (1987).
143. T. M. Wood and C. A. Wilson, *Can. J. Microbiol.*, *30*, 316 (1984).
144. R. M. Gardner, K. C. Doerner, and B. A. White, *J. Bacteriol.*, *169*, 4581 (1987).

145. D Pétré, J. Millet, R. Longin, P. Béguin, H. Girard, and
      J.-P. Aubert, *Biochimie*, *68*, 687 (1986).
146. G. Joliff, P. Béguin, M. Juy, J. Millet, A. Ruyter, R. Poljak,
      and J.-P. Aubert, *Biotechnology*, *4*, 896 (1986).
147. O. Grépinet and P. Béguin, *Nucleic Acids Res.*, *14*, 1791
      (1986).
148. G. Joliff, P. Béguin, and J.-P. Aubert, *Nucleic Acids Res.*,
      14, 8605 (1986).
149. F. Mayer, M. P. Coughlan, Y. Mori, and L. G. Ljungdahl,
      Appl. Env. Microbiol., *53*, 2785 (1987).
150. L. M. Robson and G. H. Chambliss, *J. Bacteriol.*, *169*, 2017
      (1987).
151. K.-S. Au and K.-Y. Chang, *J. Gen. Microbiol.*, *133*, 2155
      (1987).
152. Y. Koide, A. Nakamura, T. Uozumi, and T. Beppu, *Agr.
      Biol. Chem.*, *50*, 233 (1986).
153. K. Nakamura, T. Uozumi, and T. Beppu, *Eur. J. Biochem.*,
      164, 317 (1987).
154. F. Fukumori, T. Kudo, Y. Narahashi, and K. Horikoshi, *J.
      Gen. Microbiol.*, *132*, 2329 (1986).
155. O. M. Poulsen and L. W. Petersen, *Biotechnol. Bioeng.*, XXIX,
      799 (1987).
156. C. Bagnara, C. Gaudin, and J.-P. Belaich, *Biochem. Bio-
      phys. Res. Commun.*, *140*, 219 (1986).
157. D. J. Whittle, D. G. Kilburn, R. A. J. Warren, and R. C.
      Miller, Jr., *Gene*, *17*, 139 (1982).
158. N. R. Gilkes, M. L. Langsford, D. G. Kilburn, R. C. Miller,
      Jr., and R. A. J. Warren, *J. Biol. Chem.*, *259*, 10455 (1984).
159. W. K. R. Wong, B. Gerhard, Z. M. Guo, D. G. Kilburn,
      R. A. J. Warren, and R. C. Miller, Jr., *Gene*, *44*, 315
      (1986).
160. G. P. O'Neill, S. H. Goh, R. A. J. Warren, D. G. Kilburn,
      and R. C. Miller, Jr., *Gene*, *44*, 325 (1986).
161. R. A. J. Warren, C. F. Beck, N. R. Gilkes, D. G. Kilburn,
      M. L. Langsford, R. C. Miller, Jr., G. P. O'Neill, M. Scheu-
      fens, and W. K. R. Wong, *Proteins*, *1*, 335 (1986).
162. M. L. Langsford, N. R. Gilkes, B. Singh, B. Moser, R. C.
      Miller, Jr., R. A. J. Warren, and D. G. Kilburn, *FEBS Lett.*,
      225, 163 (1987).
163. M. H. Boyer, B. Cami, J.-P. Chambost, M. Magnan, and
      J. Cattanéo, *Eur. J. Biochem.*, *162*, 311 (1987).
164. B. R. Wolff, T. A. Mudry, B. R. Glick, and J. J. Pasternak,
      Appl. Env. Microbiol., *51*, 1367 (1986).
165. H. J. Gilbert, G. Jenkins, D. A. Sullivan, and J. Hall, *Mol.
      Gen. Genet.*, *210*, 551 (1987).

166. N. Moldoveanu and D. Kluepfel, *Appl. Env. Microbiol.*, *46*, 17 (1983).
167. C. R. MacKenzie, D. Bilous, and K. G. Johnson, *Can. J. Microbiol.*, *30*, 1171 (1984).
168. R. Nakai, S. Horinouchi, T. Uozumi, and T. Beppu, *Agr. Biol. Chem.*, *51*, 3061 (1987).
169. R. Bernier and F. Stutzenberger, *Appl. Env. Microbiol.*, *53*, 1743 (1987).
170. R. E. Calza, D. C. Irwin, and D. B. Wilson, *Biochemistry*, *24*, 7797 (1985).
171. E. Lin and D. B. Wilson, *Appl. Env. Microbiol.*, *53*, 1352 (1987).
172. R. Lamed, J. Naimark, E. Morgenstern, and E. A. Bayer, *J. Bacteriol.*, *169*, 3792 (1987).
173. M. McGavin and C. W. Forsberg, *J. Bacteriol.*, *170*, 2914 (1988).
174. J. P. Owolabi, P. Béguin, D. G. Kilburn, R. C. Miller, Jr., and R. A. J. Warren, *Appl. Env. Microbiol.*, *54*, 518 (1988).
175. T. Fujino, J. Sukhumavasi, T. Sasaki, K. Ohmiya, and S. Shimizu, *J. Bacteriol.*, *171*, 4076 (1989).
176. L. Huang, C. W. Forsberg, and D. Y. Thomas, *J. Bacteriol.*, *170*, 2923 (1988).
177. L. Huang and C. W. Forsberg, *Appl. Env. Microbiol.*, *54*, 1488 (1988).
178. W. W. Wakarchuk, D. G. Kilburn, R. C. Miller, Jr., and R. A. J. Warren, *Mol. Gen. Genet.*, *205*, 146 (1986).
179. K. Bronnenmeier and W. L. Staudenbauer, *Appl. Microbiol. Biotechnol.*, *28*, 380 (1988).
180. M. L. Patchett, R. M. Daniel, and H. W. Morgan, *Biochem. J.*, *243*, 779 (1987).
181. R. Lamed and E. A. Bayer, *Meth. Enzymol.*, *160*, 472 (1988).
182. L. G. Ljungdahl, M. P. Coughlan, F. Mayer, Y. Mori, H. Hon-nami, and K. Hon-nami, *Meth. Enzymol.*, *160*, 483 (1988).
183. L. Huang and C. W. Forsberg, *Appl. Env. Microbiol.*, *53*, 1034 (1987).
184. J. Gong, R. Y. C. Lo, and C. W. Forsberg, *Appl. Env. Microbiol.*, *55*, 132 (1989).
185. D. R. Love and M. B. Streiff, *Biotechnology*, *5*, 384 (1987).
186. D. R. Love, R. Fisher, and P. L. Bergquist, *Mol. Gen. Genet.*, *213*, 84 (1988).
187. K. Ohmiya, K. Nagashima, T. Kajino, E. Goto, A. Tsukada, and S. Shimizu, *Appl. Env. Microbiol.*, *54*, 1511 (1988).
188. H. Honda, T. Saito, S. Iijima, and T. Kobayashi, *Enzyme Microb. Technol.*, *10*, 559 (1988).

189. G. Pérez-Martinez, L. González-Candelas, J. Polaina, and
      A. Flors, J. Ind. Microbiol., 3, 365 (1988).
190. K. Ohmiya, T. Fujino, J. Sukhumavasi, and S. Shimizu,
      Appl. Env. Microbiol., 55, 2399 (1989).
191. K. Bronnenmeier and W. L. Staudenbauer, Appl. Microbiol.
      Biotechnol., 27, 432 (1988).
192. W. H. Schwarz, S. Schimming, and W. L. Staudenbauer,
      Appl. Microbiol. Biotechnol., 29, 25 (1988).
193. W. H. Schwarz, S. Schimming, K. P. Rücknagel, S. Burg-
      schwaiger, G. Kreil and W. L. Staudenbauer, Gene, 63, 23
      (1988).
194. J. Hall, G. P. Hazlewood, P. J. Barker, and H. J. Gilbert,
      Gene, 69, 29 (1988).
195. F. Gräbnitz and W. L. Staudenbauer, Biotechnol. Lett., 10,
      73 (1988).
196. M. P. Romaniec, K. Davidson, and G. P. Hazlewood, Enzyme
      Microb. Technol., 9, 474 (1987).
197. S. Kadam, A. Demain, J. Millet, P. Béguin, and J.-P. Aubert,
      Enzyme Microb. Technol., 10, 9 (1988).
198. F. Gräbnitz, K. P. Rücknagel, M. Seiβ and W. L. Stauden-
      bauer, Mol. Gen. Genet., 217, 70 (1989).
199. F. Jin, K. Yamasato, and K. Toda, Int. J. Syst. Bacteriol.,
      38, 279 (1988).
200. F. Jin and K. Toda, J. Ferment. Bioeng., 67, 8 (1989).
201. M. P. Tucker, A. Mohagheghi, K. Grohmann, and M. E.
      Himmel, Biotechnology, 7, 817 (1989).
202. M. Takase and K. Horikoshi, Appl. Microbiol. Biotechnol.,
      29, 55 (1988).
203. S. Kawai, H. Okoshi, K. Ozaki, S. Shikata, K. Ara, and
      S. Ito, Agr. Biol. Chem., 52, 1425 (1988).
204. S. Ito, S. Shikata, K. Ozaki, S. Kawai, K. Okamoto,
      S. Inoue, A. Takei, Y. Ohta, and T. Satoh, Agr. Biol.
      Chem., 53, 1275 (1989).
205. H. Kim and M. Y. Pack, Enzyme Microb. Technol., 10, 347
      (1988).
206. N. R. Gilkes, R. A. J. Warren, R. C. Miller, Jr., and
      D. G. Kilburn, J. Biol. Chem., 263, 10401 (1988).
207. C. Curry, N. R. Gilkes, G. P. O'Neill, R. C. Miller, Jr.,
      and N. Skipper, Appl. Env. Microbiol., 54, 476 (1988).
208. N. Misawa, T. Okamoto, and K. Nakamura, J. Biotechnol.,
      7, 167 (1988).
209. J. Hall and H. J. Gilbert, Mol. Gen. Genet., 213, 112 (1988).
210. R. Nakai, S. Horinouchi, and T. Beppu, Gene, 65, 229
      (1988).

211. D. Lupo and F. Stutzenberger, *Appl. Env. Microbiol.*, *54*, 588 (1988).

212. R. Bernier and F. Stutzenberger, *Lett. Appl. Microbiol.*, *7*, 103 (1988).

213. R. Bernier and F. Stutzenberger, *Lett. Appl. Microbiol.*, *8*, 9 (1989).

214. P. Beguin, G. Joliff, M. Juy, A. G. Amit, J. Millet, R. J. Poljak, and J.-P. Aubert, *Meth. Enzymol.*, *160*, 355 (1988).

215. M. P. M. Romaniec, N. G. Clarke, and G. P. Hazlewood, *J. Gen. Microbiol.*, *133*, 1297 (1987).

216. E. Soutschek-Bauer and W. L. Staudenbauer, *Mol. Gen. Genet.*, *208*, 537 (1987).

217. T. M. Wood, *Meth. Enzymol.*, *160*, 216 (1988).

218. L. M. Robson and G. H. Chambliss, *J. Bacteriol.*, *165*, 612 (1986).

219. A. Lejeune, S. Courtois, and C. Colson, *Appl. Env. Microbiol.*, *54*, 302 (1988).

220. A. Lejeune, V. Dartois, and C. Colson, *Biochim. Biophys. Acta*, *950*, 204 (1988).

221. H. Honda, H. Naito, M. Taya, S. Iijima, and T. Kobayashi, *Appl. Microbiol. Biotechnol.*, *25*, 480 (1987).

222. F. Stutzenberger, *Lett. Appl. Microbiol.*, *5*, 1 (1987).

223. F. Stutzenberger, *Appl. Microbiol. Biotechnol.*, *28*, 387 (1988).

224. D. B. Wilson, *Meth. Enzymol.*, *160*, 314 (1988).

225. G. S. Ghangas and D. B. Wilson, *Appl. Env. Microbiol.*, *54*, 2521 (1988).

226. G. P. O'Neill, D. G. Kilburn, R. A. J. Warren, and R. C. Miller, Jr., *Appl. Env. Microbiol.*, *52*, 737 (1986).

227. K. Nakamura and K. Kitamura, *Meth. Enzymol.*, *160*, 211 (1988).

228. W. W. Wakarchuk, N. M. Greenberg, D. G. Kilburn, R. C. Miller, Jr., and R. A. J. Warren, *J. Bacteriol.*, *170*, 301 (1988).

23

Plant Cellulases and Their Role in Plant Development

GORDON MACLACHLAN   *McGill University, Montreal, Quebec, Canada*

SEAN CARRINGTON   *University of the West Indies, Bridgetown, Barbados*

I. INTRODUCTION

"Cellulase activity" has been found in extracts of many plant tissues, particularly those that are growing or senescing (1). The activity has most commonly been assayed by measuring the rate of loss of viscosity of solutions of carboxymethylcellulose (CMC) which results from the action of endo-1,4-β-glucanase (EC 3.2.1.4). Exo-1,4-β-glucanase (EC 3.2.1.74) and cellobiohydrolase (EC 3.2.1.91), though they may be essential for the effective hydrolysis of crystalline cellulose microfibrils in "cellulase complexes" from microorganisms (2), have seldom been tested for or reported in plants. Mixed-linkage 1,3/1,4-β-glucanase (EC 3.2.1.73) in monocots is usually capable of hydrolyzing endo-1,4-β linkages but only when they are adjacent to a 1,3-β linkage (3,4). Accordingly, this chapter is confined to endo-1,4-β-glucanase (CMCase or "C_x" activity) in plants. *Methods in Enzymology* is due to publish a volume shortly on procedures for purifying and characterizing plant hydrolases, including endo-1,4-β-glucanases, and this aspect will not be dealt with here except insofar as kinetic data bear on function. The potential functions of these enzymes in plants were last reviewed in 1982 (1); thus more recent advances are emphasized in this chapter.

A variety of indirect approaches have been used in order to investigate the functions of enzymes in vivo, most of which have been employed in tests with one or another plant 1,4-β-glucanase. Methods include measurements of enzyme kinetics and substrate specificities in vitro in order to define probable limits to reaction rates and directions and to the range of potential substrates in vivo. Likewise, histochemical studies of β-glucanase distribution, particularly those using immunochemical methods, can specify precisely where in the cell this particular enzyme accumulates and thereby focus attention on potential substrates at that site. Such indirect observations are useful for defining how 1,4-β-glucanase could or might function in vivo, but they usually lack the cause-and-effect element needed for rigorous demonstration of function in the living cell.

More direct approaches, such as examination of the effects of added purified enzymes on living systems, suffer in the case of 1,4-β-glucanase from difficulties in ensuring penetration to the presumed sites of action at the inner cell wall, which is the first structure that endogenously generated 1,4-β-glucanase would encounter upon secretion in vivo. Specific inhibitors of an enzyme activity have been usefully employed in vivo to test for metabolic interference of presumed effects in other biological systems, but despite years of searching for a simple and specific cellulase inhibitor, none has been found.

Nevertheless, from time to time naturally occurring polypeptide inhibitors of 1,4-β-glucanase action are reported (5), and specific antibodies to plant cellulases have been prepared (6-8). These are inhibitory if added to the enzyme in vitro, but few of them have been tested on living systems, perhaps because of doubts about their ability to penetrate the wall. This was overcome in one successful approach where antibodies to a *Phaseolus* cellulase were injected into the abscission zone of bean explants and shown to prevent subsequent cell separation (9).

Another effective approach is based on observations that induction of plant 1,4-β-glucanase activity by application of growth regulators may lead to dramatic and apparently specific increases in enzyme level in vivo. Together with selective inhibitors of the induction process, β-glucanase activity levels may be altered up or down while physiological effects in vivo are examined. In this way, time course experiments may be conducted at different manipulated, 1,4-β-glucanase levels in order to vary the potential cause unilaterally and assay the putative effects (e.g., see Refs. 10 and 11).

This chapter is divided into two main sections: (1) a summary of current knowledge of plant 1,4-β-glucanase location and kinetic characteristics bearing on potential substrates, and (2) a summary of results of more direct tests for functions carried out on particular tissues.

II. LOCATION AND PROPERTIES OF PLANT ENDO-1,4-β-GLUCANASE

A. Distribution

1. *Tissues*

Table 1 provides a summary with examples of the many plant tissues that have been reported to contain or develop endo-1,4-β-glucanase activity. There are definite stages or conditions during the development or maturation of many plant cell types when 1,4-β-glucanase is elaborated. Three of these may be singled out where relatively high β-glucanase activity occurs and where it may be required for

TABLE 1 Distribution of Endo-1,4-β-glucanase Activity in Selected Plant Tissues

Tissues	Part	Examples (Refs.)
Growing	Epicotyl	Pea (12, 13)
	Hypocotyl	Soybean (14)
	Coleoptile	Corn (15), oat (16)
	Fiber	Cotton (17)
Differentiating	Laticifer	Rubber (18, 19)
	Root nodule	Pea (1, 20)
	Xylem, phloem	Sycamore (21), barley (22)
Aging	Ripening fruit	Avocado (23, 24)
		Tomato (25, 26), papaya (27)
	Germinating seed	Nasturtium (28, 29)
		Tamarind (30)
	Abscission zone	Bean (31, 32, 33)
		Citrus (34, 35)
	Senescing flower	Morning glory (36)

morphogenic changes that are known or suspected to occur during those periods. These are:

1. Young growing tissues where new cells and cell walls are formed with relatively thin primary walls and only a modest interwoven network of cellulose microfibrils. The endo-1,4-β-glucanase activity associated with walls may have a function in promoting the continual process of wall relaxation or "loosening" that appears to be essential for continued growth (37). When such cells or tissues cease growing and begin depositing secondary walls, their β-glucanase levels decline. In most mature tissues, 1,4-β-glucanase activity is not detectable.
2. Certain aging, ripening, senescing, or abscising tissues where cell walls visibly disintegrate (31). The implication is that the enzyme helps to promote the wall-softening or cell separation associated with such specialized tissues.
3. Instances during specific tissue differentiation where highly localized cell wall dissolution can be seen to take place as part of the process of developing specific functions, e.g., dissolution of end walls in files of cells leading to formation of vessels, sieve tubes, or articulated laticifers.

2. Subcellular Localization and Regulation

Treatment of growing pea epicotyl apices with high concentrations of auxin (indoleacetic acid) or an auxin herbicide leads to lateral cell expansion (swelling) and a massive increase in level of extractable 1,4-β-glucanase activity. Details of this process (38) can be summarized as follows: The activity is due to two endoglucanases, one buffer-soluble (0.01 M phosphate, pH 6), the other salt-soluble (buffer containing 1 M NaCl). The buffer-soluble enzyme is concentrated in ER and ER vesicles, while the buffer-insoluble, salt-soluble enzyme is bound to the inner surfaces of cell walls (12). The mRNA to buffer-soluble cellulase has been isolated and shown to increase following auxin treatment of the tissue before increases in enzyme activity can be detected (8). Ethylene treatment also causes swelling in pea epicotyls but it does not evoke any change in 1,4-β-glucanase activity and in fact inhibits its development in response to auxin (11). Accordingly, this auxin response is not brought about indirectly via an effect of auxin-induced ethylene but rather is a specific stimulation by auxin of transcription leading to mRNA for pea 1,4-β-glucanase.

In contrast, in several nongrowing but distinctly maturing plant tissues, ethylene gas, either produced by the tissue itself or applied externally, leads to an acceleration of the aging process and, at the same time, to an increase in the levels of 1,4-β-glucanase activity

(33). Examples are given in Table 1. In two of these instances, the abscission zone of kidney bean petioles and ripening avocado fruit, the endo-1,4-β-glucanases that are generated have been extensively characterized and the mechanisms of their regulation studied.

In the case of the bean petiole abscission zone, following ethylene treatment and a lag period the level of a particular isozyme of 1,4-β-glucanase (buffer-insoluble, salt-soluble, pK_I 9.5) increases markedly (7) while other isozymes are not affected. A transient form of the enzyme that is relatively inactive (latent) is bound to plasma membrane (39). Using specific monoclonal antibodies to the purified enzyme it was shown histochemically (32) that the enzyme is highly localized in the walls of cortical tissue in a thin layer that is only two cells wide through the abscission zone fracture plane.

In ripening avocado fruit, where prodigious levels of endo-1,4-β-glucanase activity have been detected particularly after ethylene treatment, the main cellulolytic activity has been purified and shown to be due to a single isozyme (23). This enzyme is precipitated by the antibody to bean petiole 1,4-β-glucanase (6), indicating common antigenic determinants. By use of a cDNA probe, avocado cellulase mRNA was shown to increase at the start of the process of fruit ripening (6). Unlike the pea endo-1,4-β-glucanase, the avocado enzyme is a glycoprotein. It is first formed as a "proenzyme" in ER, from which a signal peptide is removed during processing (24). It becomes heavily glycosylated in microsomal fractions and then is trimmed of part of the carbohydrate before being secreted and accumulated extracellularly.

These studies underline the fact that expression of β-glucanase genes following hormone treatment can be a highly localized phenomenon, sometimes confined to only a few cells or even parts of cells, and subject to various processing events before being secreted. In the case of developing laticifers, for example, 1,4-β-glucanase has been localized cytochemically only at end walls, where it is presumably involved in the process of cell wall perforation (19). This implies that these "target cells" (40) must have developed susceptibilities (hormone receptors?), different from their neighbors', that lead to uniquely enhanced levels of 1,4-β-glucanase. The mechanisms behind the differential sensitivities or receptor localization are unknown, but the systems seem ideal for more intensive studies of control points.

It is evident that ethylene and auxin have mutually antagonistic effects, not only on β-glucanase induction by auxin in growing pea stems (11,41), but on ethylene-evoked β-glucanase expression in abscising tissues as well (31,33). Abscisic acid may also evoke increases in 1,4-β-glucanase activity, but this appears to act via mediation of abscisic acid-induced ethylene formation in neighboring tissues (35). There are also several points during transcription, translation, processing, and secretion of cellulase where hormones can probably exert regulatory controls that lead to opposite end effects. The general fact that endo-1,4-β-glucanase is growing tissues is auxin-induced while that in aging tissues is ethylene-controlled, with each hormone antagonistic to the other under the two circumstances, is now established but not explained.

B. Substrate Specificity

1. β-Glucans and Derivatives

The plant endo-1,4-β-glucanases have all been detected viscometrically using as substrate CMC with a limited degree of substitution (e.g., 0.4−0.7 acetyls per glucose). It is presumed that the enzyme hydrolyzes this derivative at internal linkages between glucose units that are not substituted. Since carbon 6 is the most readily carboxymethylated, it is implied that unsubstituted or only partially substituted adjacent terminal hydroxyls are a minimal requirement for enzyme affinity and cleavage.

The pea (42) and the avocado (43) 1,4-β-glucanases have no capacity to hydrolyze cellobiose or the 1,4 linkage adjacent to the reducing end of cellodextrins. They both hydrolyze soluble cellodextrins at increasing rates as the degree of polymerization increases. The pea β-glucanases hydrolyze cellohexaose even more rapidly than cellopentaose (44), which implies a recognition site extending to at least five 1,4-β-glucosyl linkages. Internal linkages are preferentially attacked so that fragments with 3-, 4-, or 5-glucose units are generated first in reaction mixtures before being themselves hydrolyzed. Final products are cellobiose and glucose. The pea (but apparently not avocado) 1,4-β-glucanase is also capable of hydrolyzing cellulose itself (Avicel, Whatman powder, or pea microfibrils), particularly if it is first swollen with acid or alkali to reduce the degree of crystallinity. The rate of production of reducing groups from cellulose is only a fraction of that from soluble cellohexaose or CMC.

These enzymes also hydrolyze soluble mixed-linkage 1,3/1,4-β-glucans at relatively rapid rates, provided the glucan contains stretches of consecutive 1,4 linkages (4,42). No 1,3-β linkages are cleaved. There is only one report (36) that purified plant endo-1,4-β-glucanase can hydrolyze 1,4-β-xylan, which may be an isolated example since most purified plant β-glucanases do not have β-xylanase activity (e.g., see Ref. 27). Authentic plant β-xylanases do not hydrolyze β-glucan (45).

2. Xyloglucans

This natural hemicellulose in plant cell walls is readily hydrolyzed by endo-1,4-β-glucanases because all xyloglucans contain

unsubstituted glucose units in their 1,4-β-glucan backbone which
are susceptible to attack. Xyloglucans are ubiquitously distributed
in higher plants, particularly dicots, as a constituent of primary
cell walls and a minor (diminishing) component of most maturing
walls (11). In growing dicot tissues, xyloglucan is the major hemicellulose, where it is deposited in amounts comparable to that of
cellulose. Xyloglucan is also a major constituent (e.g., 30% by
weight) of many dicot seed reserves (Table 1).

All xyloglucans contain a backbone of 1,4-β-linked glucose units
which are substituted at intervals with 1,6-α-linked xylose. Some
also contain galactose, fucose, and/or arabinose units in side chains.
Because of the backbone, xyloglucan binds very strongly to cellulose
in the wall to form an insoluble macromolecular complex (46,47) from
which xyloglucan can only be solubilized by strong alkali (e.g., 24%
KOH) or other reagents which break hydrogen bonds. If primary
cell walls are extracted with dilute (4%) alkali, the residue consists
almost entirely of a xyloglucan–cellulose complex, which retains the
form of the cell wall, i.e., wall ghosts (47,48). By specific staining or radioautographic techniques, the xyloglucan component of
ghosts can be seen distributed over the whole wall, both on and between cellulose microfibrils (49). In cross-sections (50), xyloglucan
can be seen using an immunogold technique as distributed throughout the thickness of the entire wall and middle lamella. Thus it is
a true matrix polysaccharide (see Fig. 1).

The xyloglucans from pea epicotyl (47), soybean hypocotyl (51),
and sycamore cell culture (52) appear to possess the same structure,
with repeating subunits of a nonasaccharide ($Glc_4 Xyl_3 Gal Fuc$) and
a heptasaccharide ($Glc_4 Xyl_3$). Each of these subunits has a cellotetraose backbone with one unsubstituted glucose at the reducing
end. Treatment of xyloglucans with endo-1,4-β-glucanase leads to
hydrolysis at the unsubstituted glucose units. Evidently substitution of C_6 of the nonreducing glucose of the cellobiose subunit in
this backbone with xylose is no impediment to plant 1,4-β-glucanase
action in hydrolyzing the 1,4-β linkage adjacent to the reducing end.

The action of purified plant 1,4-β-glucanases on natural xyloglucan from the same tissue has been studied most extensively in pea
epicotyl (48,49,53,54). These glucanases rapidly reduce the viscosity of xyloglucan solutions and their average molecular weight, indicating that endohydrolysis does take place. The final products
after complete hydrolysis of purified pea or soybean xyloglucan are
the nona- and heptasaccharides only. K_m values for pea β-
glucanases acting on xyloglucan are the same as those for cellohexaose and CMC but V_{max} values are lower (48). Similar observations were reported for purified avocado cellulase acting on soybean xyloglucan (43). There have also been reports that extracts
from soybean cell walls contain an endo-1,4-β-glucanase that hydrolyzes soybean xyloglucan to defined fragments (14). Thus plant
xyloglucan must be regarded as an alternative substrate to cellulose

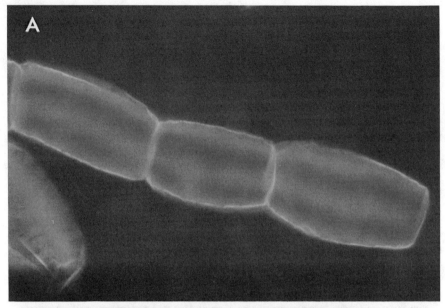

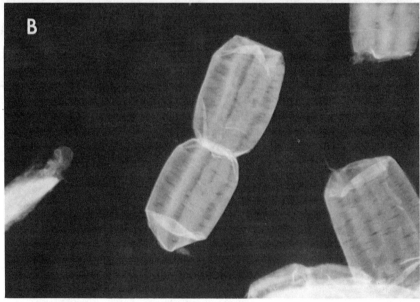

FIGURE 1   Pea stem cell wall "ghosts" stained for xyloglucan (A)
or cellulose (B). In frame A, tissue was extensively extracted with
water and 4% KOH to remove all cell wall contents except the

TABLE 2   Physical Properties of Selected 1,4-β-Glucanases

Source	Location	Mol. wt. (kD)	pH optimum	pK_I
Bean petiole	Growing tissue	70	5.1	4.5
	Plasma membrane	?	5-6	4.8
	Abscising wall	51	8.1	9.5
Pea epicotyl	Growing tissue	20	6-6.5	5.2
	Growing wall	70	6-6.5	6.9
Avocado fruit	Ripe tissue	45-49	4.5-6	4.7
		53	4.5-6	5.6-6.2
		54-56.5	4.5-6	—

for plant endo-1,4-β-glucanase, and it may be misleading to refer to the enzyme as cellulase.

It would be equally inappropriate to refer to a plant nonspecific endo-1,4-β-glucanase that hydrolyzed xyloglucan more readily than cellulose as a "xyloglucanase." However, the existence of a truly specific endoxyloglucanase, which does not appear to hydrolyze CMC, mixed-linkage β-glucan, or cellodextrins, was recently reported in *Nasturtium* seeds (29). These seeds, like those of tamarind (55), contain xyloglucan which is less fully substituted than that in growing cells.

---

xyloglucan—cellulose complex. It was stained with the fucose-specific *Ulex* lectin complexed with FITC and examined by fluorescence microscopy. The stain is distributed throughout the wall and across gaps between the microfibrillar framework. In frame B, tissue was further extracted with 24% KOH to remove xyloglucan and stained with Calcofluor to show the cellulose microfibril orientation. The cells are approx. 20 × 40 μm. (Photos supplied courtesy of Dr. T. Hayashi.)

C. Kinetic and Physical Properties

It is impossible to compare reported V_{max} and K_m values for the
plant 1,4-β-glucanases in any meaningful manner because the carboxymethylcellulose which has been universally used as substrate
has varied in degree of substitution, a factor which will affect enzyme affinity and reaction velocity.

Table 2 records properties of certain plant 1,4-β-glucanases
which are independent of substitution variations, namely, molecular
weight, pH optimum, and pK_I values. All but one of the plant glucanases isolated to date are acidic proteins with acidic pH optima
corresponding to the pH of the walls in which they are presumed to
act. The exception is the bean abscissic zone enzyme (pK_I 9.5),
which may bind to negatively charged wall components and act
in situ despite its basic pH optimum (8.1) when isolated. The molecular weights of the 1,4-β-glucanases active in growing bean and
pea tissues are about 70 kD, while those in aging tissue (abscising
bean, ripening avocado) are close to 50 kD.

III. FUNCTIONS OF PLANT
ENDO-1,4-β-GLUCANASE ACTIVITY

A. "Loosening" of Wall Components
During Growth

1. General Observations

It is widely accepted that plants must partially weaken or selectively loosen the structural integrity of primary cell walls, even
while adding to their substance by biosynthesis, in order for irreversible cell expansion to take place. Evidence that cell wall
turnover does indeed occur in growing tissues was reviewed by
Labavitch (37), who points out the difficulty in specifying which
wall components are subject to partial hydrolysis when at the same
time they are also being laid down in amounts that result in a net
increase. The question is whether 1,4-β-glucanase has a role to
play here.

There is no question that plant 1,4-β-glucanase in growing tissues (Table 1) is secreted into cell walls (13,14). It is inducible
by auxin, at least in legumes, although the lag period between
auxin application and detectable translation of cellulase mRNA, let
alone its appearance in the wall (8), precludes any casual role for
the enzyme in so-called rapid-growth responses to the hormone.
Selective wall loosening for long-term growth is another matter.

2. Cellulose

It has been suggested (38,56) that limited action of 1,4-β-glucanase
vs. microfibrils themselves may be needed in order for the tightly

woven framework of the primary wall (44; see also Fig. 1) to relax
sufficiently to permit ongoing wall expansion during growth. The
problem is that new cellulose microfibrils are continually deposited
in growing tissues, particularly after auxin treatment, and the average molecular weight of the total cellulose population continues to
increase (57). It is difficult with current methods for measuring
molecular weight distribution of cellulose to be sure that relatively
small components that may be detected (e.g., 58) are really 1,4-β-
linked cellodextrins. It must be concluded that firm evidence for
turnover of cellulose itself in growing tissue has not been obtained.

This is not to imply that plant 1,4-β-glucanases cannot hydrolyze plant microfibrils, given an accessible fibril or enough time for
other degradative reactions to open a pathway. Wholesale dissolution of plant microfibrils in vivo under specific developmental conditions has been observed repeatedly (Table 1). No such event
occurs during growth, but limited endohydrolysis with wall-loosening effects, overshadowed by continued biosynthesis, remains a
theoretical possibility.

3. Xyloglucan

In contrast to cellulose, turnover of xyloglucan in growing cell walls
has been well established through the use of a variety of investiga-
tive techniques. When pea epicotyl segments are pulse-labeled with
the relevant sugars, an auxin-dependent loss of prelabeled fucose,
xylose, or glucose from the wall takes place during chase (59−61).
The solubilized product is polymeric and can be digested enzymically
to the characteristic xylosyl-glucose dimer of xyloglucan (isoprime-
verose). Methylation analysis confirms that xyloglucan is indeed a
target for turnover in these tests (62). Low-speed centrifugation
of the living tissue was also used to demonstrate that auxin treatment leads to release of xyloglucan from the wall into the extracellular free space of pea epicotyl (63) and pine hypocotyl (64) sections. Carefully isolated pea cell walls have the capacity to carry
out autolytic reactions, and among the solubilized products are polysaccharides containing xylose and glucose (65). Enzyme solutions
extracted from the wall are capable of hydrolyzing such products
from boiled walls. Finally, an auxin-dependent marked decline in
the molecular weight of xyloglucan was demonstrated to occur in pea
epicotyl (48,49), soybean hypocotyl (14), azuki bean epicotyl (54),
Avena coleoptile (16), and pine hypocotyl (66). In the pea system,
this decline is accompanied by demonstrably higher levels of endo-
1,4-β-glucanase activity, as measured by the ability of enzyme extracts to reduce the viscosity of either CMC or xyloglucan solutions.
None of these changes takes place during ethylene-induced swelling
in peas, i.e., they are auxin-specific (11).

It may be concluded that wall xyloglucan is indeed a substrate
for β-glucanase action in young plant tissues under conditions where

growth is promoted by auxin and glucanase levels are elevated. In-deed, xyloglucan is probably the preferred target substrate, since it is degraded first when cell wall ghosts (xyloglucan —cellulose com-plex) are incubated with pea or fungal endo-1,4-β-glucanases (48, 49). This is not because these enzymes have a higher affinity for xyloglucan than cellulose but is a consequence of a structure whose microfibril surfaces are coated with xyloglucan (49). The latter is therefore more accessible to enzymic attack. The most susceptible regions of xyloglucan are presumably those that stretch between microfibrils to form the interfibrillar matrix (Fig. 1). On the as-sumption that this is a component that contributes to wall rigidity (46,52), its partial endohydrolysis by 1,4-β-glucanase would be ex-pected to contribute to loosening of the wall framework.

4. Mixed-Linkage Glucan

In many grasses an auxin-dependent loss of glucose from noncellu-losic fractions of the growing wall can be demonstrated. The glu-can which is solubilized is mixed 1,3/1,4-β-linked and a major wall component in grasses (67). There is no information on whether or not such a β-glucan contributes to wall rigidity. Nevertheless, its solubilization is due to hydrolysis to smaller fragments during growth (3,68). The main enzymes responsible appear to be a specific mixed-linkage β-glucanase and an exo-1,3-β-glucanase (69). There is no evidence that endogenous 1,4-β-glucanase also plays a role in turn-over of this β-glucan, although the possibility should be considered since the enzyme is often present in such tissues and mixed-linkage β-glucan is a good substrate for it.

B. Wall Disintegration During Differentiation

During the development of vertical conducting elements in stems and roots, localized cell wall dissolution at the ends of cells permits di-rect passage of translocates through plasma membranes while cells remain alive (sieve plates), or through unimpeded passages after cell death (vessels, laticifers). High 1,4-β-glucanase activity has been detected in such cells while wall dissolution is visibly taking place (Table 1).

Tissue hypertrophy is commonly observed as a rapid plant re-action to infection by many microorganisms. It appears to result from swelling of host cells in response to the ethylene evoked upon wounding by an invasive organism (33). This may or may not be a reaction effected through the wall-loosening action of induced 1,4-β-glucanase, since some susceptible tissues generate the en-zyme after ethylene exposure but many do not. In contrast, the symbiotic bacterium *Rhizobium* in root nodules does not actually penetrate host cells or provoke wound reactions but instead appears to be welcomed and invaginated in a vesicle of host plasma membrane. The swelling of legume host cells that is observed in root nodules is due to the fact that the bacterium secretes high concentrations of indoleacetic acid and host cells respond by generating $1,4$-β-glucanase activity (1).

C. Wall Metabolism During Aging

1. "Softening" During Fruit Ripening

Pectic materials are the most prominent components of the walls of ripening fruit that are depolymerized and solubilized during the process. In ripening pears (70), for example, very considerable losses occur in wall galactose, arabinose, and galacturonic acid, but there is no indication of any decline in the total amount of noncellulosic glucose or xylose, or in the cellulose fraction. The visible (electron microscopic) pattern of progressive wall degradation in ripening tomato (26) could be duplicated by treating cut sections of unripe fruit with tomato polygalacturonase, which dissolves the middle lamellae, but not with added cellulase. To be sure, fruit softening is often accompanied by increases in both cellulase and pectinase activities, and these may result in partial depolymerization of structural components without solubilization, as appears to take place in the hemicellulose fractions of ripening strawberry and tomato cell walls (25,71). Paradoxically, the fruit-ripening enzyme that increases in activity most dramatically, the avocado endo-$1,4$-β-glucanase (72), has been highly purified (23), its mRNA and DNA prepared (6), and details of its processing after biosynthesis elucidated (24), all without knowing what its in vivo substrate may be.

2. "Separation" During Abscission

The cell biology of the abscission process has been reviewed (31), as has the stimulatory role of ethylene (33). It is clear that visible dissolution of the middle lamella of separating cells is often preceded by an increase in levels of endopolygalacturonases and certain isozymes of $1,4$-β-glucanase before the breakstrength of the separation zone starts to decline (73). But there are numerous exceptions to this observation. In the *Begonia* flowering bud, endopolygalacturonase activity appears to rise only after abscission is complete (74). In bean petiole abscission, some β-glucanase isozymes change in amount with no apparent relation to the process of wall dissolution. In citrus petiole explants, the levels of cellulase and polygalacturonase activity increase greatly after ethylene treatment when there is no resulting sign of abscission reactions (75).

In a reexamination of early experiments, Trewavas et al. (76) concluded that it was the act of isolating the plant organ from the rest of the plant that initiated abscission, not ethylene, although ethylene accelerated the process. In explant systems, "isolation" results from excision while in the intact plant it may be a break in vascular communication between the abscising organ and stem.

D. Wall Fragmentation to Functional Products

1. *Primer Formation*

One view of the mechanism of cell growth by cell expansion identifies the main driving force and limiting factor as the rate of biosynthesis and incorporation of new structural components that extend the wall. This process could in theory depend on the availability of acceptor chain ends that can be elongated by addition of new saccharides (77). Primer requirements are well established for the biosynthesis of α-linked polysaccharides in higher plants and there is preliminary evidence for chain-lengthening reactions in pea xyloglucan (61) and tobacco β-glucan (78) biosynthesis. Recent reports (79) that β-glucan biosynthesis employs glucoprotein as primer in membranes from the alga *Prototheca zopfii* require confirmation and demonstration that similar requirements exist in other plants.

Purified pea endo-1,4-β-glucanase was added back to the surfaces of cut pea cells plus radioactive UDP-glucose in order to determine if partial hydrolysis of 1,4-β-glucan linkages could actually promote biosynthesis. The results (80) showed that with a very brief exposure to hydrolytic action, there was a significant enhancement of incorporation of glucose into β-glucan. Longer exposure to β-glucanase inhibited net incorporation, as expected. It is unlikely that these results could have been due to chain end exposure and lengthening of the xyloglucan backbone because biosynthesis of xyloglucan appears to be completed in the Golgi before excretion (61) and because there is evidence that the surface glucosyltransferase that generates 1,4-β linkages in the presence of Mg^{2+} is not the same as the Golgi glucosyltransferase that lengthens xyloglucan in a Mn^{2+}- and UDP-xylose-dependent reaction (49).

The possibility remains open, therefore, that one function of endo-1,4-β-glucanase is to cooperate with cellulose in the plasma membrane synthase in order to generate primers in preexisting wall cellulose or to excise completed cellulose from its membrane-bound initiation site. Evidence for cooperative action between hydrolase and synthase was recently reported for the biosynthesis of 1,3-β-glucan by yeast plasma membranes (81).

2. "Oligosaccharin" Production

The term "oligosaccharin" was coined by Albersheim and colleagues (82,83) to refer to fragments of plant cell wall polysaccharides that possessed biological activity such as fungicidal powers or flowering promotion. Since there is a wealth of specific chemical information inherent in the precise configurations of cell wall components, it is reasonable to look for secondary functions for the great diversity of cell wall polysaccharide subunits beyond their roles as building blocks for wall polymers.

The evidence to date for an effect of the products of plant endo-1,4-β-glucanase action on the metabolism of growing plants is based on a set of observations by Albersheim and colleagues (84) that specific xyloglucan fragments (nonasaccharides) generated by 1,4-β-glucanase action were strong inhibitors at very low concentrations (10^{-9} M) of auxin-evoked pea stem elongation. These observations were independently confirmed by repeated tests with pea stem segments (85). A curious and unexplained aspect of the results was the fact that at higher concentrations of applied nonasaccharide, the antiauxin properties disappeared. Nevertheless, the argument was advanced that since auxin treatment in peas induces 1,4-β-glucanase activity (Sec. II) which may well "loosen" the wall and provoke growth (Sec. III.A) as well as generate xyloglucan fragments, the inhibition of growth by such fragments may represent a natural feedback control system. Although these observations need further analysis and extension to other auxin-regulated growth phenomena, they represent the first evidence for a radically different function for 1,4-β-glucanase activity, namely, growth regulation by the specific products generated.

3. Nutrition

It has been proposed that the very high concentrations of xyloglucan which exist in seeds of certain tropical plants represent a "reserve" polysaccharide which, in due course, is degraded to constituent sugars and used as a source of nutritive translocate to the developing seedlings. Seed xyloglucans are in fact degraded eventually (28,30) and the enzymes required to do so have been found in germinated seeds (29). The implication is that the 1,4-β-glucanase in these seeds functions to assist in production of nutrient for the seedling.

The problem with this suggestion is that xyloglucan degradation (e.g., in seeds of *Tamarindus*; see Table 3) begins many days after the seedling is well established with its own photosynthetic capacity (30). It is not clear that the seedling would need xyloglucan sugars

TABLE 3 Changes in Xyloglucan Content and Fresh and
Dry Weight of Tamarind Seed During Germination[a]

Germination (days)	Fresh wt. (mg/cot)	Dry wt. (mg/cot)	Xyloglucan (mg/cot)
1	570	260	49
3	600	240	49
7	680	230	48
10	600	210	44
13	570	160	28
15	540	110	26
20	460	110	20
25	380	70	13
30	270	35	3

[a]Dry seeds were soaked in water and germinated in
vermiculite in a greenhouse. At intervals, the average
fresh and dry weights of the cotyledons from 10 seedlings were measured, they were extracted with hot water
and 4% KOH followed by 24% KOH, and the xyloglucan
content measured in the strong alkali extract by assaying its iodine complex spectrophotometrically (49).

as a major nutrient when it can fix its own carbon. However, it
does seem possible that xyloglucan in the seeds has a major role as
a hydrophilic reserve. The dry, hard seeds of tamarind, for example, upon wetting, swell dramatically, split their testa, and then
germinate. Since the plant grows and sheds its fruit in dry, sandy
soils in climates where rain is often sporadic, the xyloglucan "reserve" may represent an adaptation that assists the seed to adsorb
and retain whatever water comes available. The function of $1,4-\beta-$
glucanase or xyloglucanase which is generated in cotyledons much
later could be as much to release adsorbed water (epigeal cotyledons
shrivel markedly on the seedling once xyloglucan degradation begins;
Table 3) as to generate sugar nutrients.

IV. CONCLUSIONS

This chapter summarized a variety of approaches from cell to molecular biology which, in the process, have uncovered a variety of

potential functions for the substrates and products of endo-1,4-β-glucanase activity and also for related hydrolases. It has been a provocative and productive field of enquiry, and promises to continue in this vein.

Questions or approaches raised by this enquiry that remain unresolved or underemployed can be summarized briefly as a series of queries for the future:

1. Do specific inhibitors of 1,4-β-glucanase activity (polypeptides, antibodies) inhibit presumed biological effects of the enzyme?
2. Do highly localized hormone receptors develop in the very limited regions where 1,4-β-glucanase activity is induced but not elsewhere? If so, what controls the focused receptivity?
3. Why are tissues that generate 1,4-β-glucanase in response to auxin prevented from doing so by ethylene, and vice versa?
4. Does limited cellulose endohydrolysis occur during growth followed by chain elongation at newly generated primer initiation points?
5. Does purified plant 1,4-β-glucanase added to plant tissues or cells duplicate any known physiological reaction?
6. Do the products of 1,4-β-glucanase action have specific biological effects because of their unique configurations?

V. ADDENDUM

Since November 1987, when this review was concluded, there have been several important research advances and relevant reviews published which deal with plant cellulases and/or their functions:

Reviews. A volume of *Advances in Enzymology* has now been published which includes reviews of methods for purifying the auxin-induced *Pisum sativum* cellulases (86) and the three main classes of cellulase found in *Phaseolus vulgaris* (87), namely, the abscission zone basic cellulase which is ethylene-induced, the acidic cellulases in growing tissues which are auxin-induced, and the plasma membrane-bound cellulases, presumably in process of secretion. Several major reviews have also appeared on xyloglucan (88 – 90) that discuss its susceptibility to hydrolysis by endo-1,4-β-glucanases and the functions of the oligosaccharide products so produced. A recent symposium on cell walls (91) contains numerous contributions that elaborate on the roles of cellulase, particularly via hydrolysis of xyloglucan, which is increasingly regarded as its most accessible endogenous substrate.

Major research advances. Fry (91) and colleagues added further confirmatory evidence to the report of York et al. (84) that low concentrations (10^{-9} M) of the fucose-containing nonasaccharide, derived from stem xyloglucan by cellulolytic hydrolysis, do indeed inhibit auxin-induced growth in peas. A terminal α-fucosyl unit is essential for antiauxin activity (92,93), but fucosylated fragments smaller than nonasaccharides are still effective (94). In all of these studies, higher concentrations (10^{-6} M+) of nonasaccharide were no longer inhibitory to auxin-stimulated growth. This may be explained by the observation (95) that higher concentrations (μM) of several xyloglucan oligosaccharides strongly *activate* pea cellulase in vitro. A feedback activation of cellulase in situ by the products of its own action could only serve to enhance the wall-loosening effects of the enzyme and counteract the antiauxin effects of low concentrations of nonasaccharide. Evidence has also been reported by Fry et al. (91) that nonfucosylated xyloglucan oligosaccharides, such as the hepta-saccharide subunit, were able to promote pea stem growth at micro-molar concentrations in the absence of auxin. This too could help reverse the antiauxin effects of low levels of oligosaccharide.

The experimental evidence is therefore more and more convincing that the natural target for secreted plant cellulases in growing dicots is xyloglucan and that hydrolysis of this wall matrix polymer leads to wall loosening and growth. In this event, any treatment that protected xyloglucan from cellulase attack would be expected to inhibit growth. Hoson and colleagues showed (91,96) that fucose-binding lectins and antibodies specific for xyloglucans do indeed inhibit growth and the depolymerization of xyloglucan that accompanies it in bean epicotyls.

With respect to the role of cellulases in growing monocots, where xyloglucan levels are much lower than in dicots (90), it has been shown that xyloglucan also depolymerizes during growth especially after auxin treatment, e.g., in oat and rice coleoptiles (16,97). However, growing monocot walls also contain mixed 1,3/1,4-linked β-glucan which is susceptible to hydrolysis by plant cellulase (80) and specific mixed-linkage hydrolases (15,42). Antibodies to such β-glucans suppress the autodigestion of corn coleoptile β-glucan which normally occurs during growth and inhibit auxin-induced elongation (98). Thus, the hydrolysis of 1,4-β linkages in xylo-glucan and/or mixed-linkage glucan may well by key reactions that loosen cell walls in monocots during auxin-evoked growth (99).

ACKNOWLEDGMENTS

This survey was conducted with support from the Natural Sciences and Engineering Research Council in Canada. Thanks are due to the Bellairs Research Institute of McGill University in Barbados for providing facilities and Mr. Collis Barrow for assistance in obtaining data quoted in Table 3.

REFERENCES

1. D. P. S. Verma, V. Kumar, and G. Maclachlan, In *Cellulose and Other Natural Polymer Systems* (R. M. Brown, ed.), Plenum Press, New York, 1982, p. 459.
2. E. T. Reese, *Rec. Adv. Phytochem.*, *11*, 311 (1977).
3. D. J. Nevins, R. Hatfield, and Y. Kato, In *Structure, Function and Biosynthesis of Plant Cell Walls* (W. M. Dugger and S. Bartnicki-Garcia, eds.), Waverley Press, Baltimore, 1984, p. 167.
4. J. R. Woodward and G. B. Fincher, *Eur. J. Biochem.*, *121*, 663 (1982).
5. T. H. D. Jones and M. Gupta, *Biochem. Biophys. Res. Commun.*, *102*, 1310 (1981).
6. R. E. Christoffersen, M. L. Tucker, and G. G. Laties, *Plant Mol. Biol.*, *31*, 385 (1984).
7. M. L. Durbin, R. Sexton, and L. N. Lewis, *Plant Cell Env.*, *4*, 67 (1981).
8. D. P. S. Verma, G. A. Maclachlan, H. Byrne, and D. Ewings, *J. Biol. Chem.*, *250*, 1019 (1975).
9. R. Sexton, M. L. Durbin, L. N. Lewis, and W. W. Thompson, *Nature*, *283*, 873 (1980).
10. D. F. Fan and G. A. Maclachlan, *Plant Physiol.*, *42*, 1114 (1967).
11. T. Hayashi and G. A. Maclachlan, *Plant Physiol.*, *75*, 739 (1984).
12. E. Davies and G. A. Maclachlan, *Arch. Biochem. Biophys.*, *128*, 595 (1968).
13. A. K. Bal, D. P. S. Verma, H. Byrne, and G. A. Maclachlan, *J. Cell Biol.*, *69*, 97 (1976).
14. T. Koyama, T. Hayashi, Y. Kato, and K. Matsuda, *Plant and Cell Physiol.*, *22*, 1191 (1981).
15. D. J. Huber and D. J. Nivens, *Plant Physiol.*, *65*, 768 (1980).
16. M. Inouhe, R. Yamamoto, and Y. Masuda, *Plant Cell Physiol.*, *25*, 1341 (1984).
17. P. Bucheli, M. Durr, A. J. Buchala, and H. Meier, *Planta*, *166*, 530 (1985).
18. C. L. Nessler and B. Mahlberg, *Am. J. Bot.*, *68*, 730 (1981).
19. A. R. Sheldrake and G. F. J. Moir, *Physiol. Plant.*, *23*, 267 (1970).
20. D. P. S. Verma, V. Zogbi, and A. K. Bal, *Plant Sci. Lett.*, *13*, 137 (1978).
21. A. R. Sheldrake, *Planta*, *95*, 167 (1970).
22. L. Taiz and W. A. Honigman, *Plant Physiol.*, *58*, 384 (1976).

23. M. Awad and L. N. Lewis, *J. Food Sci.*, *45*, 1625 (1980).
24. A. B. Bennett and R. E. Christoffersen, *Plant Physiol.*, *81*, 830 (1986).
25. D. M. Pharr and D. B. Dickinson, *Plant Physiol.*, *51*, 577 (1973).
26. P. R. Crookes and D. Grierson, *Plant Physiol.*, *72*, 1088 (1983).
27. R. E. Paull and N. J. Chung, *Plant Physiol.*, *72*, 382 (1983).
28. M. Edwards, I. C. M. Dea, P. V. Bulpin, and J. S. G. Reid, *Planta*, *163*, 133 (1985).
29. M. E. Edwards, I. C. M. Dea, P. V. Bulpin, and J. S. G. Read, *J. Biol. Chem.*, *261*, 9489 (1986).
30. D. Reis, B. Vian, D. Darzens, and J. G. Rolland, *Planta*, *170*, 60 (1987).
31. R. Sexton and J. A. Roberts, *Ann. Rev. Plant Physiol.*, *33*, 133 (1982).
32. R. Sexton, M. L. Durbin, L. N. Lewis, and W. W. Thompson, *Protoplasma*, *109*, 335 (1981).
33. R. Sexton, L. N. Lewis, A. J. Trewavas, and P. Kelly, In *Ethylene and Plant Development* (J. A. Roberts and G. A. Tucker, eds.), Butterworths, London, 1985, p. 173.
34. J. Greenberg, R. Goren, and J. Riov, *Physiol. Plant*, *34*, 1 (1975).
35. G. K. Rasmussen, *Plant Physiol.*, *56*, 765 (1975).
36. T. Kanda, K. Wakabyashi, and K. Nisizawa, *J. Biochem.*, *79*, 989 (1976).
37. J. Labavitch, *Plant Physiol.*, *32*, 385 (1981).
38. G. A. Maclachlan, In *The New Frontiers and Future Perspectives of Plant Biochemistry* (T. Akasawa, T. Asahi, and H. Imaseki, eds.), Nijhoff/Junk, The Hague, 1983, p. 83.
39. D. E. Koehler, R. T. Leonard, W. J. Van Der Woude, A. E. Linkins, and L. N. Lewis, *Plant Physiol.*, *58*, 324 (1976).
40. D. J. Osborne, *Sci. Hort.*, *30*, 31 (1979).
41. R. Ridge and D. J. Osborne, *Nature*, *223*, 318 (1969).
42. Y.-S. Wong, G. B. Fincher, and G. A. Maclachlan, *J. Biol. Chem.*, *252*, 1402 (1977).
43. R. Hatfield and D. J. Nevins, *Plant Cell Physiol.*, *27*, 541 (1986).
44. G. A. Maclachlan and Y. S. Wong, *Adv. Chem. Series*, *181*, 347 (1979).
45. Y. Lienart, J. Comtat, and F. Barnoud, *Plant Sci.*, *41*, 91 (1985).
46. B. S. Valent and P. Albersheim, *Plant Physiol.*, *54*, 105 (1974).

47. T. Hayashi and G. A. Maclachlan, *Plant Physiol.*, 75, 596 (1984).
48. T. Hayashi, Y.-S. Wong, and G. A. Maclachlan, *Plant Physiol.*, 75, 605 (1984).
49. T. Hayashi and G A. Maclachlan, In *Cellulose: Structure, Modification and Hydrolysis* (R. A. Young and R. Rowell, eds.), Wiley-Interscience, New York, 1986, p. 67.
50. P. J. Moore, A. G. Darvill, P. Albersheim, and L. A. Staehelin, *Plant Physiol.*, 82, 787 (1986).
51. T. Hayashi, Y. Kato, and K. Matsuda, *Plant Cell Physiol.*, 21, 1405 (1980).
52. P. Albersheim, In *Plant Biochemistry* (J. Bonner and J. E. Varner, eds.), Academic Press, New York, 1976, p. 225.
53. R. Guiguard and R. E. Pilet, *CR Acad. Sci. Paris*, 286, 855 (1978).
54. K. Nishitani and Y. Masuda, *Plant Cell Physiol.*, 24, 345 (1983).
55. J. S. G. Reid, In *Biochemistry of Plant Cell Walls* (C. T. Brett and J. R. Hillman, eds.), Cambridge University Press, Cambridge, 1985, p. 259.
56. G. A. Maclachlan, *Trends Biochem. Science*, 2, 226 (1977).
57. F. S. Spencer and G. A. Maclachlan, *Plant Physiol.*, 49, 58 (1972).
58. W. Blaschek, H. Koehler, U. Semler, and G. Franz, *Planta*, 154, 550 (1982).
59. J. M. Labavitch and P. M. Ray, *Plant Physiol.*, 53, 669 (1974).
60. J. M. Labavitch and P. M. Ray, *Plant Physiol.*, 54, 499 (1975).
61. A. Camirand, D. Brummell, and G. A. Maclachlan, *Plant Physiol.*, 84, 753 (1987).
62. S. Fry, *Planta*, 169, 443 (1986).
63. M. E. Terry and B. A. Bonner, *Plant Physiol.*, 66, 321 (1980).
64. M. E. Terry, D. McGraw, and R. L. Jones, *Plant Physiol.*, 69, 323 (1982).
65. B. Dopico, E. Labrador, and G. Nicholas, *Plant Sci. Letts.*, 44, 155 (1986).
66. E. P. Lorence, J. L. Acebes, and I. Zarra, In *Cell Walls, 86, Proc. 4th Cell Wall Meeting, Paris* (B. Vian, D. Reis, and R. Goldberg, eds.), Presse de l'Ecole Normale Superieure, Paris, France, 1986, p. 270.
67. B. Stone, In *Structure, Function and Biosynthesis of Plant Cell Walls* (W. M. Dugger and S. Bartnicki-Garcia, eds.), Waverley Press, Baltimore, 1984, p. 52.

68. Y. Kato and D. J. Nevins, *Plant Physiol.*, *75*, 740 (1984).
69. D. G. Luttenegger and D. J. Nevins, *Plant Physiol.*, *77*, 175 (1985).
70. A. E. Ahmed and J. M. Labavitch, *Plant Physiol.*, *65*, 1009 (1980).
71. D. J. Huber, In *Structure, Function and Biosynthesis of Plant Cell Walls* (W. M. Dugger and S. Bartnicki-Garcia, eds.), Waverley Press, Baltimore, 1984, p. 449.
72. E. Pesis, Y. Fuchs, and G. Zauberman, *Plant Physiol.*, *61*, 416 (1978).
73. F. T. Addicott, *Abscission*, University of California Press, Berkeley, 1982, p. 153.
74. T. C. Hanish, C. H. Van Netter, J. F. Dortland, and J. Bruinsma, *Physiol. Plant.*, *33*, 276 (1975).
75. M. Huberman and R. Goren, *Physiol. Plant.*, *45*, 189 (1979).
76. A. J. Trewavas, R. Sexton, and P. Kelly, *J. Embryol. Exp. Morph.*, *1983 Supplement*, 179 (1984).
77. G. A. Maclachlan, In *Cellulose and Other Natural Polymer Systems* (R. M. Brown, ed.), Plenum Press, New York, 1982, p. 327.
78. W. Blaschek, D. Haass, H. Koehler, U. Semler, and G. Franz, *Z. Pflanzenphysiol. Bd.*, *111*, 357 (1983).
79. L. A. Rivas and R. Pont Lezica, *Eur. J. Biochem.*, *163*, 129 (1987).
80. Y.-S. Wong, G. B. Fincher, and G. A. Maclachlan, *Science*, *195*, 679 (1977).
81. E. Andaluz, A. Guillen, and G. Larriba, *Biochem. J.*, *240*, 495 (1986).
82. P. Albersheim, A. G. Darvill, M. McNeil, B. S. Valent, J. K. Sharp, E. A. Nothnagel, K. R. Davis, N. Yamazaki, D. J. Gollin, W. S. York, W. F. Dudman, J. E. Darvill, and A. Dell, In *Structure and Function of Plant Genomes*, Plenum Press, New York, 1983, p. 293.
83. P. Albersheim and A. G. Darvill, *Sci. Am.*, *253*, 58 (1985).
84. W. S. York, A. G. Darvill, and P. Albersheim, *Plant Physiol.*, *75*, 295 (1984).
85. G. J. McDougall and S. C. Fry, In *Food Hydrocolloids*, *1*, 505 (1987).
86. G. A. Maclachlan, *Meth. Enzymol.*, *160*, 382 (1988).
87. M. L. Durbin and L. N. Lewis, *Meth. Enzymol.*, *160*, 342 (1988).
88. D. A. Brummel and G. A. Maclachlan, In *Plant Cell Wall Polymers: Biogenesis and Biodegradation* (N. G. Lewis and M. G. Paice, eds.), Am. Chem. Soc. Symp. Series, Vol.. 399, 18 (1989).

89. S. C. Fry, *J. Exp. Bot.*, *40*, 1 (1989).
90. T. Hayashi, *Ann. Rev. Plant Physiol. Mol. Biol.*, *40*, 139 (1989).
91. S. C. Fry, C. T. Brett, and J. S. G. Reid, *Abstracts 5th Cell Wall Meeting*, Edinburgh, UK, Sept. 1989.
92. G. J. McDougall and S. C. Fry, *Planta*, *175*, 412 (1988).
93. G. J. McDougall and S. C. Fry, *Plant Physiol.*, *89*, 833 (1989).
94. G. J. McDougall and S. C. Fry, *J. Exp. Bot.*, *40*, 233 (1989).
95. V. Farkas and G. A. Maclachlan, *Carboh. Res.*, *184*, 213 (1988).
96. T. Hoson and T. Masuda, *Physiol. Plantarum*, *71*, 1 (1987).
97. G. Revilla and I. Zarra, *J. Exp. Bot.*, *38*, 1818 (1987).
98. T. Hoson and D. J. Nevins, *Plant Physiol.*, *90*, 1353 (1989).
99. S. C. Fry, *Physiol. Plantarum*, *75*, 532 (1989).

24

Molecular Cloning of Cellulase Genes into Noncellulolytic Microorganisms

A. LEJEUNE and C. COLSON   *Catholic University of Louvain, Louvain-la-Neuve, Belgium*

D. E. EVELEIGH   *Rutgers University, New Brunswick, New Jersey*

I. INTRODUCTION

The study of microbial cellulose degradation is relevant to both ecological and biotechnological concerns. Cellulose is a major component of vegetation, and the bioconversion of this polymer to other forms is an important aspect of the carbon cycle and thus merits special attention. Likewise, the widespread interest in the use of biomass as a feedstock for the production of energy, chemicals, food, and feedstuffs has focused further studies on cellulases.

Extensive information is available on the characteristics and mode of action of fungal cellulases, especially those of *Trichoderma reesei* (See Chap. 21). Studies on bacterial cellulases have developed more recently, and it clearly appears that some bacteria possess cellulase systems that do not resemble the fungal model. In particular, reports on cellobiohydrolases in bacteria are rare and synergistic actions between bacterial cellulase components have only recently been clarified (see Chap. 22). The molecular cloning of cellulase genes into noncellulolytic microorganisms is even more recent with a first report in 1978 (1). Research in this field has expanded tremendously over the past few years and has been predominantly devoted to prokaryotic rather than eukaryotic genes.

The reasons for cloning cellulase genes are many fold (Fig. 1). Of importance are such fundamental questions as the molecular

624

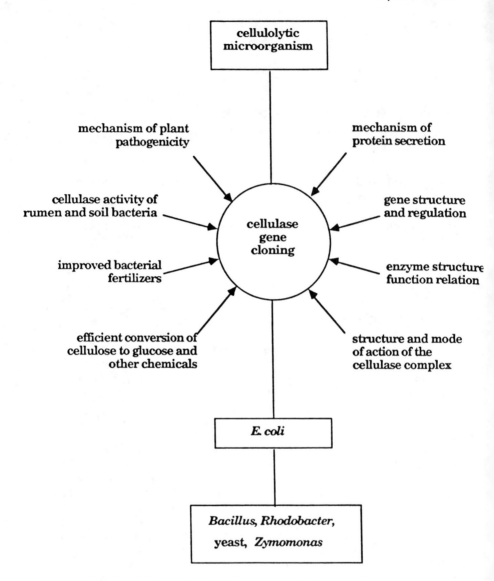

FIGURE 1   Objectives in cellulase gene cloning.

study of gene structure, the relation between enzyme structure and
function, protein secretion, the regulation of gene expression, and
the molecular relation between multiple enzyme forms and single or
multiple genes. Also under study is the evolution of genes and enzymes. The purification of the cellulase components of certain organisms by classical biochemical techniques has also proven problematic because they occur as, and form during purification of, high
molecular weight aggregates. In these instances, the cloning approach has been essential in the characterization of individual enzymes.

More applied concerns have been invoked for the cloning of cellulase genes: studying the genetic and physiological mechanisms of
attack by plant pathogens and detailing the activities of soil and
rumen bacteria. Efficient conversion of cellulosic materials of industrial, agricultural or domestic wastes into useful products is an
important goal. The reconstruction of highly active, extracellular
cellulase complexes in readily grown microorganisms containing cloned
cellulase genes, possibly of different origins, is a trend toward this
goal. Thus, microbes with special attributes have been chosen as
hosts in cellulase gene cloning in order to achieve the direct conversion of cellulose to ethanol or other chemicals, or even to improve
the quality of bacterial fertilizers. Other glycosidase genes have
been cloned, such as those having applications in the brewing and
paper industries.

Cellulase is commonly considered as comprising cellobiohydrolase
(1,4-β-D-glucan cellobiohydrolase, EC 3.2.1.91) and endoglucanase
(1,4-β-D-glucan glucanohydrolase, EC 3.2.1.4) activities. In this
chapter the molecular cloning of genes coding for these enzymes is
reviewed, irrespective of the ability for growth of the donor organism on crystalline cellulose. Although enzymes active toward cellobiose (β-glucosidases: β-D-glucoside glucohydrolase, EC 3.2.1.21,
and cellobiose phosphorylase, EC 2.4.1.20) are not usually defined
as being in the cellulase complex, their action is critical in the overall conversion of cellulose to glucose. Therefore, the cloning of
such genes is also included in this chapter and the term "cellulase"
refers to all these types of enzymes.

The next section considers the methodology of gene cloning.
Standard genetic engineering techniques are not described; rather
the methodological aspects specifically related to cellulase gene cloning are discussed. In the third section the focus is on the characterization and expression of bacterial cellulase genes and the characterization of the corresponding cellulases in *Escherichia coli*, the
host most widely used for primary cloning experiments. The primary, homologous cloning of cellulase genes in *Streptomyces*

lividans is also included in this section. Section IV is a report on
the cloning of eurkaryotic cellulase genes in *E. coli*. Section V describes the expression of both prokaryotic and eukaryotic cellulase
genes in non-*E. coli* hosts chosen for specific applications. The
cloning of related genes (endo-β-1,3/1,4-glucanase genes, xylanase,
and β-xylosidase genes and ligninase genes) is briefly reviewed in
Sec. VI, followed by a final section of general comments and perspectives.

II.   THE CLONING OF CELLULASE GENES:
THE METHODOLOGY

A general and simplified scheme of the gene-cloning procedure is
outlined in Fig. 2. Chromosomal DNA is extracted from the donor
(a bacterium in Fig. 2, any organism in general) and cleaved with
a restriction enzyme. The plasmid vector purified from *E. coli* is
linearized with the same enzyme. The vector generally carries two
genetic markers (antibiotic resistances), one of which is disrupted
by the linearization event. The chromosomal fragments and the linearized plasmid are mixed, ligated, and the products used to transform *E. coli* to the antibiotic resistance encoded by the plasmid.
The *E. coli* colonies harboring recombinant vs. recircularized plasmids are selected by controlling the inactivation of the second plasmid genetic marker. In a final step, the gene bank obtained is
screened for the gene of interest either by expression of this gene
or by DNA/DNA hybridization. The same scheme applies when bac-
teriophage or cosmid vectors are used instead of plasmids, except
that transformation is replaced by in vitro encapsidation into phage
particles followed by infection of the *E. coli* cells.

An overview of the methodology used for the cloning of cellulase
genes from both prokaryotes and eukaryotes is outlined in Table 1
and critical points are discussed in more detail in this section.

A.   Cloned Cellulases

If the cellulase complex is considered as comprising the enzyme activities of endoglucanase, cellobiohydrolase (exoglucanase), cellobiase, and cellobiose phosphorylase, then it is noteworthy that the
genes coding for all these enzymes have been cloned. The most
numerous cellulase genes cloned to date are the endoglucanases.
This is probably attributable to the fact that (1) endoglucanases
are the most commonly found among cellulolytic microorganisms,
(2) there are often several different endoglucanases which are found

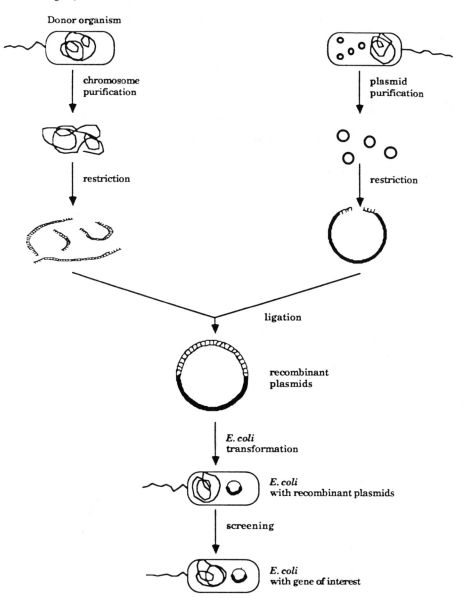

FIGURE 2   General procedure for cloning in *E. coli*.

in a cellulolytic microorganism, and (3) cloned endoglucanase genes
are easy to detect if expressed in the host organism.

While cellobiohydrolase enzymes are clearly present in many cellulolytic eukaryotes, they are reported rather infrequently in prokaryotes, e.g., in *Microbispora bispora* and *Thermomonospora fusca*.
Cloned cellobiohydrolases include those from the fungus *Trichoderma
reesei* and the bacteria *Clostridium thermocellum* and *Cellulomonas
fimi* (Table 1).

The degradation of cellulose by endoglucanases and cellobiohydrolases results mainly in cellobiose and cellulodextrins. These
products are further degraded by cellobiases and/or cellobiose phosphorylases. The former can be more generally classified with the
β-glucosidases, but one must bear in mind that not all β-glucosidases active on arylglucosides are active toward cellobiose (true cellobiases). Whenever possible, the specific terms cellobiase or cellobiose phosphorylase have been used in Table 1. If the relative substrate specificity of such enzymes is not reported but they are active
on cellobiose, then this has been explicitly cited in Table 1.

B. Donor Organisms

The vast majority of organisms from which cellulase genes have been
cloned are bacteria. An obvious reason for this is that since the
cloning generally proceeds in *E. coli*, the use of screening methods
based on the expression of the cloned bacterial cellulase genes does
not make it necessary to go through a complementary DNA (cDNA)
construction step to remove possible introns.

The choice of the donor organism is much determined by the
particular aspect of cellulose degradation under study. Thus, *Erwinia chrysanthemi* was chosen to study the role and importance of
cellulases in its pathogenicity toward plants, while *Fibrobacter succinogenes* and *Ruminococcus* spp. were studied as significant rumen
bacteria. When the efficiency of conversion of cellulose to useful
products is the primary goal, microbes are chosen whose cellulase
properties include high activity along with an extracellular and/or
thermostable nature, or with a lack of end-product inhibition, e.g.,
fungi include *Schizophyllum commune* and *Trichoderma reesei* and
the bacteria *Agrobacterium* sp., *Caldocellum saccharolyticum*, *Cellulomonas fimi*, *C. uda*, *Clostridium thermocellum*, *Microbispora bispora*, *Pseudomonas* spp., *Thermomonospora fusca*, and others.

More basic research on the structure and expression of glycosidase
genes was conducted with the yeasts *Candida pelliculosa* and *Kluyveromyces fragilis*, the fungus *Aspergillus niger*, and various bacteria including *Bacillus* spp., *Clostridium acetobutylicum*, *Escherichia adecarboxylata*, and *Pseudomonas* sp. The discrimination between basic and applied research is conceptual and both approaches
coexist.

C. Hosts for Primary Cloning

Escherichia coli is the organism for which the most numerous and
versatile genetic tools are available, e.g., many cloning, expression,
and sequencing vectors, different strains with various genetic properties, and good transformation capability. Thus, it has thus far
been the host used in the first stage of virtually all cloning experiments involving cellulase genes. Two other bacteria have been used
as hosts for primary homospecific cloning of endoglucanase genes,
namely, *Bacillus subtilis* (7) and *Streptomyces lividans* (31). In
some cases, subsequent experiments led to the cloning of cellulase
genes in other hosts exhibiting specialized properties (see Sec. V).

D. Vectors Used for Primary Cloning

The vectors used for primary cloning experiments fall into two categories according to the in vitro and in vivo cloning procedures used.

1. Vectors for In Vitro Cloning

These vectors are mainly plasmids but also include phages and
cosmids.
      The most widely used plasmid vectors are classical, high-copy-
number, narrow-host-range *E. coli* plasmids, such as pBR322.
Cloning-sequencing plasmids of the pUC series have also been used
(12,33), as well as a positive-selection plasmid, where only *E. coli*
cells transformed with recombinant plasmids (but not with recircularized plasmids) will grow (18,30). The use of a *B. subtilis/
E. coli* shuttle plasmid (6) and a mobilizable, broad-host-range
plasmid, allowing the transfer of the cloned genes to virtually all
gram-negative bacteria (27), is another useful approach. When
B. subtilis or *S. lividans* rather than *E. coli* was the primary host
for cloning, specific plasmids for these bacteria were used (7,31).
      The advantage of phage vectors over plasmids is that large
DNA fragments can be cloned and the efficiency of in vitro encapsidation vs. transformation is higher. Furthermore, as lysis of the
host occurs, the release of the glycosidases into the medium can
facilitate the screening procedure. Positive selection phages where
only recombinants will grow and expression phages have been used
successfully (14,20,24,41).
      Cosmids have some of the advantages of both plasmid and phage
vectors, although recombinants sometimes show a high structural
instability. Both narrow- and broad-host-range cosmids have been
used (19,22).

2. Vectors for In Vivo Cloning

These vectors are based on broad-host-range plasmids harboring an
R-prime formation, or chromosome-mobilizing, ability. Very large

TABLE 1 The Cloned Cellulase Genes

Enzyme	Donor organism	Host–vector system	First recombinant plasmid or strain	Gene designation	Detection of clones	Ref.
Cellobiohydrolase	Cellulomonas fimi ATCC 484	E. coli/plasmid pBR322	pDW1	cex	Immunoassay	2
			pEC1	cex	Immunoassay	3
	Clostridium thermocellum NCIB 10682	E. coli/plasmid pACYC184	pCT401 pCT1200 pCT1300		Fluorescence under UV light on plates with MUC as the substrate	4
	F7	E. coli/plasmid pBR322	pCU402		Activity on amorphous cellulose but not on CMC	5
Endoglucanase	Bacillus subtilis DGL	E. coli/plasmid pPL1202	pLG4000		Immunoassay	6
	IFO 3034	B. subtilis/plasmid pBD64	pBC5		Congo red assay on plates with CMC	7
	PAP115	E. coli/plasmid pUC8	pC6.5			8

Organism	Host/vector	Clone	Gene	Detection method	No.
Bacillus sp. N-4	*E. coli*/plasmid pBR322	pNK1 pNK2		Shallow craters on plates with CMC	9
1139	*E. coli*/plasmid pBR322	pFK1		Shallow craters on plates with CMC	10
—	*E. coli*/plasmid pUC8	pBS1 pBS1		Congo red assay on plates with CMC	11
Fibrobacter succinogenes ATCC 19169	*E. coli*/plasmid pUC8	BC14 and 5 others unnamed			12
Butyrivibrio sp. A46	*E. coli*/phage λgt11	BA46		Congo red assay on plates with CMC	13
Caldocellum saccharolyticum	*E. coli*/phage λ1059	pNZ1008		Congo red assay on plates with CMC	14
Cellulomonas fimi ATCC 484	*E. coli*/plasmid pBR322	pEC2	Portion of cenA	Immunoassay	
		pEC3		Immunoassay	
		pcEC2	cenA	Congo red assay on plates with CMC	15

TABLE 1   (Continued)

Enzyme	Donor organism	Host–vector system	First recombinant plasmid or strain	Gene designation	Detection of clones	Ref.
Endoglucanase (cont)	Cellulomonas uda CB4	E. coli/plasmid pAT325	pCM41		Congo red assay on plates with CMC	16
	Cellvibrio mixtus UQM2294	E. coli/cosmid pHC79	pJP549		Congo red assay on plates with CMC	17
	Clostridium acetobutylicum P262	E. coli/plasmid pEcoR251	pHZ100		Congo red assay on plates with CMC	18
	Clostridium thermocellum NCIB 10682	E. coli/cosmid pHC79	pCT182 (pCT1)	celA	Viscometric assay	19
			pCT262 (pCT2)	celB		
		E. coli/plasmid pACYC184	pCT113	celA	Congo red assay on plates with CMC	4
			pCT210	celB		
			pCT301	celC		
			pCT500			

Organism/host	Clone	Gene	Detection method	Reference
E. coli/ phage λ1059	pCT600	*celD*		
	pCT700, pCT800			
	CEL16	*celA*	Congo red assay on plates with CMC	20
	LIC7		Congo red assay on plates with lichenan	
E. coli/ plasmid pBR322	pRV1.1, pRV1.5, pRV1.6, pRV1.7, pRV2.1, pRV2.2, pRV2.3, pRV2.4, pRV2.5, pRV2.6, pRV2.9, pRV2.10, pRV2.18		Congo red assay on plates with CMC	21
F7	pCU101	*cel1*	Congo red on plates with CMC	5
E. coli/ plasmid pBR322	pUC103	*cel2*		
	pUC104	*cel3*		
Erwinia chrysanthemi 3665	pFB293	*celY*	Congo red assay on plates with CMC	22
E. coli/ cosmid pMMB34				23
E. coli/ plasmid pUC18	pMH13	*celZ*	Hybridization with a DNA fragment of *E. chrysanthemi* 3937 carrying	23
		celZ		

TABLE 1 (Continued)

Enzyme	Donor organism	Host–vector system	First recombinant plasmid or strain	Gene designation	Detection of clones	Ref.
Endoglucanase (cont)	*Erwinia chrysanthemi* 3937	*E. coli*/phage λL47.1	pMH17	*celY*	Congo red assay on plates with CMC	24
			pMH7	*celZ*		23
	B374	*E. coli*/plasmid pULB113		*cel*	Congo red assay on plates with CMC	25
	Microbispora bispora	*E. coli*/plasmid pBR322	pCC1528, pCC1539, pCC3622, pCC5188, pCC6704		Congo red assay on plates with CMC	26
	Pseudomonas fluorescens var. *cellulose* NCIB 10462	*E. coli*/plasmid pSUP104	pRUCL100		Congo red assay on plates with CMC	27
		E. coli/plasmid pRB322	pPFC4		Congo red assay on plates with CMC	28
	Pseudomonas fluorescens AC501	*E. coli*/plasmid XYL-K				1

Source organism	Host / plasmid	Plasmid	Gene	Assay	Ref.
Pseudomonas sp. NCIB 8634	*E. coli* / plasmid pBR322	pPC71		Congo red assay on plates with CMC	28
Ruminococcus albus SY3	*E. coli* / plasmid pBR322	RA1 RA2		Congo red assay on plates with CMC	13
F-40	*E. coli* / plasmid pBR322	pRA1		Congo red assay on plates with CMC	19
Ruminococcus flavefaciens FD1	*E. coli* / plasmid pEcoR251	pMEB200		Congo red assay on plates with CMC	30
Streptomyces lividans 1326	*S. lividans* / plasmid pIJ702	pIAF9 pIAF74		Congo red assay on plates with CMC	31
Thermomonospora fusca YX	*E. coli* / plasmid pBR322	pD316 pD365		Reducing sugar assay	32
Agrobacterium sp. ATCC 21400	*E. coli* / plasmid pUC13	pABG1		DNA/DNA hybridization	33
Cellobiase					
E. coli CSH62CT	*E. coli* / plasmid pULB113	pUF520	*celH*	Growth on mineral medium with cellobiose	34

TABLE 1 (Continued)

Enzyme	Donor organism	Host–vector system	First recombinant plasmid or strain	Gene designation	Detection of clones	Ref.
Cellobiose phosphorylase	E. adecar-boxylata ATCC 23216	E. coli/ plasmid pBR322	pBA2		Growth on mineral medium with cello-biose as the sole carbon source	36
Glucosidase active towards cellobiose	Caldocellum saccharoly-ticum	E. coli/ phage λ1059	pNZ1001		Blue plaques on plates with Xglu	37
	Cellulomonas uda CB4	E. coli/ plasmid pAT325	pCC1		Blue colonies on plates with Xglu	38
	Cellvibrio mixtus UQM2234	E. coli/ cosmid pHC79	pJP549		Browning of medium in plates with esculin	17
	Clostridium acetylbutyl-icum P262	E. coli/ plasmid pEcoR251	pHZ100		Growth on mineral medium with cello-biose as the sole carbon source	18
	Clostridium thermocellum NUB 10682	E. coli/ phage λ1059	LIC42 MUG16		In vitro activity on cellobiose fluores-cence on plates with MUG under UV light	20

	Organism/strain	Host/vector	Plasmid	Detection	Ref.
	Erwinia chrysanthemi 3665	E. coli/cosmid pMMB34	clb		22
	B374	E. coli and E. chrysanthemi plasmid pUBL113	pBEC2	clb	Growth on mineral medium cellobiose as the sole carbon source	39
Aryl-β-glucosidase[a]	Cellulomonas uda CB4	E. coli/plasmid pAT325	pCG1	Blue colonies on plates with Xglu	38
	Clostridium thermocellum NCIB 10682	E. coli/plasmid pBR322		Blue colonies on plates with Xglu	40
	F7	E. coli/plasmid pBR322			5
	Microbispora bispora	E. coli/plasmid pBR322		Fluorescence on plates with MUG under UV light	26

[a]Not tested on cellobiose.

DNA fragments (up to about 200 kb) can be transferred by conjugation, but their use is restricted to transfers between gram-negative bacteria (1,25,34,39).

E. Cloning and Screening Techniques

Chromosomal DNA has been prepared by classical purification methods and then partially restricted with restriction enzymes if large DNA fragments were to be cloned (indispensable for cosmid or phage vectors and sometimes used for plasmid vectors). The screening techniques used to detect clones containing cellulase genes can be divided into two groups, depending on the expected expression or not of the cellulase gene.

1. Detection of Positive Clones by DNA Hybridization

DNA hybridization has been used in a few cases when the expression of the cloned gene was not achieved. Two reports concern the cellobiohydrolase I gene of *Trichoderma reesei*. The detection of the *E. coli* clones carrying this gene was based on the differential hybridization of the individual clones of the gene bank with two radioactive cDNA probes made from mRNA preparations of cellulase-induced and uninduced cultures of *T. reesei* (42,43). A similar technique was used to detect the cloned endoglucanase I gene from the same organism (44,45). The cellobiase gene of *Agrobacterium* sp. is not expressed in *E. coli*. Thus screening consisted of purification of the cellobiase, partial determination of its amino acid sequence, and then construction of the corresponding synthetic radioactive oligonucleotide probe which was used to screen the gene bank (33). The cloned cellobiohydrolase II gene from *T. reesei* was isolated analogously (46).

2. Detection of Positive Clones by Gene Expression

Direct expression of the cloned cellulase genes has been obtained both naturally or by using expression vectors where the cloned gene is transcribed and/or translated from initiation sites present in the vector.

Endoglucanase genes are routinely detected by the Congo red assay procedure, which is based on the binding of the dye to carboxymethylcellulose (CMC). The latter is a soluble derivative of cellulose and is readily and specifically hydrolyzed by endoglucanases. The colonies containing recombinant plasmids (or the plaques in the case of phage vectors) are either grown on a medium containing CMC (47) or covered after growth with an overlay of soft agar containing CMC (48). After incubation, the plates are flooded with a Congo red solution and then rinsed with saline. A clear halo around certain colonies on an otherwise red background is indicative of the production of endoglucanase. In two cases, endoglucanase genes were detected by the formation of shallow craters around active colonies on plates with CMC (9,10).

Immunoassays have been based on the use of crude concentrated culture supernatants or of purified enzyme to prepare antisera from rabbits. After lysis of the gene bank colonies, they were successively incubated with the antiserum and with ^{125}I-labeled protein A that binds to the antibodies. The positive clones are selected as being radioactive (3). Alternatively, the lysed colonies can be incubated with the antiserum and then with goat anti-rabbit IgG which can be enzymatically detected (enzyme-linked immunosorbent assay, ELISA) (6). Selection can also be performed on crude extracts of individual clones that are incubated with a CMC solution. Either a decrease in viscosity (19) or the production of reducing sugars (32) is then monitored. However, in the latter case the "endo"-acting enzyme must be confirmed in that cellobiohydrolase may also produce reducing sugars, while not significantly lowering viscosity.

Cellobiohydrolase genes have been detected in two ways: (1) by radioimmunoassay with antiserum prepared from concentrated crude culture supernatant obtained from the donor organism, as mentioned above (2,3) and (2) by incubating the cells with differential substrates, methylumbelliferyl-β-D-cellobioside (MUC), a substrate initially thought to be specifically cleaved by some cellobiohydrolases (49), and with CMC, indicative of endoglucanases. Clones that are MUC-positive (by showing a blue fluorescence under ultraviolet light) and CMC-negative (by the Congo red assay procedure) were initially considered as expressing cellobiohydrolase (4). However, MUC is now realized not to be specific for cellobiohydrolases and J.-P. Aubert's group at the Pasteur Institute has isolated an endoxylanase clone based on the above criteria (unpublished observations).

β-Glucosidase genes have been detected both by positive selection, i.e., the growth of the recombinant clones on a mineral medium with cellobiose as the sole carbon source (only those clones carrying a β-glucosidase active towards cellobiose will grow), and by the use of chromogenic substrates such as methylumbelliferyl-β-D-glucose (MUG) (138) or 5-bromo-4-chloro-3-indolyl-β-D-glucopyranoside (Xglu).

Since Congo red binds not only to CMC but to other polysaccharides as well (50,51), this dye-binding procedure has also been used to detect cloned genes for endo-β-1,3/1,4-glucanases (20,52,53) and endo-β-1,3-glucanase (20). Even without the addition of a dye, the appearance of clearing zones around colonies on plates

containing lichenan, a β-1,3/1,4-glucan, was taken advantage of
to detect the production of endo-β-1,3/1,4-glucanase (54), and the
appearance of clearing zones on plates containing xylan was used
to detect the production of xylanases. The cloned *Clostridium
thermocellum celE (xylY)* gene also yielded such clearing.

3. Discussion

The cloning of genes encoding enzymes of the cellulase complex has
been approached in a direct manner, i.e., active genes have been
cloned. One of the authors (A. Lejeune, unpublished results),
working with a wild-type cellulolytic *Pseudomonas* sp. isolated from
a compost, proposed a different strategy. The rationale was to
isolate a cellulase-negative mutant of the *Pseudomonas* by transposon
mutagenesis and to clone the transposon marker and thus the inactivated gene into the wild-type *Pseudomonas* where general recombination would lead to an exchange between the mutated gene on
the plasmid and the active gene in the chromosome. This newly
in vivo-generated plasmid with the active gene would then be genetically analyzed. This strategy is currently limited by the unavail-
ability of truly cellulase-negative mutants. This lack is likely due
to the multiplicity of cellulase genes (see Sec. III) among other
reasons. In contrast, Mondou et al. (55) isolated a cellulase-negative mutant *Streptomyces lividans* and Barras et al. (39) obatined a
cellobiase-negative mutant of *Erwinia chrysanthemi*, each by chemical mutagenesis, which were used as hosts for the cloning of the
active, unmutated genes (31,39). Somewhat analogously, an endoglucanase gene from *Bacillus subtilis* was cloned into another *B. subtilis* strain that already produced an endoglucanase (7). The cloned
gene was detected by the Congo red assay procedure by the larger
clearing zones that it caused around the colonies, compared to the
untransformed *Bacillus* host.

      Several points are pertinent in the screening procedures de-
scribed above. Methods that involve DNA hybridization, though extremely sensitive are more complex and also delicate to use. Thus
screening by direct expression may be preferable. Immunoassays
are also senstitve but require the prior purification of the enzyme
whose gene is to be cloned. This cannot always be achieved, e.g.,
in the case of *C. thermocellum* (4) where proteins involved in the
degradation of cellulose from high molecular weight aggregates during purification. Antisera were also prepared from unpurified enzyme solutions (2,5). In this case, "false positive" clones may appear as the solution used to raise the antiserum also contains proteins not involved in cellulose degradation.

      The Congo red assay procedure (47,48) for the detection of en-
doglucanases is facile, can be rapidly used to screen very large
gene banks, gives more clear-cut results than some other in situ

assays, and is specific to endoglucanases (50). Cellobiohydrolases also attack CMC at the nonreducing termini that contain a few unsubstituted residues, but this does not decrease the cellulose molecule chain length significantly and thus will not prevent dye binding.

To screen using the Congo red assay procedure, bacterial colonies containing an endoglucanase gene must degrade CMC in the medium i.e., must release at least part of the enzyme. Thus phage-cloning vectors that lyse the cells or an *E. coli* host lysogenic for a phage (25,39) have sometimes been advocated. In the latter case, the genes have been cloned with a plasmid or a cosmid and the prophage induced to its lytic cycle prior to screening the bank. However, this proved unnecessary because of the high sensitivity of the Congo red assay. Even the release of minuscule amounts of enzyme (e.g., by autolysis of some cells in stationary phase) not detected by routine reducing sugar tests can be visualized via the Congo red protocol. The detection of cloned endoglucanase genes by viscometric measurements is analogous in its action to detection via the Congo red procedure, with initial reaction being toward unsubstituted oligomeric sections. However, although the viscometric assay is more laborious than the Congo red procedure, it may be even more sensitive. Thus Shareck et al. (31) selected endoglucanase-positive clones in *Streptomyces lividans* by viscometry using culture filtrates but gained no clearing zone with Congo red.

The use of MUC to detect clones with cellobiohydrolase activity deserves comment. As Van Tilbeurgh et al. (49) report, MUC is cleaved by cellobiohydrolase I of *Trichoderma reesei* but not by an endoglucanase of the same fungus. However, several endoglucanases of *C. thermocellum* whose genes were cloned in *E. coli* were shown to be active on MUC (4,56,57). Millet et al. (1985) used both CMC and MUC as substrates to screen a gene bank of *C. thermocellum* in *E. coli* for the presence of cellobiohydrolase genes, looking at a CMC-MUC phenotype. Though easy and convenient, this procedure is now realized not to be an absolute test. As noted, the Pasteur Institute cellulase group also isolated a xylanase gene by these criteria. Thus confirmation of the cellobiohydrolase activity must be gained by other methods.

When growth on a mineral medium with cellobiose as the sole carbon source is used to isolate β-glucosidase recombinant clones, one can be sure that the enzyme is active toward cellobiose. This procedure has been used to isolate several genes but in one instance the gene most likely encodes a cellobiose phosphorylase (36). Chromogenic substrates (p-nitrophenyl-β-D-glucose or MUG) have often been used for their convenience and give visually clear results. Here again, the activity of the selected enzymes toward cellobiose must be confirmed, as not all aryl-β-glucosidases are active against cellobiose.

III. CHARACTERIZATION AND EXPRESSION OF PROKARYOTIC CELLULASE GENES IN *E. COLI*

This section reviews prokaryotic cellulase genes cloned in *E. coli* and cellulase enzyme products. It is arranged in subsections according to cellulolytic donor organisms, the last subsection discussing distinctions obtained with the various donors.

A. *Agrobacterium* sp.

Wakarchuk et al. (33) cloned a cellobiase gene from *Agrobacterium* sp. into *E. coli*. Their final objective was to reconstruct an optimized cellulase complex for the production of glucose from cellulose, with the other cellulase components being obtained from *Cellulomonas fimi* cloned genes (see Sec. III.E). This particular gene was chosen, among other reasons, because the high affinity of the corresponding enzyme for cellobiose and the possible synergy between this enzyme and the *C. fimi* cellulases, as the two organisms were originally isolated in coculture and synergistic action was noted in this natural state. The gene was first cloned in *E. coli* on a recombinant plasmid pABG1 directing the synthesis of low levels of intracellular cellobiase of 50 kD MW. The transcription promoter was located in the cloned DNA fragment. Determination of the direction of transcription of the gene and trimming of the DNA fragment allowed the expression of the gene from the *E. coli lacZ* gene promoter with, upon induction, an increase in production of more than 100-fold compared with the original enzyme level produced by the wild *Agrobacterium* sp.

B. *Bacillus* spp.

Sashihara et al. (9) cloned two endoglucanase genes from an alkalophilic *Bacillus* sp. N-4 on recombinant plasmids pNK1 and pNK2 encoding thermostable enzymes (75°C for 10 min) of MW 58 and 50 kD, respectively. Fukomori et al. (10) cloned an endoglucanase gene from another alkalophilic *Bacillus* sp. 1139 (plasmid pFK1; MW of the enzyme = 94 kD). Though there are certain differences, these enzymes have an optimal pH for activity between 8.0 and 11.0. In *E. coli*, the enzymes encoded by pNK1 and pFK1 are mostly periplasmic (74 and 68% of total activity, respectively) whereas the enzyme produced by pNK2 is periplasmic (37%) as well as intracellular (49%). Only minor amounts of enzymes were released into the culture supernatant. All three enzymes were extracellular in the original bacilli (58,59). However, Kato et al. (60) constructed an excretion vector which resulted in more than 60% of the enzyme activity being in the *E. coli* culture supernatant. The genes were sequenced (10,61) and the two of *Bacillus* sp. N-4 were strongly homologous. However, a direct repeat of 60 amino acids near the C-terminal end in one of the proteins of *Bacillus* sp. N-4 was absent in the other and was also shown to be unnecessary for the enzymatic activity (61). A partial homology exists between the genes of *Bacillus* sp. N-4 and *Bacillus* sp. 1139, corresponding to a conserved N-terminal portion of the proteins (61). Interestingly, only the N-terminal half of the polypeptide chain of the endoglucanase of *Bacillus* sp. 1139 was necessary for enzyme activity (62).

A cloned endoglucanase gene from another *Bacillus* sp. yielded enzyme distributed in periplasmic and intracellular locations, and was produced throughout the growth phase in *E. coli*. This gene was expressed from its own promoter (11).

Several *B. subtilis* endoglucanase genes have been studied. An endoglucanase from *B. subtilis* IFO 3034 was cloned directly into another *B. subtilis* strain and then into *E. coli* on plasmid pBRC5 (7). The enzyme is extracellular and produced throughout the growth phase in *B subtilis*, while it is in the extracellular, periplasmic, and intracellular fractions (24, 45, and 31%, respectively) of *E. coli*. All forms of the enzyme are stable at 60°C for 1 hr. At the amino acid level, this endoglucanase showed a partial homology with the endoglucanase of an alkalophilic *Bacillus* sp. 1139, especially in the proximal half from the amino terminus (63). The endoglucanase gene from another *B. subtilis* isolate, strain DLG, was cloned in *E. coli* and expressed from its own promoter throughout the growth phase (6). The enzyme is extracellular in *B. subtilis*, but mostly intracellular in *E. coli* (63%) with the remainder loosely associated with the cells but not typically periplasmic. Two active proteins appeared in *E. coli*, with MW 51 and 39 kD, respectively. Initial biochemical analyses led to the hypothesis that the first was a "preprotein" with a "pro" sequence (different from the signal sequence for secretion), which is cleaved in the mature form. The sequence of the gene revealed the existence of a typical export signal peptide in the protein (64). This study also showed that glucose dramatically stimulated the amount of β-1,4-glucanase mRNA in vivo. This feature is unusual as many cellulase genes are repressed by glucose. From the sequence of another endoglucanase gene from *B. subtilis* PAP115 (which was also used as a donor for the cloning of xylanase genes; see Sec. VI), the derived amino acid sequence of the protein showed a short region of relative similarity with the endoglucanase EGB of *C. thermocellum* (8). The calculated molecular weight of the enzyme (52, 211 for the mature form) was larger than the measured molecular weight in both *E. coli* and the native *B. subtilis*, suggesting [as also observed by Robson and Chambliss (6)], that processing of the primary translation product occurred beyond signal peptide cleavage. All *B. subtilis* endoglucanase genes were constitutively expressed in *E. coli*.

C. *Bacteroides succinogenes*

One of six cellulase genes from *Fibrobacter succinogenes* cloned in
E. coli (12,65) has been analyzed. It was cloned on a 1.9-kb DNA
fragment and mapped by transposon mutagenesis. Transcription
started at its own promoter, which directed the synthesis of a periplasmic enzyme in *E. coli*. The latter was repressed by glucose.
The enzyme was not only active against CMC but also against p-
nitrophenylcellobioside and lichenan. It did not bind to acid-swollen
cellulose. The cloned enzyme was more sensititve to detergents,
heat, metal ions, had a narrower pH range of activity, and showed
a different mode of action against CMC than the cellulase complex of
F. succinogenes. The cloned enzyme might be a minor component
of this cellulase complex (66).

D. *Caldocellum saccharolyticum*

The gene for β-glucosidase from the thermophile *Caldocellum saccharolyticum* was expressed in *E. coli* from a promoter located inside
the vector, although its own promoter was probably present (37).
The enzyme was mostly intracellular, was active on cellobiose, and
had a temperature optimum of 85°C. Its half-life at 80°C was 14 hr
(37). At least two cloned endoglucanase genes were also reported
from this bacterium, one being expressed from its own promoter
(14).

E. *Cellulomonas* spp.

The cellulase genes of *Cellulomonas fimi* have been more extensively
characterized by the Canadian group at the University of British
Columbia. A cellobiohydrolase gene was initially cloned (2), along
with two endoglucanase genes on plasmids pEC2 and pEC3, and
again on plasmid pEC1 (3). The biochemical properties of the enzymes produced in *E. coli* showed features typical of endoglucanases
and cellobiohydrolases (67). However, there was a lack of synergy
between the two types of enzymes. The endoglucanase gene of pEC2
was expressed from initiation sites present in the plasmid vector, as
the 5' portion of the gene was absent on that plasmid (15). The
complete gene (*cenA*) was cloned on plasmid pcEC2, but its promoter as well as that of the cellobiohydrolase gene (*cex*) were poorly recognized by *E. coli* RNA polymerase (15,68). The two genes
were sequenced and show good consensus sequences for ribosome-
binding sites. The C-terminal but not the N-terminal domain of the
endoglucanase was essential for enzyme activity. The two proteins
have export signal sequences yet a lower calculated molecular weight
than the *C. fimi* proteins, probably due to the lack of glycosylation.

They show no homology except for a sequence of approximately 20 amino acids of only proline and threonine that closely correspond in both proteins (15,68).

The endoglucanase encoded by pEC2 is cytoplasmic or membrane-bound (87%) and periplasmic (13%) whereas 54% of the activity encoded by pcEC2 (complete *cenA* gene containing the portion corresponding to the signal peptide) is periplasmic (15). This enzyme (as well as the cellobiohydrolase) is extracellular in *C. fimi* (67,69). A leaky mutant of *E. coli* was isolated by chemical mutagenesis that under certain circumstances exported up to 40% of the endoglucanase into the culture medium (70). The signal sequence of the cellobiohydrolase encoded by the *cex* gene of pEC1 when modified by the replacement of its four N-terminal amino acids with the six N-terminal amino acids of the intracellular β-galactosidase did not alter the secretion pattern of 80% in the periplasm (71). By putting the *cex* gene under the control of different strong promoters, and using synthetic translation initiation sites, more than 2×10^4-fold increase in cellobiohydrolase-specific activity was obtained (72). In another construct, up to 20% of the total *E. coli* protein was cellobiohydrolase but it was no longer secreted into the periplasm and appeared instead as insoluble granules in the cytoplasm.

The cellulase system of *Cellulomonas uda* was also cloned. A mainly periplasmic endoglucanase (66% of the total activity) with a temperature optimum of $55-60°C$ was constitutively expressed in *E. coli* (16). Two further genes encoding for β-glucosidases were noted, at least one being active on cellobiose (38). The *E. coli* clones appear to produce more enzyme than the original *C. uda*, the production again being constitutive.

F. *Cellvibrio mixtus*

A gene bank of *Cellvibrio mixtus* was constructed in *E. coli* with a cosmid vector. Degradative genes with products active on CMC, chitin, cellobiose, starch, and pectin were located on a single 94.1-kb segment of chromosomal DNA. The first four genes were clustered on a 9.8-kb DNA fragment. A mutation rendering *C. mixtus* unable to degrade microcrystalline cellulose was complemented by a 36.9-kb cloned fragment encompassing this 9.8-kb fragment but none of the *E. coli* clones were able to degrade microcrystalline cellulose (17).

G. *Clostridium* spp.

Clostridium thermocellum and *Cellulomonas fimi* (see Sec. III.E) are the best characterized cellulolytic bacteria. Several endoglucanases have been cloned, along with two β-glucosidase genes encoding

enzymes active on cellobiose (20) and possibly other different
β-glucosidase genes (5,40) besides the claims for the cloning of
cellobiohydrolases (5).

Cloned endoglucanases have received the greatest focus (4,5,
19–21) and perhaps up to 15 such genes have been cloned from
strain NCIB 10682 (73). Those studied in greater detail (genes
celA, *B*, and *C*) were constitutively expressed from their own promoter, but *celD* was expressed principally from the chloramphenicol
acetyltransferase promoter of the vector (74). The genes *celA* and
celB are not contiguous on the chromosome and the enzymes they
encode (endoglucanase A and B, EGA and EGB) are mainly periplasmic and cytoplasmic, respectively (75). EGC and EGD are also
intracellular (56,57) whereas all four EGs are extracellular in
C. thermocellum cultures. The molecular weights derived for the
purified enzymes or calculated from the amino acid sequence via de-
duction from the nucleotide sequence of the genes are in good agreement and generally slightly lower than the values obtained from
C. thermocellum native enzymes. This may be due to glycosylation
in *C. thermocellum* or proteolytic cleavage in *E. coli*. A heat treatment purification step of the cloned enzyme was particularly effective, causing most *E. coli* proteins to denature and floculate (56,57,
76,77). Indeed, certain of the enzymes, after heating in sodium
dodecyl sulfate (SDS), can renature and remain active thus allowing for zymogram detection even after SDS electrophoresis. EGA
had no activity on crystalline cellulose (Avicel), was resistant to
the detergent SDS, was stable for 2 hr at 60°C, and had a temperature optimum of 75°C. It was not inhibited by cellobiose (0.1 M),
glucose (0.25 M), or ethanol (10%) (77). EGB and EGC had not
been previously found in *C. thermocellum* supernatants, but were
subsequently identified in crude clostridial filtrates by radioimmuno-
assays with antisera raised against the purified cloned enzymes of
E. coli (57,76). EGC, a monomeric enzyme, besides attacking CMC,
showed activity on substrates such as MUC, cellotriose, cellotetraose, and cellopentaose, characteristics often found in cellobiohydrolases, and was not inhibited by 0.32 M glucose with CMC as the
substrate (57). However, it does not act as a cellobiohydrolase
toward amorphous cellulose. When the gene *celD* for EGD was put
under the control of a β-galactosidase promoter, very high levels
of a fused protein were produced, resulting also in insoluble granules in the cytoplasm. However, almost 80% of the endoglucanase
activity remained soluble in the cytoplasm. The protein could be
conveniently purified from the granules by separation through differential centrifugation combined with a heat denaturation step,
leading to purification resulting in crystallization. High-resolution
X-ray diffraction studies to determine the catalytic and binding
domains of the protein may allow rational modification to yield enhanced

binding and/or activity via site-directed mutagenesis (56,78). Several of the clones isolated by Romaniec et al. (21) also exhibited activity toward MUC.

The *celA, celC*, and *celD* genes do not cross-hybridize with each other or with other *C. thermocellum* DNA sequences, and are thus distinct gene families (56,57,75). The gene *celB* hybridizes to only one other DNA sequence in the *C. thermocellum* chromosome. *CelA* or *celB* genes hybridize to the DNA of a few apparently distantly related bacteria such as *Azospirillum brazilense* and *Pseudomonas solanacearum*, but not to the DNA of *Trichoderma reesei* (79). The nucleotide sequences of *celA, celB*, and *celD* have been determined (74,80,81), and it is apparent that codon usage resembles more that of *Bacillus* than that of *E. coli*. Typical translation initiation and transcription termination sites as well as sequences for protein signal peptides were observed, but *Bacillus* or *E. coli* promoter consensus sequences were not always present. Mapping of the mRNA of *celA* (82) showed that transcription starts at several positions in *C. thermocellum* but only at one position in *E. coli*. This suggests some regulatory phenomenon. A striking feature of the amino acid sequence of EGA, EGB, and EGD deduced from the nucleotide sequence of the genes is the presence, near the C-terminal end of the protein, of a duplicated sequence of 23 or 24 amino acids linked by 8 or 12 residues. These reiterated regions are similar in all three proteins and have been suggested to be involved in the binding of the enzymes to two adjacent glucose residues of the cellulose molecule (74,80,81).

An endoglucanase gene from *Clostridium acetobutylicum* cloned in *E. coli* yielded an enzyme that was mostly periplasmic (75% of total activity), inactive toward crystalline cellulose (Avicel), and had a temperature optimum of 50°C (18). Its temperature and pH optima appear to be quite different from those of an endoglucanase of a strain derived from the donor *C. acetobutylicum*. Thus, there are perhaps other endoglucanase genes in this organism. The cloned 4.9-kb DNA fragment also allowed *E. coli* to grow on a mineral medium with cellobiose as the sole carbon source and thus contained a β-glucosidase gene. This is one of the few examples of clustered cellulase genes.

H. *Erwinia chrysanthemi*

Studies on the plant pathogenic *Erwinia chrysanthemi* have been directed more toward pectate lyase than the cellulase genes. However, several endoglucanase genes from three different strains of *E. chrysanthemi* strains have been cloned. The low level of expression of an endoglucanase gene from strain 3665 was attributed to the possible existence of positive regulatory genes that were not

cloned in *E. coli* (22). A cellulase gene from strain 3937 encoded
a protein of the same molecular weight as the major extracellular enzyme of *E. chrysanthemi* (24). Later the gene bank constructed
from this strain was shown to contain two distinct endoglucanase
genes, *celY* and *celZ*, that were homologous to *celY* and *celZ*, respectively, cloned from *E. chrysanthemi* strain 3665 (23). The antigenic determinants of the corresponding endoglucanase EGY, and
those of EGZ, were identical in both strains (23). A genetic map
for *E. chrysanthemi* was initiated via mapping of the endoglucanase
cel gene of strain B374 with respect to other genes (25).

In a genetic study on cellobiose metabolism in *E. chrysanthemi*,
three derived *E. coli* clones were able to grow on cellobiose (39).
The genes responsible for this phenotype were mapped on the
E. chrysanthemi chromosome in a region corresponding to the *bgl*
region of *E. coli*, responsible for the utilization of aryl-β-glucosides.
One of the recombinant plasmids contained two genes, for cellobiose
hydrolysis and for specific transport of cellobiose via the phosphotransferase system (83). These genes, probably in an operon, were
regulated by a cAMP-CRP complex and their expression was reduced
by the *gyrA* mutation. The latter decreases the expression of certain catabolic operons (84).

I. *Escherichia* spp.

Armentrout and Brown (36) cloned an *Escherichia adecarboxylata*
gene that was constitutively expressed in *E. coli* and allowed it to
grow on cellobiose. The membrane-bound enzyme is likely to be
β-glucoside phosphorylase and requires an unspecified cytoplasmic
factor in order to function.

Cryptic genes for the utilization of cellobiose in *E. coli* are
noteworthy and include a hydrolase gene and a gene for a transport system which are organized in an operon (35). In vivo cloning into *E. coli* was followed by subcloning on a low-copy-number
plasmid. Insertional inactivation by transposon mutagenesis allowed
the identification of the transport and hydrolase genes (34).

J. *Microbiospora bispora*

The actinomycete *Microbispora bispora* produces a thermostable cellulase complex. Cloning into *E. coli* indicated five endoglucanase
and two β-glucosidase genes (26). The sequence of the endoglucanase gene *MBICEL1* showed a severely restricted codon usage.
The position of a native promoter could not be absolutely assigned.
The deduced protein sequence showed a 33-amino-acid region rich
in proline and serine comparable with the proline-threonine box of
the cellulases of *C. fimi*. This region is essential for enzyme activity (85).

K. *Pseudomonas* spp.

The first recorded cloning of cellulase was of unspecified gene(s) from a *Pseudomonas fluorescens* strain into *E. coli* (1). This study was apparently not followed up as regulations concerning the bio-hazard status were then in formation.

Two endoglucanase genes that hybridize with each other were cloned from a *Pseudomonas* sp. and *P. fluorescens* var. *cellulosa*, and localized by deletion analysis (28). Another endoglucanase gene from *P. fluorescens* var. *cellulosa*, likely to be nonhomologous with the first one, was also cloned by Lejeune et al. (27) and expressed from its own promoter. The position of this gene was determined by transposon mutagenesis and the enzyme had MW 40 kD (141).

L. *Ruminococcus* spp.

An endoglucanase gene from *Ruminococcus flavefaciens* was recently cloned into *E. coli* (30). The enzyme was not active toward crys-talline substrates such as Avicel or filter paper. Its expression was not regulated by glucose. Though in this study the enzyme was periplasmic in *E. coli*, in contrast the endoglucanase of *R. albus* F-40 was almost equally distributed between the cytoplasm and peri-plasm of *E. coli* (29). This gene was not expressed from its own promoter.

A brief report records the cloning of two endoglucanase genes from *R. albus* SY3 and also mentions the cloning of another cellu-lase from another rumen bacterium, *Butyrivibrio* sp. A46 (13).

M. *Streptomyces lividans*

Two endoglucanase-negative mutants were obtained by chemical mu-tagenesis of *Streptomyces lividans* (31,55) and used for the homolog-ous cloning of two distinct endoglucanase genes (31). The two mutants containing the recombinant plasmids showed dramatic dif-ferences in the level of endoglucanase production, with one of them producing up to 850 times more endoglucanase than the wild-type *S. lividans*. This suggests that different loci in the two mutants were affected by mutation and that these loci are involved in the regulation of the endoglucanase gene expression in conjunction with the cloned sequences (31).

N. *Thermomonospora fusca*

An endoglucanase gene of *Thermomonospora fusca* was expressed in *E. coli* from its own promoter, and the thermostable enzyme was se-lectively excreted with up to 30% of the activity found in the super-natant of an *E. coli* culture (32). The activity was much lower than

in the original *T. fusca* and also very much strain-dependent in
E. coli. It would be significantly increased by inserting the gene
in an expression vector. The enzyme had molecular weight of 94
kD in *T. fusca* and 70 kD in *E. coli*.

O. Discussion

From these studies certain salient features emerge. Despite the
many cellulase genes that have been cloned, none has been reported
to be on a plasmid in the original bacterium. This is noteworthy
especially for pseudomonads where many catabolic genes are on plasmids, and as the cellulolytic character is not shared by all the Pseudomonads, one might have expected this additional genetic feature
to be plasmid-borne.

A further point is that no cellulase operons have yet been reported even though several cellulase genes coexist in cellulolytic
bacteria. Two exceptions concerning genes involved in cellobiose
metabolism are noted (35,83). In addition, of all the cellulase genes,
only those of *Cellvibrio mixtus* (17) and *Clostridium acetobutylicum*
(18) appear to be clustered on the chromosome. This implies that
in most cases the coordinated expression of multiple cellulase genes
is not mediated at the level of transcription of the DNA into a poly-
cistronic mRNA and its translation into proteins but perhaps occurs
via the interaction between a common regulatory protein and the individual genes.

A further important point is the multiplicity of endoglucanase
genes (Table 1). That multiple cellulase enzymes exist in cell016-
tic microbes has been known for a long time. It is now clear that
this multiplicity results not only from multiple genes but also of the
more generally known posttranslational modifications of a unique polypeptide. This observation can be questioned from the viewpoint of
the energy economy of the cell. However, it is likely, as has been
noted for the endoglucanases of *Clostridium thermocellum*, that each
enzyme has a different yet poorly understood specific action in the
process of cellulose degradation (79). Unique specificity is clearly
illustrated with polysaccharases that act on β-1,4 linkages, those
that act on β-1,3/1,4 linkages and act on β-1,4-xylan, and yet not
on the β-1,3-linked polysaccharides. Several reports state a similar multiplicity of genes involved in xylan degradation, e.g., in
Bacillus sp. (86) and *Aeromonas* sp. (87), and in pectin degradation by *Erwinia chrysanthemi* (24,25).

Few cellobiohydrolase genes have been cloned. However, it is
pertinent to note that so far cellobiohydrolases have rarely been detected in bacteria. The only bacterial cloned gene positively known
to encode a cellobiohydrolase is that of *Cellulomonas fimi* (2,3,67,71,
72,88). Other candidates were reported from *C. thermocellum*

(4,5). Indeed it could be that bacterial cellulase complexes can be subdivided into two broad categories: one analogous to the fungal systems with cellobiohydrolases and endoglucanases, and the other consisting simply of interacting endoglucanases (139). Cellulases from actinomycetes fall into the first category while those from pseudomonads (91) appear to fit the second.

The mode of action of cloned cellobiose-utilizing enzymes is yet to be detailed. The "β-glucosidase" gene from *Escherichia adecarboxylata* encodes a cellobiose phosphorylase (36). In *E. chrysanthemi* the two cloned genes responsible for cellobiose uptake and hydrolysis are perhaps arranged in an operon (83). The only other operon example at present concerns the cryptic genes for cellobiose utilization in *E. coli* (35).

Several bacterial cellulases have been shown to be glycoproteins: *Cellulomonas fimi* (69), *Clostridium thermocellum* (89), *Pseudomonas fluorescens* var. *cellulosa* (90), and *Pseudomonas* sp. (91). The glycosylation is not required for activity, as unglycosylated enzymes produced in *E. coli* are still active. However, carbohydrate moieties may play an important functional role in the binding of cellulases onto cellulose (69); such an endoglucanase of *Fibrobacter succinogenes* produced in *E. coli* no longer attached to acid-swollen cellulose (66). Likewise, the proteolytic removal of heavily glycosylated segments of *Trichoderma reesei* cellulases has been shown to decrease their binding to cellulose (92). Similarly, posttranslation modification or environmental factors do not appear to be required for the thermostability of the enzymes, as cellulases produced in *E. coli* from genes of thermophilic bacteria (*Bacillus* sp., *Caldocellum saccharolyticum*, *Cellulomonas uda*, *Clostridium thermocellum*, *Thermonomospora fusca*) remain thermostable. However, additional glycosyl residues on the endoglucanase EGI of *T. reesei* produced in yeast, compared to the native protein, rendered it less sensitive to thermal inactivation (45). Thus, thermostability is perhaps not solely determined by the primary structure of the protein (see also Chap. 20).

The location of the cellulase enzymes produced in recombinant *E. coli* cultures to that in the donor organisms is compared in Table 2. Results of experiments designed specifically to achieve export such as the use of *E. coli* leaky mutants or excretion vectors are not cited. Most extracellular enzymes from gram-positive donors are cell-bound in the gram-negative *E. coli*. Cytoplasmic location in this bacterium indicates that the signal sequence for protein secretion was not totally "recognized," although recognition of this sequence can result in the periplasmic location of the enzyme, as expected for gram-negative bacteria. The *F. succinogenes* endoglucanase (12) and the *E. adecarboxylata* β-glucosidase (36) were specifically mentioned as being membrane-bound in *E. coli*, although the

TABLE 2　Location of the Cellulases Produced in *E. coli* with Recombinant Plasmids

Donor organism	Gram stain of donor	Location of enzyme in donor			Recombinant plasmid or strain	Location of enzyme in *E. coli*			Ref.
		E	P	I		E	P	I	
Bacillus subtilis	+								
DGL		+			pLG4001a			+	6
IFO3034		+			pBRC5	+	+	+	7
Bacillus sp.	+								
N-4									
		+			pNK1			+	9
		+			pNK2		+	+	9
1139		+			pFK1		+		10
–		+			pBS1		+	+	11
Fibrobacter succinogenes ATCC 19169	–	+			BC14			+	12
Cellulomonas fimi ATCC 484	+								
		+			pUC13-1.1 *cex*		+		72
		+			pcEC2		+	+	15
Clostridium thermocellum NCIB 10682	+								
		+			pCT1		+		75
		+			pCT2			+	75
		+			pCT303	+	or	+	57
		+			pCT600	+	or	+	56
Thermomonospora fusca XY	±	+			pD316	+	+	+	32

Abbreviations: E, extracellular; P, periplasmic; I, intracellular.

enzyme activity of membrane preparations has not always been reported. A portion of the *B. subtilis* strain DLG endoglucanase produced in *E. coli* was loosely associated with the membrane (6). Significant amounts of the enzymes were found in the supernatant of an *E. coli* culture in three cases: two *B. subtilis* endoglucanases (7,8) and a *T. fusca* endoglucanase (32). This extracellular location was not due to cell lysis, but no other explanations were given for these observations.

Although quantitation of cellulase activity in *E. coli* is reported, comparison is not attempted due to the variety of cellulase assays employed. In particular, the lack of citation of the degree of substitution of the CMC poses problems because the enzyme activity is highly dependent on this factor (93). A range of methods for reducing sugar determinations (94) are employed which can be broadly interpreted as relating to the degree of hydrolysis that the investigators are studying. Thus the diverse substrates and variety assay conditions yield few absolute comparisons. However, certain generalizations can be made. Principally, many CMCs used have a degree of substitution of about 0.7, and this permits rough comparisons of relative activities. The specific activities of cloned endoglucanases produced by *E. coli* show an extremely wide range (Table 3) even taking into account the degrees of purification. Those for purified enzymes from clostridial genes are high compared with those of purified enzymes from native organisms (generally lower than 100 IU/mg protein). It should be emphasized again that purification of cellulase enzymes from *Clostridium thermocellum* has proven problematic (see Chap. 18) and that it has only been through gene-cloning techniques that purification of individual enzymes has become readily feasible.

With few exceptions, the synthesis of cellulases is repressed by glucose, although at present no regulatory genes acting in either a positive or a negative manner have been characterized. Endoglucanase genes cloned in two endoglucanase-negative mutants of *Streptomyces lividans* were affected at different regulatory loci (31).

Remarkably, though several cellulase genes have been sequenced, in general little homology is found between them. Exceptions include the homology between the two endoglucanase genes of *Bacillus* sp. N-4 and between these genes and a gene of *Bacillus* sp. 1139 (61). Both bacteria are inhabitants of alkaline environments. At the protein level, amino acid sequences that closely resemble each other can be found in the endoglucanases and cellobiohydrolase of *Cellulomonas fimi* (15) and *Microspora bispora* (85), the three endoclucanases EGA, EGB, and EGD of *Clostridium thermocellum* (74,80,81), and in EGB and the endoglucanase of *B. subtilis* PAP115 (8). These conserved sequences, although not always present, presumably have a specific role in the process of cellulose degradation, probably by substrate

TABLE 3 Specific Activity of Endoglucanases Obtained from *E. coli* with Recombinant Plasmids

Donor organism	Plasmid designation	Specific activity (mU/mg protein)[a]	Degree of substitution of CMC substrate	Ref.
Fibrobacter succinogenes ATC 19169	pRE3	0.35[b]	0.7	66
Cellumononas fimi ATCC 484	pEC2	778[c]	?	67
	pEC3	27[c]	?	67
Cellulomonas uda CB4	pCM41	12[c]	0.7	16
Clostridium acetobutylicum P262	pHZ100	1594[c]	0.7	18
Clostridium thermocellum NCIB 10682	pWS1 (EGA)	580,000[d]	0.7	77
	pCT207 (EGB)	59,500[d]	0.7	76
	pCT301 (EGC)	27,000[d]	0.7	57
	pCT603 (EGD)	428,000[d]	0.7	56

F7	pCU104	111^c	?	5
Erwinia chrysanthemi 3665	pMH13	340^c	0.7	84
Pseudomonas fluorescens var. *cellulosa* NCIB 10462	pRUCL200R	900^c	0.7	Lejune (unpublished results)
Ruminococcus flavefaciens FD1	pMEB200	16^b	0.7	30
Thermomonospora fusca XY	pD370	302^c	0.4	32

[a] One unit of enzyme corresponds to the liberation of 1 μmol of reducing sugars (as glucose) per min.
[b] Periplasmic preparation.
[c] Crude cell extract.
[d] Purified enzyme.

recognition and/or binding. Conservation of amino acid sequences
is higher between different cellulases of a given microorganism, irrespective of their function, than between enzymes with the same
function in different microorganisms. Thus, it seems that the pools
of cellulases of different microorganisms may have evolved independently (74). Complete elucidation of such evolutionary relationships
and the determination of functionally important regions will require
more protein and/or gene structure data as well as studies on the
activity of mutated genes (8). In this regard, Fukumori et al. (61)
showed that only the N-terminal half of the endoglucanase of *Bacillus*
sp. 1139 produced in *E. coli* was necessary for enzyme activity.

IV. CHARACTERIZATION AND EXPRESSION OF CLONED EUKARYOTIC CELLULASE GENES

As is the case for prokaryotic genes, most of the cellulase genes
of eukaryotic origin were first cloned and characterized in *E. coli*.
However, they generally are not expressed in this bacterial host
and more routinely expression is only achieved in yeast, sometimes
with the aid of further genetic constructions. Thus, for eukaryotic
cellulase genes, yeast is more often used as a host for fundamental
studies than as a host chosen for some particular applied objective.
Comparison is therefore made of the cloning of eukaryotic cellulase
genes in both *E. coli* and in yeast.

A. *Aspergillus niger*

Using a cosmid vector, a gene bank of *Aspergillus niger* was constructed in *E. coli* and then transferred into *Saccharomyces cerevisiae* (95). No activity was detected in the bacterial host though
one yeast cloned showed a weak activity, which was increased by
subcloning experiments. Other tested genes of *A. niger* were not
expressed in yeast although both organisms are lower eukaryotes.

B. *Candida pelliculosa*

The β-glucosidase gene of the yeast *Candida pelliculosa* was not expressed in *E. coli*, but allowed a *S. cerevisiae* strain with increased
membrane permeability to grow on cellobiose as the sole carbon
source. In *S. cerevisiae* the enzyme was located periplasmically.
The cellobiase was repressed by glucose (96), while the absence
of introns in the gene was revealed by DNA sequencing (97).

C. *Kluyveromyces fragilis*

The β-glucosidase activity obtained in *S. cerevisiae* with a cloned
gene from *Kluyveromyces fragilis* was 500-fold higher than in the

native yeast. However, *S. cerevisiae* was not able to grow on cellobiose because the enzyme was intracellular and the disaccharide
was not imported into the cells (98). As in *K. fragilis*, the β-
glucosidase was a nonadaptative, glucose-repressed enzyme in
S. cerevisiae (99).

D. *Schizophyllum commune*

A few hundred DNA fragments related to growth adaptation on cellulose were isolated from the fungus *Schizophyllum commune*. By
switching to a λ-expression vector and using a radiommunoassay,
several clones were shown to carry partial sequences for an endoglucanase and a β-glucosidase (100). The amino acid sequence of
the β-glucosidase, deduced from the nucleotide sequence, showed a
region of homology with the sequence of the β-glucosidase of *Candida pelliculosa*, suggesting the possibility of a common catalytic
mechanism (101).

E. *Trichoderma reesei*

The cellulase complex of *Trichoderma reesei* is probably the best
characterized cellulase system to date (see Chap. 21). Molecular
cloning programs have been initiated and detailed data on the cellulase genes are available.

The *cbh1* gene for cellobiohydrolase CBHI, the major cellulase
of *T. reesei*, was cloned in *E. coli* and shown to contain two short
introns and a promoter region analogous to comparable regions of
highly expressed yeast genes (42,43).

Although a strong homology at the amino acid level exists between the endoglucanase EGI and CBH I, the corresponding genes
egl1 and *cbh1* are not homologous and the position of the introns
is not conserved (44,45). Additional glycosyl residues are found
in EGI produced in yeast when compared to native *T. reesei* enzyme. They apparently have no effect on specific activity but render the protein less sensitive to thermal inactivation (45). The
cbh2 gene for cellobiohydrolase CBH II contains three introns and
its structural features indicate that it is probably regulated by the
intracellular cAMP level (46,102). Despite the functional similarities
between CBH I and II, the primary structure of these two enzymes
does not show any similarity. Analysis of cDNAs suggest that two
evolutionary distinct cellulase families, composed of CBH I/EG I and
CBH II/endoglucanase EG III, exist in *T. reesei* (102). Significantly,
however, two short consecutive blocks of strong amino acid homology
are found in all four *T. reesei* cellulases studied so far. These occur at the N terminus of CBH II and EG III but at the C terminus
of CBH I and EG I. One of these blocks is heavily glycosylated and
there is experimental evidence that the carbohydrate moieties might
have a functional role in substrate binding of cellulases (46,102).

V. EXPRESSION OF CELLULASE GENES IN NON-*E. coli* HOSTS

Examples of the expression of cellulase genes, of both prokaryotic and eukaryotic origin, in non-*E. coli* hosts are on the increase, and hosts have been chosen either for the development of basic knowledge or, more often, potential application (Table 4).

A. Expression in *Aspergillus nidulans*

Filamentous fungi of the genus *Aspergillus* are major sources of industrial enzymes. They secrete large quantities of certain proteins that are posttranslationally modified.

These features have been used to advantage to express the endoglucanase gene *cenA* of *Cellulomonas fimi* in *A. nidulans*. The expression was regulated by strong controllable promoters of *Aspergillus* and the secreted protein was biologically active. Cloned enzyme production at industrial levels appears feasible with only few improvements to the system (103).

TABLE 4   Cloning of Cellulase Genes in non-*E. coli* Hosts

Host	Special attribute	Objective
Aspergillus nidulans	Secretion and glycosylation of proteins	Posttranslational modificatons of extracellular cellulases
Bacillus subtilis	Secretion of proteins	Extracellular cellulase
Rhodobacter capsulatus	Nitrogen fixation, photosynthesis	Enhanced cellulolysis by decreasing the carbon/nitrogen ratio
Saccharomyces cerevisiae	Glycosylation, ethanologen	Posttranslational modifications of cellulases, ethanol from cellulose
Zymomonas mobilis	Ethnologen	Ethanol from cellulose
Rumen bacteria	Lignocellulolysis in the rumen	Improved forage utilization for beef and milk production

B. Expression in *B. subtilis*

Bacillus species are well known for their ability to secrete considerable amounts of enzyme and hence have been considered as useful recombinant hosts.

Cellulase enzymes of *B. subtilis* genes cloned in *B. subtilis* were predominately extracellular, as in the donor strains from which they were cloned (6,7). All of the endoglucanase synthesized in *B. subtilis* from the cloned *celA* gene from *Clostridium thermocellum* was extracellular and amounted to 50% of the total secreted proteins of *B. subtilis* (104). To date, few other cellulase genes have been cloned in *B. subtilis*. The *Caldocellum saccharolyticum* β-glucosidase gene is expressed, although at a low level, from its own promoter (this promoter is not recognized in *E. coli*) (37), as is an endoglucanase gene from the same bacterium (14).

C. Ethanol from Cellulose

Several reports have appeared on the cloning of cellulase genes in yeast. The *Clostridium thermocellum* endoglucanase *celA* gene was expressed in *Saccharomyces cerevisiae* from its bacterial promoter at a level corresponding to 10% of the extracellular EGA of *C. thermocellum*. The positive cloned were detected by the Congo red assay procedure, although the enzyme was apparently completely cytoplasmic (105). A portion of the endoglucanase *C. fimi cenA* gene on a yeast expression plasmid was used to study the excretion by yeast on the enzyme fused in various combinations with the extracellular toxin K1 polypeptide (106). One of the clones produced and secreted 10 times as much endoglucanase as *E. coli* with plasmid pEC2 (*cenA* with the tetracycline resistance gene promoter). The cellobiohydrolases CBH I and II and the endoglucanase EG I of *T. reesei* were also secreted by yeast with the help of their native signal sequences (45,107). The endoglucanase gene from *Clostridium acetobutylicum* was also cloned in *S. cerevisiae* and directed the synthesis of an intracellular enzyme (D. T. Jones, pers. commun.).

Synthesis of an intracellular endoglucanase with the gene of *Pseudomonas fluorescens* var. *cellulosa* was obtained in the ethanol-fermenting bacterium *Zymomonas mobilis* (140). Programs are currently in progress to design *Z. mobilis* plasmid expression vectors for the cloning of useful genes (108,109).

D. Expression of Cellulase Genes in Nitrogen-Fixing Bacteria

Biomass cellulosic substrates have an extremely high carbon/nitrogen ratio and microbial degradation of them is more efficient if a supply

source of nitrogen is available. Thus a program has also been initiated to increase the efficiency of *Azotobacter* as a bacterial fertilizer by enhancing its growth potential in soil via inserting cellulase genes into it. The objective is to provide large amounts of
metabolizable glucose from soil cellulosics thus increasing the *Azotobacter* population while also keeping respiration at a high level, and
thus suppressing the oxygen inhibition of nitrogenase (110).
Analogously, to gain an alternate supply of reduced nitrogen, the
cellulase genes of *C. fimi* were cloned into the nitrogen-fixing,
photosynthetic bacterium *Rhodobacter capsulatus* (88). A portion
of the *cenA* gene of *C. fimi* was fused with the initial codons of the
gene of a membrane-bound, light-harvesting protein of *R. capsulatus*, and an active endoglucanase was obtained via *R. capsulatus*
transcription and translation initiation sequences. The *C. fimi* exoglucanase *cex* gene and another endoglucanase gene, originally on
plasmid pEC3, showed expression modulated by aeration of the cultures when placed under control of *R. capsulatus* plasmid expression vectors.

E. Improved Lignocellulose Degradation in the Rumen

Ruminants rely on microbial fermentation of lignocellulose in the rumen for their nutrition (see Chap. 15). It has been speculated that
recombinant DNA techniques could be used to modify rumen bacteria
to improve rumen function (111). Numerous problems exist in this
approach, from the scarcity of basic genetic knowledge of rumen
bacteria, to the stability of recombinant bacteria in such a competitive environment.

Protoplast fusion may provide a means to modify bacteria when
suitable cloning vehicles are not available. Indeed fusion of two
rumen bacteria, *Fusobacterium varium* and *Enterococcus faecium*,
yielded fusants capable of degrading lignin-related compounds (112).
One of the hybrids was further fused with *Ruminococcus albus* and
the new fusants exhibited a lignin/cellulose degradation phenotype
derived from the three parents. These stable fusants have yet to
be evaluated in the rumen.

VI. RELATED GLYCOSIDASE GENES

Glycosidase genes are considered whose enzymes are active toward
biomass substrates including β-1,3/1,4 glucans (barley) as well as
polymers often in close association with cellulose (e.g., hemicellulose
and lignin).

A. Endo-β-1,3/1,4-glucanase Genes

Focus on these genes and in particular their expression in yeast
was initiated because of their importance to the brewing industry,
as barley β-glucan retards filtration and also can promote the formation of gels and hazes in beer. An endo-β-1,3/1,4-glucanase
gene from *B. subtilis* NCIB2117 that was first cloned in *E. coli* was
expressed from its own promoter, and directed the synthesis of a
25-kD enzyme (52). From sequencing, typical promoter and ribo-
some-binding site sequences were found (113). High levels of expression in laboratory strains of *S. cerevisiae* were achieved by
placing the gene under the control of the yeast ADH1 promoter
(114), and new plasmid constructs allowed the export of significant
amounts of β-1,3/1,4-glucanase by yeast (115). The gene has also
been expressed in brewing yeast with no detectable change in fermentation rates or beer flavor (116).

An endo-β-1,3/1,4-glucanase gene from *B. amyloliquefasciens*
yielded expression from both its own and the β-lactamase promoter
of the vector (54). It has been mapped by transposon mutagenesis.
Twenty to thirty percent of the enzyme produced by *E. coli* was extracellular, and the sequence of its gene (117) revealed about 90%
homology with the endo-β-1,3/1,4-glucanase gene of *B. subtilis* sequenced by Murphy et al. (113). There are further reports on the
cloning of possibly different endo-β-1,3/1,4-glucanase genes from
B. subtilis (53,118). In the former case, the gene was expressed
in *E. coli* where the enzyme was mostly periplasmic (55% of the activity) with significant amounts (approximately 25%) found in the
culture supernatant. The enzyme was inactive toward homopoly-
meric β-1,3-glucans or β-1,4-glucans. In the laboratory yeast
strains, this gene was expressed from its own promoter but at a
low level, and the enzyme was completely intracellular (119). A low
level of expression was observed when the gene was transformed to
brewing yeast. Better expression and excretion, and higher plasmid stability, are necessary prior to large-scale application (120).

B. Xylanase and β-Xylosidase Genes

Xylanases are of interest to the pulp and paper industry for the
gentle and selective removal of residual lignins by cleavage of the
hemicellulose−lignin complexes in the preparation of primary quality
paper. Construction of microbes able to degrade xylan is also useful in projects aimed at the utilization of agricultural wastes, as
xylan is often a major hemicellulose component (30−40% in corn
cobs).

A xylanase and a β-xylosidase gene from *Bacillus pumilus* cloned
in *E. coli* were constitutively expressed, the enzymes being

cytoplasmic in *E. coli* (121). In spite of this nonperiplasmic location, the signal peptide of the xylanase was correctly processed in
E. coli. In contrast, the enzyme was secreted when its gene was
cloned in *B. subtilis* (122). The gene was sequenced but no promoter consensus sequence was found, which could explain the low
activity observed in *E. coli* (123).

A xylanase gene from *B. subtilis* PAP115 cloned in *E. coli*
yielded mainly periplasmic enzyme (124). However, 45% of the enzyme was secreted into the supernatant when an *E. coli* leaky mutant was used (125). By DNA sequencing, a typical *Bacillus* promoter was seen to occur in front of the gene and the derived amino
acid sequence showed a 50% homology with that of the *B. pumilus*
xylanase (126). A β-xylosidase gene from the same *B. subtilis*
strain was also cloned (127).

The gene for xylanase L of *Aeromonas* sp. cloned in *E. coli*
gave a resulting 80-fold increase in xylanase production. The xylanase was periplasmic and cytoplasmic (87) but became 60% extracellular when the gene was cloned on an excretion vector (60). The
same group reported the cloning of the xylanase gene of *Bacillus*
sp. C-125, the enzyme possessing the same properties in *E. coli*
as in *Bacillus*, and 80% was secreted in the former host (86). With
this plasmid *E. coli* also secreted 68% of its normally periplasmic
β-lactamase but not its periplasmic alkaline phosphatase. Thus it
was proposed that selective permeability of the *E. coli* outer membrane is regulated through this plasmid (128). A multiplicity of
nonhomologous xylanase genes also occur in *Aeromonas* sp. No 212
(87) and *Bacillus* sp. C-125 (86), but are yet to be elucidated.
This is in contrast with *Streptomyces lividans* where a single xylanase gene apparently exists (55). It has been cloned in an xy-
lanase-negative mutant of the same organism and considerably increased extracellular xylanase activity was obtained. The xylanase
was constitutively expressed and not subject to catabolite repression
(129).

C. Genes Involved in Lignin Degradation

Lignin is a polymer of substituted phenylpropanoid units, and is
often associated with hemicellulose and cellulose where it forms "lignocellulose." As lignin enshrouds cellulose, a pretreatment of the
substrate to displace lignin is needed to gain efficient cellulose hydrolysis. Lignin itself is a potential source of chemical feedstocks,
leading to numerous studies on ligninolysis. These studies fall into
two spheres: conversion of macromolecular lignocellulose, and transformation of lignin-derived molecules. Degradation of unmodified
lignocellulose appears restricted to white rot fungi (129). In contrast, a wide variety of bacteria can modify lignin oligomers (130).

Studies of ligninolysis are progressing rapidly following the discovery of enzymes (lignin peroxidases) that degrade lignin (for review, see Ref. 131). Focus has been especially on *Phanerochaete chrysosporium*. The genes specifically expressed during the lignolytic (secondary metabolic) phase of *P. chrysosporium* were cloned by Broda et al. (132). Complementary DNA to at least two lignin peroxidase genes of the same organism was cloned and partially characterized by Zhang et al. (133) and Tien and Tu (134), who demonstrated that the synthesis of lignin peroxidase is regulated at the mRNA level. Both groups suggest the existence of multiple ligninase genes. Molecular biological studies in this area are new and few data are available, but they will expand very much in the coming years (130).

VII. CONCLUSION AND PERSPECTIVES

With the current major interest in biomass utilization, a tremendous effort has emerged in research aimed at the cloning and expression of cellulase genes from a variety of microorganisms in bacteria and yeast. We have reviewed the current results obtained in this field and the methodology for the cloning of cellulase genes, and in particular the screening techniques involved, have been discussed. The cloned genes are principally or prokaryotic origin. Both fundamental and applied objectives have motivated the investigations.

These studies have permitted deeper insights in such basic issues as gene expression and organization, and protein secretion. The multiplicity of genes and their products clearly indicates the complexity of the apparently simple process of cellulose hydrolysis. Understanding how the different enzymes relate to each other and cooperate in their action toward cellulose is now a principal research focus.

At present, no cellulase complex possessing high efficiency of cellulose degradation has been reconstructed utilizing cloned genes. Synergism between cellulases from different organisms is now well documented (135; see also Chapts. 21 and 22). Furthermore, by putting several genes, possibly of different origin, under the control of promoters of different strengths or with appropriate gene regulation, such complexes with an optimal balance of the different enzyme activities could be obtained — a goal that is approaching realization. The powerful molecular biology techniques also offer improvements, e.g., via site-directed mutagenesis to increase thermal stability or specific activity of the enzymes, or via genetic construction to gain enhanced secretion. The problems encountered in cellulose utilization due to its association with other compounds, such as hemicellulose and lignin, are also being addressed at the molecular level with the cloning of genes for xylanase, β-xylosidase, and lignin-solubilizing enzymes.

In addition to the in vitro saccharification of cellulose by "reconstructed" cellulase complexes, a single-step conversion of cellulose to fuels and chemicals can be envisioned by expressing the genes in fermentative microorganisms. For instance, the synthesis of cellulase by yeast or other homofermentative organisms can lead to direct ethanol production in high yield and concentration. Similarly, the problematic use of cocultures could be circumvented by molecular cloning approaches.

Although a strong impetus has been given to investigations by invoking immediate application, longer term goals are now evident. However, the use of isolated cellulases to effect limited yet controlled hydrolysis of cellulose with the intent to alter the structural and functional properties of cellulosic materials for the food industry may be closer to fruition (136; J.-C. de Troostenberg, pers. commun.). Practical difficulties such as plasmid stability and scale-up problems, will need attention, as will compliance with regulations of governmental agencies on the use of genetically modified strains or their products.

Finally, large amounts of cellulosic materials, mainly as agricultural residues, are produced, often constituting two-thirds or greater of the crop biomass. This resource is often collected and could be well used in developing countries as feed for animals and chemical production. Here, the biotechnology of cellulose has potential if more efficient procedures and relatively small scale fermentation plants can be developed (137).

ACKNOWLEDGMENTS

The authors thank Drs. P. L. Bergquist, D. R. Love, and M. B. Streiff, G. P. Hazlewood, D. Kluepfel, D. J. McConnell, J. A. Thomson, G. Velikodvoskaya, and M. Mogutov for communicating results prior to publication. The authors' studies with cellulase have been supported through the N. J. Agriculture Experiment Station (K-01111-88) and through the U.S. Department of Energy contracts XX-4-04150-1 and DE-AS05-83ER13140.

REFERENCES

1. A. M. Chakrabarty and J. F. Brown, In *Genetic Engineering* (A. M. Chakrabarty, ed.), CRC Press, 1978, p. 185.

2. D. J. Whittle, D. G. Kilburn, R. C. Miller, Jr., and R. A. J. Warren, *Gene*, *17*, 139 (1982).

3. N. R. Gilkes, D. G. Kilburn, M. L. Langsford, R. C. Miller, Jr., W. W. Wakarchuk, R. A. Warren, D. J. Whittle, and W. K. R. Wong, *J. Gen. Microbiol.*, *130*, 1377 (1984).

4. J. Millet, D. Pétré, P. Béguin, O. Raynaud, and J.-P. Aubert, *FEMS Microbiol. Lett.*, *29*, 145 (1986).

5. M. A. Mogutov, G. A. Velikodvorskaya, T. A. Pushkarskaya, M. Z. Yerjev, V. L. Mett, and E. S. Piruzan, In *Proc. Soviet-Finnish Seminar on Microbial Degradation of Lignocellulose Raw Materials* (L. A. Golovleva, ed.), USSR Academy of Sciences, Pushchino, 1985, p. 45.

6. M. L. Robson and G. H. Chambliss, *J. Bacteriol.*, *165*, 612 (1986).

7. A. Y. Koide, A. Nakamura, T. Ouzumi, and T. Beppu, *Agr. Biol. Chem.*, *50*, 233 (1986).

8. R. M. MacKay, A. Lo, G. Willick, M. Zuker, S. Baird, M. Dove, F. Moranelli, and V. Seligy, *Nucl. Acid Res.*, *14*, 9159 (1986).

9. N. Sashihara, T. Kudo, and K. Horikoshi, *J. Bacteriol.*, *158*, 503 (1984).

10. F. Fukumori, T. Kudo, Y. Narahashi, and K. Horikoshi, *J. Gen. Microbiol.*, *132*, 2329 (1986).

11. S. H. Park and M. Y. Pack, *Enzyme Microb. Technol.*, *8*, 725 (1986).

12. W. L. Crosby, B. Collier, D. Y. Thomas, R. M. Tether, and J. D. Erfle, In *5th Canadian Bioenergy R&D Seminar* S. Hasnlan, ed.), Elsevier, New York, 1984, p. 573.

13. M. P. M. Romaniec, N. G. E. Clarke, C. G. Orpin, and G. P. Hazlewood, *14th Int. Congress Microbiology*, Manchester, England, 1987, Abs. P.B.17-8, p. 101.

14. M. B. Streiff, D. R. Love, L. Chamley, and P. L. Berquist, *Proc. Austr. Biotechnol. Mg.*, 1986, p. 179.

15. W. K. R. Wong, B. Gerhard, Z. M. Guo, D. G. Kilburn, R. A. J. Warren, and R. C. Miller, Jr., *Gene*, *44*, 315 (1986).

16. K. Nakamura, N. Misawa, and K. Kitamura, *J. Biotechnol.*, *3*, 247 (1986).

17. E. C. Wynne and J. M. Pemberton, *Appl. Env. Microbiol.*, *52*, 1362 (1986).

18. H. Zappe, D. T. Jones, and D. R. Woods, *J. Gen. Microbiol.*, *132*, 1367 (1986).

19. P. Cornet, D. Tronik, J. Millet, and J.-P. Aubert, *FEMS Microbiol. Lett.*, *16*, 137 (1983).

20. W. Schwarz, K. Bronnenmeier, and W. L. Staudenbauer, *Biotechnol. Lett.*, *9*, 169 (1985).

21. M. P. M. Romaniec, N. G. Clarke, and G. P. Hazlewood, *J. Gen. Microbiol.*, *133*, 1297 (1987).

22. F. Barras, J.-P. Chambost, and M. Chipaux, *Mol. Gen. Genet.*, *197*, 513 (1984).

23. M.-H. Boyer, B. Cami, A. Kotoujansky, J.-P. Chambost, C. Frixon, and J. Cattaneo, *FEMS Microbiol. Lett.*, *41*, 351 (1987).

24. A. Kotoujansky, A. Diolez, M. Boccara, Y. Bertheau, T. Andro, and A. Coleno, *EMBO J.*, *4*, 781 (1985).

25. F. van Gijsegem, F. A. Toussaint, and E. Schoonejans, *EMBO J.*, *4*, 787 (1985).

26. Z. P. Shalita, M. D. Yablonsky, M. M. Dooley, S. E. Buchholz, and D. E. Eveleigh, Biochemical Conversion Program, Solar Energy Res. Institute, Golden, CO., SERI/CP-231-2988, p. 213.

27. A. Lejeune, C. Colson, and D. E. Eveleigh, *J. Ind. Microbiol.*, *1*, 79 (1986).

28. B. R. Wolff, T. A. Mudry, B. R. Glick, and J. J. Pasternak, *Appl. Env. Microbiol.*, *51*, 1357 (1986).

29. S. Kawai, H. Honda, T. Tanase, M. Taya, S. Iijima, and T. Kobayashi, *Agr. Biol. Chem.*, *51*, 59 (1987).

30. M. E. C. Barras and J. A. Thomson, *J. Bacteriol.*, *169*, 1760 (1987).

31. F. Shareck, F. Mondou, R. Morosoli, and D. Kluepfel, *Biotechnol. Lett.*, *9*, 169 (1987).

32. A. Collmer and D. B. Wilson, *Bio/Technology*, *1*, 594 (1983).

33. W. W. Wakarchuk, D. G. Kilburn, R. C. Miller, Jr., and R. A. J. Warren, *Mol. Gen. Genet.*, *205*, 146 (1986).

34. B. G. Hall, P. W. Betts, and M. Kricker, *Mol. Biol. Evol.*, *3*, 389 (1986).

35. M. Kricker and B. G. Hall, *Genetics*, *115*, 419 (1987).

36. R. W. Armentrout and R. D. Brown, *Appl. Env. Microbiol.*, *41*, 1355 (1981).

37. R. D. Love and M. B. Streiff, *Biotechnology*, *5*, 384 (1987).

38. K. Nakamura, N. Misawa, and K. Kitamura, *J. Biotechnol.*, *3*, 239 (1986).

39. F. Barras, J.-P. Chambost, and M. Chipaux, *Mol. Gen. Genet.*, *197*, 486 (1984).

40. M. P. M. Romaniec, G. P. Hazlewood, and N. G. E. Clarke, *14th Int. Congress Microbiology*, Manchester, England, 1987, Abs. P.B.17-7, p. 101.

41. R. D. Love, R. Fisher, and P. L. Bergquist, In *FEMS Symposium No. 43. Biochemistry and Genetics of Cellulose*

Degradation (J.-P. Aubert, P. Beguin, and J. Millet, eds.), Academic Press, Orlando, 1988.

42. S. Shoemaker, V. Schweickart, M. Ladner, D. Gelfand, S. Kwok, K. Myambo, and M. Innis, *Bio/Technology*, *1*, 691 (1983).

43. T. Teeri, I. Salovuori, and J. Knowles, *Bio/Technology*, *1*, 696 (1983).

44. J. Knowles, P. Lehtovaara, M. E. Penttila, T. Teeri, A. Karkki, and I. Salovuori, *Antonie van Leuweenhoek*, *57*, 335 (1987).

45. J. N. Van Arsdell, S. Kwok, V. L. Schweickart, M. B. Ladner, D. H. Gelfand, and M. A. Innis, *Bio/Technology*, *5*, 60 (1987).

46. C. M. Chen, M. Gritzali, and D. W. Stafford, *Bio/Technology* *5*, 274 (1987).

47. P. J. Wood, *Carboh. Res.*, *94*, C19 (1981).

48. R. M. Teather and P. J. Wood, *Appl. Env. Microbiol.*, *43*, 777 (1982).

49. H. Van Tilbeurgh, M. Claeyssens, and C. K. de Bruyne, *FEBS Lett.*, *149*, 152 (1982).

50. P. J. Wood, *Carboh. Res.*, *85*, 271 (1980).

51. A. G. Williams, *FEMS Microbiol. Lett.*, *20*, 253 (1983).

52. B. A. Cantwell and D. J. McConnell, *Gene*, *23*, 211 (1983).

53. E. Hinchliffe, *J. Gen. Microbiol.*, *130*, 1285 (1984).

54. R. Borris, H. Bäumlein, and J. Hofemeister, *Appl. Microbiol. Biotechnol.*, *22*, 63 (1985).

55. F. Mondou, F. Shareck, R. Morosoli, and D. Kluepfel, *Gene*, *49*, 323 (1987).

56. G. Joliff, P. Béguin, M. Juy, J. Millet, A. Ryter, R. Poljak, and J.-P. Aubert, *Bio/Technology*, *4*, 896 (1986).

57. D. Pétré, J. Millet, R. Longin, P. Béguin, H. Girard, and J.-P. Aubert, *Biochimie*, *68*, 687 (1986).

58. K. Horikoshi, M. Nakao, Y. Kurono, and N. Sashibara, *Can. J. Microbiol.*, *30*, 774 (1984).

59. F. Fukumori, T. Kudo, and K. Horikoshi, *J. Gen. Microbiol.*, *131*, 3339 (1985).

60. C. Kato, T. Kobayashi, T. Kuso, and K. Horikoshi, *FEMS Microbiol. Lett.*, *36*, 31 (1986).

61. F. Fukumori, N. Sashibara, T. Kudo, and K. Horikoshi, *J. Bacteriol.*, *168*, 479 (1986).

62. F. Fukumori, T. Kudo, and K. Horikoshi, *FEMS Microbiol. Lett.*, *40*, 311 (1987).

63. A. Nakamura, T. Uozumi, and T. Beppu, *Eur. J. Biochem.*, *164*, 317 (1987).

64. M. L. Robson and G. H. Chambliss, *J. Bacteriol.*, *169*, 2017 (1987).

65. W. L. Crosby, B. Collier, D. Y. Thomas, R. M. Teather, and J. D. Erfle, *DNA*, *3*, 184 (1984).

66. K. A. Taylor, B. Crosby, M. McGavin, C. W. Forsberg, and D. Y. Thomas, *Appl. Env. Microbiol.*, *53*, 41 (1987).

67. N. R. Gilkes, M. L. Langsford, D. G. Kilbrun, R. C. Miller, Jr., and R. A. J. Warren, *J. Biol. Chem.*, *259*, 10455 (1984).

68. G. O'Neill, S. H. Goh, R. A. J. Warren, D. G. Kilburn, and R. C. Miller, Jr., *Gene*, *44*, 325 (1986).

69. M. L. Langsford, N. R. Gilkes, W. W. Wakarchuk, D. G. Kilburn, R. C. Miller, Jr., and R. A. J. Warren, *J. Gen. Microbiol.*, *130*, 1367 (1984).

70. N. R. Gilkes, D. G. Kilburn, R. C. Miller, Jr., and R. A. J. Warren, *Bio/Technology*, *2*, 259 (1984).

71. G. P. O'Neill, R. A. J. Warren, D. G. Kilburn, and R. C. Miller, *Gene*, *44*, 331 (1986).

72. G. P. O'Neill, D. G. Kilburn, R. A. J. Warren, and R. C. Miller, Jr., *Appl. Env. Microbiol.*, *52*, 737 (1986).

73. G. P. Hazlewood, M. P. N. Romaniec, N. G. E. Clarke, and K. Davidson, In *FEMS Symposium No. 43. Biochemistry and Genetics of Cellulose Degradation* (J.-P. Aubert, P. Beguin, and J. Millet, eds.), Academic Press, Orlando, 1988, Abstract P3-14.

74. G. Joliff, P. Béguin, and J.-P. Aubert, *Nucl. Acid Res.*, *14*, 8605 (1986).

75. P. Cornet, J. Millet, P. Béguin, and J.-P. Aubert, *Bio/Technology*, *1*, 589 (1983).

76. P. Béguin, P. Cornet, and J. Millet, *Biochimie*, *65*, 495 (1983).

77. W. Schwarz, F. Grabnitz, and W. L. Staudenbauer, *Appl. Env. Microbiol.*, *51*, 1293 (1986).

78. G. Joliff, P. Béguin, J. Millet, J.-P. Aubert, P. Alzari, M. Juy, and R. J. Poljak, *J. Mol. Biol.*, *189*, 249 (1986).

79. D. Pétré, P. Béguin, J. Millet, and J.-P. Aubert, *Ann. Microbiol. Inst. Pasteur*, *136B*, 113 (1985).

80. P. Béguin, P. Cornet, and J.-P. Aubert, *J. Bacteriol.*, *162*, 102 (1985).

81. O. Grépinet and P. Béguin, *Nucl. Acid. Res.*, *14*, 1791 (1986).

82. P. Béguin, M. Rocancourt, M.-C. Chebroux, and J.-P. Aubert, *Mol. Gen. Genet.*, *202*, 251 (1986).

83. F. Barras, M. Lepelletier, and M. Chipaux, *FEMS Microbiol. Lett.*, *30*, 209 (1985).

84. F. Barras, M. Lepelletier, and M. Ghipaux, *J. Bacteriol.*, *166*, 346 (1986).

85. M. D. Yablonsky, T. Bartley, K. O. Elliston, S. K. Kahrs, Z. P. Shalita, and D. E. Eveleigh, In *FEMS Symposium No. 43. Biochemistry and Genetics of Cellulose Degradation* (J.-P. Aubert, P. Béguin, and J. Millet, eds.), Academic Press, Orlando, 1988, p. 249.
86. H. Honda, T. Kudo, and K. Horikoshi, *J. Bacteriol.*, *161*, 784 (1985).
87. T. Kudo, A. Ohkoshi, and K. Horikoshi, *J. Gen. Microbiol.*, *131*, 2825 (1985).
88. J. A. Johnson, W. K. R. Wong, and J. T. Beatty, *J. Bacteriol.*, *167*, 604 (1986).
89. M. P. Coughlan, K. Hon-Nami, H. Hon-Nami, L. G. Ljungdahl, J. J. Paulin, and W. E. Rigsby, *Biochem. Biophys. Res. Commun.*, *130*, 904 (1985).
90. K. Yamane, H. Suzuki, and K. Nisizawa, *J. Biochem.*, *67*, 19 (1970).
91. K. Ramasamy and H. Verachtert, *J. Gen. Microbiol.*, *117*, 181 (1980).
92. H. Van Tilbeurgh, P. Tomme, M. Claeyssens, R. Bhikhabhai, and G. Petterson, *FEBS Lett.*, *204*, 223 (1986).
93. M. G. Wirick, *J. Polym. Sci.*, *6*, 1965 (1968).
94. C. Breuil and J. N. Saddler, *Enzyme Microb. Technol.*, *7*, 327 (1985).
95. M. Penttilä, K. M. H. Nevalainen, A. Raynal, and J. K. C. Knowles, *Mol. Gen. Genet.*, *194*, 253 (1986).
96. C. Kohchi and A. Toh-e, *Mol. Gen. Genet.*, *203*, 89 (1986).
97. C. Kohchi and A. Toh-e, *Nucl. Acid Res.*, *13*, 6273 (1985).
98. A. Raynal and M. Guérineau, *Mol. Gen. Genet.*, *195*, 108 (1984).
99. M. Leclerc, P. Chemardin, A. Arnaud, R. Ratomahenina, P. Galzy, C. Gerbaud, and A. Raynal, *Arch. Microbiol.*, *146*, 115 (1986).
100. V. L. Seligy, M. J. Dove, C. Roy, M. Yaguchi, G. E. Willick, J. R. Barbier, L. Huang, and F. Moranelli, In *5th Canadian Bioenergy R&D Seminar* (S. Hasnian, ed.), Elsevier, New York, 1984, p. 5577.
101. F. Moranelli, J. R. Barnier, M. J. Dove, R. M. MacKay, V. L. Seligy, M. Taguchi, and G. E. Willick, *Biochem. Int.*, *12*, 905 (1986).
102. T. T. Teeri, P. Lehtovaara, S. Kauppinen, I. Salovuori, and J. Knowles, *Gene*, *51*, 43 (1987).
103. D. I. Gwynne, F. P. Buton, S. A. Williams, S. Garven, and R. W. Davies, *Bio/Technology*, *5*, 713 (1987).
104. E. Soutschek-Bauer and W. L. Staudenbauer, *Mol. Gen. Genet.*, *208*, 537 (1987).

105. M. Sacco, J. Millet, and J.-P. Aubert, *Ann. Microbiol. (Inst. Pasteur)*, *135A*, 485 (1984).
106. N. Skipper, M. Sutherland, R. W. Davies, D. G. Kilburn, R. C. Miller, Jr., A. Warren, and R. Wong, *Science*, *230*, 958 (1985).
107. J. K C. Knowles, M. Penttilä, T. T. Teeri, L. Andre, I. Salovuori, and P. Nehtovaara, In *Proc. 20th European Brewery Convention Congress*, Helsinki, Finland, 1985, p. 251.
108. S. E. Buchholz, M. M. Dooley, and D. E. Eveleigh, *Trends Biotechnol.*, *5*, 199 (1987).
109. T. Conway, M. O.-K. Byun, and L. O. Ingram, *Appl. Env. Microbiol.*, *53*, 235 (1987).
110. B. R. Glick, J. J. Pasternak, and H. E. Brooks, The development of *Azotobacter* as a bacterial fertilizer by the introduction of exogenous cellulase genes, In *Proc. First Bioenergy Specialists Meeting on Biotechnology*, National Research Council of Canada, Ottawa, Ontario, 1985, p. 180.
111. C. G. Orpin, In *FEMS Symposium No. 43. Biochemistry and Genetics of Cellulose Degradation* (J.-P. Aubert, P. Béguin, and J. Millet, eds.), Academic Press, Orlando, 1988, p. 171.
112. K. Ohmiya, W. Chen, K. Nagashima, T. Kajins, S. Shimizu, and H. Kawakami, In *FEMS Symposium No. 43. Biochemistry and Genetics of Cellulose Degradation* (J.-P. Aubert, P. Béguin, and J. Millet, eds.), Academic Press, Orlando, 1988, Abs. P4-10.
113. N. Murphy, D. J. McConnell, and B. A. Cantwell, *Nucl. Acid Res.*, *12*, 5355 (1984).
114. B. A. Cantwell, G. Brazil, N. Murphy, and D. J. McConnell, *Curr. Genet.*, *11*, 65 (1986).
115. B. A. Cantwell, P. M. Sahrp, E. Gormley, and D. J. McConnell, In *FEMS Symposium No. 43. Biochemistry and Genetics of Cellulose Degradation* (J.-P. Aubert, P. Béguin, and J. Millet, eds.), Academic Press, Orlando, 1988, p. 181.
116. B. Cantwell, G. Brazil, J. Hurley, and D. McConnell, In *Proc. 20th European Brewery Convention Congress*, IRL Press, Oxford, 1985.
117. J. Hofemeister, A. Kurtz, R. Borris, and J. Knowles, *Gene*, *49*, 177 (1986).
118. W. E. Lancashire, *The Brewer*, p. 345 (1986).
119. E. Hinchliffe and W. G. Box, *Curr. Genet.*, *8*, 471 (1984).
120. E. Hinchliffe and W. G. Box, In *Proc. 20th European Brewery Convention Congress*, IRL Press, Oxford, 1985, p. 267.
121. W. Panbangred, T. Kondo, S. Negoro, A. Shinmyo, and H. Okada, *Mol. Gen. Genet.*, *192*, 335 (1983).

122. W. Panbangred, E. Fukusaki, E. C. Epifanio, A. Shinmyo, and H. Okada, *Appl. Microbiol. Biotechnol.*, *22*, 259 (1985).
123. E. Fukusaki, W. Panbangred, A. Shinmyo, and H. Okada, *FEBS Lett.*, *171*, 197 (1984).
124. R. Bernier, Jr., H. Driguez, and M. Desrochers, *Gene*, *26*, 59 (1983).
125. R. Bernier, Jr., D. Rho, Y. Arcand, and M. Desrochers, *Biotechnol. Lett.*, *7*, 795 (1985).
126. M. G. Paice, R. Bourbonnais, M. Desrochers, L. Jurasek, and M. Yaguchi, *Arch. Microbiol.*, *144*, 201 (1986).
127. R. Bernier, Jr. and M. Desrochers, *J. Gen. Appl. Microbiol.*, *31*, 513 (1985).
128. H. Honda, T. Kudo, and K. Horikoshi, *Agr. Biol. Chem.*, *49*, 3011 (1985).
129. A. Sipat, K. A. Taylor, R. Y. C. Lo, C. W. Forsberg, and P. J. Krell, *Appl. Env. Microbiol.*, *53*, 477 (1987.
130. T. K. Kirk and R. L. Farrell, *Ann. Rev. Microbiol.*, *41*, 465 (1987).
131. T. K. Kirk, In *FEMS Symposium No. 43. Biochemistry and Genetics of Cellulose Degradation* (J.-P. Aubert, P. Béguin, and J. Millet, eds.), Academic Press, Orlando, 1988, p. 315.
132. P. Sims, A. Brown, C. James, U. Raeder, A. Schrank and P. Broda, In *FEMS Symposium No. 43. Biochemistry and Genetics of Cellulose Degradation* (J.-P. Aubert, P. Béguin, and J. Millet, eds.), Academic Press, Orlando, 1988, p. 365.
133. Y. Z. Zhang, G. J. Zylstra, R. H. Olsen and C. A. Reddy, *Biochem. Biophys. Res. Commun.*, *137*, 649 (1986).
134. M. Tien and C. P. D. Tu, *Nature*, *326*, 520 (1987).
135. M. Szakacs-Dobozi, A. Halasz, L. Vamos-Vigyazo, and G. Szakacs, *Br. Polym. J.*, *19*, 83 (1987).
136. S. P. Shoemaker, The Cellulase system of *Trichoderma reesei: Trichoderma* strain improvement and expression of *Trichoderma* cellulases in yeast. *Am. Chem. Soc. Ann. Mtg.*, Philadelphia, PN, 1984.
137. A. A. Klyosov, *Appl. Biochem. Biotechnol.*, *12*, 249 (1986).
138. M. Claeyssens, In *FEMS Symposium No. 43. Biochemistry and Genetics of Cellulose Degradation* (J.-P. Aubert, P. Béguin, and J. Millet, eds.), Academic Press, Orlando, 1988, p. 393.
139. M. P. Coughlan and L. G. Ljungdahl, In *FEMS Symposium No. 43. Biochemistry and Genetics of Cellulose Degradation* (J.-P. Aubert, P. Béguin, and J. Millet, eds.), Academic Press, Orlando, 1988, p. 11.
140. A. Lejeune, D. E. Eveleigh, and C. Colson, *FEMS Microbiol. Lett.*, *49*, 363 (1988).
141. A. Lejeune, V. Dartois, and C. Colson, *Biochim. Biophys. Acta*, *950*, 204 (1988).

Symbols and Abbreviations

Ara	arabinose
ATCC	American Type Culture Collection
cAMP	cyclic-3'-5'-adenosine-5'-monophosphate
CBF	cellulose-binding factor
CBH	cellobiohydrolase
c-di-GMP	bis-3'-5'-cyclic diguanylic acid
cel$^-$	cellulose synthesis negative
cel$^+$	cellulose synthesis positive
CHAPS	3-[3-chol(amidopropyl)dimethylammonio]-1-propane-sulfonate
CI	Colour Index
CMC	carboxymethylcellulose
CMCase	carboxymethylcellulase
CP/MAS	cross-polarization/magic angle spinning
DCB	dichlorobenzonitrile
DCPA	dichlorophenylazide
DEAE	diethylaminoethyl
DMD	dry matter digestibility

DMSO	dimethylsulfoxide
DNS	dinitrosalicylic acid
DP	degree of polymerization
DP_n	number average degree of polymerization
DP_w	weight average degree of polymerization
DS	degree of substitution
EDTA	ethylenediaminetetraacetic acid
EF	exoplasmic fracture
EG	endoglucanase
EGTA	ethylene glycol-bis-CB-aminoethylether)-N,N,N',N'-tetraacetic acid
FDA	flavin adenine dinucleotide
FBA	fluorescent brightening agent
FDP	fructose-1,6-diphosphate
FPLC	fast protein liquid chromatography
FPU	filter paper unit
Fuc	fucose
$(Glc)_\eta$	cellodextrins, where η = number of glucosyl units
Glcn	glucosamine
Glc	glucose
Glc-1-P	glucose-1-phosphate
GMP	guanidine-5'-monophosphate
GS	glucan synthase
^3H	tritium
HEC	hydroxyethylcellulose
HPLC	high-pressure liquid chromatography
HRGPs	hydroxyproline-rich glycoproteins
IEF	isoelectric focusing
K_a	association constant
K_I	inhibition constant

K_{inact}	inactivation constant
K_m	Michaelis constant
LSC	liquid scintillation counting
Man	mannose
MF	microfibril
MPN	most probable number
MT	microtubule
MUC	methylumbelliferyl-β-cellobroside
MUG	methylumbelliferyl-β-glucopyramoside
MW	molecular weight
NDF	neutral detergent fiber
NMR	nuclear magnetic resonance
oNPGlc	*ortho*-nitrophenyl-β-D-glucospyranoside
P	phosphate
PAGE	polyacrylamide gel electrophoresis
PATAg	periodic acid-thiocarbohydrazide-silver proteinate
PDE	phosphodiesterase
PEG	polyethylene glycol
PF	protoplasmic fracture
PHMB	parahydroxymercuriphenyl sulfonic acid
P_i	orthophosphate
pNPGlc	*para*-nitrophenyl-β-D-glucopyranoside
$pNP(Glc)_2$	*para*-nitrophenyl-β-cellobioside
pNPL	p-nitrophenyl-β-D-lactoside
PPA	3-phenylpropanoic acid
SDS	sodium dodecyl sulfate
SEM	scanning electron microscopy
TC	terminal complex
TEM	transmission electron microscopy
TNBS	trinitrobenzylsulfonic acid

TLC	thin-layer chromatography
TPN	triphenylnitro
UDP	uridine-5'-diphosphate
UDPG	uridine diphosphate glucose
UMP	uridine-5'-monophosphate
UTP	uridine-5'-triphosphate
V_{max}	maximal reaction velocity
VFA	volatile fatty acid
Xglu	5-bromo-4-chloro-3-inddyl-β-glucopyranoside
Xyl	xylose
$[\eta]$	intrinsic viscosity
η_{sp}	specific viscosity

Organism Index

Acanthamoeba, 12, 115, 116
Acer pseudoplatanus, 79
Acetabularia, 9
Acetivibrio cellulolyticus, 356, 359–367, 380, 537, 538, 554, 560
Acetivibrio cellulosolvens, 360, 362
Acetobacter xylinum, 12, 99–104, 107–112, 115, 119, 120, 136–138, 140, 172, 178, 182, 189, 202, 207, 209, 214, 219–240, 245, 246, 516
Acetogenium kivui, 400
Achlya, 12
Acidothermus cellulolyticus, 479, 586
Actinomycetes, 460–476
Actyostelium leptosomum, 12
Adiantum capillus veneris, 78
Aerobacter aerogenes, 414
Aeromonas, 650, 662

Aeschyonomene americana, 320
Agrobacterium, 13, 238, 240, 628, 635, 642
Alcaligenes, 120
Alcaligenes faecalis, 432
Algae:
  brown (see Phaeophyta)
  red (see Rhodophyta)
  green (see Chlorophyta)
  golden (see Chrysophyta)
  yellow–green (see Chrysophyta)
  dinoflagellates (see Pyrrophyta)
  stoneworts (see Charophyta)
  blue–green (see Cyanophyta)
Allium, 79, 151
Anabaena, 13
Anolis aeneus, 312
Anolis richardi, 312
Anolis roquet, 312
Apjohnia, 9
Apodachlya, 12

Subject Index

o-Nitrophenylglucoside, 419, 420, 422
p-Nitrophenylglucoside, 276, 416, 417, 419, 422, 537, 641

Oligosaccharins, 613
Osmometry, 270

Pectins, 27, 28, 173
Phosphodiesterases, 220, 234–236, 240
Plasma membrane:
  character of, 71–72
  effects of disruption on cellulose synthesis, 186, 187
  membrane potential of, 186
Plasmids of *Acetobacter xylinum*, 248–251
Pollen tubes:
  microtubules in, 152
  rosettes in, 79
  stress-induced particle arrays in, 88
Protoplasts:
  cell division in, 117
  regeneration into plants of, 115
  wall alteration in, 115
  wall regeneration in, 59, 62, 186
Protuberances (*see* Vesicles)

Raman spectroscopy, 294
Reducing sugar measurement, 277, 278, 281, 282
Rumen:
  anatomy, 328, 329
  microorganisms of, 331–342, 357
  physical environment, 329, 330, 356

Sclereids, 34, 35
Sclerocytes, 35
Single cell protein, 431, 432
Stone cells, 34, 35
Sucrose synthase, 184

Terminal complexes:
  biochemical composition of, 234
  fixed vs. mobile types of, 137
  linear type, 40, 41, 51–67
  assembly of, 58, 139
  changes by inhibitors, 60–66
  evidence against role in cellulose synthesis, 66, 67
  evidence for role in cellulose synthesis, 59–67
  microfibril size regulated by, 119, 120, 136, 137
  movement regulated by, 65, 73, 93, 94, 140, 155–157, 173
  ordered granule hypothesis, 54, 73
  rosette/globular type, 40, 41, 71–96
  cellulose synthesis in vitro related to, 94, 96, 180, 193
  evidence for role in cellulose synthesis, 77–82, 180
  function, 92–94, 120
  Golgi transport of, 82, 139, 172, 188
  hexagonal pattern of, 87–91
  lability of, 77, 180, 186
  microfibril width correlated with, 81
  model for function of, 93